Bashlor
2845 Dutton Dr
Titusville,
269-2085

probability and random processes

an introduction for
applied scientists and engineers

probability
and

random processes

an introduction for applied scientists and engineers

WILBUR B. DAVENPORT, JR.

department of electrical engineering
and center for advanced engineering study
massachusetts institute of technology

McGRAW-HILL BOOK COMPANY

new york / st. louis / san francisco / düsseldorf
london / mexico / panama / sydney / toronto

probability and random processes

an introduction for applied scientists and engineers

Library of Congress Catalog Card Number 70-115141

07-015440-6

34567890KPKP7987654

to joan

and

mark

and

sally

to joan
and
mark
and
sally

preface

This book, intended as a text for a first course in probability and random processes, can be used either for self-study or in a formal classroom setting. In the classroom context, the first eight chapters could form a one-semester subject on probability with the last six chapters (plus possible supplementary material at the discretion of the instructor) following as a semester course on random processes. Alternatively, most of the key ideas in both probability and random processes could be covered in one semester by a judicious selection of sections. The prerequisites are calculus for the first eight chapters and calculus plus a prior, or concurrent, introduction to Fourier theory for the remainder.

Over the past ten years or so I have taught this material to such disparate groups as first-year graduate students in electrical engineering and engineers and applied scientists trained in a variety of technical fields who have returned to the academic world for study some ten to twenty years after graduating from college. As that teaching progressed, I became increasingly convinced of the need for a text which, while directed towards students mainly interested in applications, would present to the student the underlying mathematical issues in a readable and technically honest way. In particular, I felt the need for a text which would give an indication as to how the theoretical complexity of the more advanced mathematical treatments arises from the logical difficulties inherent in the subject and is not due simply to the natural (or unnatural) perversity of the mathematician. I hope that this is such a text.

Since I believe strongly that each of us learns by doing rather than by passive observation, I have attempted to get the reader to develop a significant fraction of the key concepts and results by including them in the exercises. The exercises in this book are thus an integral part of the text and should *all* be worked out by the reader (except possibly the ones designated as supplementary). With this fact in mind, and since one of my main objectives was to write a text which could be used for self-study, complete solutions to all of the exercises are available in a supplementary book by Professor Amedeo Odoni and myself. In view of the wide availability of other books in the field, however, it did not seem necessary to also include a set of problems without solutions for class quizzes and home problems.

Since the spring of 1968 a project has been underway at the MIT Center for Advanced Engineering Study to develop a course on this material for use in industrial locations. That course is based in part upon this book and in part upon a set of videotape lectures made by Professor Harry L. Van Trees, Jr. Further information about the lecture tapes and an accompanying study guide may be obtained from the Director of the MIT Center for Advanced Engineering Study.

Of the many people who contributed to this book a few get my special thanks: my students; the secretaries of the MIT Center for Advanced Engineering Study (in particular, Mrs. Elizabeth Liljegren Borken) for the typing of the various versions of the manuscript; Richard H. Lee of WGBH, Boston, for critically reviewing most of the early chapters; John T. Fitch of the MIT-CAES for critically reviewing the entire manuscript; Professor Harry L. Van Trees, Jr., for his comments and suggestions for changes; Professor Amedeo Odoni for working up the final version of the solutions to the exercises; and last, but not least, Professor Harold S. Mickley, Director of the MIT-CAES, for both his critical comments and his generous support of the project.

Wilbur B. Davenport, Jr.

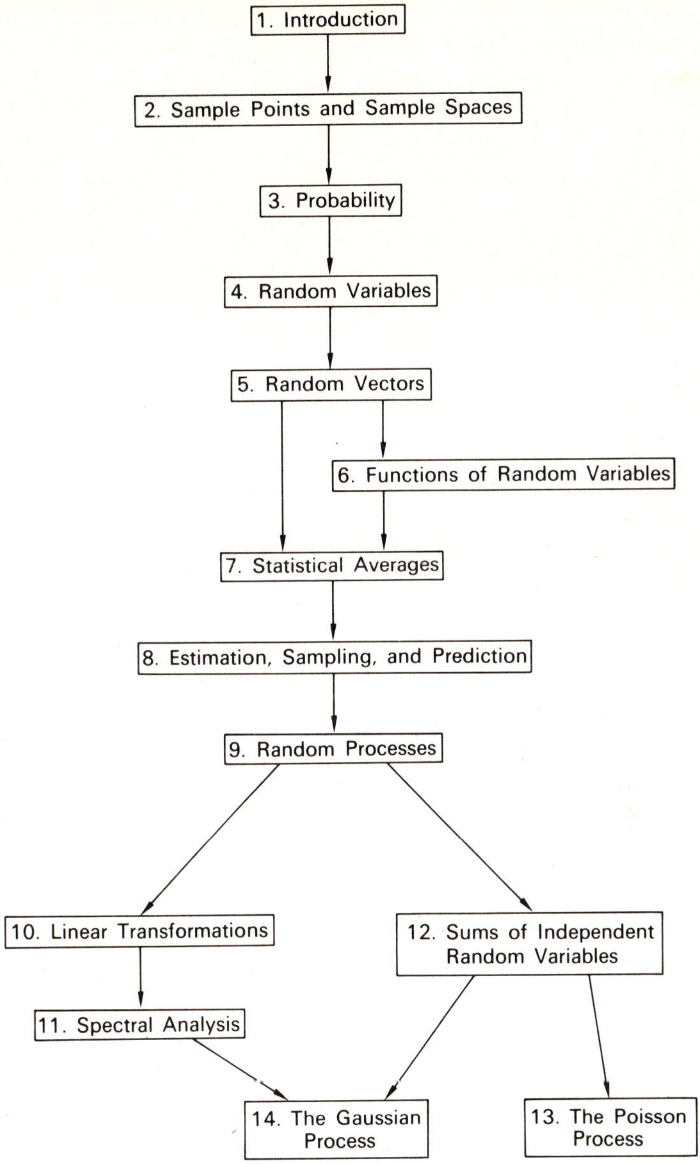

Interdependence of Chapters

to the reader

The section is the key to the numbering system used in this book; sections are single-numbered throughout a given chapter. Equations, exercises, figures, and tables are all double-numbered within a given section, and references to these items are given accordingly: "Eq. (2.4)" (Equation 4 in Section 2). There are occasional cross-references to elements in other chapters, and for this purpose the following conventions are used: (*a*) a cross-reference to a section in another chapter would be double-numbered, for example, "see Sec. 10.3" (Section 3 of Chapter 10); (*b*) cross-references to other items are then triple-numbered, for example, "cf. Fig. 12.4.2" (the second figure in Section 4 of Chapter 12).

Certain sections, subsections, and exercises throughout the work are of a supplementary nature and may be omitted or scanned without loss of continuity. The *headings* for such supplementary materials are printed in a lighter type in a manner that sets them off from the main outline of the work, that is, **Exercise 1** versus Exercise 1 (supplementary).

The symbol ▪ is used to denote the end of a proof, while the symbol □ denotes the end of an exercise.

contents

* *Supplementary section*

contents

*** *Supplementary section***

probability and random processes

an introduction for
applied scientists and engineers

1
Introduction

1 RANDOMNESS AND AVERAGES

In many physical problems of interest, there is an element of uncertainty, or unpredictability, or randomness; no matter how much we might know of the past history of a given phenomenon, we are essentially unable to predict its future behavior precisely. For example, no matter how many times we flip a coin and observe the outcome, we are unable to predict exactly the outcome of the next flip. The reasons for our inability to predict are varied: We may not know all of the causal forces at work; we may not have enough data about the initial conditions in the problem; the forces at work may be so complicated that a calculation of their combined effect is not feasible; or there actually may be some basic indeterminacy at work in the problem at hand. We shall call such unpredictable phenomena *random phenomena*.

As an example of a random phenomenon, consider the queuing problem of "customers" waiting to be served at a serving station, in particular, data blocks in buffer storage in a data processing system.

Here the time of arrival of a given customer at the entrance to the queue (the waiting line or buffer store) may well be unpredictable. Further, the time required to service a customer may also vary from customer to customer in an apparently random manner. Questions then as to the number of customers in the queue at some future time, their length of time in the queue, the number to be served in a given time interval, etc., are all incapable of being precisely answered ahead of time.

Next, let us consider the problem of mechanical vibrations set up in a spacecraft structure by variations in the thrust generated by a solid-fuel rocket engine. Here the mechanical and chemical inhomogeneities in the fuel cause variations in the burning rate and hence in the thrust generated. In any practical case then, we have a situation in which the precise behavior is far too complex to predict in detail and, accordingly, we act as though the resultant vibrations are random in nature.

Finally, let us consider the problem of measuring seismic disturbances at some point on the ocean bottom. In particular, suppose that we wish to accomplish this purpose by placing a seismometer on the ocean bottom and communicating its output to a distant point by means of acoustic waves transmitted through the ocean itself. Now acoustic transmissions through the ocean are perturbed in many ways, for example: multipath effects due to acoustic reflections from both the ocean bottom and surface, acoustic refraction effects due to variations of the water temperature from point to point, acoustic scattering effects due to the presence of plankton, schools of fish, etc. Thus we have here not only the unpredictability of the basic seismic data itself, but also the uncertainties created by the acoustic transmission path.

While in each of the above examples we are unable to predict the future in detail, we often do find experimentally that certain *average* properties exhibit a useful regularity. In the acoustic transmission case, the received energy averaged over seconds does not vary greatly over minutes, and the received energy averaged over a period of a month does not differ greatly from that averaged over the same month a year previously. Such a regularity of averages is an experimentally verifiable phenomenon in many different physical situations which supposedly involve randomly varying quantities. Since averages often do exhibit such a regularity, and are thus reasonably predictable, it seems desirable to make a study of a calculus of averages. This is the domain of the mathematical theory of probability and statistics.

We shall, in the remainder of this chapter, first study the problem of calculating averages of known data and thereby introduce the concept of *relative frequency;* we shall then look at the stability of empirical averages such as relative frequency when measured over longer and

longer sequences of data; finally, we shall introduce *probability* as a model for the limiting value of a relative frequency and *statistical average* as a model for the limiting value of an empirical average.

2 EMPIRICAL AVERAGES

Let us now consider in some detail the problem of calculating the average of a particular set of data and, in fact, of calculating averages of various functions of the given data. We call such averages *empirical averages.*

Suppose, for example, that we have the situation pictured in Fig. 2.1. Here we have blocks of data being supplied at a regular rate by a data transmission system to a data processor. Now in many cases the processing to be done on any one data block is determined by the results of the processing of the preceding data blocks and, hence, the data processing rate may well be irregular. In order to overcome the lag between the regular rate of data transmission and the irregular rate of data processing, a buffer storage unit is connected between the transmission system and the processor so that the incoming data blocks are stored until they can be processed.

In one such system, using a buffer storage unit which could store up to eight data blocks at any one time, measurements were made of the number of data blocks stored in the buffer during each of 100 successive data transmission intervals; the results of these measurements are listed in Table 2.1. Suppose now that we wish to determine the average number of data blocks in storage throughout those 100 intervals, the average of the square of the number of data blocks in storage, and the fraction of the intervals in which the buffer storage unit was more than half-full.

The *arithmetic average*, or *mean*, of the number of data blocks in storage is, using the data in Table 2.1,

$$\begin{aligned}\bar{x} &= \tfrac{1}{100}(5 + 8 + 0 + 1 + 5 + \cdots + 0 + 8 + 2)\\ &= 3.9\end{aligned}$$

That is, there were *on the average* about four data blocks in buffer storage throughout the observation period. Similarly, the *mean square* of the

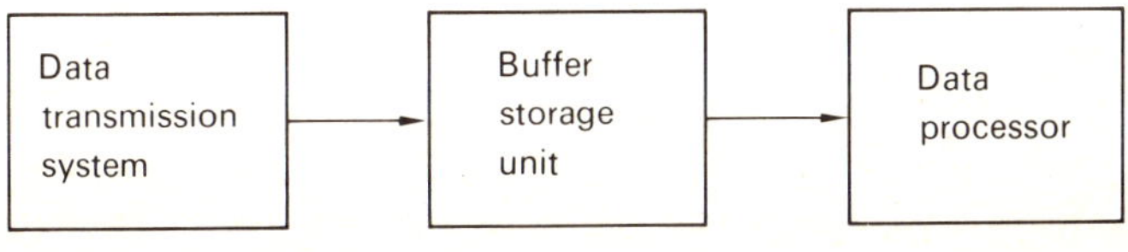

Figure 2.1 A data input system.

Table 2.1 Buffer storage raw data [i = time index; $x(t_i)$ = number of blocks in storage at time t_i]

i	$x(t_i)$	i	$x(t_i)$	i	$x(t_i)$	i	$x(t_i)$	i	$x(t_i)$
1	5	21	6	41	2	61	3	81	0
2	8	22	4	42	8	62	8	82	4
3	0	23	1	43	6	63	5	83	7
4	1	24	8	44	4	64	7	84	3
5	5	25	2	45	3	65	5	85	6
6	4	26	1	46	8	66	4	86	8
7	4	27	0	47	0	67	4	87	0
8	3	28	0	48	2	68	8	88	1
9	4	29	2	49	0	69	2	89	7
10	6	30	5	50	3	70	8	90	4
11	8	31	4	51	2	71	6	91	3
12	8	32	3	52	0	72	7	92	3
13	1	33	6	53	8	73	6	93	0
14	0	34	0	54	8	74	3	94	8
15	1	35	2	55	3	75	2	95	0
16	3	36	1	56	0	76	6	96	6
17	4	37	4	57	5	77	1	97	7
18	0	38	2	58	6	78	7	98	0
19	8	39	6	59	7	79	5	99	8
20	0	40	7	60	7	80	0	100	2

number of blocks in storage (that is, the arithmetic average of the square of the number of blocks in storage) is

$$\overline{x^2} = \tfrac{1}{100}\underbrace{(25 + 64 + 0 + 1 + 25 + \cdots + 0 + 64 + 4)}_{100 \text{ terms}}$$
$$= 29.0$$

Finally, we may calculate the *relative frequency* of occurrence of the buffer being more than half-full (the fraction of the intervals in which the buffer is more than half-full) as follows: For each entry in Table 2.1, we indicate whether or not the buffer is more than half-full by writing 1 if the Table 2.1 entry is 5 through 8, and by writing 0 if the entry is 0 through 4; this leads to the sequence of 0s and 1s shown in Table 2.2. If now we add the entries in Table 2.2 and divide by 100, we get the desired relative frequency. On denoting that relative frequency by n_A/n, we then get

$$\frac{n_A}{n} = \tfrac{1}{100}\underbrace{(1 + 1 + 0 + 0 + 1 + \cdots + 0 + 1 + 0)}_{100 \text{ terms}}$$
$$= 0.42$$

Table 2.2 Buffer storage occupancy data, more-than-half-full case

11001/00001/11000/00010/
10010/00001/00100/00011/
01100/10000/00110/01111/
01111/00101/11100/10110/
00101/10010/00010/11010/

that is, the buffer storage unit was more than half-full 42 percent of the time during the 100 observations.

The above formulas for the mean, mean square, and relative frequency can, of course, be generalized to enable the calculation of those respective averages for any other similar set of data. Thus, for example, if we denote the ith entry in a table of measurement data by $x(t_i)$, and if we denote the total number of measurements by n ($n = 100$ in the above buffer storage example), we may then write

$$\bar{x} = \frac{1}{n}[x(t_1) + x(t_2) + \cdots + x(t_n)]$$
$$= \frac{1}{n}\sum_{i=1}^{n} x(t_i) \tag{2.1}$$

as a formula for the mean of the data. Similarly, we may write

$$\overline{x^2} = \frac{1}{n}\sum_{i=1}^{n} x^2(t_i) \tag{2.2}$$

as a formula for the mean-square value of the data.

The formula given above for the relative frequency of occurrence of the buffer being more than half-full may be generalized as follows: Suppose that we are particularly interested in some specific event concerning the measured data (for example, the event that the buffer is more than half-full) and that we wish to determine the relative frequency of occurrence of that event; the procedure which we shall follow parallels that used in the buffer storage example. The operation of going from entries in a table of measurement data, such as Table 2.1, to entries in a table, such as Table 2.2, may be represented formally by introducing an *event-indicator function* which has the value 1 if the event of interest occurs and the value 0 if not. That is, if we denote the event indicator of an event A by I_A, then

$$I_A[x(t_i)] = \begin{cases} 1 & \text{if } x(t_i) \text{ has a value which corresponds to} \\ & \quad \text{the occurrence of the event } A \\ 0 & \text{if not} \end{cases} \tag{2.3}$$

The relative frequency of occurrence n_A/n of the event A is then given by the formula

$$\frac{n_A}{n} = \frac{1}{n} \sum_{i=1}^{n} I_A[x(t_i)] \tag{2.4}$$

paralleling the result in the buffer storage example.

A moment's reflection upon formulas (2.2) and (2.4) shows in each case that we are concerned with the determination of the average of some function of the measured data. In general, if we denote the function of interest by g, then we may write

$$\overline{g(x)} = \frac{1}{n} \sum_{i=1}^{n} g[x(t_i)] \tag{2.5}$$

as a formula for the average of that function of the data. Formulas (2.2) and (2.4) are then special cases of (2.5).

3 RELATIVE FREQUENCY

A review of the calculations made in the preceding section shows that as more and more complicated functions of the given data are to be determined, the detailed calculations become more and more tedious; a simpler calculation technique is therefore desired.

Inspection of the various formulas used in the buffer storage example shows that while there are 100 terms being added in each case, those terms have, in fact, only the 9 different values 0, 1, 2, . . . , 8. With this fact in mind, let us consider evaluating an average by first grouping together all the terms which have the same value. For example, we may evaluate the mean number of data blocks in storage by first multiplying each possible value by the number of times that that value occurs in the data set, then adding the resultant products, and finally dividing by 100. Thus we get

$$\begin{aligned} \bar{x} &= \tfrac{1}{100}(17 \times 0 + 8 \times 1 + 10 \times 2 + \cdots + 15 \times 8) \\ &= 3.9 \end{aligned}$$

Similarly, the mean square of the number of data blocks in the buffer store is

$$\begin{aligned} \overline{x^2} &= \tfrac{1}{100}(17 \times 0 + 8 \times 1 + 10 \times 4 + \cdots + 15 \times 64) \\ &= 29.0 \end{aligned}$$

In order to calculate averages in this way then, we need to know the number of times that each possible value occurs in a given set of data.

A plot of such numbers (called a *histogram*) is shown by Fig. 3.1 for the data given in Table 2.1.

In general, if we denote the possible values of the data $x(t_i)$ by x_k (where $k = 1, 2, \ldots, K$),† and, in addition, if we denote the number of times that the value x_k occurs in the given data by n_k, then the general

† $K = 9$ in the buffer storage example.

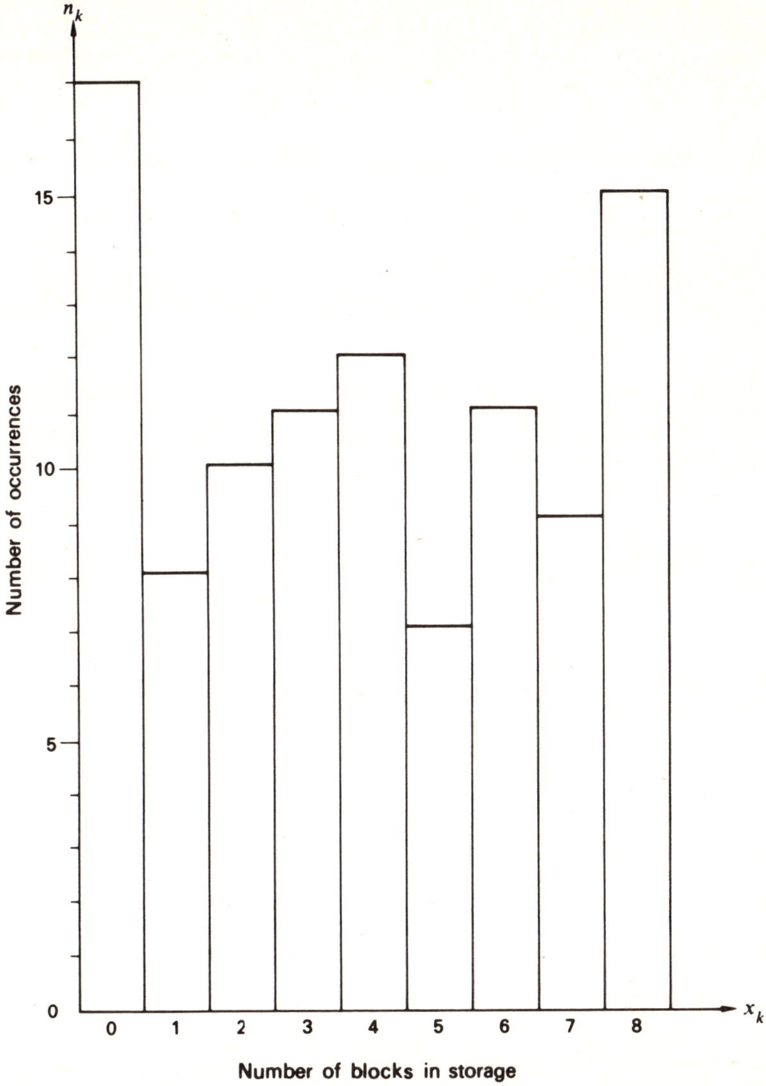

Figure 3.1 Histogram of buffer storage occupancy.

formula for the mean value of the data becomes

$$\begin{aligned}\bar{x} &= \frac{1}{n} \sum_{k=1}^{K} n_k x_k \\ &= \sum_{k=1}^{K} x_k \frac{n_k}{n}\end{aligned} \tag{3.1}$$

That is, we may calculate the mean value of the data by first weighting each possible data value x_k by the *relative frequency* of occurrence of that value, n_k/n, and then adding the results. Equation (3.1) gives the desired alternative formula for the calculation of the mean value of a set of data.

In an entirely similar manner, it follows that if we wish to calculate the average value of some function g of the given data, then we may use the formula

$$\overline{g(x)} = \sum_{k=1}^{K} g(x_k) \frac{n_k}{n} \tag{3.2}$$

Thus, for example, if we wish to calculate the mean-square value of the data, then we may write

$$\overline{x^2} = \sum_{k=1}^{K} x_k^2 \frac{n_k}{n}$$

In summary, if we wish to calculate the averages of a number of different functions of a given set of data, then it is often most convenient to calculate first the relative frequencies of occurrence of the various possible data values (as done in the preceding section) and then use those results in turn in (3.2) in order to calculate the desired averages. Averages which have been calculated in this way are often called *distribution averages* as their calculation makes use of the distribution of relative frequencies of occurrence of the various possible data values.

Properties of relative frequency

In view of the utility of relative frequency suggested by the above results, it seems worthwhile to determine a few of its basic properties.

First of all, we note that the number of times n_k that the outcome x_k can occur in n observations is bounded between zero and n:

$$0 \le n_k \le n$$

for $k = 1, 2, \ldots, K$. It therefore follows that relative frequencies are bounded between zero and one

$$0 \le \frac{n_k}{n} \le 1 \qquad k = 1, 2, \ldots, K \tag{3.3}$$

Further, if the K possible outcomes x_k are mutually exclusive (that is, if only one can occur at a time), and if they completely specify all the possible outcomes of an observation, then it follows that the numbers n_k must add up to n; that is,

$$\sum_{k=1}^{K} n_k = n$$

and hence that the corresponding relative frequencies must add up to unity:

$$\sum_{k=1}^{K} \frac{n_k}{n} = 1 \tag{3.4}$$

Finally, suppose that A and B are mutually exclusive sets of possible outcomes of an observation; for example, let A denote the event that an odd number of data blocks is in buffer storage and let B denote the event that no data blocks are in buffer storage. In a given sequence of n observations then, the number of times that either the event A or the event B occurs, say n_C, is equal to the number of times that A occurs, n_A, plus the number of times that B occurs, n_B; that is,

$$n_C = n_A + n_B$$

It therefore follows that the relative frequency of occurrence of outcomes in either A or B is given by the formula

$$\frac{n_C}{n} = \frac{n_A}{n} + \frac{n_B}{n} \tag{3.5}$$

when the events A and B are mutually exclusive.

The results stated in (3.3), (3.4), and (3.5) are the basic properties of relative frequency.

4 STABILITY

We discussed at some length in the preceding two sections various aspects of the problem of calculating the average value of a specific data record. Throughout that discussion no assumptions were made as to randomness or unpredictability of future behavior; we were concerned only with the properties of a given data record defined over a particular set of observations. While such a situation is often of interest, we are generally much more interested in our ability to use such results as a basis for predicting future average behavior. In order to do so, however, we must be able to assume that the laws or regularities or patterns of behavior which obtained in the past will persist into the future. In particular, we must be able to assume that we have a situation in which averages are stabilized.

That an assumption of stability of averages is at all reasonable is based purely and simply upon the fact that such a stability has been observed empirically by many different investigators in numerous physical experiments. For example, a plot is given in Fig. 4.1 of the results of calculating the relative frequency of occurrence of our previously discussed buffer storage unit being more than half-full using measurements made over a total of 1000 data transmission intervals. We see from that figure that while the relative frequency of being more than half-full fluctuates wildly for less than, say, 70 data transmission intervals, it does indeed also tend to stabilize in value as the number of observations becomes larger and larger.

It must not be forgotten in our subsequent theoretical developments that it is due entirely to the fact that empirical averages such as relative frequencies do so stabilize in many practical cases that a mathematical model based upon the assumption of stability of averages is at all useful in applications.

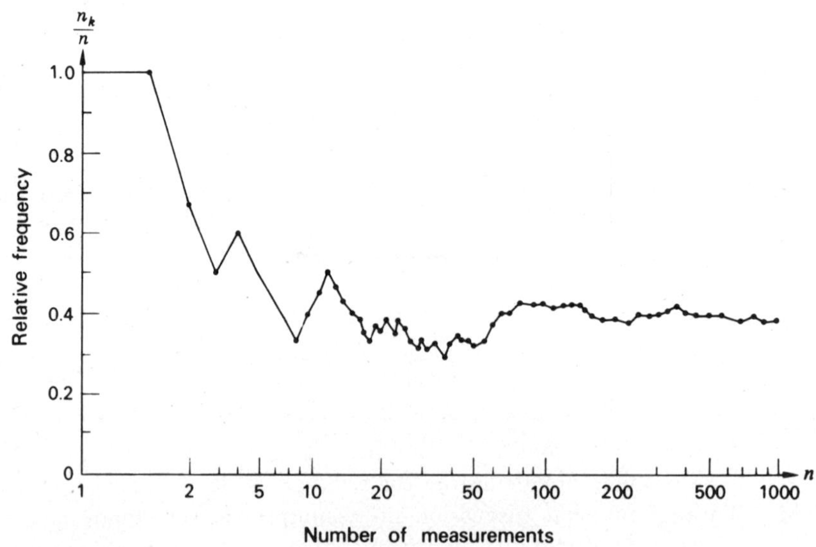

Figure 4.1 Observed relative frequencies for the buffer-storage-occupancy example.

5 PROBABILITY AND STATISTICAL AVERAGES

We now wish to introduce a mathematical model for the empirical averages discussed above. In view of the utility of expressing empirical averages in terms of relative frequencies (coupled with the fact that a relative frequency is itself an empirical average), we shall proceed by first modeling relative frequency by a function called probability, and then using probability in modeling empirical average by a function called statistical average.

The relation of our mathematical model to the physical world will be as follows:

a. First, an experiment must be defined, for example, the counting of the number of data blocks in a buffer storage unit or the measurement of the amplitude of vibration of a mechanical system. We shall assume that the given experiment can be repeated (at least conceptually) an arbitrarily large number of times under essentially identical conditions.

b. Second, all the possible outcomes of the given experiment must be specified. For example, if the capacity of the storage unit is eight blocks, then the measured number of data blocks in storage can assume any integer value between 0 and 8, while the amplitude of vibration of the given mechanical system might assume any real value between certain limits.

c. Next, to each possible outcome (or sets of possible outcomes) a non-negative number shall be assigned, called the probability of the outcome, which is intended to model the empirical relative frequency of occurrence of the outcome.

d. We then assert that an empirical relative frequency of occurrence of an outcome will, with practical certainty, differ by less and less from the corresponding probability as the number of repetitions of the defining experiment becomes larger and larger. It is in this sense that the probability of occurrence of an outcome is a model of the relative frequency of that outcome.

e. Finally, if the given experiment has K possible outcomes with the values $x_1, x_2, \ldots, x_K$, respectively, and if we denote by p_k the probability that the outcome of value x_k occurs, then we say that the statistical average of experimental outcomes, or expected value (which we denote by $E[X]$) is given by the formula

$$E[X] = \sum_{k=1}^{K} x_k p_k \tag{5.1}$$

Following (*d*), this statistical average models the empirical average

$$\bar{x} = \sum_{k=1}^{K} x_k \frac{n_k}{n}$$

in the sense that their values should differ by less and less as n becomes larger and larger.

It should be noted that while we have discussed the relation of a mathematical model (called probability theory) to the physical world, we have not yet defined probability. We shall, in fact, define probability here only by specifying its basic properties. This is, of course, no different from the usual way of defining other mathematical concepts such as *continuous function* or *linear operator*.

Since we wish to have probability model relative frequency, it seems reasonable to hypothesize that it should have the following basic properties:

(1) Let A be an arbitrary outcome (or collection of outcomes) of the defining experiment; we call A an *event*. The probability P of the event A is to be bounded by zero and one; that is, we require that

$$0 \leq P[A] \leq 1 \tag{5.2}$$

(2) Let $\mathcal{S}$ be the set of all possible outcomes of the defining experiment; $\mathcal{S}$ is then an event. Since some one of the possible outcomes must occur at each performance of the defining experiment, so must the event $\mathcal{S}$. We therefore call $\mathcal{S}$ the *sure* (or *certain*) event, and set

$$P[\mathcal{S}] = 1 \tag{5.3}$$

(3) Let A and B be *mutually exclusive* events, that is, events such that in a given performance of the defining experiment either the event A or the event B might occur, but not both. The probability that either A *or* B occurs must then be equal to the sum of their respective probabilities:

$$P[A \text{ or } B] = P[A] + P[B] \tag{5.4}$$

when A and B are mutually exclusive events. A comparison of these hypotheses with the results derived in Sec. 3 above shows that probability and relative frequency do indeed have similar properties.

We have made here an axiomatic definition of statistical probability; in other words, we have defined probability by specifying its basic properties and, in addition, we have asserted that probability models relative frequency. This approach is the one most commonly taken by those concerned with the application of probability to the physical sciences and engineering.

6 SUMMARY AND PREVIEW

We pointed out at the start of this chapter that while many physical phenomena of interest are unpredictable in detail, it is also true that such phenomena often do have properties which are reasonably predictable on the average. We then noted that it is therefore desirable to develop a mathematical theory of averages, and it was in this context that we introduced probability as a mathematical model of relative frequency and statistical average as a mathematical model of empirical average.

Now whenever an attempt is made to apply probability theory in a study of a physical system, a basic difficulty arises from the fact that probability theory tells us only that if we assign certain numbers as probabilities, then those probabilities must have certain properties and, given those probabilities, the theory tells us how to derive other probabilities and statistical averages. Unfortunately, however, the theory does not tell us how to assign the basic probabilities in the first place.

The situation before us now is essentially the same as arises in any other case of making a mathematical model of a physical system and is basically that shown schematically in Fig. 6.1. We start with a physical system which we wish to model mathematically so that we may predict the average performance of the system. The description of the physical system determines the structure of the mathematical model, but it does not tell us what values to assign to certain basic parameters in the model (such as probabilities). The basic probabilities in the model might be hypothesized on the basis of the system description (for example, we generally feel that the probability of a head occurring in a flip of a coin ought to be $\frac{1}{2}$), or they might instead be taken to be the values of the corresponding relative frequencies as measured in the physical system. In either case, the predicted performance of the system must be checked against the actual system performance so that the mathematical model (including the basic probabilities) may be adjusted accordingly if deemed necessary.

We shall develop the basic details of the theory of probability and statistical averages in the next six chapters and then we shall return to the question of the relation between a statistical average and an empirical average for further study in Chap. 8, Estimation, Sampling, and Prediction. Following that, we shall introduce the concept of a random process, study the effect of operating on random phenomena with linear systems, and, finally, study in some detail the problem of building up a probabilistic model of a physical phenomenon such as shot noise.

Because our objective for this text is limited to the providing of an introduction to the theory, however, there are many probability topics which we shall not touch upon; the reader is accordingly encouraged to explore these further in the available literature. A good introduction to

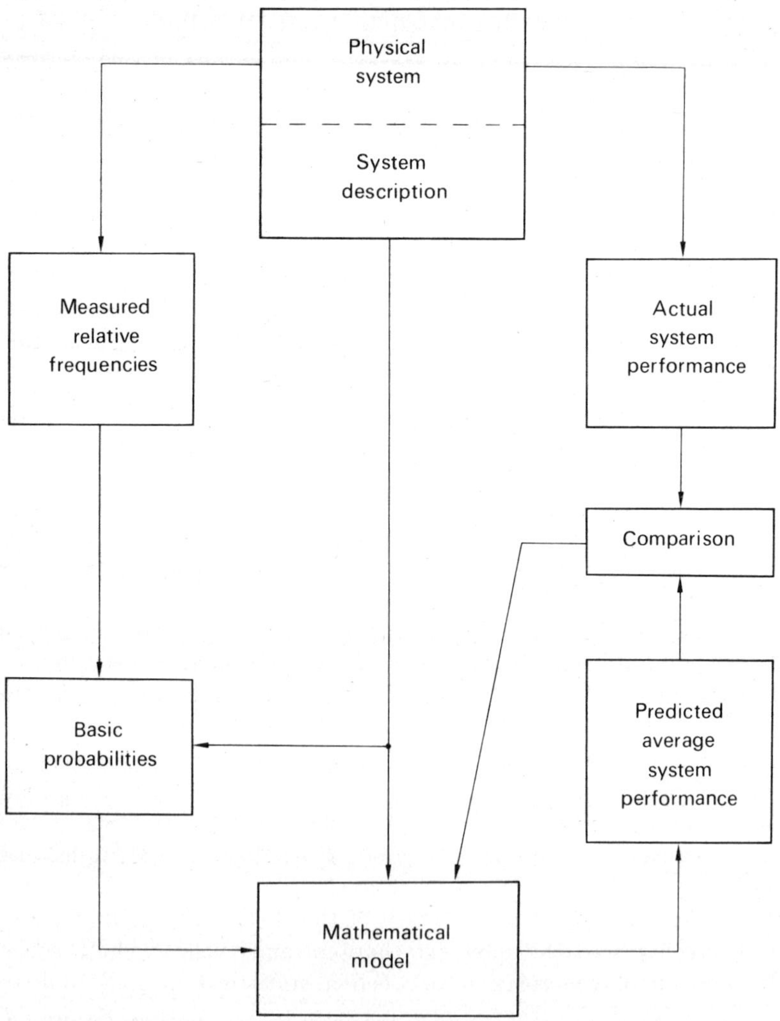

Figure 6.1 A physical system and its mathematical model.

probability theory and its application in estimation and inference problems is given in Cramér, while an excellent detailed introduction to the theory of probability and random processes is given in Feller (vol. I). Good intermediate level treatments may be found in the books by Gnedenko, Parzen, and Pfeiffer. Excellent advanced presentations of the mathematical foundations are given by Kolmogorov, Krickeberg, and Neveu, while standard advanced treatises include the works of Feller (vol. II) and Loève.

REFERENCES

Cramér, Harald: "The Elements of Probability Theory and Some of Its Applications," John Wiley & Sons, Inc., New York, 1955.

Feller, William: "An Introduction to Probability Theory and Its Applications," 3d ed., vol. 1, 1968, vol. 2, 1966, John Wiley & Sons, Inc., New York.

Gnedenko, Boris Vladimirovich: "Theory of Probability," Chelsea Publishing Company, New York, 1962.

Kolmogorov, Andrei N.: "Foundations of the Theory of Probability," Chelsea Publishing Company, New York, 1956. (English translation of 1933 German edition.)

Krickeberg, Klaus: "Probability Theory," Addison-Wesley Publishing Company, Inc., Reading, Mass., 1965.

Loève, Michel: "Probability Theory," 3d ed., D. Van Nostrand Company, Inc., Princeton, N.J., 1963.

Neveu, Jacques: "Mathematical Foundations of the Calculus of Probability," Holden-Day, Inc., Publisher, San Francisco, 1965.

Parzen, Emanuel: "Modern Probability Theory and Its Applications," John Wiley & Sons, Inc., New York, 1960.

Pfeiffer, Paul E.: "Concepts of Probability Theory," McGraw-Hill Book Company, New York, 1965.

2

Sample Points and Sample Spaces*

1 INTRODUCTION

We pointed out in the preceding chapter that whenever we wish to introduce the use of probability, we must first define an experiment and then specify all of the possible outcomes of that experiment. Now the outcomes of many experiments are often best described numerically. For example, the output of a given acoustical transducer at the time of measurement might be 2.5 millivolts; there might be six customers in a queue at the instant of observation; at a given hour on a given day, the temperature at Logan Airport in Boston might be 17° Fahrenheit and the pressure might be 30.48 inches of mercury. In such situations, an outcome of the given experiment can obviously be described geometrically by a point in some space (as shown in Fig. 1.1); the acoustical transducer output could be represented by a point on a line; whereas the Logan Airport temperature and pressure readings could well be represented by

* This entire chapter may be skimmed over rapidly if the reader is already quite familiar with the elements of set theory.

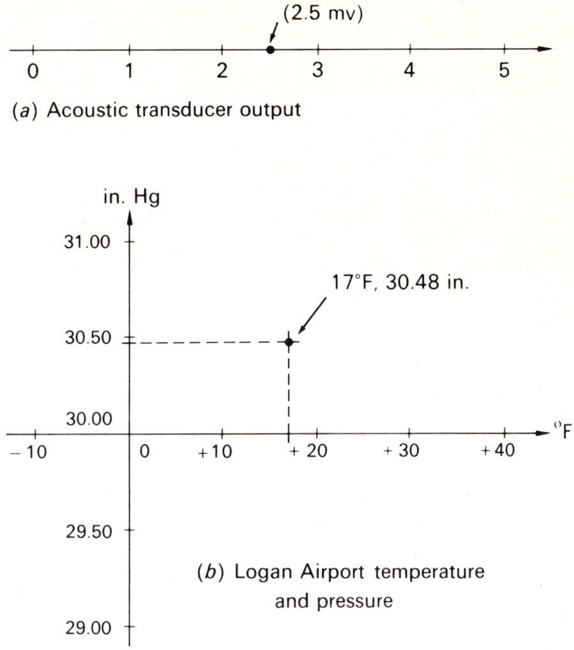

Figure 1.1 Plots of experimental results.

points in a plane. In general, if n numbers are required to specify the output of a given experiment, then that output could be represented by a point in an n-dimensional space.

While the outcomes of many experiments can be described numerically in a more or less obvious way, there are many other experiments for which a numerical description of the outcome is not particularly natural. Thus, for example, the possible results of a flip of a coin are head and tail; the seventh letter of an English language teletype message might be "a" or "b" or any other of the 26 letters of the English alphabet. Although in these cases one can, in fact, usually describe the outcome of the experiment numerically (for example, represent head by one and tail by zero), there is often no a priori way of doing so. Nonetheless, the idea of geometrically representing the outcome of an experiment as a point in some space can still be used. Thus, as shown in Fig. 1.2a, a space describing the possible outcomes of a flip of a coin could consist of two points, one labeled H and the other labeled T. Similarly, a space describing the possible seventh letters of English language teletype messages could be that consisting of 26 points labeled respectively "a," "b," "c," and continuing through "z," as shown in Fig. 1.2b. While such points

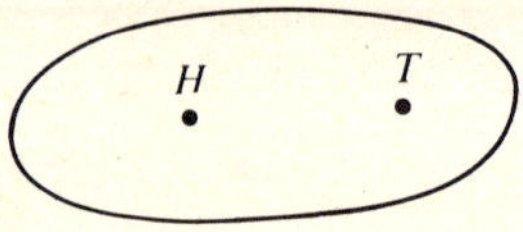

(*a*) Flip-of-a-coin sample points

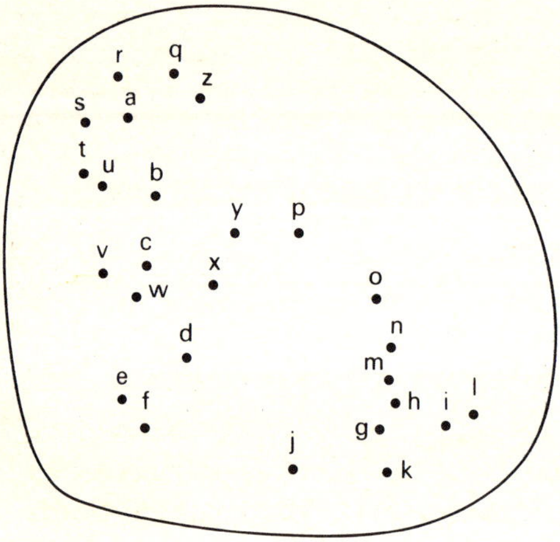

(*b*) Alphabet sample points

Figure 1.2 Some nonnumerical sample spaces with their sample points.

and spaces may, in some sense, be thought to be more abstract than those in the numerical cases, they are still useful.

In summary, we shall often find it convenient in our subsequent discussions to visualize experiments and their outcomes in terms of spaces and points (regardless of whether the appropriate space is concrete or abstract). We shall call points which represent the outcomes of an experiment *sample points*. The totality of such sample points (which represents the set of all possible outcomes of a given experiment) will be called the *sample space* of that experiment.

It is perhaps worth emphasizing at this point that, *given a particular physical situation, there is not necessarily a single unique sample-point and sample-space representation*. For example, consider the experiment of measuring the 6 A.M. temperature at Logan Airport. If the desired result of the measurement is the actual value of the temperature, then an appropriate sample space would be the set of real numbers representing all possible temperature readings. If, on the other hand, the only information desired is whether the temperature recorded was above or below the freezing point, then an appropriate sample space would be that consisting of only two sample points, one labeled *above freezing* and one labeled *below freezing*. Thus we see that, given a particular physical situation, different sample-point–sample-space representations of experimental outcomes are possible; which is most useful depends entirely upon what questions are to be asked.

Exercise 1.1 Sketch an appropriate sample space to represent the possible outcomes of the throw of a single die. □

Exercise 1.2 Suppose that the thermometer used to measure the temperature at Logan Airport has a maximum reading of 120°F and a minimum reading of −30°F, while the barometer used to measure the pressure has a maximum reading of 31.50 in. and a minimum reading of 27.50 in.

Sketch an appropriate sample space to represent the possible outcomes of the measurement of temperature and pressure. □

2 EVENTS

We previously defined in Chap. 1 an *event* to be an outcome, or a collection of outcomes, of a defining experiment. In a sample-point–sample-space representation then, an event is represented either by a single sample point or by a collection of sample points. We call such a collection or grouping of points a *set*; thus, an event may be represented by a set of sample points in a sample space. In particular, the event that *any one* of the possible outcomes of an experiment occurs (that is, the sure event), is represented by the set of *all* sample points (that is, by the sample space itself). The representation of events by sets of sample points will be found to be very useful in our subsequent discussions and derivations. We shall, in fact, find it most convenient from now on to employ the term event to denote *either* a collection of possible experimental outcomes *or* the corresponding set of sample points which represents that collection of outcomes. We shall therefore find it correspondingly convenient to denote an event by a single symbol whether we are referring to a collection of outcomes or to a set of sample points.

Let us, for example, consider the experiment of observing the number of data blocks stored at a particular time instant in a storage unit which has a capacity of 1024 blocks. Now since the number of blocks observed can be any integer between 0 and 1024, we can choose the sample space $\mathcal{S}$ to consist of those numbers and write

$$\mathcal{S} = \{0, 1, 2, \ldots, 1024\} \tag{2.1}$$†

† We shall generally use braces $\{\cdot\}$ to denote a set or collection of objects.

If the event A is the event that no blocks are in storage, then

$$A = \{0\} \tag{2.2}$$

that is, A is the set which consists of the single sample point 0. Similarly, if

$$B = \{513, 514, \ldots, 1024\} \tag{2.3}$$

then B is the event that the storage unit is more than half-full. The sample space $\mathcal{S}$ and the events A and B are shown pictorially in Fig. 2.1.

As a second example, consider the experiment consisting of measuring at some given instant the voltage output v from an acoustical transducer the maximum and minimum output of which are $+10$ and -10 volts, respectively. A suitable sample space would then be

$$\mathcal{S} = \{v: -10 \leq v \leq +10\} \tag{2.4}$$

the set of sample points v such that the value of v is greater than or equal to -10 and less than or equal to $+10$. In this case if

$$A = \{v: 0 < v \leq +10\} \tag{2.5}$$

then A is the event that the transducer output voltage is positive. This sample space and event are pictured in Fig. 2.2.

An examination of these two examples shows that an event (set)

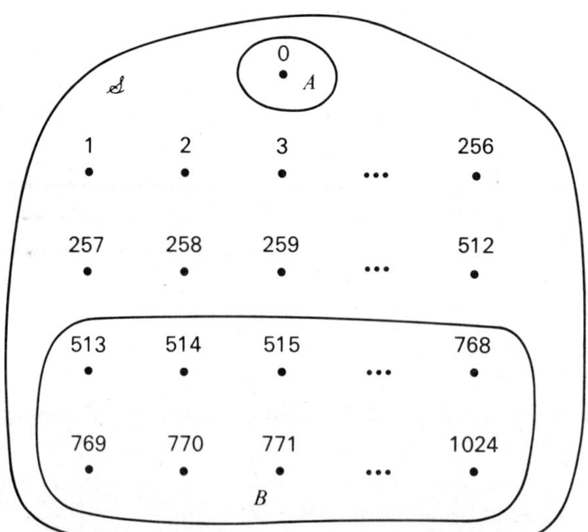

Figure 2.1 Events A and B in the storage unit sample space $\mathcal{S}$.

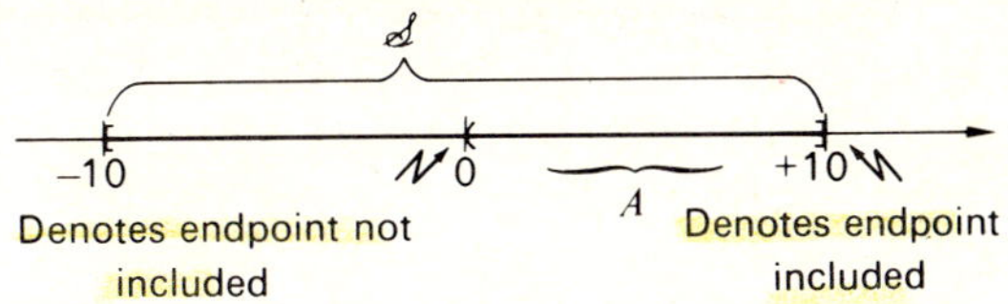

Figure 2.2 Transducer output event A and sample space $\mathcal{S}$.

may be defined either by *listing* [by enumeration of its elements as in (2.1) through (2.3)], or by *description* [by specification of the properties of its elements as in (2.4) and (2.5)]. It should be noted that definition by listing is possible only if the number of sample points in the event is finite or countably infinite;[1] definition by description is possible in any case, but is required if there is an uncountable number of sample points in the event (for in that case there are too many elements to list—either directly or by induction).

Now we say that two events A and B are *equal* (or *identical*) and write

$$A = B$$

if and only if A and B have the same elements, that is, if and only if A and B are simply two different names for the same set of elements. With this fact in mind, we note that it is often possible to define a given event either by listing or by description. Thus we may define an event A consisting of the two elements $+1$ and -1 either by

$$A \triangleq \{-1, +1\} \qquad \text{or by} \qquad A \triangleq \{x: x^2 = +1\}$$

since both expressions define the same elements.[2] It also follows from the definition of event equality that any two events defined by description are identical if the two descriptions are logically equivalent. Thus the events

$$B \triangleq \{k: k \text{ is a nonnegative, even integer}\}$$

and

$$D \triangleq \{k: k = 2j, j = 0, 1, 2, \ldots\}$$

are in fact one and the same event.

[1] A set has a *countably infinite* number of elements if and only if the elements (which are infinite in number) can be indexed by the positive integers. (Compare with chap. 2, sec. 12 of Kemeny et al.)

[2] We shall use the symbol $\triangleq$ to denote the relation "is defined to be" whenever we wish to emphasize the fact that we are defining something and not just dealing with an equality or an identity.

A second important relation between events is that of *implication*, or *inclusion*. Suppose, for example, that we consider a buffer storage unit with a capacity of 1024 data blocks and that we define the event A to be the event that "there are either two or four data blocks in storage" and the event B to be the event that "there is an even number of data blocks in storage," that is,

$$A \triangleq \{2,4\}$$

and

$$B \triangleq \{0, 2, 4, 6, \ldots, 1022, 1024\}$$

A moment's reflection upon these definitions should show that every sample point of A here is also a sample point of B. It therefore follows that if the event A occurs, then so also must the event B; we say in this case that the event A *implies* the event B and we write

$$A \subset B$$

It is also common in such a case to say that *A is included in B*, or that *A is a subset of B*.

Now we shall from here on denote the fact that a sample point s is an element of an event A by writing

$$s \in A$$

With our above specific example in mind, we shall then say in general that the event A *implies* the event B, that is,

$$A \subset B$$

if and only if for every

$$s \in A \tag{2.6}$$

it follows that

$$s \in B$$

It is often convenient to represent the relation of implication or inclusion schematically as shown in Fig. 2.3. There we have shown a sample

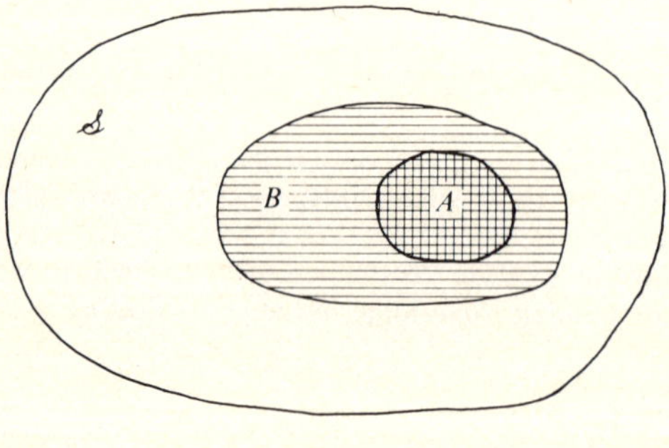

Figure 2.3 Venn diagram representing the inclusion relation $A \subset B$.

space $\mathcal{S}$, an event A (represented by the area shaded by vertical lines), an event B (represented by the area shaded with horizontal lines), and indicated the fact that A is included in B. Such a pictorial representation of abstract events (sets) and their relations in a sample space is commonly called a *Venn diagram*, named after the English logician John Venn (1834–1923), who introduced such diagrams to help visualize relations and operations in logic.

Exercise 2.1 Consider again the experiment of measuring the voltage output v from an acoustical transducer. Suppose that the maximum magnitude of the output voltage is 10 volts, that A is the event that the output voltage is positive, that B is the event that the magnitude of the output voltage is less than or equal to 1 volt, and that C is the event that the output voltage is greater than or equal to -1 volt.

a. Define the events $\mathcal{S}$, A, B, and C by description, for example, as in (2.4) and (2.5).
b. Sketch sample-point–sample-space representations, that is, Venn diagrams, of the events $\mathcal{S}$, A, B, and C.
c. Determine the implication relations between those events. □

It should be noted (as shown by the various examples above) that even if every element of an event A is also an element of an event B, it does *not* necessarily follow that every element of B is also an element of A. That is, the relation

$$A \subset B$$

does not necessarily imply the relation

$$B \subset A$$

However, if it *is* true that both $A \subset B$ and $B \subset A$, that is, if it is true that every element of A is an element of B and every element of B is an element of A, then it follows that the events A and B have the same elements and hence that A and B are equal. Thus we see that the events A and B are equal, that is,

$$A = B$$

if and only if both

$$A \subset B \tag{2.7}$$

and

$$B \subset A$$

As we shall see in the next section, this result provides a very useful tool for proving that two events are equal.

Finally, let us determine some further general properties of the event-inclusion relation: First, we note from our above discussion of equality that the inclusion relation is *antisymmetric;* that is, a result that $A \subset B$ implies $B \subset A$ if and only if $A = B$. Second, we note that it follows immediately from the definition of inclusion, (2.6), that the inclusion relation is *reflexive;* that is, that for every event A,

$$A \subset A \tag{2.8}$$

In other words, every event implies itself. An additional property is to be shown in the exercise that follows.

TRANSITIVITY

Exercise 2.2 Show that the inclusion relation is *transitive;* that is, show that if

$$A \subset B \quad \text{and} \quad B \subset C$$

then it follows that (2.9)

$$A \subset C$$

That is, if the event A implies the event B and if the event B implies the event C, then A implies C.† □

An examination of these general properties shows that the event-inclusion relation is an *order* relation which has properties essentially similar to those of the arithmetic relation $\leq$, that is, to the arithmetic relation "less than or equal to." [It should be noted that while the arithmetic relation $<$ (less than) is also antisymmetric and transitive, it is *irreflexive* as it is not true that $x < x$ for any real number x.]

3 ALGEBRA OF EVENTS

Many times in our subsequent discussions we will meet a situation in which we have defined certain events of interest and then find that we are also interested in various combinations of those events. For example, let us reconsider the experiment of counting the number of data blocks in a storage unit with a capacity of 1024 blocks. Suppose, for the moment, that we have defined A to be the event that the storage unit is more than half-full,

$$A \triangleq \{513, 514, \ldots, 1023, 1024\}$$

† However intuitively obvious this result might be, it nonetheless should be shown by the reader that it does, in fact, logically follow from the definition of inclusion as given by (2.6).

and B to be the event that there are an even number of blocks in storage,

$$B \triangleq \{0, 2, 4, \ldots, 1022, 1024\}$$

It might then happen that we are also interested in the event C that *either* the unit is more than half-full *or* that there is an even number of blocks in storage:

$$C \triangleq \{0, 2, 4, \ldots, 512, 513, 514, \ldots, 1023, 1024\}$$

In addition, we might also be concerned with the event D that the unit is more than half-full *and* that there is an even number of blocks in storage:

$$D \triangleq \{514, 516, \ldots, 1022, 1024\}$$

Finally, suppose that we are also interested in the event E that the unit is at most half-full:

$$E \triangleq \{0, 1, 2, \ldots, 511, 512\}$$

Now a comparison of the events A through E shows that C contains all the elements which are either in A or in B or in both, that D contains only those elements which are in both A and B, while E contains those elements of $\mathcal{S}$ which are not in A. Such combination events will occur over and over again and, accordingly, it will be convenient to give them special names—*union*, *intersection*, and *complement*, respectively—and to study their particular properties. This is the domain of the algebra of events.

The union

We call the combination event consisting of those elements which are either in an event A or in an event B or in both the *union* of A and B and will denote it by $A \cup B$. Thus we have:

$$A \cup B \triangleq \{s: s \in A \text{ or } s \in B \text{ or both}\} \tag{3.1}$$

This situation is represented schematically by the Venn diagram in Fig. 3.1, where the union event is indicated by shading. (It should be noted that the points which are common to both A and B are to be counted only once.)

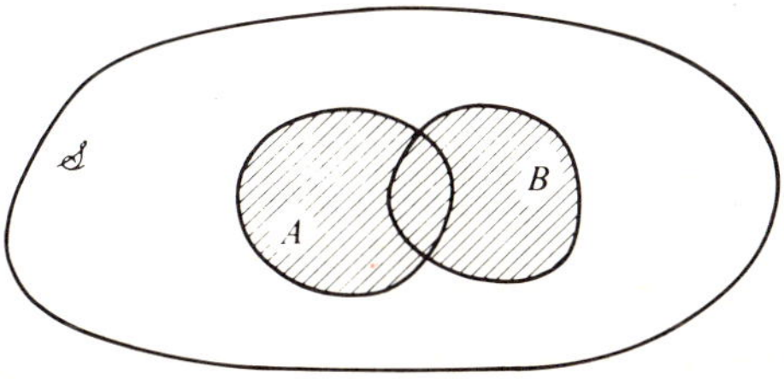

Figure 3.1 Venn diagram representing the union $A \cup B$.

Exercise 3.1 Consider again the experiment of measuring the voltage output v from an acoustical transducer as previously discussed in Exercise 2.1. Let the events A, B, and C be defined as in that exercise.

Show that it then follows that

$$C = A \cup B$$

by using the technique of proving event equality outlined in (2.7). □

Exercise 3.2 Spacecraft are to be landed at some given point 0 on the moon's surface, and we wish to observe the position of an actual landing point relative to the aiming point 0. Assume that the landing errors are small enough so that it is reasonable to measure them in the tangent plane to the moon's surface which touches at 0. Let us further establish a local rectangular coordinate system in that tangent plane which is centered on 0 and which is oriented so that the x axis points east and the y axis points north.

a. Let A be the event that the landing point is less than 5 km from the aiming point. Define the region of the xy tangent plane which corresponds to A, and sketch.

b. Let N, E, S, and W be the events that the landing point is to the north, east, south, or west, respectively, of the aiming point. Define the regions of the xy tangent plane which correspond to N, E, S, and W, and sketch.

c. Let the event B be defined by

$$B \triangleq A \cup N$$

Define the region of the xy tangent plane which corresponds to B, and sketch. □

It should be noted that it follows directly from the definition of the union that, given any two events A and B, it follows that

$$A \subset A \cup B$$

and further that (3.2)

$$B \subset A \cup B$$

An examination of the Venn diagram for the union should make this result intuitively obvious. The following theorem further connects the concepts of inclusion and union.

THEOREM 3.1 Let A and B be arbitrary events. Then

$$A \subset B$$

if and only if $\qquad$ (3.3)

$$A \cup B = B$$

Proof We have two statements to prove here:

a. The "if" statement. That is, we must show that *if* $A \cup B = B$, *then* it follows that $A \subset B$.

b. The "only if" statement. That is, we must show that, conversely, *if* $A \subset B$, *then* it follows that $A \cup B = B$.

Let us consider first the "if" statement. Suppose that

$$s \in A$$

It then follows from (3.2) and (2.6) that

$$s \in A \cup B$$

It further follows from the definition of event (set) equality that since $A \cup B = B$ here by hypothesis,

$$s \in B$$

Thus we have shown that if $A \cup B = B$, then $s \in A$ implies $s \in B$, that is, that

$$A \subset B$$

Next let us consider the "only if" statement; in particular, let us assume that $A \subset B$, and try to prove that $A \cup B = B$. Here we must remember that in order to prove equality, we must show that the two events in question have the same elements. One way of accomplishing this is to make use of the two-way inclusion idea in (2.7).

One-half of the proof of equality follows immediately from (3.2), for in any case

$$B \subset A \cup B$$

Suppose now that

$$s \in A \cup B$$

It then follows from the definition of union, (3.1), that

$$s \in A \qquad \text{or} \qquad s \in B$$

Suppose $s \in A$. Since by hypothesis $A \subset B$ here, it follows from (2.6) that then

$$s \in B$$

Thus we see that, in either case, if $s \in A \cup B$, then $s \in B$; using (2.6),

$$A \cup B \subset B$$

As we have thus proved inclusion both ways, we have established the desired equality; that is, we have shown that if $A \subset B$, then $A \cup B = B$. This completes the proof of the "only if" part of the theorem and thus Theorem 3.1 is proved in its entirety. ∎

We have deliberately gone through a theorem-proof argument in detail here in order to show just what is involved. First, a theorem states an important and useful result or conclusion and, in addition, gives in a concise and precise manner just which hypotheses are necessary for the conclusion. A proof is a sequence of statements which starts *from* the hypothesis and proceeds through a logical train of deductions *to* the conclusions. The theorem-proof style is particularly useful whenever it is desired to emphasize a conclusion or result and, further, to make certain that the conditions of validity of that result are clearly stated.[1] While we shall seldom get this formal again in our subsequent derivations, it must be remembered that we will be dealing throughout with a mathematical theory and, therefore, that results are valid only if they can be proved deductively. It should also be realized, however, that new results are most commonly first obtained by intuition or by inductive reasoning and then proved by deduction.

Exercise 3.3* Let A and B be arbitrary events. Show that

$$A \cup B = B \cup A \tag{3.4}$$

that is, that the union is a *commutative* operation. Show further that

$$A \cup (B \cup C) = (A \cup B) \cup C \tag{3.5}$$

that is, that the union is an *associative operation*. □

The intersection

The combination event consisting of those elements of two events A and B which are common to both is called the *intersection* of A and B and is denoted by $A \cap B$. Thus

$$A \cap B \triangleq \{s: s \in A \text{ and } s \in B\} \tag{3.6}$$

[1] A much more detailed but still very readable discussion of the theorem-proof style may be found in chap. 6 of Kershner and Wilcox.

* Supplementary exercise. The headings for various optional exercises and sections are printed in lighter type throughout.

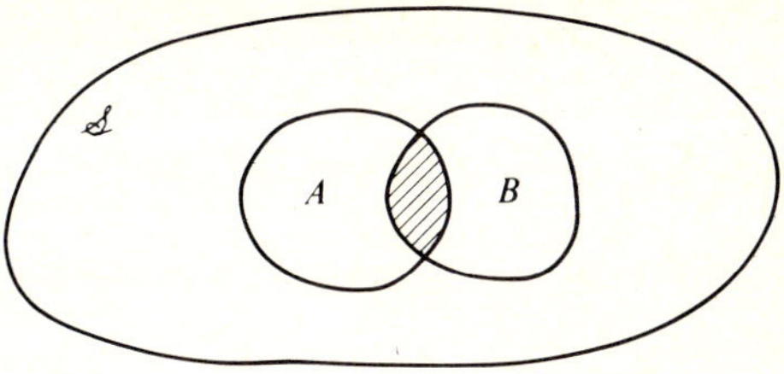

Figure 3.2 Venn diagram representing the intersection $A \cap B$.

Schematically, the situation is that pictured in Fig. 3.2 where the intersection is indicated by shading.

Exercise 3.4 Consider again the spacecraft landing-error measurement problem discussed in Exercise 3.2, and let the events A, N, and E be defined as in that exercise.

a. Define the region of the xy tangent plane which corresponds to the event $A \cap N$, and sketch.
b. Define the region of the xy tangent plane which corresponds to the event $N \cap E$, and sketch. □

From the definition of the intersection we see that, given any two events A and B, it follows that

$$(A \cap B) \subset A$$

and further that (3.7)

$$(A \cap B) \subset B$$

An examination of the Venn diagram for the event intersection should make this result more or less obvious. A further connection between the inclusion relation and the intersection operation is given by the next exercise.

Exercise 3.5 Let A and B be arbitrary events. Show that

$$A \subset B$$

if and only if (3.8)

$$A \cap B = A$$

(*Hint:* Parallel the proof of Theorem 3.1.) □

Exercise 3.6 Further basic properties of the intersection are as follows: Let A and B be arbitrary events.

a. Show that

$$A \cap B = B \cap A \tag{3.9}$$

that is, that intersection is a commutative operation.

b. Show in addition that

$$A \cap (B \cap C) = (A \cap B) \cap C \tag{3.10}$$

that is, that intersection is an associative operation. □

A review of the above definitions and results for the operations union and intersection shows that their properties are similar to those of the ordinary arithmetic operations of addition and multiplication, respectively. It should not be too surprising, therefore, to sometimes find the union called *sum* and the intersection called *product*. With this in mind, it should also not be too surprising to find that the result in the following exercise holds true.

Exercise 3.7 Let A, B, and C be arbitrary events. Show that

$$A \cap (B \cup C) = (A \cap B) \cup (A \cap C) \tag{3.11}$$

that is, that the operation of intersection is *distributive* over the union operation. □

On the other hand, it might not be expected that the following result is also true.

Exercise 3.8 Let A, B, and C be arbitrary events. Show that

$$A \cup (B \cap C) = (A \cup B) \cap (A \cup C) \tag{3.12}$$

that is, that the operation of union is distributive over the intersection operation. □

Exercise 3.9 Consider the relay network shown in Fig. 3.3. Let us denote by A the event that the relay A is closed, by B the event that the

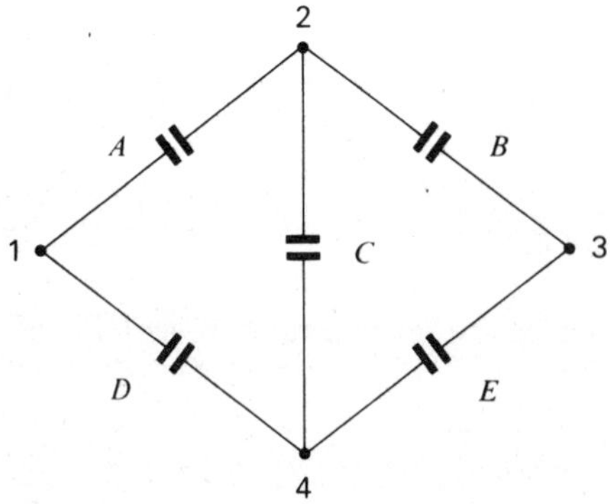

Figure 3.3 A relay network.

relay B is closed, etc. Let us also denote by K_{13} the event that there is a *closed* path between terminals 1 and 3.

a. Show by inspection of Fig. 3.3 that

$$K_{13} = (A \cap B) \cup (D \cap E) \cup (A \cap C \cap E) \cup (B \cap C \cap D)$$

b. Show also that

$$K_{13} = \{A \cap [B \cup (C \cap E)]\} \cup \{D \cap [E \cup (B \cap C)]\}$$

by applying the distributive laws of Exercises 3.7 and 3.8 to the formula in (*a*). □

The complement

Let us next consider the set of sample points which are *not* elements of a specified event. Given a particular experimental outcome (sample point), either that outcome is in a specific event A or it is not; in the latter case we write

$$s \notin A$$

The set of sample points of the corresponding sample space which are not in the event A is called the *complement* of the event A and is denoted by A^c. That is,

$$A^c \triangleq \{s: s \in \mathcal{S} \text{ and } s \notin A\} \tag{3.13}$$

The complement of the set A is thus represented schematically by the shaded area in Fig. 3.4.

Exercise 3.10 Consider again the spacecraft landing-error problem discussed in Exercises 3.2 and 3.4.

a. Define the regions of the xy tangent plane which correspond to the events A^c and N^c, and sketch.
b. What is the relation between the events S and N^c? (Careful!)
c. Define the region of the xy tangent plane which corresponds to the event $(A \cap N)^c$, and sketch. □

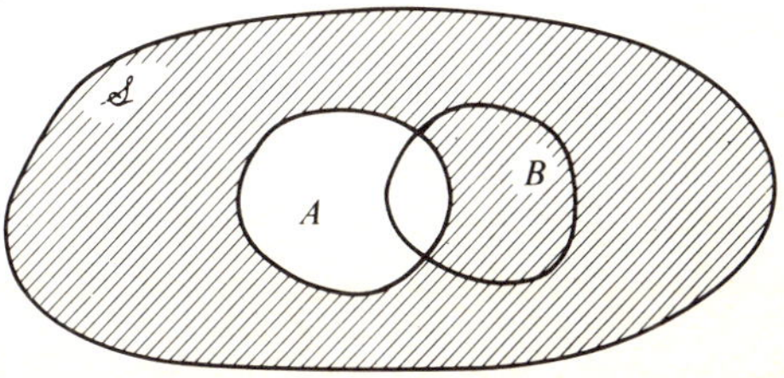

Figure 3.4 Venn diagram representing the complement A^c.

An examination of the definition of the complement shows that for any event A in $\mathcal{S}$,

$$(A^c)^c = A \tag{3.14}$$

that is, the complement of the complement of an event is that event itself.

A relation between the inclusion relation and the complementation operation is given in the next exercise.

Exercise 3.11 Let A and B be events in the sample space $\mathcal{S}$. Show that if

$$A \subset B$$

then it follows that

$$A^c \supset B^c \tag{3.15}$$

that is,

$$B^c \subset A^c$$

This statement is the event equivalent of the contrapositive argument in logic.[1] □

It should be noted that the definition of the complement requires the specification of both the sample space and the event whose complement is in question. A closely related concept to that of the complement is one in which we are not interested in all the points outside of a given event, but rather only in those outside points which are also elements of some other event. We define the *set difference* $B - A$ of the events A and B to be the set of sample points of B which are not also points of A:

$$B - A \triangleq \{s\colon s \in B \text{ and } s \notin A\} \tag{3.16}$$

Similarly we define

$$A - B \triangleq \{s\colon s \in A \text{ and } s \notin B\} \tag{3.17}$$

Now it follows from these definitions that if we take B to be the entire sample space $\mathcal{S}$, we get from (3.16)

$$\begin{aligned}\mathcal{S} - A &= \{s\colon s \in \mathcal{S} \text{ and } s \notin A\}\\ &= A^c\end{aligned} \tag{3.18}$$

Thus we see that the set difference $B - A$ is, in effect, the complement of the event A relative to the event B, rather than relative to the entire sample space $\mathcal{S}$. With this fact in mind, the set difference $B - A$ is

[1] See, for example, chap. 1, sec. 9 of Kemeny et al.

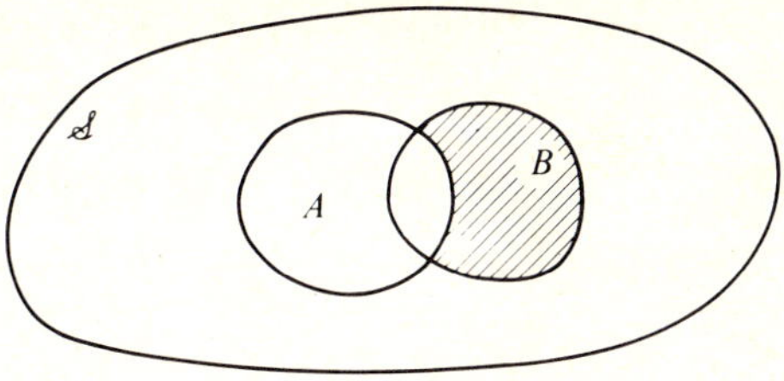

Figure 3.5 Venn diagram representing the relative complement $B - A$.

often called the *relative complement* of the event A with respect to the event B. The relative complement $B - A$ is thus that region in Fig. 3.5 indicated by the shaded area.

Exercise 3.12 Consider again the spacecraft landing-error problem discussed in Exercises 3.2, 3.4, and 3.10.

a. Define the region of the xy tangent plane which corresponds to the event $N - A$, and sketch.

b. Show that

$$N - A = N \cap A^c$$

in this case. □

The result obtained in part (*b*) of the last exercise is a special case of the general result obtained in the next exercise.

Exercise 3.13 Let A and B be arbitrary events in the sample space $\mathcal{S}$. Show that

$$B - A = B \cap A^c \tag{3.19}$$

Thus, we see that while the relative complement might seem to be a new and different operation, it is in fact expressible in terms of our previously defined intersection operation.

Even though the result here seems obvious from Fig. 3.5, it should be proved directly using the definitions of set difference, complement, and intersection. □

The null set

Let us reconsider for the moment the experiment of counting the number of data blocks in a storage unit with a capacity of 1024 blocks. Suppose here that A is the event that the storage unit is more than half-full

$$A \triangleq \{513, 514, \ldots, 1024\}$$

and that B is the event that there are 10 blocks in storage

$$B \triangleq \{10\}$$

Now we note that these two events have no elements in common and hence that their intersection contains no sample points. As this kind of situation will occur quite often in our studies to follow, it is convenient to introduce at this time some special terminology to describe this case. We shall say that two events which have no elements in common are *mutually exclusive*, or *disjoint*, events and note that if one occurs the other cannot. The event which contains no sample points we call the *null* event (or the *empty* or *vacuous* event) and denote by the symbol ϕ. Thus it follows that we have

$$A \cap B = \phi \tag{3.20}$$

for the events A and B defined in our above data storage example. It should be noted in this case that the null event ϕ and the event that no data blocks are in storage (the event $\{0\}$) are quite different events.

It follows immediately from our definitions above that *any* event and its complement are disjoint; that is,

$$A \cap A^c = \phi \tag{3.21}$$

for any event A in $\mathcal{S}$. It obviously also follows immediately that

$$\mathcal{S}^c = \phi \tag{3.22}$$

and hence that

$$\phi^c = \mathcal{S} \tag{3.23}$$

since if $\mathcal{S}$ contains all the allowable sample points, then its complement cannot contain any.

Exercise 3.14 Let A be an arbitrary event in $\mathcal{S}$ and let ϕ be the null event.

a. Show that

$$A \cup \phi = A \tag{3.24}$$

and further that

$$A \cap \phi = \phi \tag{3.25}$$

b. Show that the complementary results

$$A \cup \mathcal{S} = \mathcal{S} \tag{3.26}$$

and

$$A \cap \mathcal{S} = A \tag{3.27}$$

also hold for any event A in $\mathcal{S}$. □

It might well be expected that any event which contains no elements would be a rather slippery thing and have some unusual properties. We note, for example, that (3.25) says in essence that the null event is disjoint with *every* other event (and in fact with itself)! A further oddity results if we require the relations between event intersection and event inclusion [cf. (3.7)],

$$A \cap B \subset A \qquad \text{and} \qquad A \cap B \subset B \tag{3.28}$$

to hold for any pair of events—whether disjoint or not. If the events A and B do have elements in common, then it seems quite reasonable that (3.28) should hold true. If, however, A and B are disjoint events, i.e., if

$$A \cap B = \phi$$

and, further, if we wish (3.28) to cover this case, it necessarily follows that we must get the result that

$$\phi \subset A \qquad \text{and} \qquad \phi \subset B$$

here! However odd this result might seem to be, we would give up more than we would gain if we did not allow (3.28) to hold in the case of disjoint events. We shall therefore require from now on that, for *any* event A,

$$\phi \subset A \tag{3.29}$$

that is, the null event is "included" in every event!

Exercise 3.15 Consider a sample space consisting of a finite number of sample points, say n; that is, let

$$\mathcal{S} \triangleq \{s_1, s_2, \ldots, s_n\}$$

a. Enumerate all of the subsets of $\mathcal{S}$ when $n = 3$.
b. Show that for any n, there are 2^n subsets of $\mathcal{S}$.
[*Hint:* Do not forget (2.8) and (3.29).] □

De Morgan's rules

If we reconsider the spacecraft landing-error problem discussed in several of the preceding exercises, we note that it follows from the results of Exercise 3.10 that

$$(A \cap N)^c = A^c \cup N^c$$

where A is the event that the landing error is less than 5 km and N is the event that the landing point is to the north of the aiming point. This conclusion is a special case of one of the results in the next exercise.

Exercise 3.16 Let A and B be arbitrary events in a sample space $\mathcal{S}$. Show that

$$(A \cap B)^c = A^c \cup B^c \tag{3.30}$$

and further that

$$(A \cup B)^c = A^c \cap B^c \tag{3.31}$$

The results are often known as *De Morgan's rules*, named after the English mathematician Augustus De Morgan (1806–1871). □

It is interesting to note that (3.31) can be gotten from (3.30) simply by interchanging the operations of union and intersection. This is a special case of an important result following from De Morgan's rules called the *principle of duality*.

Any true identity formed of unions, intersections, and complements of events remains true if in it the symbols

$$\cup \quad \cap \quad \subset \quad \mathcal{S} \quad \text{and} \quad \phi$$

are replaced, respectively, by the symbols

$$\cap \quad \cup \quad \supset \quad \phi \quad \text{and} \quad \mathcal{S}$$

as long as the relation of equality and the operation of complementation are left unchanged.

Exercise 3.17 Consider again the relay network shown in Fig. 3.3 and discussed in Exercise 3.9. The event $K_{13}{}^c$ is the event that there is not a closed path between terminals 1 and 3.

a. Show that

$$K_{13}{}^c = (A^c \cap D^c) \cup (B^c \cap E^c) \cup (A^c \cap C^c \cap E^c) \cup (B^c \cap C^c \cap D^c)$$

by using the principle of duality to obtain the complement of K_{13} given in Exercise 3.9 and then applying De Morgan's rules.

b. Remembering that the event A^c corresponds to the relay A being open, etc., restudy Fig. 3.3 in the light of the result in part (*a*). □

We have tried so far in this chapter to give an introduction to those parts of the algebra of events (sets) which are necessary for our discussions and derivations to follow. Somewhat more complete elementary treatments of this material may be found, for example, in chaps. 1 and 2 of Kemeny et al., and in secs. 1 to 10 of Halmos. A very interesting and quite readable discussion of both the pertinent results and the concepts involved may be found in Fraenkel's book.

4 PARTITIONS*

A common problem in probability theory and its applications is the splitting up of a space or an event into disjoint events. Now it follows from the definition of complement that any set, say A, and its complement, A^c, are disjoint. Thus

$$A \cap A^c = \phi$$

It further follows from that definition that

$$A \cup A^c = \mathcal{S}$$

that is, that A and its complement *cover* the sample space $\mathcal{S}$ in the sense that every point of $\mathcal{S}$ is contained in the union of A and A^c. We therefore say that A and A^c *partition* the sample space $\mathcal{S}$.

In general, a collection of sets $A_1, A_2, \ldots, A_n$ is said to *partition* the sample space $\mathcal{S}$ if and only if (whether n is finite or infinite)

$$\mathcal{S} = A_1 \cup A_2 \cup \cdots \cup A_n \tag{4.1}$$

$$\mathcal{S} \triangleq \bigcup_{i=1}^{n} A_i$$

that is, if they cover the sample space, and if

$$A_i \cap A_j = \phi \qquad \text{when } i \neq j \tag{4.2}$$

that is, if they are pairwise disjoint. An example of such a partition is shown schematically in Fig. 4.1 for the case $n = 9$.

* While this section follows logically in the order given here, the motivation for introducing partitions probably will not become very obvious until after the first part of Chap. 3, Probability, has been studied.

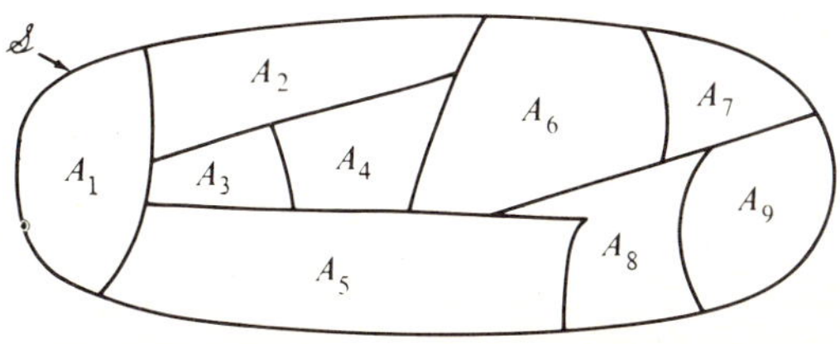

Figure 4.1 A partitioning of the sample space.

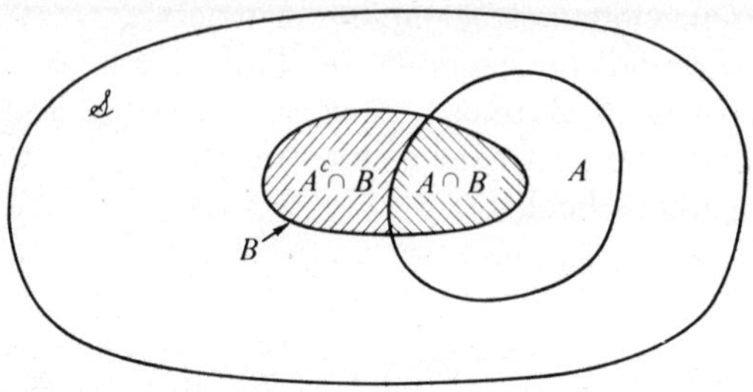

Figure 4.2 Partitioning of one event by another.

Once we have partitioned a sample space, we may always use such a partitioning to induce a partition of any event contained in that sample space. Suppose, for example, that we are given two events A and B (say, as shown in Fig. 4.2) and that we wish to partition B by A. Now it follows from the above results, and those of the preceding section, that

$$\begin{aligned} B &= B \cap \mathcal{S} = B \cap (A \cup A^c) \\ &= (B \cap A) \cup (B \cap A^c) \end{aligned}$$

that is, that the events formed by intersecting B with A and A^c cover B. It also follows that

$$\begin{aligned} (B \cap A) \cap (B \cap A^c) &= B \cap (A \cap A^c) = B \cap \phi \\ &= \phi \end{aligned}$$

and hence that the events $(B \cap A)$ and $(B \cap A^c)$ are disjoint. Since those events also cover B, we see that they therefore form our desired partition of the event B. This result is a special case of the following exercise.

Exercise 4.1 Suppose that the events $A_1, A_2, \ldots, A_n$ partition the sample space $\mathcal{S}$ and that B is an event in $\mathcal{S}$. Let

$$B_i \triangleq B \cap A_i \qquad i = 1, 2, \ldots, n \tag{4.3}$$

Show that

$$B = \bigcup_{i=1}^{n} B_i \tag{4.4}$$

and also that

$$B_i \cap B_j = \phi \qquad \text{when } i \neq j \tag{4.5}$$

that is, the events $B_1, B_2, \ldots, B_n$ formed by the intersection of B with the events A_i (which partition the sample space $\mathcal{S}$) partition the event B. □

5 SEQUENCES OF EVENTS*

Much of advanced probability theory has to do with the probabilistic properties of sequences of events, that is, with sequences of subsets of some given sample space. We wish to introduce in this section some of the basic ideas involved in a study of such sequences.

The sequence of events $\{A_1, A_2, A_3, \ldots\}$ is said to be a *monotone-increasing*[1] sequence of events if and only if each event in the sequence is contained in the immediately following one, that is, if and only if

$$A_1 \subset A_2 \subset A_3 \subset \cdots \tag{5.1}$$

For example, let the sample space $\mathcal{S}$ be the unit interval on the real line,

$$\mathcal{S} \triangleq \{x: 0 \leq x \leq 1\}$$

If the events A_i are defined by

$$A_i \triangleq \left\{x: \frac{1}{i+1} < x \leq 1\right\}$$

where $i = 1, 2, 3, \ldots$, then it follows that

$$A_1 \triangleq \{x: \tfrac{1}{2} < x \leq 1\}$$

that

$$A_2 = \{x: \tfrac{1}{3} < x \leq 1\}$$

and so forth. Thus we see that

$$A_1 \subset A_2 \subset A_3 \subset \cdots$$

here, and hence that the given sequence of events is in fact a monotone-increasing sequence of events.

Next, let us consider unions of the elements of a monotone-increasing sequence of events. Now it follows from Theorem 3.1 that, since

$$A_1 \subset A_2 \qquad \text{then } A_1 \cup A_2 = A_2$$

Further since

$$A_{n-1} \subset A_n$$

for every integer $n \geq 2$, we also have

$$A_{n-1} \cup A_n = A_n$$

* This section is supplementary and may be omitted if the reader so wishes. The material in it is somewhat more advanced and is required essentially only in our subsequent discussion of the continuity theorem of probability (Sec. 3.4).

[1] As in *monotonously*.

It therefore follows from these results that

$$\bigcup_{i=1}^{n} A_i = A_1 \cup A_2 \cup \cdots \cup A_n = A_n \tag{5.2}$$

when the events A_i form a monotone-increasing sequence.

Now our previous definition of a union of events was limited to the case of the union of a finite number of events. We may, however, easily extend that definition to cover the case of a union of a countably infinite collection of events by writing

$$\begin{aligned}\bigcup_{i=1}^{\infty} A_i &\triangleq \{s: s \in A_1 \text{ or } s \in A_2 \text{ or } s \in A_3 \text{ or } \cdots\} \\ &= A_1 \cup A_2 \cup A_3 \cup \cdots\end{aligned} \tag{5.3}$$

With this definition [and the result in (5.2)] in mind, we further define the *limit of a monotone-increasing sequence of events* to be

$$\begin{aligned}\lim_{n\to\infty} A_n &= \lim_{n\to\infty} \bigcup_{i=1}^{n} A_i \\ &= \bigcup_{i=1}^{\infty} A_i\end{aligned} \tag{5.4}$$

For example, let us reconsider the events A_i defined earlier. On using (5.2), it follows in this case that

$$\bigcup_{i=1}^{n} A_i = \left\{x: \frac{1}{n+1} < x \leq 1\right\}$$

and hence that

$$\lim_{n\to\infty} A_n = \{x: 0 < x \leq 1\}$$

It is interesting to note that for this sequence we have the result

$$\mathcal{S} = \{0\} \cup \lim_{n\to\infty} A_n$$

since the limit of the given sequence does not contain the point $x = 0$.

A second kind of monotone sequence of interest is that in which each succeeding event in the sequence is contained in the immediately preceding one, that is, a sequence of events B_i such that

$$B_1 \supset B_2 \supset B_3 \supset \cdots \tag{5.5}$$

Such a sequence of events is said to be a *monotone-decreasing sequence of events*. For example, let us again suppose the sample space $\mathcal{S}$ to be the unit interval on the real line and also that the events B_i are defined by

$$B_i \triangleq \left\{x: 0 < x < \frac{1}{i}\right\}$$

where $i = 1, 2, 3, \ldots$. Then

$$B_1 = \{x: 0 < x < 1\}$$

and

$$B_2 = \{x: 0 < x < \tfrac{1}{2}\}$$

and so forth. Thus we see that in this case

$$B_1 \supset B_2 \supset B_3 \supset \cdots$$

and hence that this sequence of events is indeed monotone decreasing.

Next let us consider intersections of the elements of a monotone-decreasing sequence of events. Now it follows from the results of Exercise 3.5 that, since

$$B_1 \supset B_2 \qquad \text{then } B_1 \cap B_2 = B_2$$

Further, since

$$B_{n-1} \supset B_n$$

for any integer $n \geq 2$, we get

$$B_{n-1} \cap B_n = B_n$$

It therefore follows from these results that

$$\begin{aligned} \bigcap_{i=1}^{n} B_i &= B_1 \cap B_2 \cap B_3 \cap \cdots \cap B_n \\ &= B_n \end{aligned} \tag{5.6}$$

when the events B_i form a monotone-decreasing sequence of sets.

Now just as our first definition of a union of events was limited to the case of a finite number of sets, so also was our previous definition of intersection limited to the case of a finite number of events. However, we may again easily extend our definition to cover the case of a countably infinite collection of events and write

$$\begin{aligned} \bigcap_{i=1}^{\infty} B_i &= B_1 \cap B_2 \cap B_3 \cap \cdots \\ &\triangleq \{s: s \in B_1 \text{ and } s \in B_2 \text{ and } s \in B_3 \text{ and } \cdots\} \end{aligned} \tag{5.7}$$

With this definition [and the result contained in (5.6)] in mind, we further define the *limit of a monotone-decreasing sequence of events* to be

$$\lim_{n\to\infty} B_n = \lim_{n\to\infty} \bigcap_{i=1}^{n} B_i = \bigcap_{i=1}^{\infty} B_i \tag{5.8}$$

For example, let us consider the events B_i defined above. It then follows in this case, on using (5.6), that

$$\bigcap_{i=1}^{n} B_i = \left\{x: 0 < x < \frac{1}{n}\right\}$$

and hence that

$$\lim_{n\to\infty} B_n = \{x: 0 < x < 0\} = \phi$$

since there is no real number x such that x is both greater than zero and less than zero.

One result of our above discussion is that we have obtained expressions for the limits of monotone[1] sequences of events. The question of defining the limit of an arbitrary sequence of events is more complicated[2] and will not be dealt with by us. Fortunately, however, whenever we do have to deal with sequences of events in our subsequent discussions, we will always be able to structure the problem at hand in such a way that we never have to deal with other than monotone sequences.

6 SUMMARY OF DEFINITIONS AND FORMULAS

Sample point

A sample point is a representation of a possible outcome of an experiment.

Sample space

A sample space is the totality of all possible sample points, that is, the representation of all possible outcomes of an experiment.

Event

An event is an outcome or a collection of outcomes. It is also defined as the corresponding sample point or set of sample points, respectively.

[1] Either monotone increasing or monotone decreasing.
[2] See, for example, chap. 1, sec. 1.3 of Loève.

Event defined by listing

$A \triangleq \{s_1, s_2, \ldots, s_n\}$ or $B \triangleq \{s_1, s_2, s_3, \ldots\}$

Event defined by description

$A \triangleq \{s\colon \text{prop }(s) \text{ is true}\}$

where prop (s) is some proposition about s; for example, $|s| < 1$.

Equality

Two events A and B are equal,

$A = B$

if and only if A and B have the same elements.

Implication or inclusion

The event A implies the event B,

$$A \subset B \tag{2.6}$$

if and only if for every $s \in A$ it follows that $s \in B$.

Equality

$$A = B \tag{2.7}$$

if and only if both $A \subset B$ and $B \subset A$.

Union

a.

$$A \cup B \triangleq \{s\colon s \in A \text{ or } s \in B \text{ or both}\} \tag{3.1}$$

b. In any case,

$$A \subset A \cup B \quad \text{and} \quad B \subset A \cup B \tag{3.2}$$

c.

$A \subset B$

if and only if (3.3)

$A \cup B = B$

Intersection

a.

$$A \cap B \triangleq \{s\colon s \in A \text{ and } s \in B\} \tag{3.6}$$

b. In any case,

$$A \cap B \subset A \qquad \text{and} \qquad A \cap B \subset B \tag{3.7}$$

c.

$$A \subset B$$

if and only if (3.8)

$$A \cap B = A$$

Distributive laws

$$A \cap (B \cup C) = (A \cap B) \cup (A \cap C) \tag{3.11}$$

and

$$A \cup (B \cap C) = (A \cup B) \cap (A \cup C) \tag{3.12}$$

Complement

a.

$$A^c \triangleq \{s\colon s \in \mathcal{S} \text{ and } s \notin A\} \tag{3.13}$$

b.

$$(A^c)^c = A \tag{3.14}$$

c. If $A \subset B$, then

$$B^c \subset A^c \tag{3.15}$$

Relative complement

a.

$$B - A \triangleq \{s\colon s \in B \text{ and } s \notin A\} \tag{3.16}$$

b.

$$B - A = B \cap A^c \tag{3.17}$$

Null set

a. The null set ϕ is the set which contains no sample points.

b.

$$A \cap A^c = \phi \tag{3.21}$$

c.

$$\mathcal{S}^c = \phi \tag{3.22}$$

d.

$$A \cup \phi = A \tag{3.24}$$

and

$$A \cap \phi = \phi \tag{3.25}$$

e.

$$\phi \subset A \text{ for any } A \subset \mathcal{S} \tag{3.29}$$

Disjoint events

The events A and B are disjoint if and only if

$$A \cap B = \phi$$

De Morgan's rules

$$(A \cap B)^c = A^c \cup B^c \tag{3.30}$$
$$(A \cup B)^c = A^c \cap B^c \tag{3.31}$$

Partitions

A collection of events $A_1, A_2, \ldots, A_n$ is said to partition the sample space $\mathcal{S}$ if and only if

$$\bigcup_{i=1}^{n} A_i = \mathcal{S} \tag{4.1}$$

and

$$A_i \cap A_j = \phi \qquad \text{for } i \neq j \tag{4.2}$$

REFERENCES

Fraenkel, Abraham A.: "Abstract Set Theory," North-Holland Publishing Company, Amsterdam, 1953.

Halmos, Paul R.: "Naive Set Theory," D. Van Nostrand Company, Inc., Princeton, N.J., 1960.

Kemeny, John G., Hazleton Mirkil, J. Laurie Snell, and Gerald L. Thompson: "Finite Mathematical Structures," Prentice-Hall, Inc., Englewood Cliffs, N.J., 1958.

Kershner, Richard B., and Lee Roy Wilcox: "The Anatomy of Mathematics," The Ronald Press Company, New York, 1950.

Loève, Michel: "Probability Theory," 3d ed., D. Van Nostrand Company, Inc., Princeton, N.J., 1963.

3
Probability

1 PROBABILITY AXIOMS

Having introduced in the preceding chapter the concepts of sample points and sample spaces to represent the possible outcomes of an experiment and the totality of such outcomes, respectively, and having also discussed events (collections of possible outcomes) in terms of sample-point representations, let us now reformulate our earlier list of desired properties for probability in terms of these new concepts. The reader is warned that our new list of hypothesized properties (which we shall call *axioms*) is somewhat different from that given in Sec. 1.4, for the aim in choosing axioms is to assume no more than is absolutely required.[1]

> *AXIOM 1* *Given an experiment, there exists a sample space $\mathcal{S}$ representing the totality of possible outcomes of that experiment, and a collection $\mathcal{A}$ of subsets A of $\mathcal{S}$, called events.*

[1] See, for example, sec. 6.1 of Kershner and Wilcox.

It should be remembered that there are generally a number of possible sample spaces which correspond to any single given experiment;[1] the choice of which sample space to use is commonly made on the basis of convenience.

AXIOM 2 *To each event A in the collection of events $\mathcal{A}$, there can be assigned a nonnegative real number $P[A]$, that is, a number such that*

$$0 \leq P[A]$$

This number is called the probability of the event A.

We call the rule which determines the assignment of a number $P[A]$ to each set (event) A in a collection of sets $\mathcal{A}$, a *set function* defined on that collection. Now Axiom 2 tells us only that a set function P *can* be defined on the sets A; it does not tell us *how* to define that function. The answer to the latter question can be determined only from a detailed knowledge of the physical situation being modeled. Thus, for example, the probability $P[6]$ that a die will show a 6 when thrown should be $\frac{1}{6}$ if it is equally likely that each of the numbers 1 through 6 will show, that is, if the die is a fair die. If, however, someone has weighted that die so that a 6 can practically never show, then it is much more reasonable to take $P[6]$ to be zero.

A comparison of Axiom 2 above with property 1, Sec. 1.5, shows that while we assumed earlier that $0 \leq P[A] \leq 1$, we require here only that $0 \leq P[A]$. Although the earlier hypothesis seems to be the more natural one in view of our desire to have probability model relative frequency, it will be shown later (in Exercise 2.3) that the statement $P[A] \leq 1$ follows as a conclusion from our present axioms and hence that it does not have to be made as an a priori hypothesis.

AXIOM 3 *For any probability assignment P, it is required that*

$$P[\mathcal{S}] = 1$$

Axiom 3 states that the probability of the sure, or certain, event is unity. It should be realized, however, that the converse statement, that is, if $P[A] = 1$, then $A = \mathcal{S}$, does not necessarily hold true. Suppose, for example, that we are given a wheel with a very low (but not zero) bearing friction, that we mark the highest point on the wheel, and that we then spin the wheel and note the final position of our mark. As we shall see later (in Exercise 2.2), the probability is unity that an unmarked point will be the highest point after spinning—even though

[1] Compare with Sec. 2.1.

the set of all unmarked points does not exhaust the set of all possible outcomes (for it is certainly possible for the marked point to end up on top!).

AXIOM 4 *If the events A and B are mutually exclusive (disjoint) events, that is,*

$$A \cap B = \phi$$

then any probability assignment P that is made must be such that

$$P[A \cup B] = P[A] + P[B]$$

Such a set function is said to be finitely additive.

As we shall see below (Theorem 2.1ff), the above four axioms suffice for all probability problems in which the given sample space has only a finite number of sample points. In order to deal with sample spaces which consist of an infinite number of sample points, however, we must extend Axiom 4 to become:

AXIOM 5 *If the events A_1, A_2, A_3, . . . are mutually exclusive events,*

$$A_i \cap A_j = \phi \qquad \text{for } i \neq j$$

then any probability assignment P that is made must be such that

$$P\left[\bigcup_{i=1}^{\infty} A_i\right] = P[A_1] + P[A_2] + P[A_3] + \cdots$$

$$= \sum_{i=1}^{\infty} P[A_i]$$

Such a set function is said to be countably additive.

A moment's reflection upon the last two axioms should show that we have defined how to determine the probability of a union of events only for those cases in which the events under consideration are mutually exclusive or disjoint. *It is for this reason that whenever we are faced in the future with a problem in which we are given an arbitrary collection of events, we will generally have to construct from them a new collection of disjoint sets.*

In summary, we note that a formal definition of probability involves the specification of a sample space $\mathcal{S}$, an appropriate collection $\mathcal{A}$ of events in that sample space, and a set function P which is defined on the elements of $\mathcal{A}$ and which has the properties specified in Axioms 2 through 5. The combination of these three entities [the triple $(\mathcal{S}, \mathcal{A}, P)$] is called a *probability space.* The probability axioms which we have given here are essentially the same as those first proposed (in 1933) by the Russian mathematician Andrei Nikolaevich Kolmogorov (1903–).[1]

2 ELEMENTARY PROPERTIES OF PROBABILITY

Let us next derive a few of the elementary properties of probability from the axioms given in the preceding section. In what follows we shall generally give (or ask for) complete details of each derivation; later on we shall be much less formal. Our motivation in being so complete in our derivations here is to demonstrate the basic techniques used in solving probability problems and to give practice in their use. We shall assume in all the general examples and exercises of this section that some appropriate probability space $(\mathcal{S}, \mathcal{A}, P)$ is given; that is, we shall assume that a sample space $\mathcal{S}$, an appropriate collection $\mathcal{A}$ of events, and a probability assignment P are all given whenever needed.

LEMMA† 2.1 COMPLEMENTS Given any event A in a sample space $\mathcal{S}$ and the probability $P[A]$,

$$\boxed{P[A^c] = 1 - P[A]} \tag{2.1}$$

that is, the probability of the complement of an event equals one minus the probability of the given event.

Proof Remembering the definition of the complement, we recall that the union of any event and its complement covers the entire sample space,

$$A \cup A^c = \mathcal{S}$$

and, further, that an event and its complement are disjoint,

$$A \cap A^c = \phi$$

It then follows from Axiom 3 that

$$P[A \cup A^c] = P[\mathcal{S}] = 1$$

Since A and A^c are disjoint, we may next use Axiom 4 to get

$$P[A \cup A^c] = P[A] + P[A^c]$$

The assertion (2.1) may now be obtained by combining the last two results and solving for $P[A^c]$. ∎

[1] Compare with chap. I (arts. 1, 2, and 3) and chap. II of Kolmogorov's monograph.
† A *lemma* is a small theorem.

The proof of Lemma 2.1 illustrates a basic technique of proof which we shall use over and over again: *an event of particular interest is decomposed, or partitioned, into a collection of disjoint events and then the additivity property of probability is used.*

Exercise 2.1 Using (2.1), show that if ϕ is the null event, then

$$P[\phi] = 0 \tag{2.2}$$

that is, that the probability of the null event is zero (as it obviously should be!). □

It should be noted that the converse to the result of Exercise 2.1 is not necessarily true; it is not necessarily true that if $P[A] = 0$, then $A = \phi$. An example of this situation is given in the next exercise.

Exercise 2.2 Consider a wheel with a very low, but not zero, bearing friction. Suppose that a mark is put at the top of the wheel, that the wheel is then spun, that the position of the mark is noted after the wheel has stopped spinning, and that the counterclockwise angle θ through which the wheel must be rotated in order to bring the marked point uppermost is noted. This is the situation shown schematically in Fig. 2.1. The results of many experiments show that it is reasonable to assume that

$$P[a \le \theta \le b] = \frac{b - a}{2\pi}$$

that is, that the probability that θ falls between two angles a and b (each

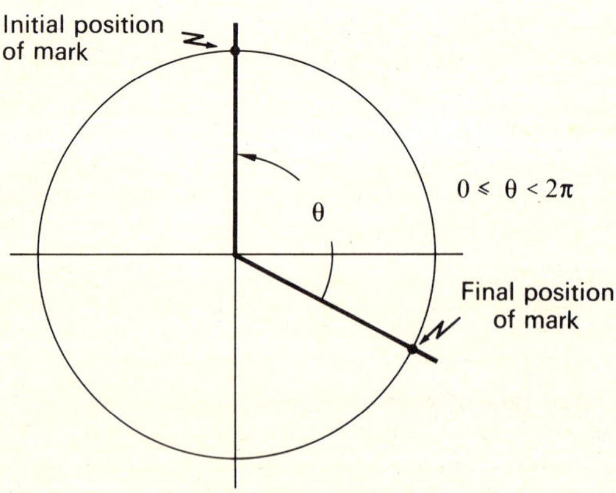

Figure 2.1 The spinning wheel problem.

of which is equal to or greater than zero and less than 2π) is equal to the difference between those angles divided by 2π.

a. Show that the probability is zero that the final position of the mark is exactly the same as the original position; that is,

$$P[\theta = 0] = 0$$

even though the event $[\theta = 0]$ can in fact happen.

b. Show that the probability is unity that the final position of the mark has any value other than its original position; that is,

$$P[\theta \neq 0] = 1$$

even though the event $[\theta \neq 0]$ does not correspond to the totality of all of the possible outcomes. □

Exercise 2.3 Using (2.1), show that for any event A in a sample space $\mathcal{S}$

$$P[A] \leq 1 \tag{2.3}$$

that is, that the probability of any event is less than or equal to unity. □

Exercise 2.3 shows that the statement "$P[A] \leq 1$, for any event A" does indeed follow from the axioms as stated in Sec. 1 of this chapter and hence that it would be redundant to include that statement in those axioms.

The following six exercises develop further elementary properties of probability. The technique of solution to be used is again that of partitioning the event of interest into disjoint subsets and then using the additivity property of probability. The use of Venn diagrams will often help in visualizing the various events under consideration.

Exercise 2.4 Let A and B be arbitrary events in a sample space $\mathcal{S}$. Show that

$$P[B - A] = P[B] - P[A \cap B] \tag{2.4}$$

that is, that the probability of the event difference $B - A$ equals the probability of the event B minus the probability of occurrence of those sample points which are common to both of the events A and B. □

Exercise 2.5 Let A and B be two events in a sample space $\mathcal{S}$. Show that if

$$A \subset B$$

then (2.5)

$$P[B - A] = P[B] - P[A]$$

that is, that the probability of an event difference equals the difference of the probabilities of the respective events when one event is included in the other.

□

Exercise 2.6 Let A and B be two events in a sample space $\mathcal{S}$. Show that if

$$A \subset B$$

then

$$P[A] \leq P[B] \tag{2.6}$$

A set function P which has this property is said to be a *monotone* function.

□

Exercise 2.7 Let A and B be two arbitrary events in a sample space $\mathcal{S}$.

a. Show that

$$P[A \cup B] = P[A] + P[B] - P[A \cap B] \tag{2.7}$$

b. Show that it therefore follows that

$$P[A \cup B] \leq P[A] + P[B] \tag{2.8}$$

Thus, we see from (2.7) that the probability of occurrence of the union of two arbitrary events is less than the sum of the respective probabilities of those events by an amount equal to the probability of occurrence of those sample points which are common to both events. The probability property expressed by (2.8) is called *subadditivity*.

□

A moment's reflection on (2.7) shows that we get the result

$$P[A \cup B] = P[A] + P[B] \tag{2.9}$$

whenever the probability of the intersection of the events A and B is zero:

$$P[A \cap B] = 0 \tag{2.10}$$

Now it follows from Axiom 4 that this condition is met whenever A and B are disjoint. It does not necessarily follow, however, that the converse is true, that is, that if two events A and B satisfy both (2.9) and (2.10), then A and B are disjoint. Consider, for example, the following exercise.

Exercise 2.8 Recall the spinning wheel discussed in Exercise 2.2 above. Let the events A and B be defined by

$$A \triangleq \left\{\theta : 0 \leq \theta \leq \frac{\pi}{4}\right\} \quad \text{and} \quad B \triangleq \left\{\theta : \frac{\pi}{4} \leq \theta \leq \pi\right\}$$

where θ is the angle defined in Exercise 2.2 and has the probability assignment given there.

a. Determine the events $A \cup B$ and $A \cap B$.
b. Determine $P[A]$, $P[B]$, $P[A \cup B]$, and $P[A \cap B]$, and show that

$$P[A \cup B] = P[A] + P[B]$$

even though A and B are not disjoint events. □

It follows from the definition of union that $A \cup B$ is the event that *at least one* of the events A and B occurs. We are also often interested in the probability that *exactly one* of a pair of events occurs. Referring to the Venn diagram in Fig. 2.2, we see that the set of sample points which represents the event that exactly one of the events A and B occurs is the set containing those points of A which are not also points of B plus those points of B which are not also points of A, that is, the event $(A - B) \cup (B - A)$. This event is often called the *symmetric difference* of the events A and B and is denoted by the symbol $A \triangle B$:

$$\boxed{A \triangle B \triangleq (A - B) \cup (B - A)} \tag{2.11}$$

is the event that exactly one of the events A and B occurs.

Exercise 2.9 Let A and B be arbitrary events in a sample space $\mathcal{S}$. Show that

$$P[A \triangle B] = P[(A - B) \cup (B - A)] = P[A \cup B] - P[A \cap B] \tag{2.12}$$

that is, that the probability that exactly one of two events occurs is equal to the probability that at least one occurs minus the probability that both occur. □

The derivations of our various results to this point have really only required that we deal with two events at a time and, therefore, have required only the use of Axioms 1 to 4 (and not 5). In order to be able to consider more complicated probability problems, however, we need to

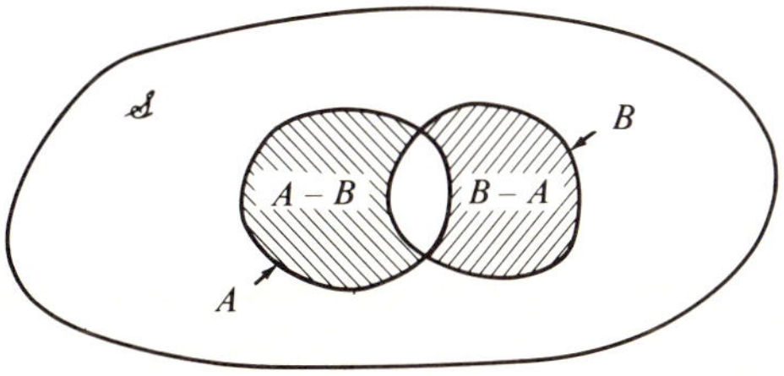

Figure 2.2 Venn diagram for the occurrence of exactly one of two events.

be able to deal with unions of more than two events. One way of accomplishing this is, of course, to employ Axiom 5, the axiom of countable additivity. That this is not in fact necessary in the case of a *finite* number of disjoint events, though, is shown by Theorem 2.1.

THEOREM 2.1 FINITE DISJOINT UNIONS Given only Axioms 1 through 4, it follows that if $A_1, A_2, \ldots, A_n$ are mutually exclusive events, then

$$P\left[\bigcup_{i=1}^{n} A_i\right] = \sum_{i=1}^{n} P[A_i] \tag{2.13}$$

that is, that the probability of a finite union of disjoint events equals the sum of the probabilities of those events.

Proof We must now show that the theorem statement holds for *any* finite, nonnegative, integer value of the parameter n.

First, it is trivial that (2.13) holds when $n = 1$, for in that case we have

$$P[A_1] = P[A_1]$$

When $n = 2$, (2.13) states that

$$P[A_1 \cup A_2] = P[A_1] + P[A_2]$$

which holds true as a consequence of Axiom 4. When $n = 3$, we have from (2.13)

$$P[A_1 \cup A_2 \cup A_3] = P[A_1] + P[A_2] + P[A_3]$$

which remains to be proven. Now we note that we may write

$$A_1 \cup A_2 \cup A_3 = (A_1 \cup A_2) \cup A_3$$

and we note further that the events $(A_1 \cup A_2)$ and A_3 are disjoint since the events A_1, A_2, and A_3 are all disjoint events by hypothesis. On using Axiom 4 twice, we obtain

$$\begin{aligned} P[(A_1 \cup A_2) \cup A_3] &= P[A_1 \cup A_2] + P[A_3] \\ &= P[A_1] + P[A_2] + P[A_3] \end{aligned}$$

which shows that (2.13) holds in the case $n = 3$.

It seems now that our above procedure could well be continued for each succeeding value of n, if it were not for the fact that we will never live long enough to complete it. If, however, we can show that assuming that (2.13) holds for some value of n, say k, automatically guarantees that it also holds for $n = k + 1$ (whatever the value of k), then we are assured that there is no logical fallacy in our line of argument and that the validity of (2.13) has thus been established for all finite values of n.

With this fact in mind, suppose that (2.13) is indeed valid for $n = k$. Now

$$\bigcup_{i=1}^{k+1} A_i = A_1 \cup A_2 \cup \cdots \cup A_k \cup A_{k+1}$$
$$= \left(\bigcup_{i=1}^{k} A_i\right) \cup A_{k+1}$$

The situation here is essentially the same as that in which we took n to be 3 in that the events $\left(\bigcup_{i=1}^{k} A_i\right)$ and A_{k+1} are disjoint since the events $A_1, A_2, \ldots, A_{k+1}$ are all disjoint by hypothesis. It therefore follows from Axiom 4 that

$$P\left[\bigcup_{i=1}^{k+1} A_i\right] = P\left[\left(\bigcup_{i=1}^{k} A_i\right) \cup A_{k+1}\right]$$
$$= P\left[\bigcup_{i=1}^{k} A_i\right] + P[A_{k+1}]$$

We have assumed already that (2.13) is valid when $n = k$, hence we may write

$$P\left[\bigcup_{i=1}^{k+1} A_i\right] = \sum_{i=1}^{k} P[A_i] + P[A_{k+1}] = \sum_{i=1}^{k+1} P[A_i]$$

thus showing that *if* (2.13) holds for any value of n, say, k, *then* it also holds for $n = k + 1$. We are thus assured that our successive evaluation procedure will always work for any finite value of n, and hence that the theorem is proved. ■

The proof of Theorem 2.1 illustrates the use of a technique called *proof by induction*, commonly employed to establish the validity of statements like (2.13), which are functions of an integer parameter, for example, n. In summary, the proof steps are:

a. Establish the validity of the given statement for some initial value of the index (say, 1 or 2).

b. Show that if the statement holds for some arbitrary value of the index, say, k, then it also holds for the next value $k + 1$.

A very interesting and well-written extensive discussion of proof by induction may be found in chap. VII of Polya's book "Induction and Analogy in Mathematics."

Having established by Theorem 2.1 the additivity of probability over finite disjoint unions, it seems quite natural that one should have the extension of additivity to infinite disjoint unions which we gave as Axiom 5. It must be realized, however, that there is no way of *proving* that Axiom 5 is a consequence of the preceding four axioms; we can only *hypothesize* its validity (take it to be true a priori). The extension of results from a finite domain to an infinite one is always fraught with difficulty and in mathematics, at least, is usually handled by adding new axioms or definitions.

The remaining exercises in this section illustrate applications of Theorem 2.1.

Exercise 2.10

a. Suppose that the events $A_1, A_2, \ldots, A_n$ partition a sample space $\mathcal{S}$. Show that

$$\sum_{i=1}^{n} P[A_i] = 1 \tag{2.14}$$

b. Suppose that the events $B_1, B_2, \ldots, B_m$ partition an event B in a sample space $\mathcal{S}$. Show that

$$P[B] = \sum_{i=1}^{m} P[B_i] \tag{2.15}$$

While these results are quite elementary conclusions drawn from Theorem 2.1, they are very important results in their own right. □

Exercise 2.11 A study was made of the occupancy of a data storage unit with a capacity of 16 data blocks which showed that any one number of data blocks in storage was just as likely as any other number (including zero). Suppose that we are concerned with the following three conditions of occupancy:

$A \triangleq$ [the unit is at least $\frac{1}{8}$ full but no more than $\frac{5}{8}$ full]
$B \triangleq$ [the unit is at least $\frac{3}{8}$ full but no more than $\frac{7}{8}$ full]

and

$C \triangleq$ [the unit is either empty or an even number of data blocks is in storage]

Determine the probability that at least one of the events A, B, and C occurs. (*Hint:* Draw a sample-point–sample-space representation). □

A generalization of Exercise 2.11 is given in the following exercise.

Exercise 2.12 Let A, B, and C be arbitrary events in a sample space $\mathcal{S}$.

a. Show that

$$\boxed{\begin{aligned} P[A \cup B \cup C] = P[A] + P[B] + P[C] - P[A \cap B] \\ - P[B \cap C] - P[C \cap A] + P[A \cap B \cap C] \end{aligned}} \tag{2.16}$$

b. Show that it therefore follows that

$$P[A \cup B \cup C] \leq P[A] + P[B] + P[C] \tag{2.17}$$

(*Hint:* Draw a Venn diagram). □

We showed earlier (in Exercise 2.7) that the probability of the union of two events was bounded above by the sum of the probabilities of those two events; the preceding exercise extends that result to the case of three events. The further generalization of these results is given in the following exercise.

Exercise 2.13 Let $A_1, A_2, \ldots, A_n$ be arbitrary events in a sample space $\mathcal{S}$. Show that

$$P\left[\bigcup_{i=1}^{n} A_i\right] \leq \sum_{i=1}^{n} P[A_i] \tag{2.18}$$

that is, that the probability that at least one of the events $A_1, A_2, \ldots, A_n$ occurs is bounded above by the sum of the probabilities of those events. (*Hint:* Prove by induction). □

3 PROBABILITY SPACES*

We pointed out in the first section of this chapter that a formal definition of probability involves the specification of:

a. A sample space $\mathcal{S}$
b. An appropriate collection of events $\mathcal{A}$ in that sample space
c. A set function P which is defined on the elements of $\mathcal{A}$ and which has the properties specified by Axioms 2 to 5

We called the triplet $(\mathcal{S},\mathcal{A},P)$ a probability space. Now a moment's reflection upon our discussions to this point should make us realize that

* Supplementary section. The material in this section is primarily of a background nature and can well be skimmed over lightly (or completely ignored) so far as most practical applications of probability theory are concerned.

while we have previously defined sample space and event, we have not as yet defined just what we mean by "an appropriate collection of events." Let us therefore digress from the mainstream of our discussion for the moment and consider this problem.

First, we note that if the given sample space is *finite* (if it consists of a finite number of sample points), then there really is no problem at all—for, as we shall see below, one can assign a probability to every possible subset of the sample space without difficulty.

Suppose that the given sample space consists of only three sample points:

$$\mathcal{S} \triangleq \{s_1, s_2, s_3\}$$

Suppose further that we assign a probability to each of the events which consists of a single sample point, in particular,

$$P[\{s_i\}] \triangleq p_i \qquad i = 1, 2, 3$$

where the p_i are nonnegative real numbers such that

$$p_1 + p_2 + p_3 = 1$$

We can now extend this assignment to the other subsets of $\mathcal{S}$ as follows: The only remaining subsets of $\mathcal{S}$ are[1] ϕ

$$\begin{aligned} \{s_1,s_2\} &= \{s_1\} \cup \{s_2\} \\ \{s_2,s_3\} &= \{s_2\} \cup \{s_3\} \\ \{s_3,s_1\} &= \{s_3\} \cup \{s_1\} \end{aligned}$$

and

$$\mathcal{S} = \{s_1,s_2,s_3\} = \{s_1\} \cup \{s_2\} \cup \{s_3\}$$

Since the events $\{s_1\}$, $\{s_2\}$, and $\{s_3\}$ are disjoint, it follows that if we set the probabilities of these remaining sets to be

$$\begin{aligned} P[\phi] &\triangleq 0 \\ P[\{s_1,s_2\}] &\triangleq p_1 + p_2 \\ P[\{s_2,s_3\}] &\triangleq p_2 + p_3 \\ P[\{s_3,s_1\}] &\triangleq p_3 + p_1 \end{aligned}$$

and

$$P[\{s_1,s_2,s_3\}] \triangleq p_1 + p_2 + p_3 = 1 = P[\mathcal{S}]$$

then our probability assignment is entirely consistent with our axioms. Thus we see that we can perfectly well assign a probability to *every* possible subset of the given sample space here (and hence to every event that could possibly be of interest to us). An "appropriate collection of

[1] Compare with Exercise 2.3.15.

events" in this case, then, would be the collection of *all* possible subsets of the sample space $\mathcal{S}$. Since the total number of subsets of any finite sample space is finite, the above procedure generalizes, as does our conclusion. That is, if the given sample space is finite, then an appropriate collection of events is always the collection of all possible subsets of that sample space.

Suppose, on the other hand, that the given sample space is not finite. In particular, consider the sample space

$$\mathcal{S} \triangleq \{\theta : 0 \leq \theta \leq 2\pi\}$$

which we discussed in the spinning wheel problem of Exercise 2.2 above. Now we assigned probabilities there to intervals, for we set the probability that the angle θ falls between any two angles a and b to be

$$P[a \leq \theta \leq b] = \frac{b - a}{2\pi}$$

so long as $a \leq b$ and both a and b have values in $\mathcal{S}$. While this probability assignment allows us to determine the probability that θ falls in any interval (or, for that matter, in any union or intersection of intervals), it does have the unhappy property that not all the possible subsets of $\mathcal{S}$ can, in fact, be assigned a probability by it.[1] If therefore we really wish to use this probability assignment, it follows that we must in some way choose our allowable collection of events to be smaller than the collection of *all* possible subsets of the given sample space. The same situation unfortunately arises in many other probability problems of interest.

In summary, we have indicated that while the collection of all possible subsets of a sample space makes a quite appropriate collection of events in the case of a finite sample space, it is too large a collection to be useful in other cases of interest. The question before us then is: Given an arbitrary sample space $\mathcal{S}$, how are we to select a suitable collection of subsets of $\mathcal{S}$ which is large enough to enable us to handle the probability problems of interest, and yet is small enough that we are able to avoid inconsistencies arising when we try to satisfy all of our axioms?

To help answer this question, let us return for the moment to the ideas introduced in Chap. 2. We pointed out there that given certain events of interest, we are also often interested in unions, intersections, complements, and differences of those events. With this fact in mind, let us now consider the class (collection) $\mathcal{A}$ of subsets of a sample space $\mathcal{S}$ such that

[1] The proof of this statement is far beyond our capabilities here. See, for example, Royden, chap. 3, sec. 4.

a. *If* $A \in \mathcal{A}$,

$$\textit{then } A^c \in \mathcal{A} \tag{3.1}$$

that is, if the event A is an element of the class $\mathcal{A}$, *then so is its complement.*

b. *If* $A \in \mathcal{A}$ *and* $B \in \mathcal{A}$,

$$\textit{then } A \cup B \in \mathcal{A} \tag{3.2}$$

that is, if the events A and B are elements of the class $\mathcal{A}$, *then so is their union.*

Such a class of events is often called a *field* or an *algebra*.

Exercise 3.1 Consider the probability space $(\mathcal{S},\mathcal{A},P)$ where the class $\mathcal{A}$ is an algebra as defined above.

a. Show that if the events A and B are elements of the algebra $\mathcal{A}$, then so is their intersection $A \cap B$.

b. Show that both the sample space $\mathcal{S}$ and the null set ϕ are elements of $\mathcal{A}$.

c. Show that if the events A and B are elements of $\mathcal{A}$, then so is the difference $B - A$. □

Exercise 3.2 Consider the probability space $(\mathcal{S},\mathcal{A},P)$. Suppose that $\mathcal{S}$ is a finite sample space and that $\mathcal{A}$ is the class of all subsets of $\mathcal{S}$.

Show that the class $\mathcal{A}$ is an algebra. □

We see from the results of Exercise 3.1 that if the class $\mathcal{A}$ of subsets of the sample space $\mathcal{S}$ is an algebra, then it is closed under the operations of complementation, taking finite unions and intersections, and taking differences.[1] Now remembering our axioms, we recall that we often also wish to be able to deal with countable unions of events. Let us, accordingly, add the requirement that the class $\mathcal{A}$ be such that

c. *If* $A_i \in \mathcal{A}$ *for* $i = 1, 2, 3, \ldots$

$$\textit{then } \bigcup_{i=1}^{\infty} A_i \in \mathcal{A} \tag{3.3}$$

that is, if the events A_i *are all elements of the class* $\mathcal{A}$, *then so is their countable union.*

A class of events which has the properties (*a*) and (*c*) above is often called a σ-field or σ-algebra.[2]

[1] A class is said to be *closed* under an operation applied to its elements if the result of that operation is also an element of the class.

[2] The symbol σ as used here denotes the extension of the union and intersection operations to countably infinite collections of events.

Now it follows (from an extension of the results of Exercise 3.1) that if a class $\mathcal{A}$ of subsets of a sample space $\mathcal{S}$ is a σ-algebra, then it is closed under the operations of complementation, taking countable unions and intersections, and taking differences. Such a class of events thus contains all the events likely to be of interest to us, and hence makes an "appropriate collection of events" if we can assign a probability to every element of the class in such a way that all our axioms are satisfied. While we are not able to prove it here, it can in fact be shown that a probability assignment which is consistent with our axioms can always be made to the elements of a σ-algebra.[1] We shall therefore assume throughout all our subsequent discussions that, for any probability space $(\mathcal{S},\mathcal{A},P)$, the class $\mathcal{A}$ of events is either an algebra if $\mathcal{S}$ is finite or a σ-algebra if $\mathcal{S}$ is infinite.

While specifying that the class $\mathcal{A}$ of events in a probability space $(\mathcal{S},\mathcal{A},P)$ is to be either an algebra or a σ-algebra does indeed ensure that we will be able to assign a probability in a consistent way to all elements of the class, it nonetheless does not uniquely specify the actual elements of that class. For example, suppose that $\mathcal{S}$ is a finite sample space with four sample points:

$$\mathcal{S} \triangleq \{s_1,s_2,s_3,s_4\}$$

Then, as shown by Exercise 3.2, the class of all subsets of $\mathcal{S}$ is an algebra. There are, however, smaller classes of subsets of $\mathcal{S}$ which are also algebras, as is shown by the next exercise.

Exercise 3.3 Consider the four-point sample space defined above. Let the class $\mathcal{A}$ of events be an algebra [let $\mathcal{A}$ have the properties specified in (*a*) and (*b*) above] and suppose that

$$\{s_1\} \in \mathcal{A} \qquad \text{and} \qquad \{s_2\} \in \mathcal{A}$$

a. List all of the subsets of $\mathcal{S}$ which *are* contained in the class $\mathcal{A}$.
b. List all of the subsets of $\mathcal{S}$ which *are not* contained in $\mathcal{A}$. □

We see, therefore, that in order to specify uniquely the class of events in a probability space, we not only must say that it is an algebra (or a σ-algebra), but also we must specify certain of its basic elements. Now in most practical applications of probability theory, an appropriate sample space will be a euclidean space;[2] one then usually takes the basic elements of the σ-algebra of events to be intervals.[3] In this case, the

[1] See, for example, Loève, chap. I, sec. 4, or Neveu, sec. I.4.
[2] A line, a plane, or the everyday three-dimensional space.
[3] *Interval* as used here denotes an interval in a one-dimensional sample space, a rectangle in a two-dimensional sample space, a rectangular parallelepiped in a three-dimensional sample space, etc.

σ-algebra of events is commonly called a *Borel field*, named after the French mathematician Emile Borel (1871–1956), who first introduced and used this class. " IF A FIELD HAS THE PROPERTY THAT THE SETS $A_1, A_2 \cdots A_N$ BELONG TO IT AND SO DO THE SETS $A_1 \cup A_2 \cup \cdots \cup A_n \cdots$ And $A_1 \cap A_2 \cap \cdots \cap A_n \cap \cdots$ THEN THE FIELD IS A BOREL FIELD.

4 THE CONTINUITY THEOREM OF PROBABILITY*

We previously introduced, in Sec. 5 of Chap. 2, monotone sequences of events and limits of such sequences. We now wish to derive one of the important limit theorems of probability, the *continuity theorem*, which is concerned with the probabilistic properties of monotone sequences. It should be noted that the proof to be given follows the usual theme of decomposition into disjoint subsets, followed by the use of the additivity property of probability.

THE CONTINUITY THEOREM Let the events B_i, $i = 1, 2, 3, \ldots$, be elements of a σ-algebra $\mathcal{A}$ of subsets of a sample space $\mathcal{S}$. If the sequence $B_1, B_2, B_3, \ldots$ is a monotone sequence of events (whether increasing or decreasing), then it follows that

$$P\,[\lim_{n\to\infty} B_n] = \lim_{n\to\infty} P\,[B_n] \tag{4.1}$$

That is, the operations of "taking the probability of" and "passing to the limit" may be interchanged when dealing with monotone sequences of events.

Proof Suppose first that the monotone sequence $B_1, B_2, B_3, \ldots$ is increasing, for example, as shown schematically in Fig. 4.1. It then follows from Eq. (2.5.4),† that

$$\lim_{n\to\infty} B_n = \bigcup_{i=1}^{\infty} B_n \tag{4.2}$$

Now since, by hypothesis, the given sequence is increasing, $B_{i-1} \subset B_i$, and hence we may decompose the event B_i into two disjoint parts by writing

$$B_i = B_{i-1} \cup (B_i - B_{i-1})$$

for all $i = 1, 2, 3, \ldots$, where we have defined

$$B_0 \triangleq \phi$$

With this fact in mind, we may re-express the union in (4.2) as a

* Supplementary section. This section may be omitted if desired. The material in it is somewhat more advanced and is required only in our subsequent discussion of the continuity properties of the probability distribution function.

† That is, (5.4) in Chap. 2.

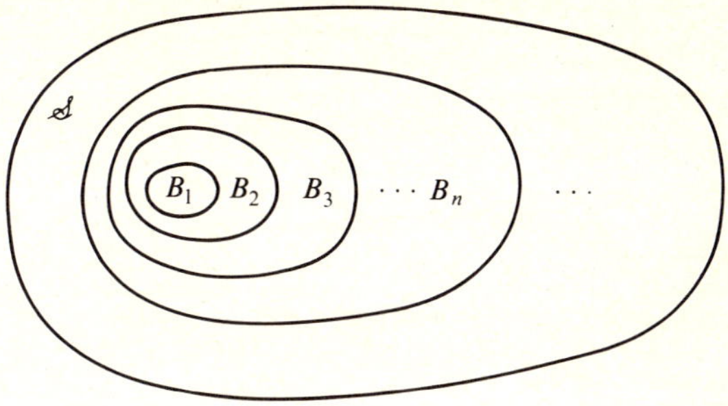

Figure 4.1 A monotone-increasing sequence of events.

union of event differences $B_i - B_{i-1}$ to obtain the result

$$\bigcup_{i=1}^{\infty} B_i = \bigcup_{i=1}^{\infty} (B_i - B_{i-1}) \tag{4.3}$$

The virtue of this procedure is that we have thus obtained a union of disjoint events since, by construction

$$(B_i - B_{i-1}) \cap (B_j - B_{j-1}) = \phi \qquad \text{for } i \neq j$$

It therefore follows that, if we substitute (4.3) into (4.2) and take the probability of both sides of the result, we get

$$P[\lim_{n\to\infty} B_n] = \sum_{i=1}^{\infty} P[B_i - B_{i-1}] \tag{4.4}$$

on using the axiom of countable additivity. Next, we note that it follows, from (2.5), that

$$P[B_i - B_{i-1}] = P[B_i] - P[B_{i-1}]$$

since $B_{i-1} \subset B_i$. On using this result in (4.4) we may obtain the formula

$$P[\lim_{n\to\infty} B_n] = \lim_{n\to\infty} \sum_{i=1}^{n} (P[B_i] \quad P[B_{i-1}]) \tag{4.5}$$

Let us now consider the finite sum on the right-hand side here. Direct expansion of this sum gives

$$\begin{aligned}\sum_{i=1}^{n} (P[B_i] - P[B_{i-1}]) &= P[B_1] - P[B_0] + P[B_2] - P[B_1] + P[B_3] \\ &\qquad - P[B_2] + \cdots + P[B_n] - P[B_{n-1}] \\ &= P[B_n] - P[B_0] = P[B_n] - P[\phi] \\ &= P[B_n]\end{aligned}$$

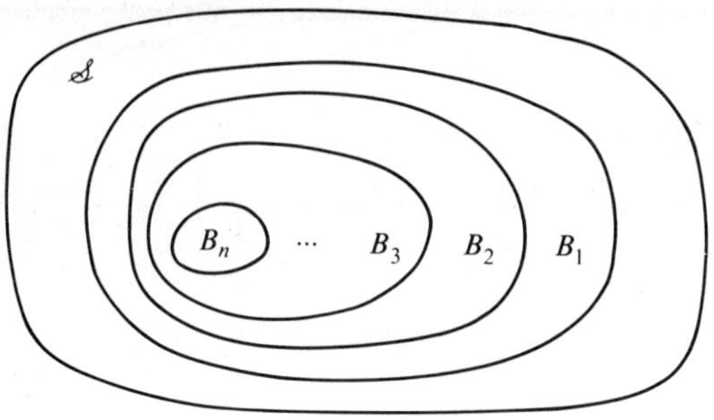

Figure 4.2 A monotone-decreasing sequence of events.

On using this result in (4.5), we thus find that if $B_1, B_2, B_3, \ldots$ is a monotone-increasing sequence of events, then it follows that

$$P[\lim_{n\to\infty} B_n] = \lim_{n\to\infty} P[B_n]$$

and half our theorem is proved.

Next let us suppose that the monotone sequence B_1, B_2, B_3, . . . is decreasing, say, for example, as shown schematically in Fig. 4.2.

Now since the given sequence is here assumed to be decreasing, it follows that the sequence of event differences $B_1 - B_i$, $i = 1, 2, 3, \ldots$, forms an increasing sequence of events. We may therefore apply the result of the first half of our proof to the sequence of these event differences and so obtain the result

$$P[\lim_{n\to\infty} (B_1 - B_n)] = \lim_{n\to\infty} P[B_1 - B_n] \tag{4.6}$$

Now since, by hypothesis, $B_n \subset B_1$ here, it follows that

$$P[B_1 - B_n] = P[B_1] - P[B_n]$$

for all $n = 1, 2, 3, \ldots$. On using this result in (4.6), we then get

$$P[\lim_{n\to\infty} (B_1 - B_n)] = P[B_1] - \lim_{n\to\infty} P[B_n] \tag{4.7}$$

Next let us consider the limit in the left-hand side of (4.7). Since the event differences $B_1 - B_n$ form an increasing sequence of events, it follows from (2.5.4) that

$$\begin{aligned}\lim_{n\to\infty} (B_1 - B_n) &= \bigcup_{n=1}^{\infty} (B_1 - B_n) \\ &= \bigcup_{n=1}^{\infty} (B_1 \cap B_n{}^c)\end{aligned}$$

the last equality following from (2.3.19). In addition, it follows from the distributive laws for unions and intersections, and from De Morgan's rules, that

$$\bigcup_{n=1}^{\infty} (B_1 \cap B_n^c) = B_1 \cap \left(\bigcup_{n=1}^{\infty} B_n^c\right) = B_1 \cap \left(\bigcap_{n=1}^{\infty} B_n\right)^c$$
$$= B_1 - \left(\bigcap_{n=1}^{\infty} B_n\right)$$

Now, by hypothesis, the events B_i here form a monotone-decreasing sequence of events, hence

$$\bigcap_{n=1}^{\infty} B_n = \lim_{n\to\infty} B_n$$

by (2.5.8). On combining these various results we get

$$\lim_{n\to\infty} (B_1 - B_n) = B_1 - \lim_{n\to\infty} B_n$$

which gives, on taking the probability of both sides,

$$P[\lim_{n\to\infty} (B_1 - B_n)] = P[B_1 - \lim_{n\to\infty} B_n]$$
$$= P[B_1] - P[\lim_{n\to\infty} B_n]$$

since $\lim_{n\to\infty} B_n \subset B_1$. On combining this result with that in (4.7), we finally see that if the monotone sequence of events $B_1, B_2, B_3, \ldots$ is decreasing, then it follows that

$$P[\lim_{n\to\infty} B_n] = \lim_{n\to\infty} P[B_n]$$

We have thus completed the proof of the continuity theorem of probability. ∎

A review of the above proof of the continuity theorem shows that it depends strongly upon the use of the countable additivity axiom. Conversely, it may be shown[1] that if the statement of the continuity theorem is taken as an axiom, then our statement of the countable-additivity axiom may be derived as a theorem. The notions of countable additivity and continuity are therefore entirely equivalent, and either may be chosen as the more basic according to personal preference. Kolmogorov, in fact, chose continuity as an axiom and derived countable additivity as a theorem in his pioneering work on the axiomatic foundation of probability theory. Today it is more common to choose countable additivity as an axiom.

[1] See, for example, Loève, pp. 84–85, or chap. II of Kolmogorov's monograph.

Quite aside from the "foundational" aspects of the continuity theorem, it is also an important theorem in applications. We shall, in fact, make use of this theorem in the next chapter when we study the continuity properties of the probability distribution function.

5 JOINT PROBABILITY

We have noted several times before that given two events A and B, we are often interested in the event that both A and B occur, that is, in the event $A \cap B$. Such a situation may arise in many different contexts. We may, for example, be concerned with an experiment in which the possible outcomes of interest refer only to a single variable, for example, number of data blocks in storage or temperature or pressure, etc., and then the two events A and B refer to the same variable. Alternatively, we may be concerned with an experiment in which the outcomes of interest refer to two, or more, variables, for example, temperature and pressure or three space coordinates or three velocity components, etc., and where the two events A and B refer to different variables.

Particularly in this latter case, it is common to refer to the event $A \cap B$ as a *joint event* and to refer to its corresponding probability $P[A \cap B]$ as a *joint probability*. In spite of the special name, however, joint probabilities are of course no different from any other probabilities discussed so far; in particular, joint probabilities satisfy the basic axioms just as do the probabilities $P[A]$ and $P[B]$ of the individual events A and B, respectively. Nonetheless, when the events of interest do refer to different variables, it is of considerable importance to determine the relations between the probabilities of such joint events and the probabilities of their corresponding individual component events. The remaining sections of this chapter will be devoted to this topic.

Exercise 5.1 Let us reconsider the spacecraft-landing problem first introduced in Exercise 2.3.2. Suppose that the probability is zero that the spacecraft lands more than 10 km from the aiming point and, further, that the probability that the spacecraft lands in some particular region (no point of which is more than 10 km from the aiming point) is directly proportional to the area of that region.

a. Let A_k be the event that the landing point is less than k km from the aiming point. Determine $P[A_k]$ for $k = 1, 3, 5$.
b. Let B be the event that a landing is more than 1 km and yet less than 3 km from the aiming point. Determine $P[B]$.
c. Let N and E be the events that the landing point is to the north and east, respectively, of the aiming point. Determine $P[N]$ and $P[E]$.

d. Determine the probability that the landing point is less than 5 km from the aiming point while also being both to the north and east of the aiming point.

Sketch an appropriate sample space and the events of interest in each case. □

A case of common interest is that in which a given sample space is partitioned both by the collection of events $\{A_1, A_2, \ldots, A_m\}$ and by the collection $\{B_1, B_2, \ldots, B_n\}$, that is, the case in which

$$\mathcal{S} = \bigcup_{i=1}^{m} A_i \qquad \text{and} \qquad \mathcal{S} = \bigcup_{j=1}^{n} B_j$$

and in which

$$A_i \cap A_k = \phi \qquad \text{for } i \neq k$$

and

$$B_j \cap B_r = \phi \qquad \text{for } j \neq r$$

If now the probabilities $P[A_i \cap B_j]$ of each of the m times n joint events $(A_i \cap B_j)$ are specified, then the probabilities of the individual events A_i and B_j may easily be calculated as follows: Since the events A_i partition the sample space $\mathcal{S}$, it follows from the partition results in Sec. 2.4 that we may express the event B_j as a union of disjoint subsets by the formula

$$B_j = \bigcup_{i=1}^{m} (A_i \cap B_j)$$

where, since the events A_i partition $\mathcal{S}$, it follows that

$$(A_i \cap B_j) \cap (A_k \cap B_j) = \phi \qquad \text{for } i \neq k$$

That is, the events $(A_1 \cap B_j), \ldots, (A_m \cap B_j)$ partition the event B_j. On applying the results of Exercise 2.10, we therefore get the result that

$$P[B_j] = \sum_{i-1}^{m} P[A_i \cap B_j] \tag{5.1}$$

That is, if the events A_i partition the sample space $\mathcal{S}$, then the probability of an event B_j equals the sum (over all values of the index i) of the probabilities of the joint events $(A_i \cap B_j)$. It similarly follows that we obtain the result

$$P[A_i] = \sum_{j=1}^{n} P[A_i \cap B_j] \tag{5.2}$$

since, by hypothesis, the events B_j also partition the sample space $\mathcal{S}$.

Exercise 5.2 Suppose that the collections of events $\{A_1, A_2, \ldots, A_m\}$ and $\{B_1, B_2, \ldots, B_n\}$ both partition a sample space $\mathcal{S}$. Show that

$$P[A_i] \geq P[A_i \cap B_j] \tag{5.3}$$

for any j and all i, and that

$$P[B_j] \geq P[A_i \cap B_j] \tag{5.4}$$

for any i and all j. □

Exercise 5.3 Let the events A_i and B_j be defined as in the preceding exercise. Show that

$$\sum_{i=1}^{m} \sum_{j=1}^{n} P[A_i \cap B_j] = 1 \tag{5.5}$$

(A hardly unexpected result!) □

It should be noted that while the above results tacitly assumed m and n to be finite, either or both may be infinite, in fact.

The results obtained above for joint events concerning two of the variables required to describe the output of an experiment can, of course, be extended to cover cases in which more than two such variables are of interest. Suppose in the following exercises that the collections of events $\{A_i, i = 1, 2, \ldots, n\}$, $\{B_j, j = 1, 2, \ldots, p\}$, $\{C_k, k = 1, 2, \ldots, q\}$, and $\{D_m, m = 1, 2, \ldots, r\}$ each partition the sample space $\mathcal{S}$ of some given defining experiment.

Exercise 5.4 For the events A_i, B_j, C_k, and D_m defined above, show that

$$P[B_j \cap C_k \cap D_m] = \sum_{i=1}^{n} P[A_i \cap B_j \cap C_k \cap D_m]$$

for all j, k, and m; that

$$P[C_k \cap D_m] = \sum_{j=1}^{p} \sum_{i=1}^{n} P[A_i \cap B_j \cap C_k \cap D_m]$$

for all k and m; and that

$$P[D_m] = \sum_{k=1}^{q} \sum_{j=1}^{p} \sum_{i=1}^{n} P[A_i \cap B_j \cap C_k \cap D_m]$$

for all values of m. □

Exercise 5.5 For the events A_i, B_j, C_k, and D_m defined above, show that

$$P[D_m] \geq P[C_k \cap D_m] \geq P[B_j \cap C_k \cap D_m] \geq P[A_i \cap B_j \cap C_k \cap D_m]$$

for all values of i, j, k, and m. □

Exercise 5.6 For the events A_i, B_j, C_k, and D_m defined above, show that

$$\sum_{m=1}^{r} \sum_{k=1}^{q} \sum_{j=1}^{p} \sum_{i=1}^{n} P[A_i \cap B_j \cap C_k \cap D_m] = 1$$

As in the two-variable case, both this and the preceding results hold when any or all of the indices n, p, q, and r are infinite. □

6 CONDITIONAL PROBABILITY

It was pointed out in the preceding section that we very often are interested in the interrelations amongst the various probabilities concerning a pair, or more, of events referring to some given defining experiment. A question which often arises in such situations is: What is the probability of occurrence of some specified event, say B, if we know that another event, say A, has occurred? Such probabilities, which give the probability of occurrence of one event subject to the hypothesis of the occurrence of another event, are called *conditional* probabilities. We shall in this section define conditional probability and determine some of its properties.

Let us, for the moment, restrict our attention to two arbitrary events, say A and B, referring to some given defining experiment with a corresponding sample space $\mathcal{S}$. Suppose now that we wish to determine the probability of the event B, given that the event A has occurred. We are then concerned with the various sets shown pictorially in Fig. 6.1. Once we insist upon the conditioning hypothesis that the event A occurs, we are no longer concerned with the probability that the sample point

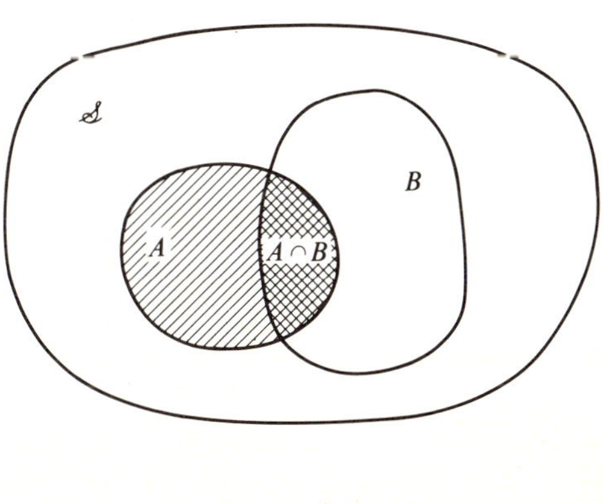

Figure 6.1 Sets concerning the conditional probability $P[B|A]$.

Experiment Number	1	2	3	4	5	6	7	8	9	10
Event *A* occurs	✓	✓	×	×	✓	×	✓	×	×	✓
Event *B* occurs	×	×	✓	×	×	✓	✓	✓	×	✓
Event $A \cap B$ occurs	×	×	×	×	×	×	✓	×	×	✓

Figure 6.2

representing a particular outcome falls anywhere in the subset B of the sample space $\mathcal{S}$. Rather, we are concerned with the probability that that sample point falls in that part, $A \cap B$, of B which is also a subset of the set of points representing the event A. In essence then, instead of dealing with a subset B of a sample space $\mathcal{S}$, we must now deal with a subset $A \cap B$ of a new and restricted "sample space" A, and our conditional probability of the event B must reflect this change in effective sample spaces.

With this fact in mind, suppose next that we perform the given defining experiment a number of times, say m, under essentially identical conditions. Suppose further that in this set of m performances of the given experiment, the event A occurs $n(A)$ times, the event B occurs $n(B)$ times, and the joint event $A \cap B$ occurs $n(A \cap B)$ times. A typical table of results might then be that shown in Fig. 6.2, where there we have $m = 10$, $n(A) = 5$, $n(B) = 5$, and $n(A \cap B) = 2$. Let us now consider the occurrence of the event B given that the event A also occurs. In this case only those $n(A)$ performances of the given experiment in which the event A occurs are of interest to us. Now the event B occurs $n(B \cap A)$ times in those $n(A)$ performances, and hence the relative frequency of occurrence of the event B given that event A has occurred is $n(B \cap A)/n(A)$. For example, we see that in the particular sequence of experimental outcomes reported in Fig. 6.2, the relative frequency of occurrence of the event B given that the event A also occurs is $n(B \cap A)/n(A) = \frac{2}{5}$.

In line with our previous view that a probability should be a mathematical model of an empirical relative frequency, the conditional probability of the event B given that the event A has occurred should be a model of the conditional relative frequency $n(B \cap A)/n(A)$. Suppose now that we divide both the numerator and the denominator of the conditional relative frequency by m, the total number of experiments performed. In this way we get

$$\frac{n(B \cap A)}{n(A)} = \frac{\dfrac{n(B \cap A)}{m}}{\dfrac{n(A)}{m}}$$

The numerator on the right-hand side here is modeled by the joint probability $P[B \cap A]$, while the denominator is modeled by $P[A]$. With this fact in mind, we define the *conditional probability* $P[B|A]$ of the event B, subject to the hypothesis of the occurrence of the event A, by the formula

$$P[B|A] \triangleq \frac{P[B \cap A]}{P[A]} \qquad \text{so long as } P[A] > 0 \tag{6.1}$$

The latter restriction is required, of course, to prevent an indeterminacy from arising on the right-hand side of the defining equation. A moment's reflection upon this defining equation, and upon our earlier discussion centered about Fig. 6.1, should show that we have taken into account the change in the effective sample space from $\mathcal{S}$ to A, due to the conditioning hypothesis, by weighting the probability with respect to the original sample space $\mathcal{S}$ of the joint event $B \cap A$ by the probability with respect to the original sample space $\mathcal{S}$ of the new "sample space" A.

Exercise 6.1 Consider again the spacecraft landing problem discussed in Exercise 5.1. Using the probability assignment made there,

a. Determine the conditional probability that the landing point is more than 1 km from the aiming point, given that it is less than 3 km from the aiming point.

b. Determine the conditional probability that the landing point is both to the north and the east of the aiming point, given that it is less than 5 km from the aiming point. □

Once the conditional probability $P[B|A]$ has been defined as in (6.1) above, it follows immediately that we may write

$$P[B \cap A] = P[B|A]\,P[A] \tag{6.2}$$

that is, the joint probability of the events A and B may be calculated by forming the product of the conditional probability of the event B given that the event A occurs and the probability of the event A. This result may easily be extended to cases involving more than two events, as shown by the following exercise.

Exercise 6.2 Let A, B, C, and D be events in a sample space $\mathcal{S}$. Show that

$$P[A \cap B \cap C \cap D] = P[D|A \cap B \cap C]P[C|A \cap B]P[B|A]P[A] \quad \square$$

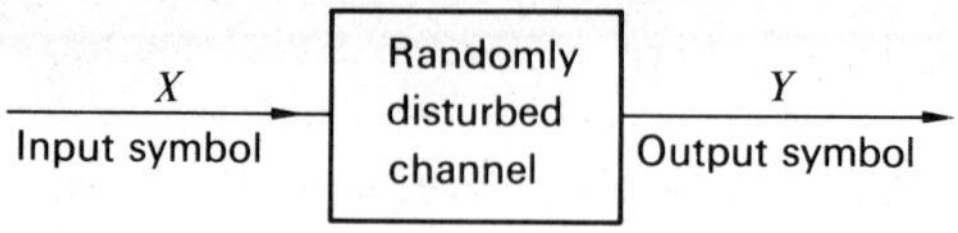

Figure 6.3 A binary communication channel.

Exercise 6.3 Suppose that the events $A_1, A_2, \ldots, A_m$ partition a sample space $\mathcal{S}$, and that $P[A_i] > 0$ for all $i = 1, 2, \ldots, m$. Show that

$$P[B] = \sum_{i=1}^{m} P[B|A_i]P[A_i] \tag{6.3}$$

where B is an arbitrary event in $\mathcal{S}$. □

A typical example of the application of conditional probability is shown by the following problem: Consider the binary communication channel shown schematically in Fig. 6.3. The channel input symbol X may assume either the state 0 or the state 1 and, similarly, the channel output symbol Y may assume either the state 0 or the state 1. The random disturbances in the channel can convert an input 1 to an output 0 or an input 0 to an output 1. Let us characterize the channel probabilistically by the numbers q_0, q_1, p_0, and p_1, where q_0 and q_1 are the *channel transition probabilities* corresponding to no errors

$$q_0 \triangleq P[Y = 0|X = 0] \qquad \text{and} \qquad q_1 \triangleq P[Y = 1|X = 1]$$

while p_0 and p_1 are thus the channel transition probabilities corresponding to errors

$$p_0 \triangleq P[Y = 1|X = 0] \qquad \text{and} \qquad p_1 \triangleq P[Y = 0|X = 1]$$

where, of course,

$$p_0 + q_0 = 1 = p_1 + q_1$$

It is common to represent a channel characterized in terms of the channel transition probabilities by a state diagram as shown in Fig. 6.4.

The observer of the channel output knows only that he sees an output symbol 0 or an output symbol 1. In order to evaluate these

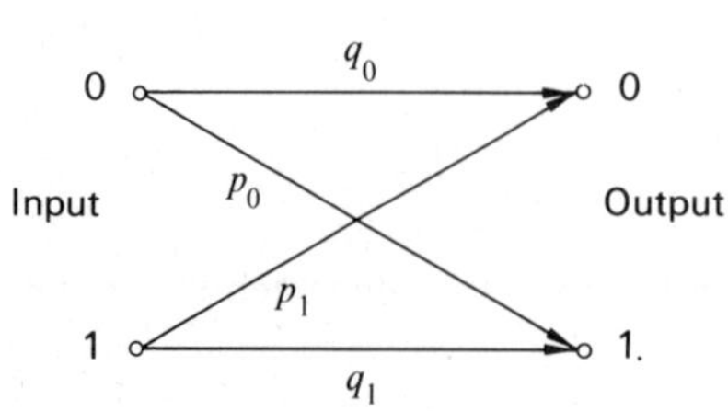

Figure 6.4 State diagram for a binary communication channel.

output symbols, he would also like to determine such probabilities as the probability that the input symbol was 1 given that the output was 1, etc.

Exercise 6.4 Suppose that a 1 appears at the input with a probability π_1 and that a 0 appears at the input with a probability π_0, where

$$\pi_0 + \pi_1 = 1$$

a. Determine the probability of occurrence of the output symbols 1 and 0 in terms of the input probabilities and the channel transition probabilities.

b. Determine the conditional probabilities

$$P[X = 1|Y = 1] \qquad P[X = 1|Y = 0]$$
$$P[X = 0|Y = 1] \qquad P[X = 0|Y = 0]$$

□

A generalization of the results in Exercise 6.4 is given by the next exercise.

Exercise 6.5 Let B be an event in a sample space $\mathcal{S}$. Suppose that the events $A_1, A_2, \ldots, A_m$ partition the sample space $\mathcal{S}$, and that $P[A_i] > 0$ for all i. Show that

$$P[A_i|B] = \frac{P[B|A_i]P[A_i]}{\sum_{j=1}^{m} P[B|A_j]P[A_j]} \tag{6.4}$$

This result is known as *Bayes' rule*, after the English philosopher Reverend Thomas Bayes (1702–1761). □

The communication problem discussed in Exercise 6.5 provides a typical example of the use of Bayes' rule: A situation arises in which one knows beforehand the conditional probabilities $P[B|A_i]$ that a given event B occurs given that certain causes A_i occur, as well as the probabilities of occurrence $P[A_i]$ of those causes. If the event B is then observed to occur, Bayes' rule may be used to determine which of the possible causes was the most likely to have given rise to the event B. The unconditional probabilities $P[A_i]$ are sometimes called *a priori probabilities* in such a situation, while the conditional probabilities $P[A_i|B]$ are then called *a posteriori probabilities.*

There has, over the years, been considerable philosophical discussion in the statistical literature of the "validity" of Bayes' rule. When reading such discussions, it is worthwhile remembering that Bayes' rule is a perfectly valid conclusion drawn from the probability axioms and the definition of conditional probability. The only possible point to be questioned is whether or not, in a given practical application, one does in fact have valid (or even reasonable) values for the various probabilities $P[A_j]$ and $P[B|A_j]$. Such a question must, of course, be raised about the application of *any* result derived from a mathematical model (such as probability) to a problem in the physical world.

Having defined conditional probability and seen some of its applications, let us next determine some bounds which may be placed upon it. As usual, we shall assume that A and B are events in some sample space $\mathcal{S}$ and, further, that $P[A] > 0$. We showed in Sec. 2 of this chapter that $0 \leq P[A] \leq 1$ for any event A. It then follows from these inequalities and the definition (6.1) of conditional probability that

$$P[B|A] \geq P[A \cap B] \tag{6.5}$$

that is, that the conditional probability of the event B given the event A is always at least as great as is the joint probability of A and B. The relation between the conditional probability $P[B|A]$ and the unconditioned probability $P[B]$ is not so simple, however, as shown by the following exercises.

Exercise 6.6 Let A and B be events in a sample space $\mathcal{S}$ such that $P[A] > 0$. Show that if A and B are mutually exclusive events, then

$$P[B|A] = 0$$

that is, that the conditional probability of the event B given the event A is zero when A and B are mutually exclusive.

It obviously then follows that

$$P[B|A] \leq P[B]$$

in this case. □

Exercise 6.7 Let A and B be events in a sample space $\mathcal{S}$ such that $P[A] > 0$. Show that if the event B implies the event A; that is, if

$$B \subset A$$

then

$$P[B|A] \geq P[B]$$

that is, that the conditional probability of the event B given the event A is at least as great as is the unconditional probability of the event B if B implies A. □

Exercise 6.8 Let A and B be events in a sample space $\mathcal{S}$ such that $P[A] > 0$. Show that if the event A implies the event B; that is, if

$$A \subset B \qquad \text{then} \qquad P[B|A] = 1$$

that is, that the conditional probability of the event B given the event A is unity if A implies B.

It is then clearly true that

$$P[B|A] \geq P[B]$$

in this case. □

The last three exercises show that the conditional probability of the event B given the event A may well be greater than, equal to, or less than the unconditional probability of the event B. We shall discuss this point further in the next section.

While it seems quite reasonable to call the quantity $P[B|A]$, defined by (6.1) above, a conditional probability, the fact remains that it is incumbent upon us to show that it does indeed warrant being called a probability, in particular, that it satisfies the axioms of probability. To this end, suppose that A is an event in a sample space $\mathcal{S}$, that $P[A] > 0$, and that we assign a real number $P[B|A]$, determined by (6.1), to each event B in the sample space $\mathcal{S}$. Now it follows from the probability axioms as given in Sec. 1 of this chapter that $P[A \cap B] \geq 0$. It therefore then follows from the defining equation (6.1) that, for any event B in $\mathcal{S}$,

$$0 \leq P[B|A] \qquad \text{(Axiom 2)}$$

since, by hypothesis, $P[A] > 0$. Thus we see that conditional probability satisfies Axiom 2. It may similarly be shown that it also satisfies the rest of the probability axioms.

Exercise 6.9 Show that

$$P[\mathcal{S}|A] = 1 \qquad \text{(Axiom 3)}$$

where the event A and the sample space $\mathcal{S}$ are defined as above. □

Exercise 6.10 Let the event A and the sample space $\mathcal{S}$ be as defined above. Show that the conditional probability satisfies the additivity axioms, that is,

a. Show that

$$P[(B \cup C)|A] = P[B|A] + P[C|A] \qquad \text{(Axiom 4)}$$

if B and C are mutually exclusive events in the sample space $\mathcal{S}$.

b. Show further that

$$P\left[\left(\bigcup_{j=1}^{\infty} B_j\right) \Big| A\right] = \sum_{j=1}^{\infty} P[B_j|A] \qquad \text{(Axiom 5)}$$

where the events $B_1, B_2, B_3, \ldots$ are mutually exclusive events in the sample space $\mathcal{S}$. □

Since it thus follows that conditional probability does indeed satisfy the probability axioms as given in Sec. 1, it further follows that a conditional probability has properties analogous to those derived in Sec. 2 of this chapter for an ordinary probability, *so long as the probability of the conditioning event is not zero.*

7 INDEPENDENT EVENTS

Suppose that A and B are events defined in some given sample space $\mathcal{S}$ such that

$$P[A] > 0 \qquad \text{and} \qquad P[B] > 0$$

It then follows from (6.2) that both of the conditional probabilities $P[B|A]$ and $P[A|B]$ are defined unambiguously. We noted in the preceding section that the conditional probability $P[B|A]$ could be greater than, equal to, or less than the unconditioned probability $P[B]$.† Suppose now that these two probabilities are in fact equal, that is, that

$$\boxed{P[B|A] = P[B]} \tag{7.1}$$

It then follows from (6.2) that

$$\boxed{P[A \cap B] = P[A]P[B]} \tag{7.2}$$

and hence, on using this result in (6.1), that

$$P[A|B] = P[A] \tag{7.3}$$

Thus we see that if (7.1) holds true, then the knowledge that either of the events A and B has occurred does not affect the probability of occurrence of the other. In a probabilistic sense then, the two events are independent of each other.

A moment's reflection upon the above results shows that if the product formula (7.2) holds, then both of the conditional probability relations (7.1) and (7.3) also hold—so long as neither $P[A]$ nor $P[B]$ is zero. For the sake of compactness then, it is common to use the product formula in a formal definition of probabilistic independence rather than the relations which directly involve conditional probabilities. Thus, if A and B are two events defined in a sample space $\mathcal{S}$, then those two events are said to be *statistically independent* (or *stochastically independent*) (or

† See Exercises 6.6 to 6.8.

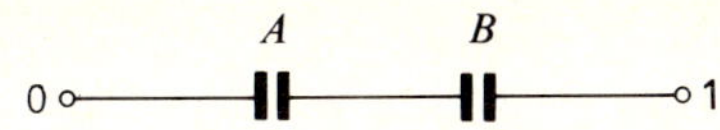

Figure 7.1 A series relay circuit.

simply *independent*) if and only if

$$P[A \cap B] = P[A]P[B] \tag{7.4}$$

that is, if and only if their joint probability factors into the product of their individual probabilities.

Consider, for example, the relay circuit shown in Fig. 7.1. Let us denote by A the event that relay A is closed, by A^c the event that relay A is open, and by B and B^c the events that relay B is closed and open, respectively. On studying this circuit, we see that there are four possible mutually exclusive events: $A \cap B$, $A \cap B^c$, $A^c \cap B$, and $A^c \cap B^c$; we may therefore represent the possible conditions of the circuit by a new four-point sample space (one point for each circuit condition), as shown in Fig. 7.2. Suppose next that, in a particular application of this circuit, it has been found that these four basic events are equally likely:

$$P[A \cap B] = P[A \cap B^c] = P[A^c \cap B] = P[A^c \cap B^c] = \tfrac{1}{4}$$

Let us now determine the probabilities $P[A]$ and $P[B]$. Consideration of the structure of our four-point sample space shows that we may write

$$A = [A \cap B] \cup [A \cap B^c]$$

and, similarly,

$$B = [A \cap B] \cup [A^c \cap B]$$

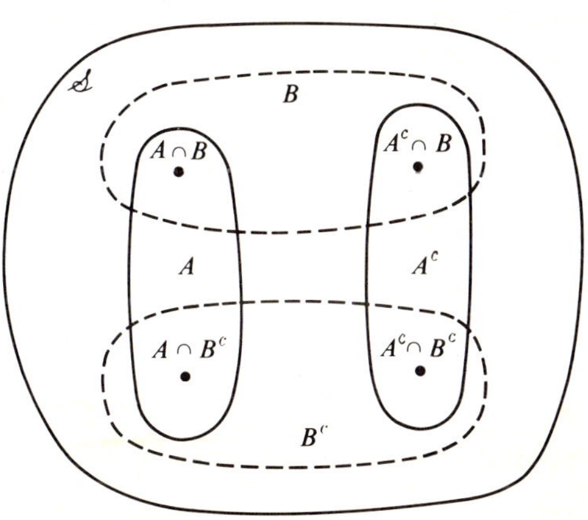

Figure 7.2 Sample space for the series relay circuit.

Since we have in each of these cases a disjoint decomposition, it follows that we get

$$\begin{aligned} P[A] &= P[A \cap B] + P[A \cap B^c] \\ &= \tfrac{1}{4} + \tfrac{1}{4} = \tfrac{1}{2} \end{aligned}$$

and

$$\begin{aligned} P[B] &= P[A \cap B] + P[A^c \cap B] \\ &= \tfrac{1}{4} + \tfrac{1}{4} = \tfrac{1}{2} \end{aligned}$$

It therefore further follows that

$$P[A \cap B] = \tfrac{1}{4} = \tfrac{1}{2} \times \tfrac{1}{2} = P[A]P[B]$$

that is, that the closings of the two relays are independent events in this case.

A further question might now arise as to the statistical relationship between any other two of the events A, A^c, B, and B^c, say, between A and B^c. Since we have already calculated the probability of the event B, it follows that the probability of its complement is given by

$$\begin{aligned} P[B^c] &= 1 - P[B] \\ &= 1 - \tfrac{1}{2} = \tfrac{1}{2} \end{aligned}$$

Now $P[A \cap B^c]$ is $\frac{1}{4}$ by hypothesis and we found $P[A]$ to be $\frac{1}{2}$. It therefore follows that

$$\begin{aligned} P[A \cap B^c] &= \tfrac{1}{4} = \tfrac{1}{2} \times \tfrac{1}{2} \\ &= P[A]P[B^c] \end{aligned}$$

that is, that the events A and B^c are also independent. This result is a special case of that in the following exercise.

Exercise 7.1 Let A and B be events defined in a sample space $\mathcal{S}$. Show that if A and B are statistically independent, then so are A and B^c, A^c and B, and A^c and B^c.

□

A common confusion arises in attempting to keep straight the relation between events being *mutually exclusive* and events being *statistically independent.* A review of the series relay circuit and its corresponding sample space (cf. Fig. 7.1 and 7.2) shows that while the events A and B are independent, they certainly are not mutually exclusive (since one can occur whether or not the other does). Similarly, we see that while the events A and A^c are mutually exclusive, they certainly are not independent (since if one occurs, the other cannot). The next two exercises develop this point further.

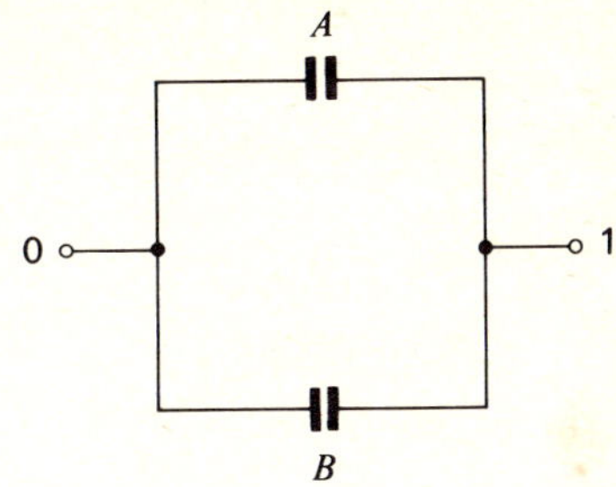

Figure 7.3 A parallel relay circuit.

Exercise 7.2 Let A and B be events defined in a sample space $\mathcal{S}$. Show that if both $P[A]$ and $P[B]$ are nonzero, then the events A and B *cannot* be both mutually exclusive and statistically independent. □

Exercise 7.3 Let A and B be events defined in a sample space $\mathcal{S}$. Show that if the events A and B are both mutually exclusive and statistically independent, then either or both of $P[A]$ and $P[B]$ must be zero. □

Exercise 7.4 Consider the relay circuit shown in Fig. 7.3. Again let A denote the event that the relay A is closed, etc. Suppose that the events A and B are independent and that

$$P[A] = p = P[B]$$

where p is some number between 0 and 1.

Determine the probability that a closed path exists between the points 0 and 1 in that circuit. □

It should be noted that, by convention, we did *not* specify that either or both $P[A]$ and $P[B]$ had to be nonzero when we set up the product formula (7.4) as a definition of the statistical independence of the two events A and B. Some rather peculiar looking results are obtainable because of this convention, as shown in the following exercise.

Exercise 7.5 Let A be an event in a sample space $\mathcal{S}$ and suppose that

$$P[A] = 0$$

Show that A is statistically independent of *any* other event B in that sample space. □

An immediate consequence of this result is the fact that the null event is statistically independent of *any* other event.

Exercise 7.6 Let A and B be arbitrary events defined in a sample space $\mathcal{S}$. Show that the events A and B are statistically independent if and only if *either*

$$P[A] = 0 \qquad \text{or} \qquad P[B] = 0$$

or

$$P[B|A] = P[B] \qquad \text{and} \qquad P[A|B] = P[A]$$

□

The above discussion of statistical independence was confined to the case of two events; when more than two events are to be considered, however, the question of independence becomes quite a bit more complicated. Consider, for example, the relay circuit shown in Fig. 7.4. We have there two relays A and B, each with multiple contacts. If we denote by A the event that relay A is energized, then the contacts labeled A are closed when event A occurs, while the contact labeled A^c is closed when relay A is not energized (when the event A^c occurs). Similarly, if we denote by B the event that relay B is energized, then the contacts labeled B are closed when event B occurs, and the contact labeled B^c is closed when the event B^c occurs. Let us further denote by C_1 the event that there is a closed path from the point 0 to the point 1, by C_2 the event that there is a closed path from 0 to 2, and by C_3 the event that there is a closed path from 0 to 3. We now wish to investigate the statistical relationships which exist among the events C_1, C_2, and C_3 under the conditions that

$$P[A] = \tfrac{1}{2} = P[B]$$

and that the events A and B are statistically independent in the sense discussed previously.

Since the operation of the circuit shown in Fig. 7.3 is basically determined by the operation of the two relays A and B, each of which may be energized or not, we can if we wish view the problem in terms of the four-point sample space shown in Fig. 7.5. Now an examination of Figs. 7.4 and 7.5 shows that we may express the events C_1, C_2, and C_3 in

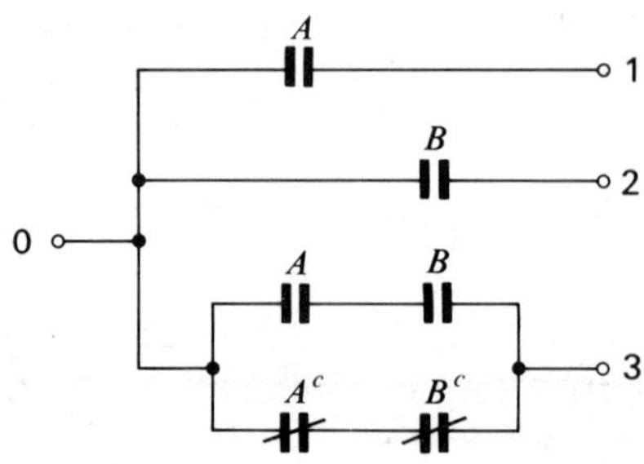

Figure 7.4 A two-relay multiple-contact circuit.

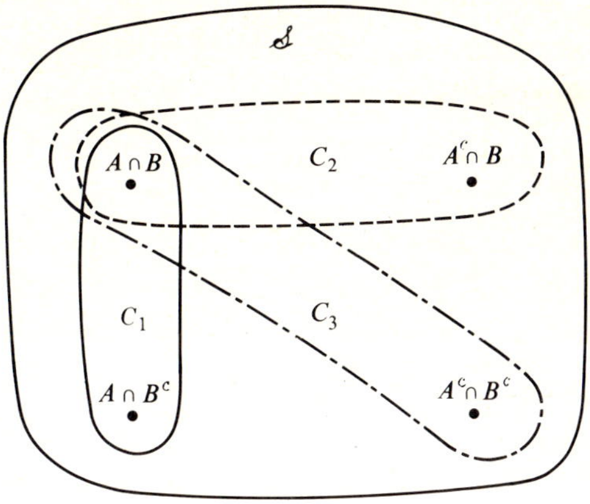

Figure 7.5 A sample-space representation of the relay circuit in Fig. 7.4.

terms of the events A and B as follows:

$$C_1 = A = (A \cap B) \cup (A \cap B^c)$$
$$C_2 = B = (A \cap B) \cup (A^c \cap B)$$

and

$$C_3 = (A \cap B) \cup (A^c \cap B^c)$$

Since we have already said that the events A and B are independent, it follows from (7.2) and the results of Exercise 7.1 that

$$P[A \cap B] = P[A]P[B] = \tfrac{1}{4}$$

and

$$P[A^c \cap B^c] = P[A^c]P[B^c] = \tfrac{1}{4}$$

On using these results with the expressions for C_1, C_2, and C_3, we then see that

$$P[C_1] = \tfrac{1}{2}$$
$$P[C_2] = \tfrac{1}{2}$$

and

$$P[C_3] = \tfrac{1}{4} + \tfrac{1}{4} = \tfrac{1}{2}$$

It further follows from the definitions of C_1, C_2, and C_3 (see Fig. 7.5) that

$$C_1 \cap C_2 = A \cap B$$
$$C_2 \cap C_3 = A \cap B$$

and that

$$C_3 \cap C_1 = A \cap B$$

It therefore further follows that

$$P[C_1 \cap C_2] = P[C_2 \cap C_3] = P[C_3 \cap C_1]$$
$$= P[A \cap B] = \tfrac{1}{4}$$

Thus we finally see that

$$P[C_1 \cap C_2] = P[C_1]P[C_2]$$
$$P[C_2 \cap C_3] = P[C_2]P[C_3]$$

and

$$P[C_3 \cap C_1] = P[C_3]P[C_1]$$

in this case. That is, the events C_1, C_2, and C_3 are here *pairwise independent* (statistically independent by pairs).

Suppose, however, that we now investigate the events C_1, C_2, and C_3 all together. We note, first, that (cf. Fig. 7.4)

$$C_1 \cap C_2 \cap C_3 = A \cap B$$

and hence that

$$P[C_1 \cap C_2 \cap C_3] = P[A \cap B] = \tfrac{1}{4}$$

in this case. The results in Exercise 7.7 therefore follow.

Exercise 7.7 Show that

$$P[C_1|C_2 \cap C_3] = 1 \neq P[C_1] = \tfrac{1}{2}$$
$$P[C_2|C_3 \cap C_1] = 1 \neq P[C_2] = \tfrac{1}{2}$$

and

$$P[C_3|C_1 \cap C_2] = 1 \neq P[C_3] = \tfrac{1}{2}$$

where C_1, C_2, and C_3 are the events defined above. □

The various results above show that while the probability of occurrence of the event C_1 *is not* changed by adding the hypothesis that C_2 has occurred (or by adding the hypothesis that C_3 has occurred), the probability of occurrence of C_1 *is* changed by adding the hypothesis that both C_2 and C_3 have occurred. Thus the events C_1, C_2, and C_3 here are *not* independent when considered all together. A moment's reflection upon Fig. 7.4 should show that this conclusion must, of course, be true.

Our above example shows that the specification that three events are pairwise independent is not sufficient to guarantee that those three events are independent when considered together; clearly some other requirement must be added. It seems intuitively obvious that if we add the requirement that the joint probability of the three events factors into the product of their respective individual probabilities, then we should accomplish our goal. That this step does in fact achieve the desired result is shown by the next exercise.

Exercise 7.8 Let A, B, and C be events in a sample space $\mathcal{S}$. Show that *if* A, B, and C are pairwise independent, *and if*

$$P[A \cap B \cap C] = P[A]P[B]P[C]$$

then

$$P[A|B \cap C] = P[A]$$
$$P[B|C \cap A] = P[B]$$

and

$$P[C|A \cap B] = P[C]$$

that is, the knowledge that either or both of the two other events occurred does not change the probability of occurrence of any one of the three events in this case. □

With these various results in mind, we are led to the following definition:

Let $A_1, A_2, \ldots, A_n$ be events defined in some given sample space $\mathcal{S}$. These n events are said to be mutually independent *events if and only if the relations*

$$\begin{aligned} P[A_i \cap A_j] &= P[A_i]P[A_j] \\ P[A_i \cap A_j \cap A_k] &= P[A_i]P[A_j]P[A_k] \\ &\cdots\cdots\cdots \\ P[A_1 \cap A_2 \cap \cdots \cap A_n] &= P[A_1]P[A_2] \cdots P[A_n] \end{aligned} \tag{7.5}$$

hold for all combinations of the indices such that $1 \le i < j < k < \cdots \le n$. The three events A, B, and C in Exercise 7.8 are therefore mutually independent.

Exercise 7.9 Show that there are $2^n - n - 1$ different equations in the system (7.5). (*Hint:* Recall how to determine the number of combinations of different things and use the binomial-expansion formula.) □

A question might now arise as to whether or not we really have not overdone it by requiring that all the equations in (7.5) be satisfied in order to have the mutual independence of n events, and whether or not it would not be sufficient just to require that the joint probability of the n events factors into the product of the n respective individual probabilities. That this is not in fact the case is shown by the following example.

Consider the relay circuit shown in Fig. 7.6. We have there three relays: A, B, and D with multiple contacts. As before, we denote by A the event that relay A is energized, then the contact labeled A is closed when the event A occurs, the contact labeled A^c is closed when the event A^c occurs, etc. Again let C_i ($i = 1, 2, 3$) denote the event that there is a closed path between the points 0 and i. We now wish to

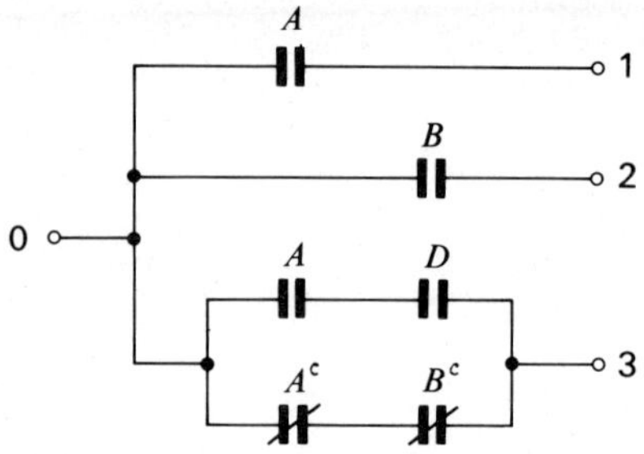

Figure 7.6 A three-relay multiple-contact circuit.

investigate the statistical relationships which exist among the events C_1, C_2, and C_3 here under the conditions that

$$P[A] = P[B] = P[D] = \tfrac{1}{2}$$

and that the events A, B, and D are mutually independent.

Exercise 7.10 Show that in this case

$$P[C_1 \cap C_2 \cap C_3] = P[C_1]P[C_2]P[C_3]$$

and that

$$P[C_1 \cap C_2] = P[C_1]P[C_2]$$

and

$$P[C_1 \cap C_3] = P[C_1]P[C_3]$$

but that

$$P[C_2 \cap C_3] \neq P[C_2]P[C_3]$$

That is, the factoring of the joint probability $P[C_1 \cap C_2 \cap C_3]$ *is not* enough here to guarantee that the events C_1, C_2, and C_3 are pairwise independent. □

8 INDEPENDENT EXPERIMENTS

Up to this point, we have tended to view each problem at hand as being concerned with the possible outcomes of some single defining experiment. At times, however, it is more convenient instead to view the defining experiment as consisting of several subexperiments. Thus, for example, it might well be more convenient to view the experiment consisting of measuring the temperature and pressure at some given location at some instant of time as a combination of two subexperiments: one consisting of measuring the temperature and the other consisting of measuring the pressure. Similarly, the experiment consisting of determining the results of eight successive flips of a coin (or of determining whether or not a pulse is present in each of eight successive time slots) is sometimes more conveniently viewed as the successive performance of eight different experiments.

One of the reasons for the convenience of introducing subexperiments is that it is often possible to convince oneself that the various subexperiments are independent in the sense that the outcome of one has no probabilistic effect upon the outcomes of another. For example, it is not hard to believe that the result of the fourth coin flip is probably independent of the outcomes of either the second or the seventh coin flips.

We now wish to extend our definitions of statistical independence to cover the case of experiments; our new definitions should, of course, be based upon and consistent with those definitions which we have already made concerning events. Consider for the moment a combined experiment consisting of two subexperiments $\mathcal{E}_1$ and $\mathcal{E}_2$. We say that the pair of experiments $\mathcal{E}_1$ and $\mathcal{E}_2$ are *statistically independent experiments* if and only if we have

$$P[A \cap B] = P[A]P[B]$$

for any pair of events A and B such that $A \in \mathcal{E}_1$ and $B \in \mathcal{E}_2$. That is, $\mathcal{E}_1$ and $\mathcal{E}_2$ are statistically independent experiments if and only if every event concerning $\mathcal{E}_1$ alone is independent of every event concerning $\mathcal{E}_2$ alone. That this definition is consistent with our previous ones is indicated by the following exercise.

Exercise 8.1 Suppose that the experiment $\mathcal{E}_1$ has the two possible outcomes A and A^c, and that the experiment $\mathcal{E}_2$ has the two possible outcomes B and B^c.

Show that the specification that the *events* A and B are independent is equivalent here to the specification that the *experiments* $\mathcal{E}_1$ and $\mathcal{E}_2$ are independent. □

The extension of the definition of independence for two experiments to the case of more than two experiments is made in a more or less obvious way:

Suppose that we are concerned with the outcomes of n different experiments $\mathcal{E}_1, \mathcal{E}_2, \ldots, \mathcal{E}_n$. *Suppose further that the sample space* $\mathcal{S}_k$ *of the kth of these n experiments is partitioned by the* m_k *events* A_{ki_k}, $i_k = 1, 2, \ldots, m_k$ (cf. Fig. 8.1). *The n given experiments are then said to be* statistically independent *if and only if the equation*

$$P[A_{1i_1} \cap A_{2i_2} \cap \cdots \cap A_{ni_n}] = P[A_{1i_1}]P[A_{2i_2}] \cdots P[A_{ni_n}] \quad (8.1)$$

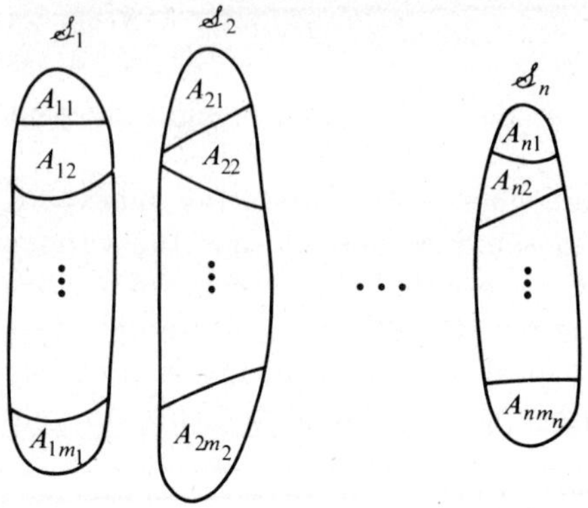

Figure 8.1 The partitioning of the n different sample spaces.

holds for every possible set of n integers $\{i_1, i_2, \ldots, i_n\}$ where the index i_k ranges from 1 to m_k.

Consider, for example, the joint experiment consisting of the observation of eight successive pulse positions on a communication link. Suppose that at each of the eight possible pulse positions there can be either a positive pulse, a negative pulse, or no pulse. Suppose also that the individual experiments which determine the kind of pulse at each possible position are independent (as defined above). Let us denote the event that the ith pulse is positive by $(x_i = +1)$, that it is negative by $(x_i = -1)$, and that it is zero by $(x_i = 0)$. Finally, let us suppose that

$$P[x_i = +1] = p$$
$$P[x_i = -1] = q$$

and

$$P[x_i = 0] = 1 - p - q$$

for $i = 1, 2, \ldots, 8$. It then follows from our definition of independent experiments that the probability that all pulses are positive is given by

$$\begin{aligned} P[(x_1 = +1) \cap (x_2 = +1) \cap \cdots \cap (x_8 = +1)] \\ = P[x_1 = +1]P[x_2 = +1] \cdots P[x_8 = +1] \\ = p^8 \end{aligned}$$

It further follows that the probability that the first four pulses are positive, that the next two are negative, and that the last two are zero is

$$\begin{aligned} P[(x_1 = +1) \cap (x_2 = +1) \cap (x_3 = +1) \cap (x_4 = +1) \\ \cap (x_5 = -1) \cap (x_6 = -1) \cap (x_7 = 0) \cap (x_8 = 0)] \\ = P[x_1 = +1]P[x_2 = +1]P[x_3 = +1]P[x_4 = +1] \\ P[x_5 = -1]P[x_6 = -1]P[x_7 = 0]P[x_8 = 0] \\ = p^4q^2(1 - p - q)^2 \end{aligned}$$

It is this ability to calculate the probability of joint events from the probabilities of individual events that makes the concept of statistically independent experiments so useful.

Exercise 8.2 Show that if n experiments are statistically independent, then any $n - 1$ of those experiments are also statistically independent. [*Hint*: If the omitted experiment is the mth, sum (8.2) over all possible outcomes of that experiment.] □

Note that it follows by induction on the result of Exercise 8.2 that if n experiments are statistically independent, then so also are any k of those experiments, where k is an integer such that $2 \leq k < n$.

Exercise 8.3 Consider the combined experiment consisting of four statistically independent flips of a coin. Suppose that on each flip, a head and a tail are equally likely outcomes.

a. Enumerate each of the possible outcomes of the combined experiment.
b. Determine the probability of occurrence of each of the possible outcomes of the combined experiment. □

Exercise 8.4 For the combined experiment of Exercise 8.3, determine (for $k = 0, 1, 2, 3, 4$) the probability $P[kH]$ that k heads occur in a single performance of the combined experiment. □

Exercise 8.5 For the combined experiment of Exercise 8.3:

a. Determine the probability that *at most* two heads occur in a single performance of the combined experiment, and
b. Determine the probability that *at least* two heads occur in a single performance of the combined experiment. □

On recalling how to calculate the number of combinations of things, the results in Exercise 8.4 may be generalized to give the result in the following exercises.

Exercise 8.6 Consider the combined experiment consisting of n statistically independent flips of a coin. Suppose that on each flip, a head and a tail are equally likely outcomes. Show that the probability $P[kH]$ that k heads occur in a single performance of the combined experiment is given by

$$P[kH] = \frac{1}{2^n}\binom{n}{k} \tag{8.2}$$

where $\binom{n}{k}$ is the *binomial coefficient*

$$\binom{n}{k} = \frac{n!}{k!(n-k)!} \tag{8.3}$$

As we shall see later (Sec. 6.2), (8.3) represents a special case of the so-called *binominal-probability distribution.* □

The last several exercises are typical of the combinatorial problems which occur in various applications of probability theory. Numerous other similar problems are discussed, for example, in Feller, vol. 1, chap. II.

9 SUMMARY OF AXIOMS, DEFINITIONS, AND FORMULAS

Set function

A rule which assigns a real number to each set in a specified collection of sets is called a set function.

Axiom 1

Given an experiment, there exists a sample space $\mathcal{S}$ representing the totality of possible outcomes of that experiment and a collection $\mathcal{A}$ of subsets A of $\mathcal{S}$, called events.

Axiom 2

To each event A in the collection of events $\mathcal{A}$, there can be assigned a nonnegative real number $P[A]$, that is, a number such that

$$0 \leq P[A]$$

This number is called the probability of the event A.

Axiom 3

For any probability assignment P, it is required that

$$P[\mathcal{S}] = 1$$

Axiom 4

If the events A and B are mutually exclusive (disjoint) events,

$$A \cap B = \phi$$

then any probability assignment P that is made must be such that

$$P[A \cup B] = P[A] + P[B]$$

Such a set function is said to be finitely additive.

Axiom 5

If the events $A_1, A_2, A_3, \ldots$ *are mutually exclusive events,*

$$A_i \cap A_j = \phi \qquad \text{for } i \neq j$$

then any probability assignment P *that is made must be such that*

$$P\left[\bigcup_{i=1}^{\infty} A_i\right] = P[A_1] + P[A_2] + P[A_3] + \cdots$$

$$= \sum_{i=1}^{\infty} P[A_i]$$

Such a set function is said to be countably additive.

Elementary properties of probability

$$P[A^c] = 1 - P[A] \tag{2.1}$$

$$P[\phi] = 0 \tag{2.2}$$

$$P[A] \leq 1 \tag{2.3}$$

$$P[B - A] = P[B] - P[A \cap B] \tag{2.4}$$

If

$$A \subset B \qquad \text{then} \qquad P[B - A] = P[B] - P[A] \tag{2.5}$$

If

$$A \subset B \tag{2.6}$$

then

$$P[A] \leq P[B]$$

In any case

$$P[A \cup B] = P[A] + P[B] - P[A \cap B] \tag{2.7}$$

$$P[A \cup B] \leq P[A] + P[B] \tag{2.8}$$

Joint probability

If the sample space $\mathcal{S}$ is partitioned both by the collection of events $\{A_1, A_2, \ldots, A_m\}$ and by the collection $\{B_1, B_2, \ldots, B_n\}$, then

$$P[B_j] = \sum_{i=1}^{m} P[A_i \cap B_j] \tag{5.1}$$

and

$$P[A_i] = \sum_{j=1}^{n} P[A_i \cap B_j] \tag{5.2}$$

Conditional probability

$$P[B|A] \triangleq \frac{P[A \cap B]}{P[A]}$$

so long as (6.1)

$$P[A] > 0$$

Bayes' rule

Let B be an arbitrary event in a sample space $\mathcal{S}$. Suppose that the events $A_1, A_2, \ldots, A_m$ partition $\mathcal{S}$ and that $P[A_i] > 0$ for all i. Then

$$P[A_i|B] = \frac{P[B|A_i]P[A_i]}{\sum_{j=1}^{m} P[B|A_j]P[A_j]} \tag{6.4}$$

Independent events

The events A and B are said to be statistically independent if and only if

$$P[A \cap B] = P[A]P[B] \tag{7.4}$$

The events $A_1, A_2, \ldots, A_n$ are said to be mutually independent if and only if the relations

$$\begin{aligned} P[A_i \cap A_j] &= P[A_i]P[A_j] \\ P[A_i \cap A_j \cap A_k] &= P[A_i]P[A_j]P[A_k] \\ &\cdots\cdots\cdots\cdots \\ P[A_1 \cap A_2 \cap \cdots \cap A_n] &= P[A_1]P[A_2] \cdots P[A_n] \end{aligned} \tag{7.5}$$

hold for all combinations of the indices such that $1 \le i < j < k < \cdots \le n$

Independent experiments

Suppose that we are concerned with the outcomes of n different experiments $\mathcal{E}_1, \mathcal{E}_2, \ldots, \mathcal{E}_n$. Suppose further that the sample space $\mathcal{S}_k$ of the kth of these n experiments is partitioned by the m_k events A_{ki_k}, $i_k = 1, 2, \ldots, m_k$. The n given experiments are then said to be statistically independent if and only if the equation

$$P[A_{1i_1} \cap A_{2i_2} \cap \cdots \cap A_{ni_n}] = P[A_{1i_1}]P[A_{2i_2}] \cdots P[A_{ni_n}] \tag{8.1}$$

holds for every possible set of n integers $\{i_1, i_2, \ldots, i_n\}$, where the index i_k ranges from 1 to m_k.

REFERENCES

Feller, William: "An Introduction to Probability Theory and Its Applications," 3d ed., vol. 1, John Wiley & Sons, Inc., New York, 1968.

Kershner, Richard B., and Lee Roy Wilcox: "The Anatomy of Mathematics," The Ronald Press Company, New York, 1950.

Kolmogorov, Andrei N.: "Foundations of the Theory of Probability," Chelsea Publishing Company, New York, 1956.

Loève, Michel: "Probability Theory," 3d ed., D. Van Nostrand Company, Inc., Princeton, N.J., 1963.

Neveu, Jacques: "Mathematical Foundations of the Calculus of Probability," Holden-Day, Inc., Publishers, San Francisco, 1965.

Polya, George: "Induction and Analogy in Mathematics," Princeton University Press, Princeton, N.J., 1954.

Roydan, Halsey L.: "Real Analysis," The Macmillan Company, New York, 1963.

Bernoulli trials

BINOMIAL EXPANSION: $\binom{n}{k} p^k q^{n-k}$

$p+q=1$

k = NUMBER OF OUTCOMES (SUCCESSFUL)
n = number of TRIALS
p = probability of success
q = probability of failure

P(EXACTLY k SUCCESSES) = $\binom{n}{k} p^k q^{n-k}$

P(OF AT LEAST m k SUCCESSES) = $\sum_{k=m}^{n} \binom{n}{k} p^k q^{n-k}$ or $1 - \sum_{k=n-m}^{n} \binom{n}{k} p^k q^{n-k}$

THIS TERM IS THE MISSING TERM'S IN FIRST EQU

example: Let n = 10, m = 3 ∴ $\sum_{k=3}^{10} \binom{n}{k} p^k q^{n-k}$

or $1 - [P(0) + P(1) + P(2)]$

P(no success's) = q^n P(AT LEAST 1 SUCCESS) = $1 - q^n$

A FAIR DIE IS ROLLED 5 TIMES. FIND THE PROB. THAT THE NUMBER 6 WILL APPEAR AT LEAST TWICE $\sum \binom{n}{k} p^k q^{n-k}$

$p = 1/6$ (SUCCESS) $q = 1-p$ (NO SIX)

P[6 APPEARS AT LEAST TWICE] = $1 - [P(0) + P(1)]$

$P(0) = \binom{5}{0} p^0 q^5$
$P(1) = \binom{5}{1} p^1 q^4$

A PAIR OF FAIR DICE IS ROLLED FOUR TIMES. WHAT IS THE PROB. THAT SEVEN WILL NOT APPEAR AT ALL?

$S = \{(11)(12) \dots (66)\}$ = 36 combinations

A SEVEN CAN OCCUR AS FOLLOWS $\{(16)(25)(34)(43)(52)(61)\}$

∴ THE PROBABILITY p OF ROLLING A SEVEN = $6/36 = 1/6$

THE PROB. q OF NOT ROLLING A SEVEN IS $q = 1-p = 5/6$

∴ THE PROB OF A SEVEN NOT APPEARING IS q^n

P{NO SEVEN} = $\binom{4}{0} p^k q^{n-k} = \binom{4}{0} p^0 q^4 = q^4$

4
Random Variables

1 RANDOM VARIABLES

One is, in practice, often interested in various numerical functions of the results of an experiment. For example, suppose that we are studying a particle moving along a straight line by measuring at some given instant of time both its position along the line, say x, and its velocity

$$v = \frac{dx}{dt}$$

and suppose, further, that we have determined in some way or other the probability distribution P for events in the two-dimensional sample space $\mathcal{S}$ of sample points (x,v). Finally, suppose that we are interested not only in the position and velocity of the particle, but also in its kinetic energy

$$K = \tfrac{1}{2}mv^2$$

where m is the mass of the particle. In other words, we wish to study probabilities of events which concern the kinetic energy of the particle as well as of events which concern only its position or its velocity.

Let us now, for the moment, focus our attention on the kinetic energy of the particle. Since to any particular outcome of our experiment, say (x_0,v_0), there corresponds one and only one value of the kinetic energy $(mv_0^2/2)$, we may if we wish act as though we are dealing with a new experiment whose possible outcomes are the single numbers K [rather than the number pairs (x,v)] and whose corresponding sample space, say $\mathcal{S}_K$, is therefore one-dimensional. The correspondence between points in this new kinetic energy sample space $\mathcal{S}_K$ and points in the original position-velocity sample space $\mathcal{S}$ is determined by the formula

$$K(x,v) = \tfrac{1}{2}mv^2$$

We say in this case that K is a function defined on the sample space $\mathcal{S}$ and that it determines a *mapping* of points (x,v) in $\mathcal{S}$ into points K in $\mathcal{S}_K$; such a mapping is pictured, for example, in Fig. 1.1.

While the kinetic energy function K may well be of particular importance to us, it is of course by no means the only function of interest which can be defined on the position-velocity sample space $\mathcal{S}$. Since, by hypothesis, we are also concerned here with the position and velocity of the particle, we are therefore also concerned with the functions X and

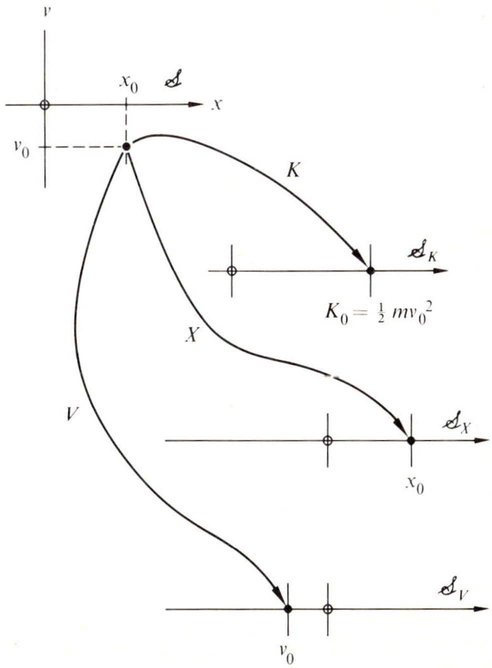

Figure 1.1 Various sample spaces for the moving particle.

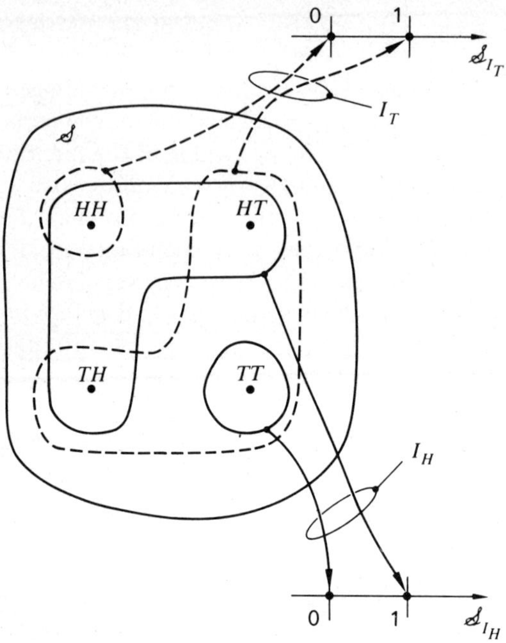

Figure 1.2 The head- and tail-indicator mappings.

V which are defined on $\mathcal{S}$ by the formulas

$$X(x,v) = x \qquad \text{and} \qquad V(x,v) = v$$

and which map points in $\mathcal{S}$ into points in new one-dimensional sample spaces, say $\mathcal{S}_X$ and $\mathcal{S}_V$ respectively. These mappings are also pictured in Fig. 1.1.

The problem remaining before us now is to determine the probabilities of events in each of the separate sample spaces $\mathcal{S}_K$, $\mathcal{S}_X$, $\mathcal{S}_V$ when we are given a probability distribution defined on events in the original sample space $\mathcal{S}$. We shall return to this point a bit later.

A situation similar to the above particle-motion example can arise, of course, when one is dealing with experiments which have nonnumerical results. For example, consider an experiment which consists of two independent flips of an unbiased coin, that is, a coin for which the probability of a head, $P[H]$, equals the probability of a tail, $P[T]$:

$$P[H] = \tfrac{1}{2} = P[T]$$

Suppose further that we are interested in whether or not at least one head occurs in the two flips. It is often convenient in studying such a situation to introduce a *set-indicator function* which has the value unity for sample points in the set of particular interest and which is zero elsewhere. Thus we define here[1]

$$I_H(s) \triangleq \begin{cases} 1 & \text{if } s \text{ corresponds to the occurrence} \\ & \quad \text{of at least one head} \\ 0 & \text{otherwise} \end{cases}$$

We then have the situation pictured in Fig. 1.2 and see that the set-indicator function I_H maps points in an abstract sample space into points on the real line. Now while the event that at least one head occurs may be important to us, we might well also be interested in other events which relate to the two coin flips. Thus, for example, if we are also interested in whether or not at least one tail occurs in the two flips, then we would introduce the set-indicator function

$$I_T(s) \triangleq \begin{cases} 1 & \text{if } s \text{ corresponds to the occurrence} \\ & \quad \text{of at least one tail} \\ 0 & \text{otherwise} \end{cases}$$

The mapping generated by I_T is also shown in Fig. 1.2. As in the preceding example, the problem remaining is again to determine the probabilities of events in a new sample space, given a probability distribution on events in an original sample space.

The two examples discussed above are concrete special cases of the abstract situation shown schematically in Fig. 1.3. In general, we start with some given sample space $\mathcal{S}$ and a probability distribution P defined

[1] We shall often use the symbol I when we wish to emphasize that a particular function is a set indicator; the reader is cautioned, however, that we shall not always do so.

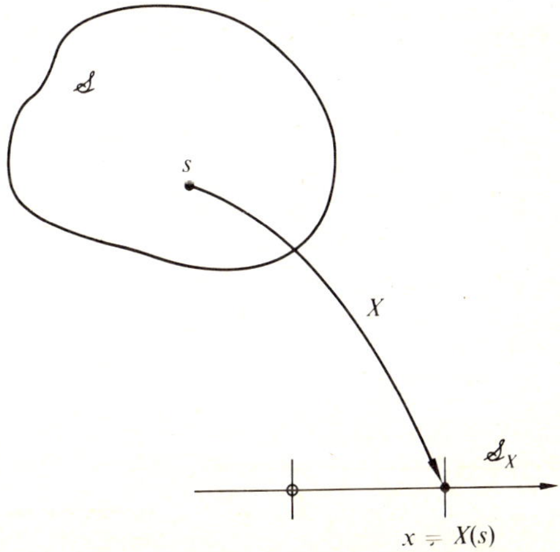

Figure 1.3 A general mapping.

on events in that sample space. A single-valued real function X is then defined on the sample space $\mathcal{S}$ so that to each point s of $\mathcal{S}$ there corresponds a single real number

$$x \triangleq X(s) \tag{1.1}$$

where $-\infty < x < +\infty$; the function X thus maps points in the sample space $\mathcal{S}$ into points on the real line.[1] The sample space $\mathcal{S}$ is called the *domain* of the function X, while the set of points on the real line which are determined by the values $X(s)$ as s takes on values throughout $\mathcal{S}$ is called the *range* of the function X. Thus, if we denote the range of X by R_X, we have

$$R_X \triangleq \{x: x = X(s) \text{ for } s \in \mathcal{S}\} \tag{1.2}$$

It therefore follows from our above discussion of the particle-motion and coin-flipping examples that the range of the kinetic energy mapping K is

$$R_K = \{K: 0 \leq K < +\infty\}$$

while the range of the head indicator I_H is

$$R_{I_H} = \{0,1\}$$

It should be noted that while to each $s \in \mathcal{S}$ there corresponds only one $x \in R_X$, it is also true that to a given $x \in R_X$ there may well correspond several values of $s \in \mathcal{S}$ such that $X(s) = x$, for example, as shown by the examples pictured in Fig. 1.2. That is, the function X is not necessarily a one-to-one mapping.[2]

Now once we have defined a function on a sample space and determined its range, it is often convenient to think of that range as a new sample space. Thus, for example, we found it convenient in our study of a particle moving in one dimension to consider the range R_K of the kinetic energy function

$$K = \tfrac{1}{2}mv^2$$

to be the sample space for a new experiment which consists of "measuring" the kinetic energy of a moving particle at some given instant of time. In order to emphasize the fact that we wish to think of the *range* R_X of a given function X as a new *sample space*, we shall from here on generally denote that range by the symbol $\mathcal{S}_X$; that is, we shall in general define

$$\begin{aligned} \mathcal{S}_X &\triangleq R_X \\ &= \{x: x = X(s) \text{ for } s \in \mathcal{S}\} \end{aligned} \tag{1.3}$$

[1] The *real line* is the set of real numbers x such that $-\infty < x < +\infty$.
[2] A *one-to-one mapping* X from a sample space $\mathcal{S}$ to the range R_X is a mapping such that to each point $s \in \mathcal{S}$ there corresponds one and only one point $x \in R_X$, while to each point $x \in R_X$ there corresponds one and only one point $s \in \mathcal{S}$.

Since the values of $X(s)$ are real numbers, a range sample space $\mathcal{S}_X$ will either be the entire real line $(-\infty < x < +\infty)$ or some part thereof.

On viewing a range as a sample space, subsets of this space become "events," and the question again arises as to how to assign probabilities to such events when our given probability assignment is made instead to events in the original domain sample space. With this question in mind, let us consider for a moment some particular set A in the range sample space $\mathcal{S}_X$ of a function X defined on a sample space $\mathcal{S}$. The set of points in $\mathcal{S}$ which map into A is called the *inverse image* of the set A under the mapping X and will be denoted by the symbol $X^{-1}(A)$:

$$X^{-1}(A) \triangleq \{s: s \in \mathcal{S}, A \subset \mathcal{S}_X, \text{ and } X(s) \in A\} \tag{1.4}$$

A typical set A and its inverse image $X^{-1}(A)$ are shown schematically in Fig. 1.4. Now even though $A \subset \mathcal{S}_X$ and $X^{-1}(A) \subset \mathcal{S}$ are defined in different sample spaces, they are nonetheless equivalent sets in that it follows from the definition of the inverse image that a point $s \in \mathcal{S}$ is in the inverse image $X^{-1}(A)$ if and only if the point $x \in \mathcal{S}_X$ is in the event A when $x = X(s)$. It follows from this that if $X^{-1}(A)$ is an *event* in $\mathcal{S}$, that is, if $X^{-1}(A)$ is one of the subsets of $\mathcal{S}$ for which a probability is defined (or is calculable), then it is entirely reasonable to assign the value of the probability of the inverse-image event $X^{-1}(A) \subset \mathcal{S}$ to be the value of the probability of the event $A \subset \mathcal{S}_X$. *We shall arbitrarily define the*

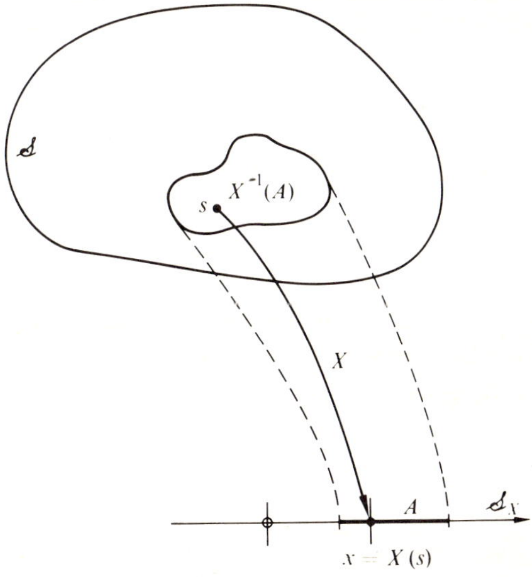

Figure 1.4 The set A and its inverse image $X^{-1}(A)$.

probability of an event A in the range sample space $\mathcal{S}_X$ *by the formula*

$$P[A \subset \mathcal{S}_X] \triangleq P[X^{-1}(A) \subset \mathcal{S}] \tag{1.5}$$

whenever the right-hand probability exists. We say then that the function X *induces a probability distribution* from its domain sample space $\mathcal{S}$ onto its range sample space $\mathcal{S}_X$ by the formula (1.5); our problem of finding the probabilities of events in the range sample space is thus solved (at least in principle).

As an example of this probability induction procedure, let us consider again the case of two independent flips of an unbiased coin. The single point 1 in the range sample space $\mathcal{S}_{I_H}$ of the head-indicator function I_H defined above has the set of sample points $\{HH,HT,TH\} \subset \mathcal{S}$ as its inverse image, while the single point 0 in $\mathcal{S}_{I_H}$ has the single point inverse image $\{TT\} \subset \mathcal{S}$. It then follows from (1.5) that

$$P[I_H = 1] = P[\{HH,HT,TH\}]$$

and hence that

$$\begin{aligned} P[I_H = 1] &= P[HH] + P[HT] + P[TH] \\ &= \tfrac{1}{4} + \tfrac{1}{4} + \tfrac{1}{4} \\ &= \tfrac{3}{4} \end{aligned}$$

since the points HH, HT, and TH are disjoint. It similarly follows that

$$\begin{aligned} P[I_H = 0] &= P[\{TT\}] \\ &= \tfrac{1}{4} \end{aligned}$$

These two probabilities are said to form the *probability distribution* induced on the range sample space $\mathcal{S}_{I_H}$ by the function I_H.

Exercise 1.1 Consider again the experiment which consists of two independent flips of an unbiased coin. Suppose now that we are interested in the event A that either a head or a tail occurs in the two flips, but not both a head and a tail occur. Define a set-indicator function I_A which corresponds to the event A and determine the probability distribution induced by it on its range sample space $\mathcal{S}_{I_A}$. □

Exercise 1.2 Consider the kinetic energy mapping K defined previously. Suppose that it has been found in a particular physical situation that a particle of mass 2 has a velocity v which is uniformly distributed over the interval $-1 \leq v \leq +1$; that is,

$$P[A] \triangleq \frac{b-a}{2}$$

when the event A is the interval $[a,b]$ with $-1 \leq a \leq b \leq +1$.

Let B be the event that the particle has a kinetic energy greater than $\frac{1}{2}$. Determine the probability of the event B. □

Whenever a function X defined on a sample space $\mathcal{S}$ is such that the probability of the inverse image $X^{-1}(A) \subset \mathcal{S}$ is defined for *each* event A in the range sample space $\mathcal{S}_X$, then that function X is said to be a *measurable function* on $\mathcal{S}$ and is called a *random variable.*[1] While a random variable is in fact a function and not a "variable" (at least in the common usage of that term), the name random variable is so deeply imbedded in the literature of probability and random processes that we shall continue its usage.

An examination of our examples and exercises in this section shows that the range of a given random variable may well be a continuum, for example, an infinite interval as in the case of the kinetic energy function K, or it may be a finite or countably infinite set of isolated points, for example, the points 0 and 1 as in the case of the head-indicator function I_H, or it may be a combination of the two. If the range is a continuum, we say that we have a *continuous random variable;* if the range consists only of isolated points, we say that we have a *discrete random variable;* and, finally, if the range is a combination of both continuum parts and isolated points, we say that we have a *mixed random variable.* Thus we see, for example, that the kinetic energy random variable K is a continuous random variable and that the head-indicator random variable I_H is a discrete random variable.

We have tried to introduce in this section the basic ideas involved in the concept of a random variable. A more detailed, though quite readable, treatment of these ideas may be found in secs. 3-1 to 3-3 of Pfeiffer, while a more advanced treatment of the subject may be found in secs. 6 and 7 of chap. I of Krickeberg.

2 PROBABILITY DISTRIBUTIONS, DENSITIES, AND DISTRIBUTION FUNCTIONS

Suppose, for the moment, that we are concerned with an experiment whose outcomes are characterized by a discrete random variable X which has the possible values $x_1, x_2, x_3, \ldots$ with the probabilities

$$P[X = x_i] = p_i$$

for $i = 1, 2, 3, \ldots$, respectively. We call the set of probabilities $P[X = x_i]$, that is, the set of numbers

$$\{p_1, p_2, p_3, \ldots\}$$

[1] Since X here is a real function, it is often called a *real random variable.* With this in mind, we define a *complex random variable* to be a function whose real and imaginary parts are separately real random variables.

the *probability distribution of the discrete random variable X*. Since all the events concerning X must be specifiable in terms of the possible outcomes x_i, the probability distribution of X completely determines the probabilistic properties of that random variable.

For example, suppose that we toss n unbiased coins independently and that we count the number of heads which result. If we characterize the outcomes of this experiment by the counting random variable X which counts the number of heads, then X has the possible values 0, 1, 2, . . . , n, and it follows (from the results of Exercise 3.8.6) that

$$P[X = k] = \frac{1}{2^n}\binom{n}{k} = \frac{1}{2^n}\frac{n!}{k!(n-k)!}$$

for $k = 0, 1, 2, \ldots, n$. The set of probabilities

$$\left\{\frac{1}{2^n}, \frac{n}{2^n}, \frac{n(n-1)}{2^n \times 2}, \ldots, \frac{n}{2^n}, \frac{1}{2^n}\right\}$$

then forms the probability distribution of this counting random variable and completely characterizes X.

Suppose next that we plot the probability distribution of the given discrete random variable versus the possible values of that random variable, say, for example, as we have done in Fig. 2.1 for our counting random variable in the cases $n = 3$ and $n = 10$. The virtue of such a plot is that it gives a compact representation of the probability distribution and, as such, shows at a glance the essential features of that distribution. Thus we see from Fig. 2.1 that the most probable value of the head-counting random variable is about $n/2$ for n coins tossed and, further, that the extreme values zero and n are the least probable outcomes.

Suppose now, on the other hand, that we are concerned with an experiment whose outcomes are characterized by a continuous random variable. For example, let us consider again the moving-particle case discussed in Exercise 1.2 above. We specified in that example that the velocity of the moving particle was uniformly distributed over the velocity interval $[-1,+1]$; in particular, we asserted that if A is any interval, say $[a,b]$, in $[-1,+1]$, then

$$P[A] = \frac{b-a}{2}$$

We would now like not only to be able to calculate the probability of any velocity event (whether represented by an interval or not), but also we

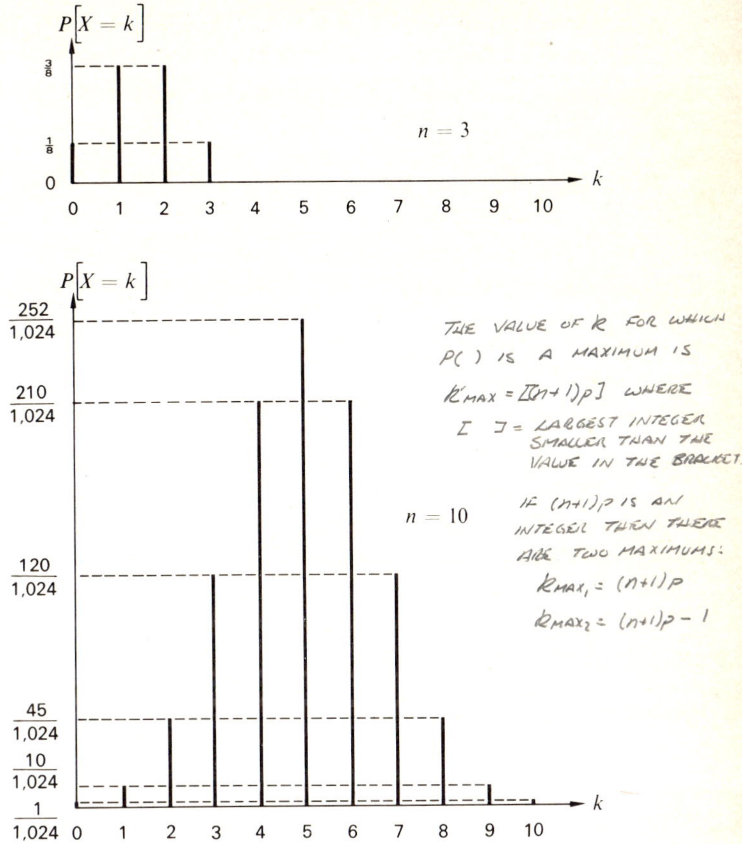

Figure 2.1 Probability distributions for the head-counting random variable.

would like to obtain a visualization of this situation which is essentially as informative as is the plot of the probability distribution in the discrete case.

Since we have stated here how to calculate the probability of an interval, one way to proceed might be to partition the velocity interval $[-1,+1]$ into m equal-length subintervals, say, by inserting the points v_i into that interval such that

$$-1 \triangleq v_0 < v_1 < v_2 < \cdots < v_{m-1} < v_m \triangleq +1$$

where

$$v_i - v_{i-1} = \frac{2}{m}$$

(since the length of the total interval is 2). It then follows from our given probability specification that each such subinterval has a probability of occurrence equal to

$$\frac{1}{2}\frac{2}{m} = \frac{1}{m}$$

and hence that a plot of the distribution of these subinterval probabilities would be as shown in Fig. 2.2*a*.

Suppose now that we are asked to calculate the probability of some velocity event B which is neither one of the above m subintervals nor a union of such intervals. While we cannot then use this particular partitioning to calculate the probability of B exactly, we can, of course, obtain an approximation to $P[B]$ by adding up the probabilities of all the subintervals contained in B; that is, we can write

$$P[B] \approx \sum_{\substack{\text{subintervals}\\ \text{in } B}} \frac{v_i - v_{i-1}}{2}$$

$$\triangleq \sum_{\substack{\text{subintervals}\\ \text{in } B}} \frac{\Delta v_i}{2}$$

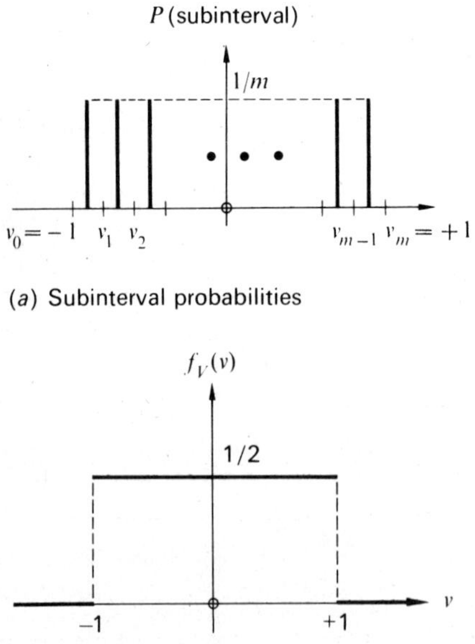

(*a*) Subinterval probabilities

(*b*) The velocity probability density

Figure 2.2 The velocity distribution.

While for any given value of m this result is not exact, we can reduce the approximation error by making the subintervals smaller and smaller (by increasing m). We may, in fact, reduce the error to zero by allowing the number of subintervals to increase without limit and hence obtain the result that

$$P[B] = \int_B \tfrac{1}{2} dv$$

Thus we see that we may calculate the probability of any event B in $[-1,+1]$ by integrating over B the function which has the value $\frac{1}{2}$ for all points in $[-1,+1]$. That is, if we define the function f_V by

$$f_V(v) \triangleq \begin{cases} \frac{1}{2} & \text{for } -1 < v < +1 \\ 0 & \text{otherwise} \end{cases}$$

then we may write

$$P[B] = \int_B f_V(v)\, dv$$

We shall call this function f_V the *probability density of the velocity random variable V*. A probability density is therefore a function which you integrate to obtain a probability and hence is similar in nature to the (perhaps) more familiar mass density; a plot of the velocity probability density f_V is given in Fig. 2.2*b*. We note that while the subinterval probability distribution changes each time we change the total number of subintervals, the probability density is fixed. We also note that the plot of f_V conveys succinctly the idea of a uniform distribution and hence that it fills the role for the continuous random variable V which the probability distribution fills for a discrete random variable.

While it is clear that a probability distribution completely specifies the probabilistic properties of a discrete random variable and that a probability density does the same for a continuous random variable, it would be very convenient on many occasions to have a single way of specifying the probabilistic properties of any random variable—whether discrete, continuous, or mixed. With this fact in mind, let us reconsider the probability density for a moment, recalling that we may calculate a probability by integrating the density.

Suppose that we are given a continuous random variable X defined on a sample space $\mathcal{S}$, that X has a probability density f_X, and, further, that we denote the indefinite integral of f_X by F_X; let us write

$$F_X(x) = \int_c^x f_X(u)\, du$$

where c is an, as yet, undetermined constant. It then follows from the fundamental theorem of calculus that[1]

$$f_X(x) = \frac{dF_X(x)}{dx} \tag{2.1}$$

that is, that the probability density is the derivative of the indefinite integral F_X. While this result is true for any value of the constant c, F_X is not uniquely specified until we choose a specific value of that constant. In order to avoid having to worry about whether x is greater or less than c, let us arbitrarily take c to be minus infinity; that is,

$$F_X(x) = \int_{-\infty}^{x} f_X(u)\,du \tag{2.2}$$

F_X is then uniquely determined by the probability density and is called the *probability distribution function* (or *cumulative distribution function*) of the random variable X.

While we introduced the probability distribution function here via the probability density, we shall next show that we can, in fact, define such a function for any random variable. Suppose that we evaluate (2.2) at some point b:

$$F_X(b) = \int_{-\infty}^{b} f_X(u)\,du \tag{2.3}$$

Now remembering the fact that the integral of the probability density f_X over a given region equals the probability that X has a value in that region, we see that the right-hand side of (2.3) equals the probability that X has a value in the semi-infinite interval $(-\infty,b]$. Thus we see that

$$F_X(b) = P[X(s) \in (-\infty,b]] \tag{2.4}$$

that is, the probability distribution function F_X evaluated at the point b equals the probability that the random variable X has a value less than or equal to b. Now it follows from the very definition of random variable that we can (in principle, at least) always determine a probability of the type shown on the right-hand side of (2.4) for any random variable, whether it has a probability density or not. With this fact in mind, we shall from now on *define* the probability distribution function of the random variable X (whether discrete, continuous, or mixed) by the formula[2]

$$\boxed{F_X(x) \triangleq P[X \leq x]} \tag{2.5}$$

where, for the sake of compactness of notation, we have introduced the

[1] See, for example, sec. 2.14, vol. I of Apostol.

[2] The definition in (2.5) is quite common. The reader is cautioned, however, that some authors define the probability distribution function to be $P[X < x]$; cf. sec. 10 of Loève. Be careful!

symbol $P[X \leq x]$ defined by

$$P[X \leq x] \triangleq P[X(s) \in (-\infty, x]] \\ = P[\{s: -\infty < X(s) \leq x, s \in \mathcal{S}\}] \tag{2.6}$$

Having defined the probability distribution function as in (2.5), we shall next determine that function for a few typical random variables. Then, in the remainder of the chapter, we shall derive the properties of the probability distribution function which follow from this definition and, finally, we shall return to the probability density defined as the derivative of the probability distribution function.

Let us now consider several specific examples of probability distribution functions. First, consider the head-indicator random variable I_H discussed in the preceding section. We pointed out there that this random variable assumed only the two possible values zero and one, and showed that it had the probability distribution

$$P[I_H = 1] = \tfrac{3}{4} \qquad \text{and} \qquad P[I_H = 0] = \tfrac{1}{4}$$

It now follows from the definition of the probability distribution function, (2.5), that

$$F_{I_H}(a) = \begin{cases} 0 & \text{for } -\infty < a < 0 \\ \frac{1}{4} & \text{for } 0 \leq a < +1 \\ 1 & \text{for } +1 \leq a < +\infty \end{cases}$$

We note that this distribution function (plotted in Fig. 2.3) is a step function; that is it is constant except at the points zero and one where it has jumps. We note also that the height of the jump at zero (or one) equals the probability that I_H equals zero (or one).

A random variable, such as I_H, which has only two possible values, and hence which has a probability distribution function of the type shown in Fig. 2.3, is called a *Bernoulli random variable* after the Swiss mathematician Jacob (James) Bernoulli (1654–1705).[1]

[1] Jacob Bernoulli's main probability work, "Ars Conjectandi," was published posthumously in 1713.

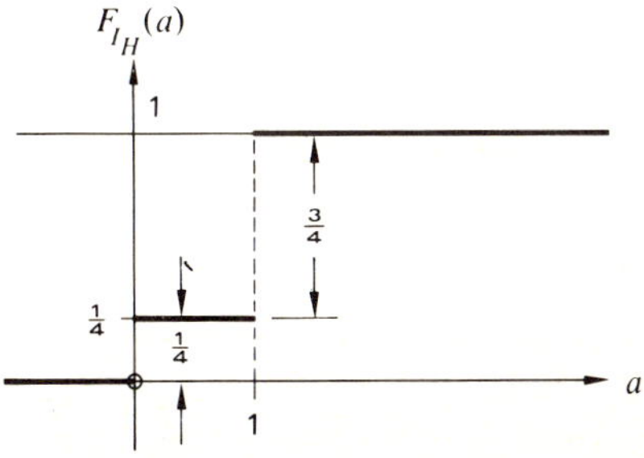

Figure 2.3 The head-indicator probability distribution function.

As a second discrete example, let us reconsider the experiment of tossing n coins independently and counting the number of heads which result. Suppose that we again define a counting random variable, say X, which counts the number of heads which occur in the n tosses. As noted earlier, the counting random variable X can assume only the values $0, 1, 2, \ldots, n$ with the probabilities

$$P[X = k] = \frac{1}{2^n}\binom{n}{k}$$

for $k = 0, 1, 2, \ldots, n$, respectively. As in the preceding Bernoulli example, the probability distribution function here is constant between the jumps which occur at the possible values of the random variable X. Thus, if we set $n = 3$, we have

$$F_X(x) = \begin{cases} 0 & \text{for } x < 0 \\ \frac{1}{8} & \text{for } 0 \le x < 1 \\ \frac{1}{2} & \text{for } 1 \le x < 2 \\ \frac{7}{8} & \text{for } 2 \le x < 3 \\ 1 & \text{for } 3 \le x \end{cases}$$

as shown in Fig. 2.4 (cf. Fig. 2.1a).

We will often have to deal with the probability distribution functions of discrete random variables in the future and a more compact representation than those given above for the head-indicator and head-counting random variables would be desirable. With this in mind, it is convenient to introduce here the use of the *unit-step function* U defined by the formula

$$U(x) \triangleq \begin{cases} 1 & \text{for } 0 \le x \\ 0 & \text{for } x < 0 \end{cases} \tag{2.7}$$

as shown in Fig. 2.5a. It follows from this definition that if we translate the unit-step function to the right by an amount a, we have

$$U(x - a) = \begin{cases} 1 & \text{for } a \le x \\ 0 & \text{for } x < a \end{cases} \tag{2.8}$$

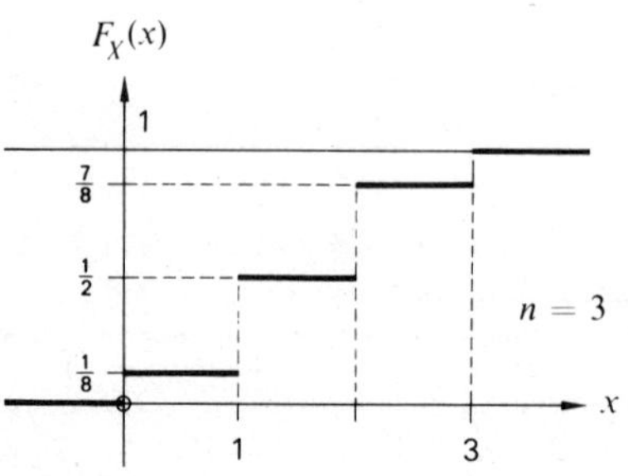

Figure 2.4 The head-counting probability distribution function.

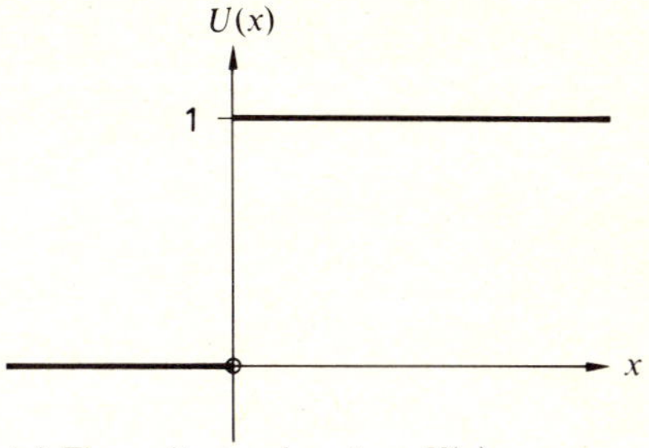

(a) The unit-step function $U(x)$

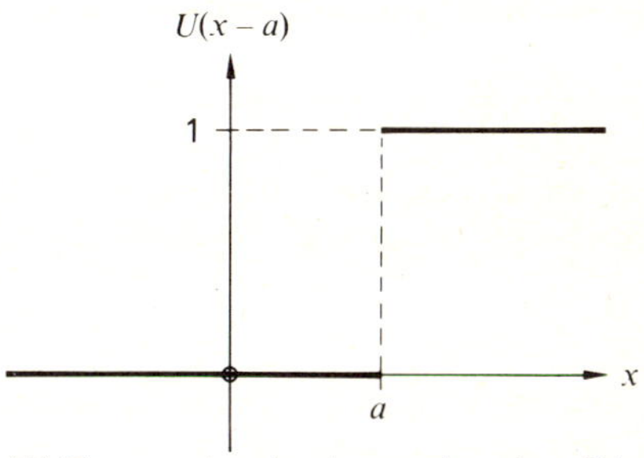

Figure 2.5 Unit-step functions. (b) The translated unit-step function $U(x - a)$

as shown in Fig. 2.5*b*. A moment's reflection upon these formulas should then show that we may write

$$F_X(x) = \sum_{k=0}^{n} \frac{1}{2^n} \binom{n}{k} U(x - k) \tag{2.9}$$

as an expression for the probability distribution function of our counting random variable X. We say that X has a *binomial*-probability distribution function in this case.

Examples involving random variables which have probability densities are given in the following two exercises.

Exercise 2.1 As in Exercise 1.2 above, suppose that it has been found that a moving particle has a velocity which is uniformly distributed over the interval $[-1,+1]$. Let V denote the random variable which characterizes the possible values of the particle velocity.

a. Determine the probability distribution function $F_V(v)$ of the velocity random variable V, and plot.

b. Calculate the derivative

$$f_V(v) \triangleq \frac{dF_V(v)}{dv}$$

of the probability distribution function, and plot (cf. Fig. 2.2*b*). Such a random variable is said to have a uniform probability density. □

Exercise 2.2 Let K be the random variable which characterizes the possible values of the kinetic energy of the moving particle and suppose that the particle velocity has the uniform probability density derived in Exercise 2.1. Assuming that the particle has a mass of 2,

a. Determine the probability distribution function $F_K(y)$ of the kinetic energy random variable, and plot.

b. Calculate the derivative

$$f_K(y) \triangleq \frac{dF_K(y)}{dy}$$

of the probability distribution function, and plot. □

We noted in the preceding section that a function defined on a sample space was a random variable if and only if the probability of each event in the range sample space could be defined. We shall show in the following section that once a probability distribution function is defined, the probability of every interval in the range sample space (and hence of all unions and intersections of such intervals) may be calculated. Now the class of subsets of a range space which consists of intervals and of all unions and intersections of intervals is generally adequate to describe events of interest.[1] It therefore follows that a function defined on a sample space is a random variable if and only if it has a probability distribution function. Hence whenever the name random variable is used from here on, there must be a corresponding probability distribution function lurking around somewhere; a random variable and its probability distribution function always go together!

3 PROPERTIES OF PROBABILITY DISTRIBUTION FUNCTIONS

Suppose that we are given a random variable X defined on a sample space $\mathcal{S}$. Having defined the probability distribution function F_X of X by the formula

$$\boxed{\begin{aligned} F_X(x) &\triangleq P[X(s) \in (-\infty, x]] \\ &= P[X \leq x] \end{aligned}} \tag{3.1}$$

[1] Such a class is a Borel field as pointed out in Sec. 3.3.

let us now determine some of the special properties of such functions.

First, we note that the probability distribution function is defined on the entire real line and *not* just on the range of the random variable in question. For example, the probability distribution function of the head-indicator random variable shown in Fig. 2.3 was plotted there for all values of the parameter a and not just for those values of a which fell in the range of the head-indicator random variable, that is, not just in $\{0,1\}$.

Next, we note that since the range sample space of any random variable is always contained within the real line, that is, since for any random variable X defined on a sample space $\mathcal{S}$ we have

$$\mathcal{S}_X = \{x: x = X(s),\ s \in \mathcal{S}\} \subset (-\infty < x < +\infty)$$

it follows that[1]

$$\begin{aligned} F_X(+\infty) &= P[X \le +\infty] \\ &= 1 \end{aligned} \tag{3.2}$$

On the other hand, since the event that the random variable X has values falling in the interval $(-\infty, -\infty]$ is the null event, it follows that

$$F_X(-\infty) = 0 \tag{3.3}$$

Thus we see that the probability distribution function $F_X(x)$ achieves its lower limit of zero as x tends to minus infinity, while it achieves its upper limit of unity as x tends to plus infinity.

Let us now study the behavior of the probability distribution function $F_X(x)$ for finite values of its argument x. On recalling that F_X is by definition a probability concerning semi-infinite intervals, let us consider two points a and b, where $a \le b$, and decompose the semi-infinite interval $(-\infty,b]$ in terms of $(-\infty,a]$ to obtain

$$(-\infty,b] = (-\infty,a] \cup (a,b] \tag{3.4}$$

where $(a,b]$ is the half-open interval

$$(a,b] \triangleq \{x: a < x \le b\}$$

The two sets on the right-hand side of (3.4) are disjoint and hence the probabilities that the random variable X has values in those sets add. Thus we get

$$P[X \in (-\infty,b]] = P[X \in (-\infty,a]] + P[X \in (a,b]]$$

which becomes

$$F_X(b) = F_X(a) + P[X \in (a,b]] \tag{3.5}$$

when we use the definition of the probability distribution function.

[1] A detailed proof of (3.2) and (3.3) is given in the following section.

Now the probability that X has values falling in any set must be non-negative (by Axiom 2). Hence it follows from (3.5) that

$$F_X(b) \geq F_X(a) \tag{3.6}$$

as long as $b \geq a$. That is, the value of the probability distribution function always increases (or, at the very least, remains constant) as the value of its argument increases from minus infinity to plus infinity. Thus we have shown that *a probability distribution function is a monotone-increasing function which is bounded between zero and one.*

The fact that the probability distribution function is a bounded monotone function places severe restrictions upon the kind of discontinuities that it can have. In particular, it can be shown that the probability distribution function F_X is continuous on the right,[1] that is, that

$$F_X(a) = F_X(a + 0) \tag{3.7}$$

where, given that $\epsilon > 0$,

$$F_X(a + 0) \triangleq \lim_{\epsilon \to 0} F_X(a + \epsilon) \tag{3.8}$$

and, further, that

$$F_X(a) = F_X(a - 0) + P[X = a] \tag{3.9}$$

where, given that $\delta > 0$,

$$F_X(a - 0) \triangleq \lim_{\delta \to 0} F_X(a - \delta) \tag{3.10}$$

That is, the probability distribution function F_X is continuous on the left at a point a if and only if the probability that X equals a is zero.

In summary, given a random variable X with its probability distribution function F_X, it follows that at each point a of the real line, *either*

$$P[X = a] > 0$$

in which case F_X has a jump discontinuity of height $P[X = a]$ at a (as shown in Fig. 3.1), *or*

$$P[X = a] = 0$$

in which case

$$F_X(a - 0) = F_X(a) = F_X(a + 0)$$

and F_X is therefore continuous at a.

[1] See Sec. 4 of this chapter.

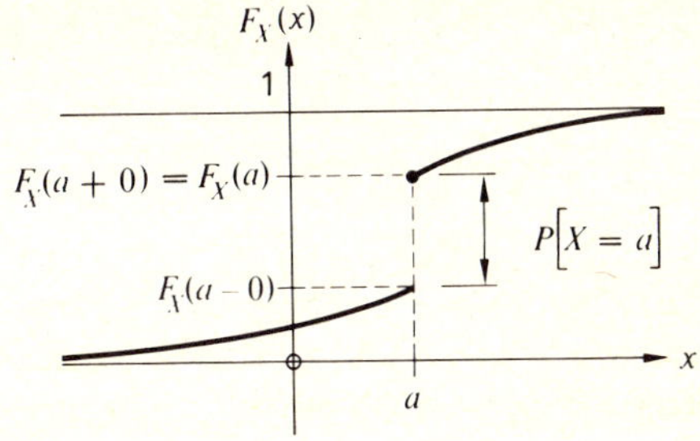

Figure 3.1 A jump discontinuity.

Now we have already seen examples of probability distribution functions which are everywhere continuous (cf. Exercises 2.1 and 2.2) and we have also already seen that the probability distribution functions of discrete random variables are step functions in that they increase their values only by jumps (cf. the Bernoulli and binomial cases discussed in the preceding section). Since we have shown above that the only discontinuities that a probability distribution function can have are jump discontinuities, a question then arises as to whether or not it would be possible to decompose an arbitrary probability distribution function into only two parts: one which is everywhere continuous and the other which is a step function. It turns out that the answer to this question is yes, since it can be shown that a probability distribution function can have at most a countable number of jump discontinuities (cf. Sec. 4). That this in fact answers our question may be shown as follows: Suppose that we are given a random variable X with a probability distribution function F_X which has no more than a countable number of jump discontinuity points, say x_i, $i = 1, 2, \ldots$, and which is not necessarily constant between jumps. Recalling that such a probability distribution function jumps by an amount $P[X = x_i]$ at the discontinuity point x_i, let us consider for the moment the step function

$$D_X(x) \triangleq \sum_{i=1}^{\infty} P[X = x_i]U(x - x_i) \tag{3.11}$$

where $U(x - x_i)$ is the unit-step function which changes value at $x = x_i$, and where the summation is taken over all possible jump-point indices (which is possible since there are at most a countable number of jumps). The step function D_X thus has exactly the same jump discontinuities as does the probability distribution function F_X. If, then, we subtract it from F_X, we obtain the function

$$C_X(x) \triangleq F_X(x) - D_X(x) \tag{3.12}$$

which is everywhere continuous (since we have removed all its possible discontinuities). Thus we see that we may express *any* probability

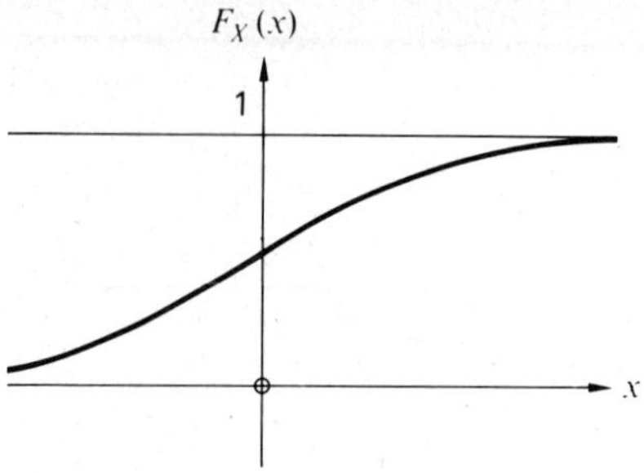

Figure 3.2 A continuous probability distribution function.

distribution function in the form

$$F_X(x) = D_X(x) + C_X(x) \tag{3.13}$$

where D_X is a step function and C_X is continuous everywhere.

Now a moment's reflection upon the above definitions of the functions D_X and C_X should show that they are both bounded monotone-increasing functions. It therefore follows that every probability distribution function is *either* everywhere continuous (say, as shown in Fig. 3.2), *or* is a step function (say, as shown in Fig. 3.3), *or* is decomposable into two parts, one of which is everywhere continuous and the other of which is a step function (say, as shown in Fig. 3.4). We call the

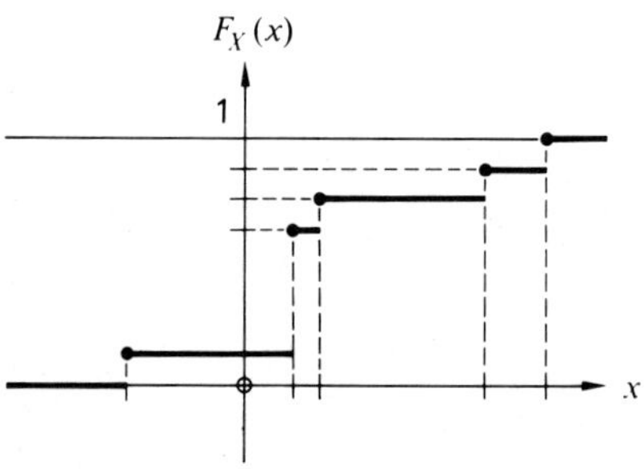

Figure 3.3 A discrete probability distribution function.

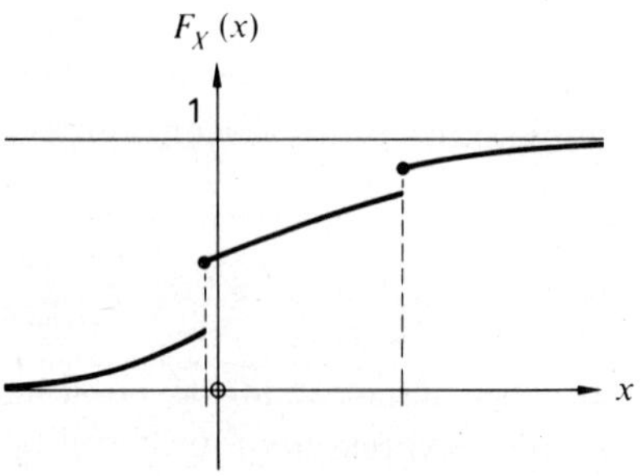

Figure 3.4 A mixed probability distribution function.

first case the *continuous* case, the second the *discrete* case, and the last the *mixed* case.

Exercise 3.1 Let us consider a version of the spinning-wheel problem introduced in Exercise 3.2.2. In particular, consider a wheel with very low, but not zero, bearing friction. Suppose that a mark is put at the top of the wheel, that the wheel is spun, that the position of the mark is noted after the wheel has stopped spinning, and that the counterclockwise angle θ through which the wheel must be rotated in order to bring the marked point uppermost is determined (see Fig. 3.5). Suppose that an indicating device is attached to the wheel which notes the value of θ and generates an output ϕ by the rule

$$\phi \triangleq \begin{cases} 0 & \text{if } 0 \leq \theta < \dfrac{\pi}{2} \\ \theta & \text{if } \dfrac{\pi}{2} \leq \theta < \pi \\ \pi & \text{if } \pi \leq \theta < \dfrac{3\pi}{2} \\ \theta & \text{if } \dfrac{3\pi}{2} \leq \theta < 2\pi \end{cases}$$

that is, the indicator reads the actual angle θ only for values of θ in the regions $[\pi/2, \pi)$ and $[3\pi/2, 2\pi)$.

Assuming that it has been determined that

$$P[a \leq \theta \leq b] = \frac{b-a}{2\pi}$$

for $0 \leq a \leq b < 2\pi$, determine and plot the probability distribution

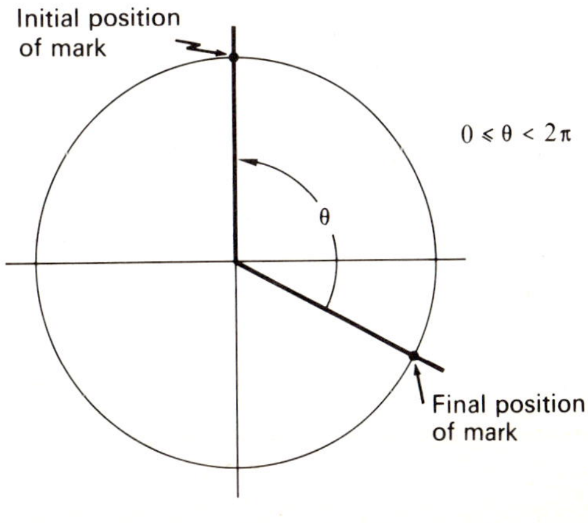

Figure 3.5 The spinning-wheel problem.

function of the random variable Φ which characterizes the readings ϕ of the indicating device. (Note that Φ is a mixed random variable.) □

Having determined various general properties of the probability distribution function, let us now see how to use that function in calculating the probabilities of various events concerning the possible values of the related random variable. First, let us determine the probability that a random variable X (with a probability distribution function F_X) has values falling in the half-open interval $(a,b]$, where $a \leq b$. If we return for the moment to (3.5) and solve that equation for $P[X \in (a,b]]$, we get the result that

$$P[X \in (a,b]] = F_X(b) - F_X(a) \tag{3.14}$$

so long as $a \leq b$. That is, the probability that a random variable has values falling in the half-open interval $(a,b]$ is given by the value of the probability distribution function of that random variable at the point b minus its value at the point a.

The above result may now be extended to other kinds of intervals as shown by the following exercise. We consider there a closed interval, say

$$[a,b] \triangleq \{x\colon\ a \leq x \leq b\}$$

an open interval, say

$$(a,b) \triangleq \{x\colon\ a < x < b\}$$

and the half-open interval

$$[a,b) \triangleq \{x\colon a \leq x < b\}$$

where, in each case, $a \leq b$.

Exercise 3.2 Let X be a random variable defined on a sample space $\mathcal{S}$, and let F_X be the probability distribution function of X. Assuming that $a \leq b$:

a. Show that

$$P[X \in [a,b]] = P[X = a] + F_X(b) - F_X(a) \tag{3.15}$$

b. Show that

$$P[X \in (a,b)] = \begin{cases} F_X(b) - F_X(a) - P[X = b] & \text{if } a < b \\ 0 \quad \text{if } a = b \end{cases} \tag{3.16}$$

c. Show that

$$P[X \in [a,b)] = \begin{cases} P[X = a] + F_X(b) - F_X(a) - P[X = b] \\ \qquad\qquad\qquad\qquad\qquad\qquad \text{if } a < b \\ 0 \qquad \text{if } a = b \end{cases} \tag{3.17}$$

Don't forget to check the case $a = b$ in part (*a*)! □

Once we know how to determine the probability that the random variable X has values which fall in any kind of interval, we can easily determine the probability that X has a value which falls in the union or intersection of two arbitrary intervals. The following exercise considers the simplest case: that in which we study half-open intervals which are open on the left.

Exercise 3.3 Let X be a random variable defined on a sample space $\mathcal{S}$ and let F_X be the probability distribution function of that random variable. Let $(a,b]$ and $(c,d]$ be two arbitrary half-open intervals on the real line.

a. Calculate the probability that X has a value falling in the union of the two intervals, first when they are disjoint and then when they overlap.

b. Calculate the probability that X has a value falling in the intersection of the two intervals, first when they are disjoint and then when they overlap.

(*Hint:* Be careful of the number of different cases of overlap.) □

Using the results of Exercise 3.2, the results of Exercise 3.3 can, of course, be extended to cover the union and intersection of any pair of intervals. Once that has been done, a further extension can be made to show that one can always calculate both the probability that the given random variable has a value falling in any arbitrary union of intervals and the probability that it has a value falling in any arbitrary intersection of intervals. That is, given a random variable and its probability distribution function, we can calculate the probability of essentially any event of interest which concerns the values of that random variable. This justifies our comment at the end of the preceding section to the effect that a function defined on a sample space is a random variable if and only if it has a probability distribution function.

Exercise 3.4 Consider again the spinning wheel and indicating device discussed in Exercise 3.1.

a. Determine the probability that the indicator reading ϕ has a value in the interval

$$A \triangleq \left[\pi, \frac{7\pi}{4}\right]$$

b. Determine the probability that the indicator reading ϕ has a value in the interval

$$B \triangleq \left[\frac{3\pi}{4}, \frac{5\pi}{4}\right]$$

c. Determine the probability that the indicator reading ϕ has a value in the union $A \cup B$.

d. Determine the probability that the indicator reading ϕ has a value in the intersection $A \cap B$. □

OMIT

4 DERIVATIONS*

Let us now derive some of the results used in the preceding section:

THEOREM 4.1 THE RIGHT-HAND LIMIT Let X be a random variable defined on a sample space $\mathcal{S}$ and let F_X be the probability distribution function of that random variable.

Then

$$F_X(+\infty) = +1 \tag{4.1}$$

Proof First of all, we note that since X assumes values only on the real line $(-\infty, +\infty)$, the symbol $F_X(+\infty)$ must be defined. We take here

$$F_X(+\infty) \triangleq \lim_{i \to +\infty} F_X(x_i) \tag{4.2}$$

where the sequence $\{x_i\}$ is a countable, monotone-increasing sequence of points x_i, that is, a sequence of points x_i, $i = 1, 2, 3, \ldots$ such that

$$x_i \leq x_{i+1} \qquad \text{and} \qquad \lim_{i \to +\infty} x_i = +\infty$$

It then follows from the definition of the probability distribution function that we may rewrite (4.2) as

$$F_X(+\infty) = \lim_{i \to +\infty} P[X \in (-\infty, x_i]]$$

* Supplementary material. The developments in this section are rather theoretical; the techniques used and the results obtained, however, are quite important.

Now in view of the assumed character of the x_i, the sequence of intervals $(-\infty, x_i]$ is monotone increasing; that is,

$$(-\infty, x_i] \subset (-\infty, x_{i+1}]$$

It therefore follows from the continuity theorem of probability derived in Sec. 3.4 that

$$\lim_{i \to +\infty} P[X \in (-\infty, x_i]] = P[\lim_{i \to +\infty} (X \in (-\infty, x_i])]$$

It further follows from the fact that our sequence of intervals is monotone increasing that

$$\lim_{i \to +\infty} (-\infty, x_i] = \bigcup_{i=1}^{\infty} (-\infty, x_i] \tag{4.3}$$

[cf. (2.5.4)]. The right-hand side of (4.3) is the real line $(-\infty < x < +\infty)$ since the x_i increase without limit, by hypothesis. On combining these various results we see that

$$F_X(+\infty) = P[X \in (-\infty, +\infty)] \tag{4.4}$$

Now it follows from the definition of random variable that the range of any random variable is contained within the real line:

$$R_X = \{x \colon x = X(s),\ s \in \mathcal{S}\} \subset (-\infty < x < +\infty)$$

Since X must assume some value in its range, the right-hand side of (4.4) therefore must equal unity. The theorem statement is thus proved. ∎

If next we define

$$F_X(-\infty) \triangleq \lim_{j \to +\infty} F_X(x_j) \tag{4.5}$$

where the sequence of points $\{x_j\}$ here is a countable, monotone-decreasing sequence of points, that is, a sequence of points x_j, $j = 1, 2, 3, \ldots$, such that

$$x_{j+1} \le x_j \quad \text{and} \quad \lim_{j \to +\infty} x_j = -\infty \tag{4.6}$$

then one can derive the corresponding left-hand limit theorem in a similar manner.

Exercise 4.1 The left-hand limit Let X be a random variable defined on a sample space $\mathcal{S}$ and let F_X be the probability distribution function of that random variable.

Show that

$$F_X(-\infty) = 0 \tag{4.7}$$

where $F_X(-\infty)$ is defined as in (4.5). □

Let us next study the continuity properties of the probability distribution function.

THEOREM 4.2 CONTINUITY ON THE RIGHT Let X be a random variable defined on a sample space $\mathcal{S}$ and let F_X be the probability distribution function of that random variable.

Then

$$F_X(a + 0) = F_X(a) \tag{4.8}$$

at any point a on the real line.

Proof Now by definition [cf. (3.7)], $F_X(a + 0)$ is the limiting value of F_X as the point a is approached from the right. Let us therefore take a monotone-decreasing sequence of numbers α_i, that is, a sequence of numbers α_i, $i = 1, 2, 3, \ldots$, such that

$$\alpha_{i+1} \leq \alpha_i$$

and (4.9)

$$\lim_{i \to +\infty} \alpha_i = 0$$

and consider the points $a + \alpha_i$; the sequence of these points then approaches the point a from the right as the index i increases without limit. We may therefore write

$$F_X(a + 0) = \lim_{i \to +\infty} F_X(a + \alpha_i)$$

It now follows using this result that

$$\begin{aligned} F_X(a + 0) - F_X(a) &= \lim_{i \to +\infty} F_X(a + \alpha_i) - F_X(a) \\ &= \lim_{i \to +\infty} [F_X(a + \alpha_i) - F_X(a)] \\ &= \lim_{i \to +\infty} P[X \in (a, a + \alpha_i]] \end{aligned}$$

the last equality following from (3.8).

Consider next the sequence of intervals $(a, a + \alpha_i]$. It follows from the definition of the α_i, (4.9), that

$$(a, a + \alpha_i] \supset (a, a + \alpha_{i+1}]$$

that is, that the sequence is monotone decreasing. It then further follows from the continuity theorem of probability that

$$\lim_{i \to +\infty} P[X \in (a, a + \alpha_i]] = P[\lim_{i \to +\infty} (X \in (a, a + \alpha_i])]$$

and, from (2.5.8), that

$$\lim_{i \to +\infty} (a, a + \alpha_i] = \bigcap_{i=1}^{\infty} (a, a + \alpha_i]$$
$$= \phi$$

the last equality coming from the fact that the α_i tend to zero.

On combining these various results, we then get

$$F_X(a + 0) - F_X(a) = P[\phi] = 0$$

Equation (4.8) then follows from this result and the theorem is proved. ∎

The arguments used above in proving continuity on the right can now be paraphrased to prove the fact that a probability distribution function F_X is continuous on the left at a point a if and only if $P[X = a]$ is zero.

Exercise 4.2 Continuity on the left Let X be a random variable defined on a sample space $\mathcal{S}$ and let F_X be the probability distribution function of that random variable.

Show that

$$F_X(a - 0) = F_X(a) - P[X = a]$$

at any point a on the real line. □

Finally, let us prove the fact that a probability distribution function can have at most a countable number of jump discontinuities:

THEOREM 4.3 COUNTABILITY OF JUMPS Let X be a random variable defined on a sample space $\mathcal{S}$ and let F_X be the probability distribution function of that random variable.

It then follows that F_X can have at most a countable number of jump discontinuities on the real line.

Proof First, we note that it follows from Theorem 4.2, and the results of Exercise 4.2, that F_X can have no discontinuities other than jumps.

Consider next an arbitrary bounded interval on the real line, say $(a,b]$. Suppose that the interval contains n discontinuity points x_i, say

$$a < x_1 < x_2 < \cdots < x_n \leq b$$

such that

$$P[X = x_i] \geq \frac{1}{m} \qquad i = 1, 2, \ldots, n \tag{4.10}$$

where m is an arbitrary positive integer. We now wish to determine whether or not there is an upper bound on the number n of such discontinuities.

It follows from the monotone character of the probability distribution function, (3.6), and from our continuity results above that

$$F_X(a) \leq F_X(x_1 - 0) < F_X(x_1) \leq F_X(x_2 - 0) \\ < F_X(x_2) \leq \cdots \leq F_X(b)$$

Now the total height of these various jumps cannot exceed the total change in F_X over the interval $(a,b]$; that is,

$$\sum_{i=1}^{n} [F_X(x_i) - F_X(x_i - 0)] = \sum_{i=1}^{n} P[X = x_i] \\ \leq F_X(b) - F_X(a) \leq 1 \tag{4.11}$$

the last inequality following from the fact that any probability distribution function is bounded between zero and one.

If next we apply (4.10) to the sum of jump probabilities, we get

$$\sum_{i=1}^{n} P[X = x_i] \geq n \frac{1}{m}$$

On using this result in conjunction with that in (4.11), we then obtain

$$n \leq m[F_X(b) - F_X(a)] \leq m$$

Thus we see that, since m was assumed to be an arbitrary positive integer (and hence is finite), the integer n must be finite, and therefore there can be at most a *finite* number of discontinuity points (in particular, at most m) in any bounded interval which have jumps of height greater than or equal to $1/m$. Since there are only a countable number of positive integers, we see (on summing over all values of m), that there can be at most a countable number of discontinuities in any bounded interval.

Finally, since a countable number of bounded intervals of the type $(a,b]$ will completely cover the real line [for example, the intervals $(n, n + 1]$ with $n = 0, \pm 1, \pm 2, \ldots$ will do so], and since a countable union of countable sets is countable,[1] it follows that a probability distribution function can have at most a countable number of jump discontinuities on the real line. The theorem is thus proved. ∎

5 PROBABILITY DENSITIES

We noted earlier (Sec. 3) that every probability distribution function can be decomposed into two parts: a step-function part and a part which is continuous everywhere. Now our previous specific examples of continuous random variables were such that the probability distribution functions of those random variables happened to be differentiable.[2] A question then arises as to whether or not this is always the case. Theoretically, the answer is no; in particular, although it is well beyond our capabilities to do so here, it can be shown[3] that every everywhere continuous probability distribution function F_c may be decomposed into two parts:

$$F_c = c_1 F_{ac} + c_2 F_s$$

where (5.1)

$$c_1 + c_2 = 1$$

where F_{ac} is an *absolutely continuous* probability distribution function, that is, one which has a derivative everywhere, and where F_s is a *singular* probability distribution function, that is, one which is a nonconstant, continuous everywhere, probability distribution function whose derivative is zero almost everywhere.[4] Fortunately, however, one seldom meets singular probability distribution functions in applications, and we shall accordingly always assume from here on that our probability distribution functions are either discrete, differentiable, or may be expressed as the sum of a discrete part and a differentiable part.

With this latter assumption in mind, we wish to investigate here the differentiable case in some detail. Suppose that we have a random

[1] See, for example, theorem 3.2 of Bartle or proposition 7, chap. 1, of Royden.

[2] See Exercises 2.1 and 2.2.

[3] See, for example, Loève, pp. 177ff.

[4] For an example of a singular probability distribution function see pp. 15–16 of Lukacs.

variable X defined on a sample space $\mathcal{S}$ and that the probability distribution function F_X of that random variable has a derivative

$$f_X(x) \triangleq \frac{dF_X(x)}{dx} \tag{5.2}$$

which is continuous everywhere[1] in the range of X; we call this derivative the *probability density* of the random variable X. That is, having previously defined the probability distribution function $F_X(x)$ to be the probability of the event $[X \leq x]$, we now *define* the probability density as the derivative of the probability distribution function. As we shall see in our subsequent developments, the solution of many applied problems is simplified considerably whenever the probability density exists. It seems worthwhile, therefore, to spend some time in investigating some of its basic properties which follow from the definition (5.2).

First, we recall that the usual definition of the derivative gives

$$\frac{dF_X(x)}{dx} = \lim_{x' \to x} \frac{F_X(x') - F_X(x)}{x' - x} \tag{5.3}$$

Now we showed previously [cf. (3.6)] that the probability distribution function is a monotonically increasing function:

$$F_X(b) \geq F_X(a)$$

whenever $b \geq a$. It therefore follows that whether $x' > x$ or $x' < x$, the numerator of the ratio on the right-hand side of (5.3) must either be zero or it must have the same sign as does the denominator of that ratio; that is, the ratio on the right-hand side of (5.3) must always be nonnegative. Thus we see that

$$f_X(x) \geq 0 \tag{5.4}$$

whenever that derivative exists. That is, *the probability density of any random variable must always be nonnegative.*

Next, we note that if the probability density is continuous everywhere on the closed interval $[a,b]$, then it follows from the fundamental theorem of calculus[2] that

$$F_X(x) - F_X(a) = \int_a^x f_X(\xi)\, d\xi \tag{5.5}$$

for every $x \in [a,b]$. If then we let the point a recede to minus infinity, we have

$$\lim_{a \to -\infty} F_X(a) = F_X(-\infty) = 0$$

[1] Everywhere, that is, except possibly at a finite number of points in any finite interval.
[2] See, for example, sec. 2.15, vol. I of Apostol.

and hence it follows that we get

$$F_X(x) = \int_{-\infty}^{x} f_X(\xi)\, d\xi \tag{5.6}$$

That is, *given the probability density of a random variable, we may calculate the corresponding probability distribution function by integration.* It should be noted that (5.6) here is a derived result which follows from the definition (5.2)—not vice versa as in our informal introduction to these ideas given in Sec. 2 above.

Once we have established the above integral relations between the probability distribution function F_X and the probability density f_X, certain other results follow immediately. For example, if we let x in (5.5) tend to plus infinity, the left-hand side of (5.5) tends to

$$\lim_{x \to +\infty} F_X(x) = F_X(+\infty) = 1$$

and thus it follows that

$$\int_{-\infty}^{+\infty} f_X(\xi)\, d\xi = 1 \tag{5.7}$$

that is, that *the area under any probability density must be unity.*

On setting $x = b$ in (5.5), it further follows immediately that

$$\begin{aligned} \int_a^b f_X(\xi)\, d\xi &= F_X(b) - F_X(a) \\ &= P[X \in (a,b]] \end{aligned} \tag{5.8}$$

that is, that the probability that a random variable assumes a value in any interval in its range may be calculated by integrating its probability density over that interval. We may then combine this result with those derived in Sec. 3 to show that whenever a given random variable has a probability density, we may calculate the probability of essentially any event of interest which concerns the values of that random variable by integrating its probability density over those points in the range of X which represent the specified event. That is, if a random variable X has a probability density f_X, then we may write

$$P[X \in A] = \int_A f_X(x)\, dx \tag{5.9}$$

where A is any event in the range of X. It should be noted that it is precisely this result which makes the probability density so important, for it enables the use of the techniques of ordinary calculus to derive probabilities.

Let us now investigate a few of the more common probability densities:

Both the velocity random variable V in Exercise 2.1 and the angular

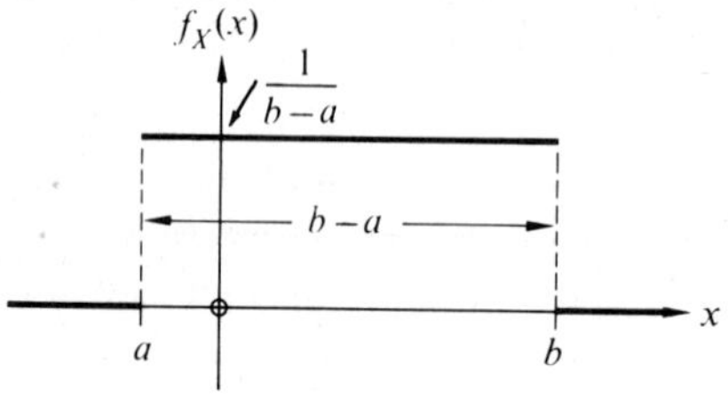

Figure 5.1 A uniform probability density.

random variable Θ which characterizes the angle θ in the spinning-wheel problem in Exercise 3.1 have probability densities of the type

$$f_X(x) = \begin{cases} \dfrac{1}{b-a} & \text{for } a \le x \le b \\ 0 & \text{otherwise} \end{cases} \tag{5.10}$$

as shown in Fig. 5.1. Such a random variable is said to have a *uniform* probability density.

Exercise 5.1 Suppose that a random variable X has a uniform probability density as given by (5.10).

a. Calculate the probability that

$$X \le \frac{2a+b}{3}$$

b. Let $a = -1$ and $b = +2$. In this case, calculate the probability that

$$|X| \le c$$

for $c = \frac{1}{2}$ and for $c = 1\frac{1}{2}$. □

All man-made devices and machines fail to work sooner or later. It has been found in practice that it is quite common for such objects to have a time to failure which obeys an exponential law. That is, if we characterize the times to failure by a random variable T, then it is quite common to find that T has the *exponential* probability density

$$f_T(t) = \begin{cases} ae^{-at} & \text{for } 0 \le t \\ 0 & \text{otherwise} \end{cases} \tag{5.11}$$

(where a is a positive number) as shown in Fig. 5.2.

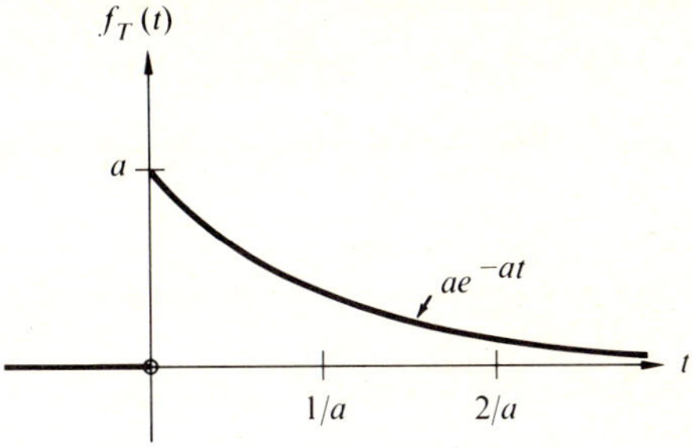

Figure 5.2 An exponential probability density.

Exercise 5.2 Determine the probability distribution function of a random variable which has an exponential probability density and plot your result. □

Exercise 5.3 Suppose that a particular class of light bulbs has been found to have an exponential failure law.

a. Measurements show that the probability that the time to failure exceeds 200 hr for light bulbs in the given class is $e^{-1}(\approx 0.368)$. Calculate the value of the parameter a in this case.

b. Using the value of the parameter a determined above, calculate the time T_0 such that the probability is 0.1 that the time to failure is less than T_0. □

One of the most common probability densities occurring in practice is the *normal* probability density function which has the form

$$f_X(x) = \frac{1}{\sqrt{2\pi}} e^{-x^2/2} \tag{5.12}$$

This probability density is also often called the *laplacian* density by the French [after the French mathematician the Marquis Pierre Simon de Laplace (1749–1827)] and the *gaussian* density by the Germans [after the German mathematician Johann Friedrich Carl Gauss (1777–1855)]. Normal is a British usage. Partly for historical reasons, probability theorists and statisticians commonly use normal, while physicists and engineers often use gaussian. We shall use all these names interchangeably.

Roughly speaking, whenever the random variable X may be expressed as a sum of a large number of independently acting random variables, then the probability density of X is approximately normal. Such is the case, for example, when X represents the fluctuations of the voltage developed across a resistor by thermal agitation.

Exercise 5.4 Show that (5.12) does in fact define a true probability density; in particular, show that

$$I \triangleq \frac{1}{\sqrt{2\pi}} \int_{-\infty}^{+\infty} e^{-x^2/2}\, dx = 1$$

(*Hint:* Consider I^2 and evaluate by converting to polar coordinates.) □

Unfortunately, the probability distribution function for the normal random variable,

$$F_X(x) = \frac{1}{\sqrt{2\pi}} \int_{-\infty}^{x} e^{-\xi^2/2}\, d\xi \tag{5.13}$$

is not expressible in terms of elementary functions (polynomials, exponentials, or trigonometric functions). Tables of the normal probability density and the normal probability distribution function are readily available,[1] however, as also are very extensive tables of the related *normal probability integral*,[2] defined by

$$\begin{aligned} P[x] &\triangleq \frac{1}{\sqrt{2\pi}} \int_{-x}^{+x} e^{-\xi^2/2}\, d\xi \\ &= 2F_X(x) - 1 \qquad \text{for } 0 \le x \end{aligned} \tag{5.14}$$

where F_X here is the normal probability distribution function defined in (5.13). Plots of the normal probability density and distribution functions are given in Fig. 5.3.

Exercise 5.5 Let X be a normal random variable with f_X given by (5.12). Show that, for $0 < x$,

$$\left(1 - \frac{1}{x^2}\right) \frac{e^{-x^2/2}}{\sqrt{2\pi}\, x} < Q(x) < \frac{e^{-x^2/2}}{\sqrt{2\pi}\, x} \tag{5.15}$$

where Q is the "tail" of the normal probability distribution function defined by

$$Q(x) \triangleq \frac{1}{\sqrt{2\pi}} \int_{x}^{+\infty} e^{-\xi^2/2}\, d\xi = 1 - F_X(x) \tag{5.16}$$

when F_X is given by (5.13). (*Hint:* Use integration by parts.) □

[1] See, for example, table 26.1 of National Bureau of Standards, "Handbook of Mathematical Functions."

[2] See, for example, National Bureau of Standards, "Tables of Normal Probability Functions."

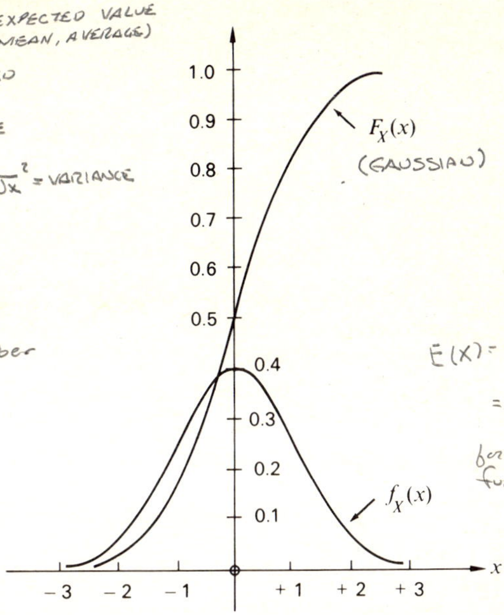

Figure 5.3 Normal probability functions.

The results of Exercise 5.5 may be extended to give

$$Q(x) \approx \frac{e^{-x^2/2}}{\sqrt{2\pi}\, x}\left(1 - \frac{1}{x^2} + \frac{1 \times 3}{x^4} - \frac{1 \times 3 \times 5}{x^6} + \cdots\right) \tag{5.17}$$

as an asymptotic expansion for the tail of the normal probability distribution function which is valid for large positive values of x.

One of the various probability densities related to the normal density is the *Rayleigh probability density* [named after the English physicist John William Strutt, Lord Rayleigh (1842–1919)] which has the form

$$f_R(r) = \begin{cases} \dfrac{r}{b}\, e^{-r^2/2b} & \text{for } 0 \le r \\ 0 & \text{otherwise} \end{cases} \tag{5.18}$$

(where $b > 0$), which is shown in Fig. 5.4. Such a probability density

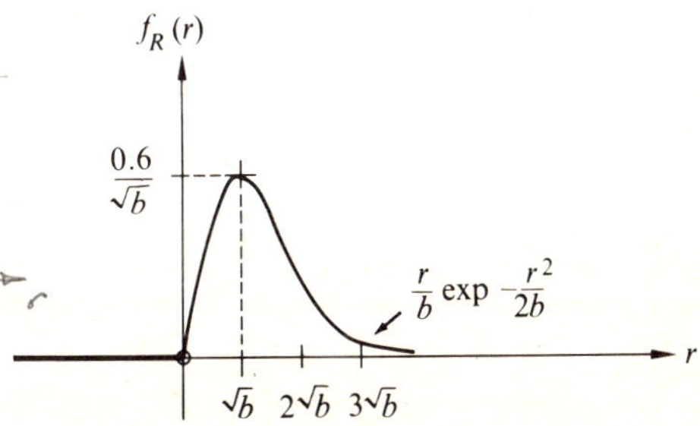

Figure 5.4 A Rayleigh probability density.

would arise, for example, in the following way. Suppose that we are studying the landing errors of a spacecraft and that the random variables X and Y which characterize the rectangular components of that error are independent, identically distributed normal random variables. It may then be shown that the radius

$$R \triangleq \sqrt{X^2 + Y^2}$$

of the landing point from the aiming point has a Rayleigh density (cf. Exercise 6.4.5).

Exercise 5.6 Let R be a random variable with the Rayleigh probability density given in (5.18).

a. Determine the corresponding probability distribution function and plot your result.

b. Calculate (in terms of the parameter b) the radius R_0 such that the probability that $R > R_0$ is $e^{-1}(\approx 0.368)$. □

Another probability density which is related to the normal density is the *Cauchy probability density* [named after the French mathematician Augustin Louis Cauchy (1789–1857)] which has the form

$$f_Z(z) = \frac{a}{\pi} \frac{1}{a^2 + z^2} \tag{5.19}$$

(where $a > 0$) and which is shown in Fig. 5.5. As an example of where a Cauchy random variable arises, consider again the spacecraft-landing-error example discussed in the Rayleigh case. If we again assume that the random variables X and Y which characterize the rectangular components of the landing error are independent, identically distributed, normal random variables, then it may be shown that the tangent Z of

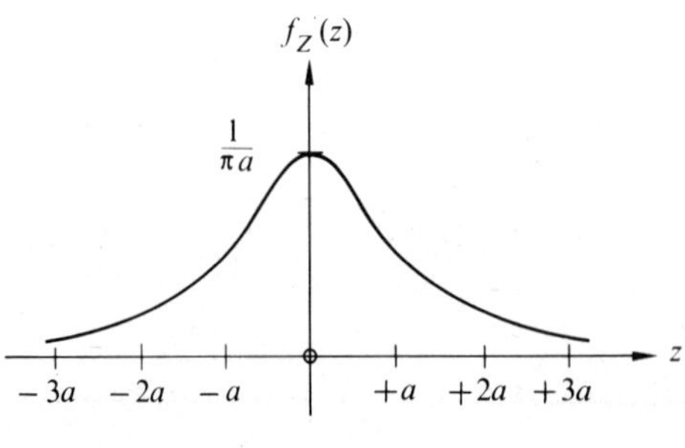

Figure 5.5 A Cauchy probability density.

the angle of the landing-error vector from the x axis,

$$Z \triangleq \frac{Y}{X}$$

has a Cauchy probability density (cf. Exercise 6.3.9).

Exercise 5.7 Let Z be a random variable with the Cauchy probability density given in (5.19). Determine the corresponding probability distribution function and plot your result. □

6 MIXED RANDOM VARIABLES*

In many problems of interest we often have to deal with a mixed random variable, that is, with a random variable whose probability distribution function has both a differentiable part and a step-function part. Now it is quite possible to handle these two parts separately and we shall, in fact, generally do so. We would on occasion, however, like to be able to handle both parts together if at all possible—simply for the sake of compactness. There are two techniques commonly used to handle this situation: one is to introduce a new kind of integral (called the Stieltjes integral) which directly involves the probability distribution function; the other is to extend the common definition of a derivative so as to be able to talk about the probability density of a discrete random variable. The Stieltjes integral approach is commonly used in theoretical developments,[1] while the generalized derivative approach is the more common in applications. We wish to introduce the latter approach in this section.

To see why it is necessary to extend the definition of the derivative, let us consider for the moment a discrete random variable X whose probability distribution function F_X has jumps of height

$$p_k \triangleq P[X = x_k] \tag{6.1}$$

at the points x_k, $k = 1, 2, 3, \ldots$. We may then write [see Eq. (2.9)]

$$F_X(x) = \sum_k p_k U(x - x_k) \tag{6.2}$$

* Supplementary material. This section may be skipped if the reader so wishes; it treats a specialized technique which, although quite powerful, is seldom used in our subsequent discussions.

[1] See, for example, chap. 4 of Pfeiffer or chap. 7 of Cramér.

where the summation is over all values of the index k and where U is the unit-step function defined by the formula

$$U(x) \triangleq \begin{cases} 1 & \text{for } 0 \leq x \\ 0 & \text{otherwise} \end{cases} \tag{6.3}$$

Let us now attempt to take the derivative of such a discrete probability distribution function in order to obtain its corresponding probability density. As pointed out in the preceding section [cf. (5.3)], the derivative of F_X is, by definition,

$$\frac{dF_X(x)}{dx} = \lim_{x' \to x} \frac{F_X(x') - F_X(x)}{x' - x} \tag{6.4}$$

Suppose that we consider this derivative at point x in the vicinity of some jump point, say x_i, as shown in Fig. 6.1a. For values of x such that $x \neq x_i$, we have (for values of x' very near x)

$$F_X(x') = F_X(x)$$

and hence, on using this result in (6.4), we get

$$\frac{dF_X(x)}{dx} = 0 \qquad \text{for } x \neq x_i$$

Suppose, on the other hand, that we take $x = x_i$. If x' is to the right

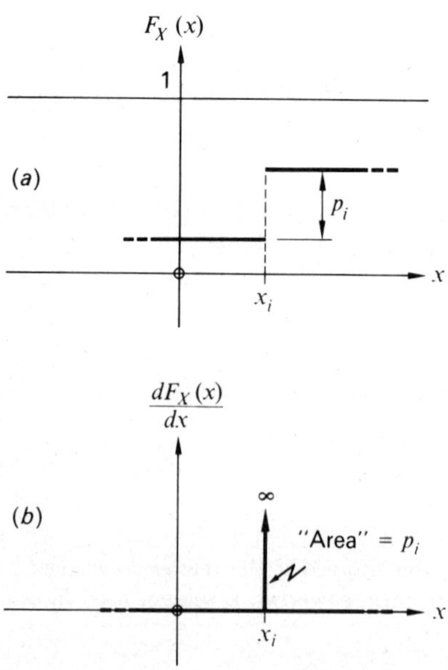

Figure 6.1 The derivative of a probability distribution function in the vicinity of a jump point.

of x_i, we then have

$$F_X(x') = F_X(x_i)$$

while if x' is to the left of x_i we have

$$F_X(x') = F_X(x_i) - p_i$$

If now we try to evaluate (6.4) at $x = x_i$, we either get an answer of zero as x' approaches x from the right, or we get

$$\left.\frac{dF_X(x)}{dx}\right|_{x=x_i} = \lim_{x' \to x_i} \frac{F_X(x_i) - p_i - F_X(x_i)}{x' - x_i}$$
$$= \lim_{x' \to x_i} \frac{p_i}{x_i - x'} = +\infty$$

as x' approaches x_i from the left. This is the situation indicated graphically in Fig. 6.1*b*. Thus we see that if the probability distribution function $F_X(x)$ is a step function, then its derivative is zero at all values of x which are not jump points and is infinite (or, more precisely, is not defined) at the jump points.

While the derivative of a discrete probability distribution function therefore does not exist in any usual sense, we would still like to be able to talk about a probability density in such a case. Somehow or other we must turn the problem around in order to get out of this dilemma. Now a moment's reflection upon the discussion of the probability density given in the preceding section should bring out the fact that we never really *use* a probability density all by itself. In fact, we either integrate it directly in order to calculate a probability, as in the formula

$$P[X \in A] = \int_A f_X(x)\, dx$$

[cf. (5.9)], or, as we shall show later (in Chap. 7), we shall use it inside a different integral in order to calculate a statistical average. That is, we shall never use a probability density in a *calculation* except through the process of integration. It is important to keep this fact in mind throughout the following discussion.

Suppose for the moment that the derivative δ of the unit-step function U did in fact exist and that we really could write

$$\delta(x) = \frac{d}{dx} U(x) \tag{6.5}$$

which would then give the inverse relation

$$U(x) = \int_{-\infty}^{x} \delta(\xi)\, d\xi \tag{6.6}$$

The derivative of the discrete probability distribution function in (6.2) could then be obtained by differentiating to get

$$f_X(x) = \sum_k p_k \delta(x - x_k) \tag{6.7}$$

which, upon using (6.6), would then lead to the inverse relation

$$\begin{aligned} F_X(x) &= \sum_k p_k \int_{-\infty}^{x} \delta(\xi - x_k)\, d\xi \\ &= \sum_k p_k U(x - x_k) \end{aligned} \tag{6.8}$$

and our problem would be solved.

Since we have already shown that the derivative δ of the unit-step function U does not really exist in any usual sense, we cannot, in fact, proceed quite as we have done above. In particular, we cannot consider the "derivative" δ of the unit-step function as an ordinary function,[1] but rather we must define it in terms of the property which we wish it to have and consider it as a more general kind of function. With this in mind, let us define the *unit-impulse function* δ, or *Dirac* δ *function* [after the English theoretical physicist Paul A. M. Dirac (1902–)], by the formula

$$\int_{-\infty}^{+\infty} g(\xi)\delta(\xi - a)\, d\xi \triangleq g(a) \tag{6.9}$$

where g is any function which is continuous at the point a. The integral in (6.8) is commonly called a *sifting integral* as its effect is to "sift" out the value of g at the point a. The "function" δ defined by (6.9) is commonly called a *generalized function*[2] or *distribution*.

Having defined the unit-impulse function by (6.9), let us next take the function g in that formula to be a translated unit-step function; in particular, let us take

$$g(\xi) = U(x - \xi)$$

We then have

$$g(a) = U(x - a)$$

and (6.9) becomes in this case

$$\begin{aligned} U(x - a) &= \int_{-\infty}^{+\infty} U(x - \xi)\delta(\xi - a)\, d\xi \\ &= \int_{-\infty}^{x} \delta(\xi - a)\, d\xi \end{aligned} \tag{6.10}$$

[1] That is, as a rule δ which gives the number $\delta(a)$ corresponding to the point a.
[2] See appendix I of Papoulis, "The Fourier Integral," for a good introductory treatment of generalized functions. A quite readable though much more advanced treatment is given in chap. I of Gel'fand and Shilov.

the last equality following from the fact that $U(x - \xi)$ is (by definition) zero for $\xi > x$ and equal to unity for $\xi < x$. Thus we see that if we define the generalized function δ by the sifting integral (6.9), then we can obtain the unit-step function U by integrating δ; we shall accordingly call δ the *generalized derivative* of the unit-step function. On using (6.10) in (6.8), we then see that we may say that a discrete probability distribution function does indeed have a corresponding discrete probability density so long as we define that density in terms of generalized derivatives. In essence, then, the problem before us has been solved.

7 SUMMARY OF DEFINITIONS AND FORMULAS

The following definitions and formulas are given essentially in the order in which they occur in the chapter.

Set-indicator function

$$I_A(s) \triangleq \begin{cases} 1 & \text{if } s \in A \\ 0 & \text{if } s \notin A \end{cases}$$

Range

The range of a function X defined on a sample space $\mathcal{S}$ is defined as:

$$R_X \triangleq \{x: x = X(s) \text{ for } s \in \mathcal{S}\} \tag{1.2}$$

Range sample space

$$\mathcal{S}_X \triangleq R_X \tag{1.3}$$

Inverse image

The inverse image of a set A in the range of a function X defined on a sample space $\mathcal{S}$ is:

$$X^{-1}(A) \triangleq \{s: s \in \mathcal{S},\ A \subseteq \mathcal{S}_X, \text{ and } X(s) \in A\} \tag{1.4}$$

Induced probability

$$P[A \subset \mathcal{S}_X] \triangleq P[X^{-1}(A) \subset \mathcal{S}] \tag{1.5}$$

Random variable

A random variable is a function X defined on a sample space $\mathcal{S}$ such that the probability of the inverse image $X^{-1}(A) \subset \mathcal{S}$ is defined for each event A in the range sample space $\mathcal{S}_X$, that is, a *measurable function* on $\mathcal{S}$. RANDOM VARIABLE X IS A FUNCTION WHOSE DOMAIN IS THE SET S OF ALL OUTCOMES OF AN EXPERIMENT AND WHOSE RANGE IS A SET OF NUMBERS.

A PARTICULAR NUMBER ASSOCIATED WITH THE OUTCOME a IS X(a)

Probability distribution function

The probability distribution function of a random variable X defined on a sample space $\mathcal{S}$ is:

$$F_X(x) \triangleq P[X \le x] \tag{2.5}$$
$$= P[\{s\colon s \in \mathcal{S},\ -\infty < X(s) \le x\}] \tag{2.6}$$

Limits on F

$$F_X(+\infty) = 1 \tag{3.2}$$

and

$$F_X(-\infty) = 0 \tag{3.3}$$

Monontonicity

If

$$b \ge a$$

then

$$F_X(b) \ge F_X(a) \tag{3.6}$$

Continuity

$$F_X(a + 0) = F_X(a) \tag{3.7}$$

where, for $\epsilon > 0$,

$$F_X(a + 0) \triangleq \lim_{\epsilon \to 0} F_X(a + \epsilon) \tag{3.8}$$

and

$$F_X(a - 0) + P[X = a] = F_X(a) \tag{3.9}$$

where, for $\delta > 0$,

$$F_X(a - 0) \triangleq \lim_{\delta \to 0} F_X(a - \delta) \tag{3.10}$$

Decomposition of distribution functions

Every probability distribution function may be decomposed into two parts:

$$F_X(x) = D_X(x) + C_X(x) \tag{3.13}$$

where D_X is a step function and hence may be expressed as

$$D_X(x) = \sum_i P[X = x_i]U(x - x_i) \tag{3.11}$$

and where C_X is continuous everywhere.

Interval probability

$$P[X \in (a,b]] = F_X(b) - F_X(a) \tag{3.14}$$

Probability density

$$f_X(x) \triangleq \frac{dF_X(x)}{dx} \tag{5.2}$$

Nonnegativeness

$$\boxed{f_X(x) \geq 0} \tag{5.4}$$

for all x for which the density exists.

Inversion

$$\boxed{F_X(x) = \int_{-\infty}^{x} f_X(\xi)\, d\xi} \tag{5.6}$$

Unit area

$$\boxed{\int_{-\infty}^{+\infty} f_X(\xi)\, d\xi = 1} \tag{5.7}$$

Calculation of probability

$$P[X \in A] = \int_A f_X(x)\, dx \tag{5.9}$$

where A is any event in the range of the random variable X.

Common densities

$$f_X(x) = \begin{cases} \dfrac{1}{b-a} & \text{for } a \leq x \leq b \\ 0 & \text{otherwise} \end{cases} \qquad \text{(uniform)} \tag{5.10}$$

$$f_X(x) = \begin{cases} ae^{-ax} & \text{for } 0 \leq x \\ 0 & \text{otherwise} \end{cases} \qquad \text{(exponential)} \tag{5.11}$$

where $a > 0$.

$$f_X(x) = \frac{1}{\sqrt{2\pi}} e^{-x^2/2} \qquad \text{(normal)} \tag{5.12}$$

$$f_X(x) = \begin{cases} \dfrac{x}{b} e^{-x^2/2b} & \text{for } 0 \leq x \\ 0 & \text{otherwise} \end{cases} \qquad \text{(Rayleigh)} \tag{5.18}$$

where $b > 0$.

$$f_X(x) = \frac{a}{\pi} \frac{1}{a^2 + x^2} \qquad \text{(Cauchy)} \tag{5.19}$$

where $a > 0$.

REFERENCES

Apostol, Tom M.: "Calculus," Blaisdell Publishing Company, New York, vol. I, 1961; vol. II, 1962.

Bartle, Robert G.: "The Elements of Real Analysis," John Wiley & Sons, Inc., New York, 1964.

Cramér, Harald: "Mathematical Methods of Statistics," Princeton University Press, Princeton, N.J., 1946.

Gel'fand, Izrail Moiseevich and Georgiĭ Evgen'evich Shilov: "Generalized Functions," vol. I, Academic Press, Inc., New York, 1964.

Krickeberg, Klaus: "Probability Theory," Addison-Wesley Publishing Company, Inc., Reading, Mass., 1965.

Loève, Michel: "Probability Theory," 3d ed., D. Van Nostrand Company, Inc., Princeton, N.J., 1963.

Lukacs, Eugene: "Characteristic Functions," Hafner Publishing Company, Inc., New York, 1960.

National Bureau of Standards: "Tables of Normal Probability Functions," Applied Mathematics Series (AMS) 23, U.S. Government Printing Office, 1953.

National Bureau of Standards: "Handbook of Mathematical Functions," AMS 55, U.S. Government Printing Office, 1964.

Papoulis, Athanasios: "The Fourier Integral and Its Applications," McGraw-Hill Book Company, New York, 1962.

Pfeiffer, Paul E.: "Concepts of Probability Theory," McGraw-Hill Book Company, New York, 1965.

Royden, Halsey L.: "Real Analysis," The Macmillan Company, New York, 1963.

DEFINITIONS: $\{x_1 \le X \le x_2\}$ IS THE SET CONSISTING OF ALL OUTCOMES a, SUCH THAT $x_1 \le X(a) \le x_2$ & $\{x_1 \le X \le x_2\} = \{X \le x_2\} - \{X \le x_1\}$

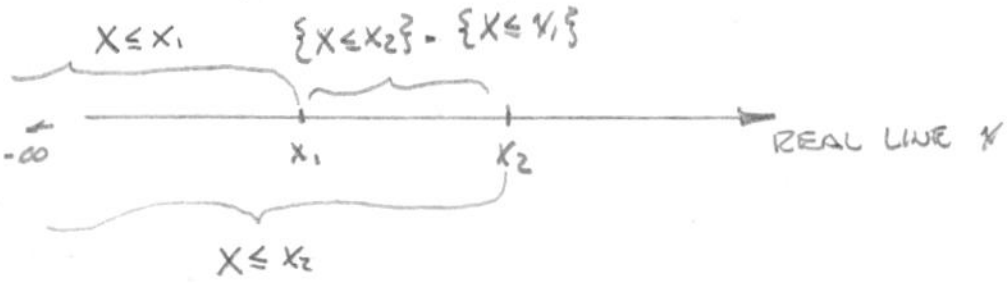

$F_X(0)=0,\ F_X(+\infty)=1,\ F_X(-\infty)=0$

FROM GRAPH: $F_X(x) = \frac{1}{3}x + \frac{1}{3}U(x-1) + (\frac{1}{6}x + \frac{1}{2})$

$f_X(x) = 1/3 \quad 0 \le x \le 1 \quad F_X(x) = \frac{1}{3}x$

UNIT IMPULSE FUNCTION $f_X(x) = \frac{1}{3}\delta(x-1) \quad x=1 \quad F_X(x) =$ UNIT STEP

$f_X(x) = 1/6 \quad 1 \le x \le 3 \quad F_X(x) = \frac{1}{6}x + \frac{1}{2}$

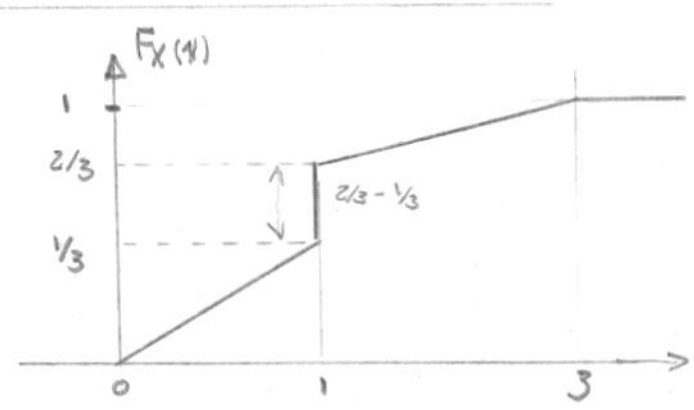

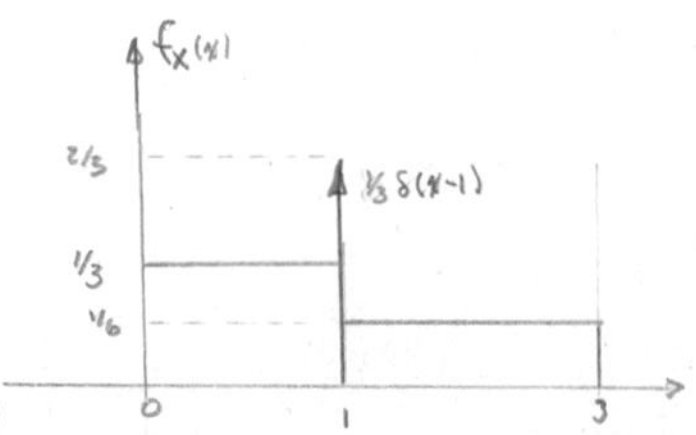

$\int_{-\infty}^{\infty} f_X(x)\,dx = \int_0^1 \frac{1}{3}dx + \int \frac{1}{3}\delta(x-1)\,dx + \int_1^3 \frac{1}{6}x\,dx$

$= 1/3 + 1/3 + 1/3 = 1$ OK

CONT $P\{0 \le X \le 1/2\} = \int_0^{1/2} \frac{1}{3}dx = 1/6$ or $= F_X(\frac{1}{2}) - F_X(0) = \frac{1}{6} - 0 = \frac{1}{6}$

NOT CONT. $P\{1 \le X \le 3\} = \int_1^3 f_X(x)\,dx = \int_1^3 [\frac{1}{3}\delta(x-1) + 1/6]\,dx = \frac{1}{3} + \frac{1}{3} = 2/3$

$= F_X(3) - F_X(1) + P\{X=1\} = 1 - \frac{2}{3} + \frac{1}{3} = 2/3$

CONT. $P\{1 < X \le 3\} = F_X(x_2) - F_X(x_1) = F_X(x_3) - F_X(x_1) = 1 - 2/3 = \frac{1}{3}$

NOTE: $P\{1 \le X \le 3\} \ne P\{1 < X \le 3\}$ — INCLUDES DIRAC FUNCTION; Does not include DIRAC function

$P\{X = 1/2\} = 0$ (NO DISCONTINUITY) AT THIS POINT)

$P\{X = 1\} = \frac{1}{3}$ (DISCON. @ THIS POINT)

$P\{\frac{1}{2} \le X \le \frac{1}{2}\} = \int_{1/2}^{1/2} f_X(x)\,dx = \int_{1/2}^{1/2} \frac{1}{3}dx = \frac{1}{6} - 1/6 = 0$

5
Random Vectors

1 RANDOM VECTORS

As we have pointed out previously on several occasions, an outcome of a given experiment may well require several numbers for its description. For example, suppose that we consider again the experiment which consists of two independent flips of a coin and, further, suppose that we are interested here both in whether or not at least one head occurs in the two flips and in whether or not at least one tail occurs in the two flips. In our discussion of this experiment in Sec. 4.1, we were also interested in both of these events and, accordingly, introduced there both the head-indicator function

$$I_H(s) \triangleq \begin{cases} 1 & \text{if } s \text{ corresponds to the occurrence} \\ & \quad \text{of at least one head} \\ 0 & \text{if not} \end{cases}$$

and the tail-indicator function

$$I_T(s) \triangleq \begin{cases} 1 & \text{if } s \text{ corresponds to the occurrence} \\ & \quad \text{of at least one tail} \\ 0 & \text{if not} \end{cases}$$

where, in each case, s is any sample point in the sample space $\mathcal{S}$ which describes the possible outcomes of the given experiment. Now it follows from our discussions in Chap. 4 that the functions I_H and I_T are both random variables defined on the same basic sample space $\mathcal{S}$. If we denote their respective range sample spaces by S_{I_H} and S_{I_T}, then the situation becomes that pictured in Fig. 1.1.

When dealing simultaneously (as we are here) with two random variables defined on the same basic sample space, it is often convenient to think of those two random variables as being the components of a two-dimensional vector. Thus, for example, we might find it desirable in the present case to introduce the vector $\mathbf{I}$ defined to have the components I_H and I_T:

$$\mathbf{I}(s) \triangleq (I_H(s), I_T(s))$$

The vector $\mathbf{I}$ is then the two-dimensional vector, defined upon the sample space $\mathcal{S}$, whose values are the number pairs (x,y), where

$$x \triangleq I_H(s) \qquad \text{and} \qquad y \triangleq I_T(s)$$

for any $s \in \mathcal{S}$.

It follows from the definitions of I_H and I_T that the only possible number pairs which can occur here are: (1,0) corresponding to the occurrence of at least one head but no tail, (0,1) corresponding to the occurrence of at least one tail but no head, and (1,1) corresponding to the occurrence of at least one head and at least one tail; the number pair (0,0) corresponding to the occurrence of neither a head nor a tail on either flip obviously cannot occur. The set of all possible pairs of values of x and y is, by

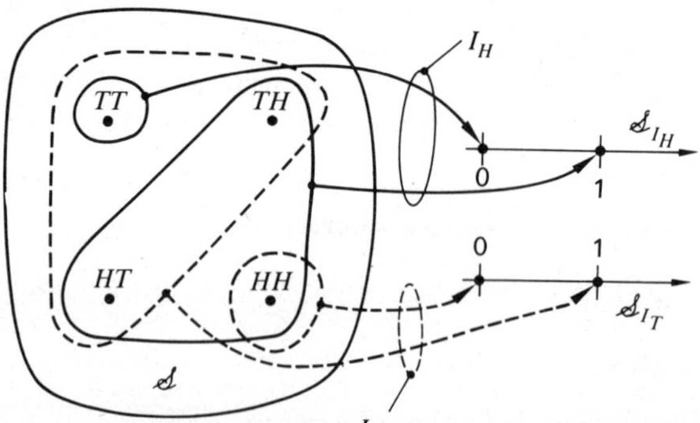

Figure 1.1 The random variables I_H and I_T.

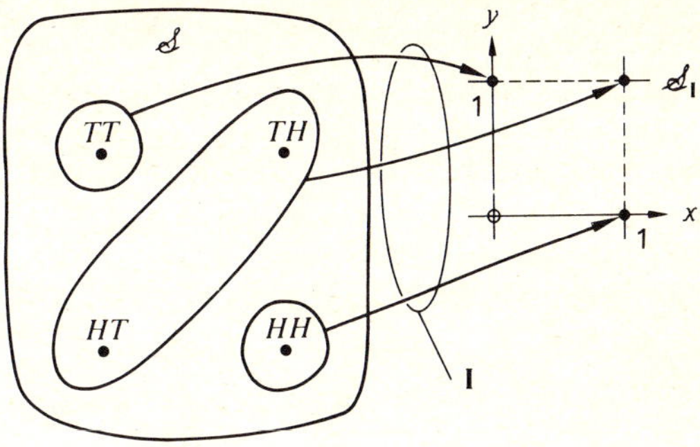

Figure 1.2 The random vector **I**.

definition, the range R_I of the vector function **I**. Thus

$$\begin{aligned} R_I &\triangleq \{(x,y): x = I_H(s),\ y = I_T(s), \text{ and } s \in \mathcal{S}\} \\ &= \{(0,1),(1,0),(1,1)\}. \end{aligned}$$

The situation here is thus that pictured in Fig. 1.2.

If now the vector function **I** defined above is such that every event A in the range R_I is measurable [if every inverse image $I^{-1}(A)$ is an event in $\mathcal{S}$ with a definite probability] then **I** meets our previous criteria placed on random variables and will, therefore, be called a *random vector*. We shall then (as in the single random variable case) think of the range R_I as a new sample space and denote it by $\mathcal{S}_I$:

$$\mathcal{S}_I \triangleq R_I$$

It then follows that

$$\mathcal{S}_I = \{(0,1),(1,0),(1,1)\}$$

Since the mapping **I** from the basic sample space $\mathcal{S}$ to the new range sample space $\mathcal{S}_I$ is so simple here, it is quite easy to find the probability distribution induced on $\mathcal{S}_I$ by **I**. Thus, on referring to Fig. 1.2, we get

$$P[\mathbf{I} = (0,1)] = P[s \in \{TT\}] = \tfrac{1}{4}$$
$$P[\mathbf{I} = (1,0)] = P[s \in \{HH\}] = \tfrac{1}{4}$$

and

$$\begin{aligned} P[\mathbf{I} = (1,1)] &= P[s \in \{HT,TH\}] \\ &= P[s \in \{HT\}] + P[s \in \{TH\}] \\ &= \tfrac{1}{4} + \tfrac{1}{4} = \tfrac{1}{2} \end{aligned}$$

This is the probability distribution shown in Fig. 1.3.

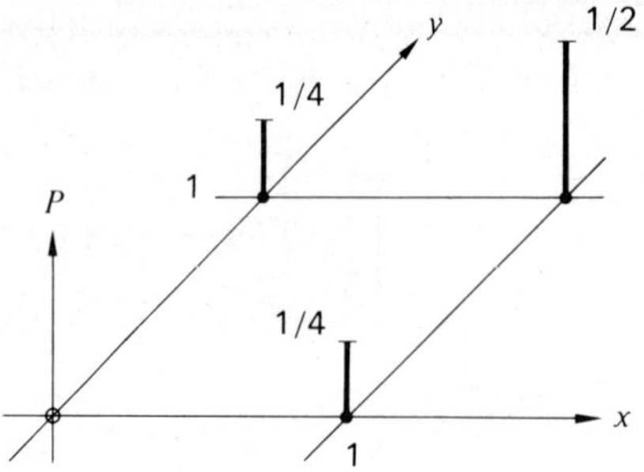

Figure 1.3 The probability distribution of the values of the random vector $\mathbf{I} = (I_H, I_T)$.

Exercise 1.1 A given communication system transmits binary digits, for example, 0 and 1, over a particular link. For convenience, the digit stream is broken into blocks (called *words*) each consisting of four successive digits, as shown in Fig. 1.4. Experience has shown that the various digits on this particular link are mutually independent and that each digit is equally likely to be 0 or 1.

Now the link receiver here counts both the number of 1s which occur in the first two digit positions in each word and the total number of 1s which occur in each word. Let N_2 be the random variable characterizing the number of 1s in the first two digit position and N_4 be the random variable characterizing the number of 1s in a word.

a. Describe a suitable basic sample space $\mathcal{S}$ upon which N_2 and N_4 may be defined.

b. Determine the range sample-spaces $\mathcal{S}_{N_2}$ and $\mathcal{S}_{N_4}$, and determine the probability distribution induced on those spaces.

Let **N** be the vector

$$N(s) \triangleq (N_4(s), N_2(s))$$

where s is any sample point in the basic sample space $\mathcal{S}$.

c. Determine the range $R_{\mathbf{N}}$ of the vector **N**.

d. Determine the probability distribution induced on $R_{\mathbf{N}}$ by **N** and sketch your result. □

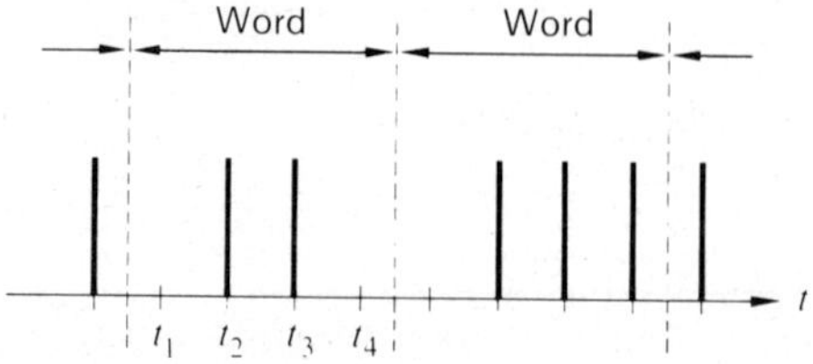

Figure 1.4 A typical stream of binary digits.

Exercise 1.2 Let the sample space $\mathcal{S}$ consist of all points s in the unit square; that is, let $\mathcal{S}$ consist of all points

$$s = (s_1,s_2)$$

where

$$0 \le s_1 \le 1 \qquad \text{and} \qquad 0 \le s_2 \le 1$$

The probability P is distributed uniformly on this square; for example, for each rectangle B with sides of length w_1 and w_2 in this square we have

$$P[s \in B] = w_1 \times w_2$$

Suppose that two random variables X_1 and X_2 are defined on this sample space by the formulas

$$X_1(s) \triangleq s_1^2 \qquad \text{and} \qquad X_2(s) \triangleq s_2^2$$

Let $\mathbf{X}$ be the random vector whose components are X_1 and X_2:

$$\mathbf{X}(s) \triangleq (X_1(s),X_2(s))$$

a. What is the range $R_\mathbf{X}$ of $\mathbf{X}$?

Let the event $A \subset R_\mathbf{X}$ be defined by

$$A \triangleq \{(x_1,x_2): x_1 = X_1(s),\ x_2 = X_2(s),\ \text{and } x_1 + x_2 \le \tfrac{1}{2}\}$$

b. Sketch $R_\mathbf{X}$, $A \subset R_\mathbf{X}$, and $\mathbf{X}^{-1}(A) \subset \mathcal{S}$.

c. Calculate the probability of the event A. □

The example and exercises above are all concrete special cases of the general situation shown schematically in Fig. 1.5. We start, in general,

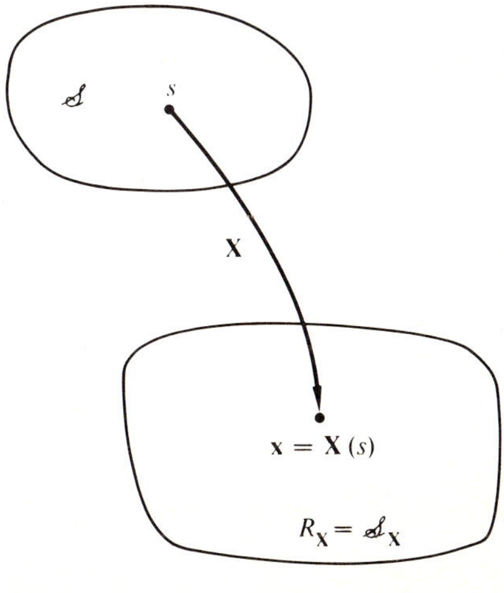

Figure 1.5 A general random vector **X**.

with some given sample space $\mathcal{S}$ and define upon it a *finite family* of single-valued real functions, say $\{X_1, X_2, \ldots, X_n\}$. For each point s in the basic sample space $\mathcal{S}$ we use these functions to calculate a set of n real numbers $\{x_1, x_2, \ldots, x_n\}$, where

$$x_i \triangleq X_i(s) \qquad s \in \mathcal{S} \tag{1.1}$$

for $i = 1, 2, \ldots, n$. Such a set of n real numbers may now be thought of as specifying the components of an n-dimensional real vector $\mathbf{X}$; that is, we may define

$$\begin{aligned}\mathbf{X}(s) &\triangleq (x_1, x_2, \ldots, x_n) \\ &= (X_1(s), X_2(s), \ldots, X_n(s))\end{aligned} \tag{1.2}$$

The set of all possible such vectors is then the range $R_{\mathbf{X}}$ of the vector function $\mathbf{X}$:

$$\begin{aligned}R_{\mathbf{X}} \triangleq \{(x_1, x_2, \ldots, x_n)\colon x_1 = X_1(s),\ x_2 = X_2(s),& \\ \ldots, x_n = X_n(s),\ s \in \mathcal{S}\}&\end{aligned} \tag{1.3}$$

As in the single random variable case, it is again convenient to think of the points in this range as new sample points, in other words, to think of the range $R_{\mathbf{X}}$ as a new n-dimensional sample space and, accordingly, to denote it by $\mathcal{S}_{\mathbf{X}}$.

The mapping of probabilities in the general n-dimensional case is the direct analog of the mapping of probabilities in the one-dimensional case: Suppose that A is some event of interest in the range sample space $\mathcal{S}_{\mathbf{X}}$. If the inverse image $\mathbf{X}^{-1}(A)$ of A under the mapping $\mathbf{X}$ is measurable, that is, if the probability of the event $\mathbf{X}^{-1}(A) \subset \mathcal{S}$ is defined for each $A \subset \mathcal{S}_{\mathbf{X}}$, then we set the probability of the event A in the range sample space $\mathcal{S}_{\mathbf{X}}$ equal to that of the event $\mathbf{X}^{-1}(A)$ in the original sample space $\mathcal{S}$. That is, we define

$$P[A \subset \mathcal{S}_{\mathbf{X}}] \triangleq P[\mathbf{X}^{-1}(A) \subset \mathcal{S}] \tag{1.4}$$

whenever the right-hand probability is defined.

If the vector-valued function $\mathbf{X}$ defined on a sample space $\mathcal{S}$ is such that the probability of the inverse image $\mathbf{X}^{-1}(A) \subset \mathcal{S}$ is defined for each event A in the n-dimensional range sample space $\mathcal{S}_{\mathbf{X}}$, then the function $\mathbf{X}$ is said to be a *vector-valued measurable function* on $\mathcal{S}$ and is called an *n-dimensional random vector* (or, sometimes, an *n-dimensional random variable*).

It therefore follows that the components $X_1, X_2, \ldots, X_n$ of the n-dimensional random vector $\mathbf{X}$ are individually random variables and hence that a random vector may be viewed as a finite family of random variables.

It is perhaps worthwhile to mention two additional points concerning random vectors at this time. First, it is sometimes desirable in applications to consider *complex* random variables. Having defined random vectors as above, a complex random variable may be considered to be a two-dimensional random vector whose components are the real and imaginary parts, respectively, of the given complex random variable. Once we have made this identification, we may proceed as though we were again dealing only with real random variables. Second, it is quite common in the mathematics literature to call the range sample space $\mathcal{S}_{\mathbf{X}}$ of the n-dimensional random vector $\mathbf{X}$ (with components X_1, X_2, . . . , X_n) an *n-dimensional product space* and to write

$$\mathcal{S}_{\mathbf{X}} = \mathcal{S}_{X_1} \times \mathcal{S}_{X_2} \cdots \mathcal{S}_{X_n} \tag{1.5}$$

where $\mathcal{S}_{X_i}$ is the range sample space of the random variable X_i. We shall discuss this point further at the end of the next section.

2 JOINT-PROBABILITY DISTRIBUTION FUNCTIONS

Now just as we found it convenient and useful to introduce the probability distribution function into the study of a single random variable, so also shall we find it useful to introduce multivariate probability distribution functions into a study of random vectors. For the sake of simplicity, we shall restrict our discussion in this section to the case of two-dimensional random vectors; the extension of our results to n-dimensional random vectors should not be too difficult for the reader.

Suppose now that $\mathbf{Z}$ is a two-dimensional random vector with components X and Y:

$$\mathbf{Z} \triangleq (X,Y)$$

where X and Y are random variables defined on a basic sample space $\mathcal{S}$. The probability distribution function $F_{\mathbf{Z}}$ of this random vector is then defined by

$$\begin{aligned} F_{\mathbf{Z}}(\mathbf{z}) &\triangleq P[\{s: -\infty < X(s) \le x, -\infty < Y(s) \le y, s \in \mathcal{S}\}] \\ &= P[X \le x, Y \le y] \end{aligned} \tag{2.1}$$

where

$$\mathbf{z} \triangleq (x,y)$$

It is sometimes convenient to make the dependence of $F_{\mathbf{Z}}$ on the random variables X and Y more evident by using the symbol $F_{X,Y}$ for $F_{\mathbf{Z}}$; thus we write

$$\begin{aligned} F_{X,Y}(x,y) &\triangleq F_{\mathbf{Z}}(\mathbf{z}) \\ &= P[X \le x, Y \le Y] \end{aligned} \tag{2.2}$$

In order to emphasize the fact that we are concerned here with more than one random variable, we shall call $F_{X,Y}$ (and hence F_Z) the *joint-probability distribution function* of the random variables X and Y.

For an example of a two-dimensional joint-probability distribution function, let us reconsider the random vector

$$\mathbf{I} = (I_H, I_T)$$

describing the results of two independent flips of an unbiased coin, which we discussed in the preceding section. It now follows from the results of Sec. 1 (cf. Fig. 1.3) that

$$F_{I_H,I_T}(x,y) = \begin{cases} 1 & \text{for } 1 \le x,\ 1 \le y \\ \frac{1}{4} & \text{for } 1 \le x,\ 0 \le y < 1 \\ \frac{1}{4} & \text{for } 0 \le x < 1,\ 1 \le y \\ 0 & \text{otherwise} \end{cases} \tag{2.3}$$

This result is sketched in Fig. 2.1.

Exercise 2.1 Determine the joint-probability distribution function of the random variables N_4 and N_2 which were defined in Exercise 1.1 above and sketch your result. □

Some of the general properties of the joint-probability distribution function follow simply and directly from its definition. Thus, for exam-

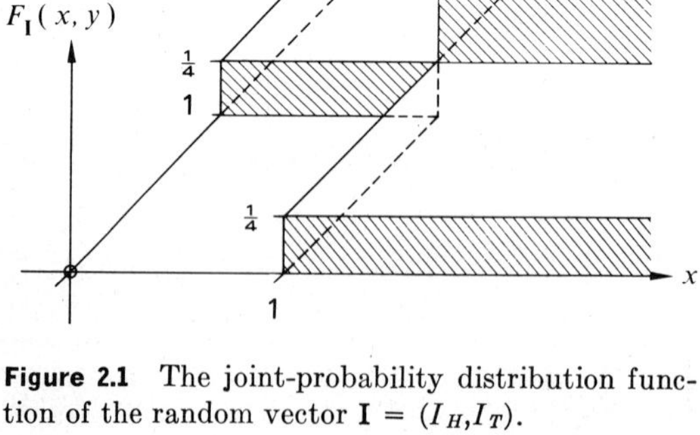

Figure 2.1 The joint-probability distribution function of the random vector $\mathbf{I} = (I_H, I_T)$.

ple, it follows from (2.2) that

$$F_{X,Y}(-\infty,-\infty) = P[X \leq -\infty, Y \leq -\infty] \\ = 0 \tag{2.4}$$

that

$$F_{X,Y}(+\infty,+\infty) = P[X \leq +\infty, Y \leq +\infty] \\ = 1 \tag{2.5}$$

and further, that $F_{X,Y}(x,y)$ is monotone increasing in x and y. Suppose next that we set only one of the variables x and y, say y, to the extreme value $+\infty$. We then get

$$F_{X,Y}(x,+\infty) = P[X \leq x, Y \leq +\infty]$$

The right-hand side here is the probability that $X \leq x$ and that Y assumes *any* value whatsoever; that is, it is simply the probability that $X \leq x$. Thus we obtain the result that

$$F_{X,Y}(x,+\infty) = P[X \leq x] \\ = F_X(x) \tag{2.6}$$

In a similar manner we can also obtain the result

$$F_{X,Y}(+\infty,y) = P[Y \leq y] \\ = F_Y(y) \tag{2.7}$$

The individual probability distribution functions $F_X(x)$ and $F_Y(y)$ of the component random variables can be gotten from their joint-probability distribution function $F_{X,Y}(x,y)$ simply by evaluating that function at $y = +\infty$ and at $x = +\infty$, respectively.

For example, it follows on applying (2.6) to (2.3) that we get

$$F_{I_H}(x) = \begin{cases} 1 & \text{for } 1 \leq x \\ \frac{1}{4} & \text{for } 0 \leq x < 1 \\ 0 & \text{for } x < 0 \end{cases}$$

as the probability distribution function of the random variable I_H (cf. Sec. 4.2).

Exercise 2.2 Let $\mathbf{Z}$ be a two-dimensional random vector with components X and Y. Suppose that the joint-probability distribution function of $\mathbf{Z}$ is

$$F_{X,Y}(x,y) = \begin{cases} 1 & \text{for } 1 \leq x,\ 1 \leq y \\ x & \text{for } 0 \leq x < 1,\ 1 \leq y \\ y & \text{for } 1 \leq x,\ 0 \leq y < 1 \\ xy & \text{for } 0 \leq x < 1,\ 0 \leq y < 1 \\ 0 & \text{otherwise} \end{cases}$$

a. Sketch $F_{X,Y}(x,y)$.
b. Determine $F_X(x)$ and plot.
c. Determine $F_Y(y)$ and plot.

3 JOINT-PROBABILITY DENSITIES

The decomposition of joint-probability distribution functions is by no means as simple in general as is the decomposition of the probability distribution function of a single random variable.[1] However, it is true that the joint-probability distribution functions of interest in practical problems are often either completely discrete (that is, are n-dimensional step functions—as in Fig. 2.1), or they are differentiable. Let us next consider the latter case. In particular, let us suppose that the joint-probability distribution function of the two-dimensional random variable $\mathbf{Z} = (X,Y)$ is continuous everywhere, and that the partial derivative

$$f_{\mathbf{Z}}(\mathbf{z}) \triangleq f_{X,Y}(x,y) \triangleq \frac{\partial^2 F_{X,Y}(x,y)}{\partial x \, \partial y} \tag{3.1}$$

exists and is continuous everywhere except possibly at points belonging to a finite number of curves.[1] We call $f_{\mathbf{Z}}$ (or $f_{X,Y}$) the *joint-probability density* of the random vector $\mathbf{Z}$ (or of the random variables X and Y). Now it follows from the usual definition of the partial derivative that (assuming positive Δx and Δy):

$$\begin{aligned} f_{X,Y}(x,y) &= \lim_{\substack{\Delta x \to 0 \\ \Delta y \to 0}} \frac{1}{\Delta x \, \Delta y} [F_{X,Y}(x + \Delta x, y + \Delta y) - F_{X,Y}(x + \Delta x, y) \\ &\qquad - F_{X,Y}(x, y + \Delta y) + F_{X,Y}(x,y)] \\ &= \lim_{\substack{\Delta x \to 0 \\ \Delta y \to 0}} \frac{1}{\Delta x \, \Delta y} P[x < X \le x + \Delta x, y < Y \le y + \Delta y] \end{aligned} \tag{3.2}$$

where the various regions of the range sample space $\mathcal{S}$ which are pertinent here are sketched in Fig. 3.1. It then follows from (3.2) that

$$f_{X,Y}(x,y) \ge 0 \tag{3.3}$$

for all values of x and y. That is, *the joint-probability density is always nonnegative.*

[1] See, for example, sec. 8.4 of Cramér.

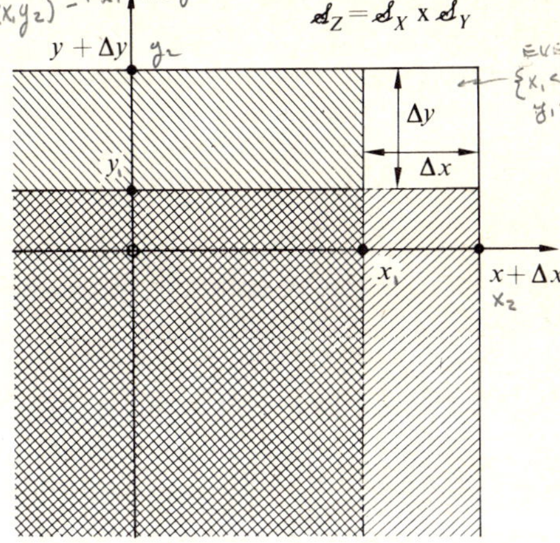

Figure 3.1 Regions pertinent to the calculation of the joint probability density $f_{X,Y}$.

Exercise 3.1 Let the two-dimensional random vector $\mathbf{Z} = (X,Y)$ have the joint-probability distribution function specified in Exercise 2.2.

Determine the corresponding joint-probability density and sketch your result. □

If the joint-probability density of the two-dimensional random vector $\mathbf{Z} = (X,Y)$ is continuous everywhere, then it follows from the two-dimensional version of the fundamental theorem of calculus that

$$F_{X,Y}(x,y) = \int_{-\infty}^{y} \int_{-\infty}^{x} f_{X,Y}(\xi,\eta)\, d\xi\, d\eta \tag{3.4}$$

Now since the joint-probability distribution function assumes its maximum value of unity when $x = +\infty$ and $y = +\infty$, it then follows from (3.4) that we get the result

$$\iint_{-\infty}^{+\infty} f_{X,Y}(\xi,\eta)\, d\xi\, d\eta = 1 \tag{3.5}$$

that is, the *total volume under the graph of the joint-probability density must be unity*. Similarly, it follows [from (2.6) and (2.7)] that

$$F_X(x) = \int_{-\infty}^{+\infty} \int_{-\infty}^{x} f_{X,Y}(\xi,\eta)\, d\xi\, d\eta \tag{3.6}$$

and that

$$F_Y(y) = \int_{-\infty}^{y} \int_{-\infty}^{+\infty} f_{X,Y}(\xi,\eta)\, d\xi\, d\eta \tag{3.7}$$

If now we differentiate both sides of (3.6) with respect to x and both sides of (3.7) with respect to y, we get

$$f_X(x) = \int_{-\infty}^{+\infty} f_{X,Y}(x,\eta)\, d\eta \tag{3.8}$$

and

$$f_Y(y) = \int_{-\infty}^{+\infty} f_{X,Y}(\xi,y)\, d\xi \tag{3.9}$$

Thus we see that the individual probability densities of the component random variables X and Y may be obtained by appropriately integrating their joint-probability density.

Exercise 3.2 Let the two-dimensional random vector $\mathbf{Z} = (X,Y)$ have the joint-probability density determined in Exercise 3.1 above.

a. Determine the probability density $f_X(x)$ and plot your result.
b. Determine $f_Y(y)$ and plot. □

Exercise 3.3 Let $\mathbf{Z} = (X,Y)$ be a two-dimensional random vector with the joint-probability density

$$f_{X,Y}(x,y) = \frac{1}{2\pi\sqrt{1-\rho^2}} \exp\left[-\frac{x^2 - 2\rho xy + y^2}{2(1-\rho^2)}\right] \tag{3.10}$$

where $|\rho| \leq 1$. Show that

$$f_X(x) = \frac{1}{\sqrt{2\pi}} e^{-x^2/2}$$

and that

$$f_Y(y) = \frac{1}{\sqrt{2\pi}} e^{-y^2/2}$$

[*Hint:* Complete the square and use the formula

$$\frac{1}{\sqrt{2\pi}} \int_{-\infty}^{+\infty} e^{-t^2/2}\, dt = 1$$

(cf. Exercise 4.5.4).] The random variables X and Y here are said to have a *two-dimensional normal* (or *gaussian*) joint-probability density. □

We showed in (3.4) above that the joint-probability distribution function $F_{X,Y}(x,y)$ of the random vector $\mathbf{Z} = (X,Y)$ could be calculated by integrating its joint-probability density $f_{X,Y}$ over the region $(-\infty < X \leq x,\ -\infty < Y \leq y)$ in the range sample space $\mathcal{S}_Z = \mathcal{S}_X \times \mathcal{S}_Y$. We then showed in (3.6) and (3.7) that the individual probability distribution functions F_X and F_Y could also be obtained by integrating the joint-probability density over appropriate regions in $\mathcal{S}_Z$. These results can now be extended as shown in the following exercises.

Exercise 3.4 Let $\mathbf{Z} = (X,Y)$ be a two-dimensional random vector with a joint-probability density $f_{X,Y}$. Consider a rectangle

$$(a < x \leq b, c < y \leq d)$$

in the range sample space $\mathcal{S}_Z = \mathcal{S}_X \times \mathcal{S}_Y$.

Show that the probability that the random vector $\mathbf{Z}$ has a value falling in that rectangle is given by the formula

$$\boxed{P[\mathbf{Z} \in (a < x \leq b, c < y \leq d)] = \int_c^d \int_a^b f_{X,Y}(x,y)\, dx\, dy} \tag{3.11}$$

[*Hint:* Consider a sketch like that in Fig. 3.1, and use Eq. (3.4).] □

The result in (3.11) can be extended in a relatively straightforward manner to show that one can calculate the probability that the random vector $\mathbf{Z}$ has a value falling in any arbitrary union or intersection of rectangles in the range sample space $\mathcal{S}_Z$ by integrating the joint-probability density of $\mathbf{Z}$ over the appropriate union or intersection. It further follows that we may calculate the probability of essentially any event of interest concerning the values of the random vector $\mathbf{Z}$ by integrating its joint-probability density over those points in the range sample space $\mathcal{S}_Z$ which represent the specified event. Thus if A is an event in the range sample space $\mathcal{S}_Z$, then we may calculate the probability that $\mathbf{Z} \in A$ by using the formula

$$P[\mathbf{Z} \in A] = \iint_A f_{X,Y}(x,y)\, dx\, dy \tag{3.12}$$

where, as usual, $\mathbf{Z} = (X,Y)$.

Let us now use the formula in (3.12) to calculate a few probabilities.

Exercise 3.5 Let the two-dimensional random vector $\mathbf{Z} = (X,Y)$ have the joint probability density determined in Exercise 3.1.

a. Sketch the region of the range sample space

$$\mathcal{S}_Z = \mathcal{S}_X \times \mathcal{S}_Y$$

for which $X \geq 2Y$.

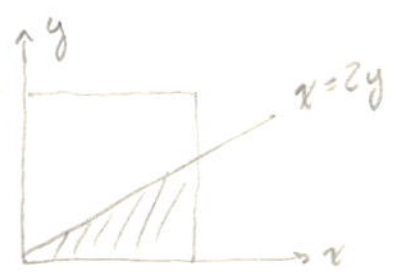

b. Calculate the probability $P[X \geq 2Y]$. □

Exercise 3.6 Let us reconsider again the spacecraft-landing problem first introduced in Exercise 2.3.2. Let the random variable X characterize the east-west component of the landing error and the random variable Y characterize the north-south component of the landing error. A much more reasonable assumption for the probability distribution of the landing error than that used in Exercise 3.5.1 is that the error components X and Y have a two-dimensional normal joint-probability density. In particular, let us suppose that experiments have indicated that

$$f_{X,Y}(x,y) = \frac{1}{2\pi} \exp\left(-\frac{x^2+y^2}{2}\right)$$

[that is, (3.10) with $\rho = 0$] where x and y are measured in kilometers.

Calculate the probability that the landing point is no more than 2 km from the aiming point. □

Exercise 3.7 A manufacturer has been using two different manufacturing processes to make 100-watt light bulbs. On defining X to be the random variable characterizing the time to failure of bulbs made by one process, and Y to be the random variable characterizing the time to failure of bulbs made by the other process, he found that X and Y had the joint-probability density

$$f_{X,Y}(x,y) = \begin{cases} ab \exp\left[-(ax+by)\right] & \text{for } 0 \le x,\ 0 \le y \\ 0 & \text{otherwise} \end{cases}$$

where $a = \frac{1}{200}$ and $b = \frac{1}{300}$ (cf. Exercise 4.5.3).

Determine the probability that the time to failure is greater for bulbs characterized by X than it is for bulbs characterized by Y. □

Exercise 3.8 Let $\mathbf{Z} = (X,Y)$ be a two-dimensional random vector with the joint-probability density:

$$f_{X,Y}(x,y) = \begin{cases} c(x+y)^2 & \text{for } -1 \le x \le 1,\ -1 \le y \le 1 \\ 0 & \text{otherwise} \end{cases}$$

where c is a constant.

a. Evaluate the constant c.
b. Determine $f_X(x)$ and plot.
c. Determine $f_Y(y)$ and plot. □

Exercise 3.9 For the random vector $\mathbf{Z}$ of Exercise 3.8:

a. Sketch the region in $\mathcal{S}_{\mathbf{Z}}$ for which $X \ge |Y|$.
b. Calculate the probability $P[X \ge |Y|]$. □

In our discussions above we were concerned primarily with events defined in the range sample space $\mathcal{S}_Z$ of the random vector $\mathbf{Z}$. In many applied problems, however, one is concerned with events defined in the separate range sample spaces of the components X and Y, that is, with events defined separately in $\mathcal{S}_X$ and $\mathcal{S}_Y$, respectively. Let us now consider this latter situation. Suppose, in particular, that we are given the events

$$A \subset \mathcal{S}_X \qquad \text{and} \qquad B \subset \mathcal{S}_Y$$

and that we are interested in the probability of the joint event $[X \in A] \cap [Y \in B]$. If now we are to apply directly our above results, we must first determine the set in $\mathcal{S}_Z$ which corresponds to this joint event. Clearly the event A in $\mathcal{S}_X$ and the event

$$[X \in A \text{ and } Y \text{ has any value}] = [X \in A, -\infty < Y \leq +\infty]$$

in $\mathcal{S}_Z$ are equivalent events. Similarly, the events B in $\mathcal{S}_Y$ and

$$[Y \in B \text{ and } X \text{ has any value}] = [Y \in B, -\infty < X \leq +\infty]$$

in $\mathcal{S}_Z$ are also equivalent events. If then we give the same name to equivalent events (even though they are defined in different sample spaces), we have a situation such as that shown schematically in Fig. 3.2. As much as anything, it is this picturization of the range sample space $\mathcal{S}_Z$ which leads to its being called a product space and denoted by $\mathcal{S}_X \times \mathcal{S}_Y$.

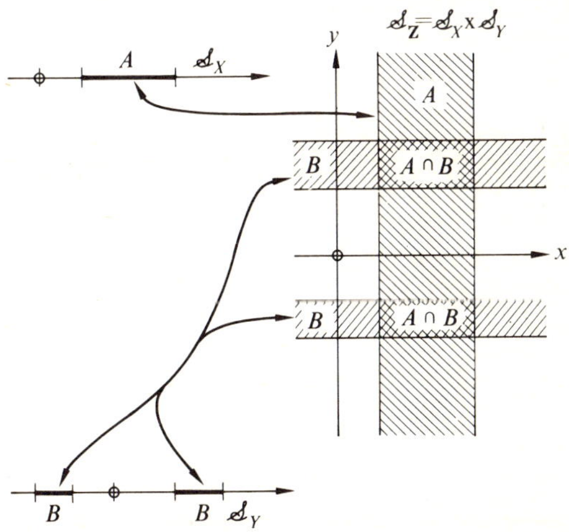

Figure 3.2 Same events in different sample spaces.

A moment's reflection upon Fig. 3.2 should then convince one that

$$[X \in A] \cap [Y \in B] = [\mathbf{Z} \in A \cap B]$$

and hence that

$$P[[X \in A] \cap [Y \in B]] = P[\mathbf{Z} \in A \cap B] \tag{3.13}$$

As we shall often have to deal with joint probabilities of this kind in the future, it is convenient to introduce at this time the somewhat simpler notation:

$$P[X \in A, Y \in B] \triangleq P[[X \in A] \cap [Y \in B]] \tag{3.14}$$

Suppose now that the joint-probability density $f_{X,Y}$ exists and is continuous almost everywhere. We may then use (3.12) to evaluate the right-hand side of (3.14). Thus it follows that we may write

$$\begin{aligned} P[X \in A, Y \in B] &= \iint\limits_{A \cap B} f_{X,Y}(x,y)\, dx\, dy \\ &= \iint\limits_{B\,A} f_{X,Y}(x,y)\, dx\, dy \end{aligned} \tag{3.15}$$

whenever we are given the joint-probability density.

Exercise 3.10 Suppose that the two-dimensional random vector $\mathbf{Z} = (X,Y)$ has the uniform joint-probability density

$$f_{X,Y}(x,y) = \begin{cases} 1 & \text{for } 0 \le x \le 1, 0 \le y \le 1 \\ 0 & \text{otherwise} \end{cases}$$

Let the events A and B be defined by

$$A \triangleq (0,\tfrac{1}{4}) \subset \mathcal{S}_X \qquad \text{and} \qquad B \triangleq (\tfrac{1}{2},1) \subset \mathcal{S}_Y$$

a. Sketch the sets in $\mathcal{S}_Z$ which correspond to the events A, B, and $A \cap B$.
b. Calculate the joint probability $P[X \in A, Y \in B]$. □

4 CONDITIONAL-PROBABILITY DISTRIBUTION FUNCTIONS

We pointed out in Sec. 3.6 that, given a pair of events, it is often of interest to ask questions about the probability of occurrence of one event given that the other event has occurred. Our earlier discussions of such conditional probabilities dealt with general events defined on some given sample space; let us now apply those ideas to the specific case of events concerning random variables. As in the preceding section, we shall restrict our discussion here to pairs of random variables, that is, to the case of two-dimensional random vectors. Our results, however, can be extended to the case of n-dimensional random vectors in a more or less obvious manner.

Suppose now that $\mathbf{Z}$ is a two-dimensional random vector with components X and Y defined upon some given sample space $\mathcal{S}$:

$$\mathbf{Z}(s) = (X(s),Y(s)) \tag{4.1}$$

for each $s \in \mathcal{S}$. Suppose further that we are interested in the events $A \subset \mathcal{S}_X$ and $B \subset \mathcal{S}_Y$. It then follows from the definition of conditional probability, (3.6.1), that the probability of the occurrence of A, under the hypothesis that B has occurred, is

$$P[X \in A|Y \in B] = \frac{P[X \in A,\ Y \in B]}{P[Y \in B]} \tag{4.2}$$

so long as $P[Y \in B] > 0$. Suppose now, in particular, that the event A is the semi-infinite interval

$$A \triangleq (-\infty,x] \subset \mathcal{S}_X$$

Equation (4.2) then becomes

$$P[X \leq x|Y \in B] = \frac{P[X \leq x,\ Y \in B]}{P[Y \in B]} \tag{4.3}$$

in this case. This conditional probability may be viewed as a point function of the parameter x; we shall call it a *conditional-probability distribution function* and write

$$F_X(x|Y \in B) \triangleq P[X \leq x|Y \in B] \tag{4.4}$$

whenever $P[Y \in B] > 0$.

As a specific example, consider the two-dimensional random vector $\mathbf{Z} = (X,Y)$, introduced in Exercise 2.2, which has the joint-probability distribution function

$$F_{X,Y}(x,y) = \begin{cases} 1 & \text{for } 1 \leq x,\ 1 \leq y \\ x & \text{for } 0 \leq x < 1,\ 1 \leq y \\ y & \text{for } 1 \leq x,\ 0 \leq y < 1 \\ xy & \text{for } 0 \leq x < 1,\ 0 \leq y < 1 \\ 0 & \text{otherwise} \end{cases} \tag{4.5}$$

and suppose that we wish to determine the conditional probability distribution function $F_X(x|Y \leq \frac{1}{2})$. The event that the random variable Y does not exceed $\frac{1}{2}$ is represented by the semi-infinite interval

$$(-\infty,\tfrac{1}{2}] \subset \mathcal{S}_Y$$

hence it follows from the definition of the probability distribution function that

$$P[Y \leq \tfrac{1}{2}] = F_Y(\tfrac{1}{2})$$

It now follows from (4.5) and the fact that

$$F_Y(y) = F_{X,Y}(+\infty,y)$$

[cf. (2.7)], that the probability distribution function of Y is

$$F_Y(y) = \begin{cases} 1 & \text{for } 1 \leq y \\ y & \text{for } 0 \leq y < 1 \\ 0 & \text{for } y < 0 \end{cases}$$

and therefore that

$$P[Y \leq \tfrac{1}{2}] = \tfrac{1}{2} \tag{4.6}$$

The joint probability $P[X \leq x, Y \leq \frac{1}{2}]$ may next be calculated by integrating the joint-probability density $f_{X,Y}$ over the appropriate region in $\mathcal{S}_X \times \mathcal{S}_Y$; thus

$$P[X \leq x,\ Y \leq \tfrac{1}{2}] = \int_{-\infty}^{1/2} \int_{-\infty}^{x} f_{X,Y}(\xi,\eta)\, d\xi\, d\eta \tag{4.7}$$

We may obtain $f_{X,Y}$ by differentiating (4.5) with respect to x and y to get

$$f_{X,Y}(x,y) = \begin{cases} 1 & \text{for } 0 \leq x < 1,\ 0 \leq y < 1 \\ 0 & \text{otherwise} \end{cases}$$

(cf. Exercise 3.1). It then follows on using this joint-probability density in (4.7) that we get

$$P[X \leq x,\ Y \leq \tfrac{1}{2}] = \begin{cases} \tfrac{1}{2} & \text{for } 1 \leq x \\ x/2 & \text{for } 0 \leq x < 1 \\ 0 & \text{otherwise} \end{cases}$$

If we then substitute this result and (4.6) into (4.3), we finally obtain

$$F_X(x|Y \leq \tfrac{1}{2}) = \begin{cases} 1 & \text{for } 1 \leq x \\ x & \text{for } 0 \leq x < 1 \\ 0 & \text{otherwise} \end{cases} \tag{4.8}$$

as the formula for our desired conditional-probability distribution function.

Exercise 4.1 Consider the light bulb time-to-failure problem discussed in Exercise 3.7 where the time-to-failure random variables X and Y have the joint-probability density

$$f_{X,Y}(x,y) = \begin{cases} ab \exp\left[-(ax + by)\right] & \text{for } 0 \leq x, 0 \leq y \\ 0 & \text{otherwise} \end{cases} \tag{4.9}$$

where $a = \frac{1}{200}$ and $b = \frac{1}{300}$.

Determine the conditional-probability distribution function of the random variable X, given that $Y \leq 300$ and plot your result. □

Next let us determine a few of the basic properties of the conditional probability distribution function.

Exercise 4.2 Let X and Y be random variables defined on the same basic sample space $\mathcal{S}$.

Show that

$$F_X(-\infty|Y \in B) = 0 \tag{4.10}$$

and that

$$F_X(+\infty|Y \in B) = 1 \tag{4.11}$$

under the assumption that the probability of the conditioning event $Y \in B$ is not zero. □

Exercise 4.3 Let X and Y be random variables defined on the same basic sample space $\mathcal{S}$ and suppose that $P[Y \in B] > 0$.

Show that

$$F_X(b|Y \in B) - F_X(a|Y \in B) = P[a < x \leq b|Y \in B] \tag{4.12}$$

when $a < b$, and hence that the conditional-probability distribution function $F_X(x|Y \in B)$ is a monotone-increasing function of the argument x. □

It follows from the result of these last two exercises that $F_X(x|Y \in B)$ does indeed deserve to be called a probability distribution function.

5 CONDITIONAL-PROBABILITY DENSITIES

We find in many applied problems that the derivative with respect to x of the conditional-probability distribution function $F_X(x|Y \in B)$ exists for almost all x. We call this derivative a *conditional-probability density* and denote it by the symbol $f_X(x|Y \in B)$; thus

$$f_X(x|Y \in B) \triangleq \frac{dF_X(x|Y \in B)}{dx} \tag{5.1}$$

For example, if $\mathbf{Z} = (X,Y)$ is the two-dimensional random vector with the joint-probability distribution function given in (4.5) above, it follows from our derivation of $F_X(x|Y \leq \frac{1}{2})$ given in the preceding section [cf. (4.8)] that

$$f_X(x|Y \leq \tfrac{1}{2}) = \begin{cases} 1 & \text{for } 0 \leq x < 1 \\ 0 & \text{otherwise} \end{cases} \tag{5.2}$$

in this case.

Exercise 5.1 Let X and Y be the light bulb time-to-failure random variables discussed in Exercises 3.7 and 4.1.

Determine the conditional probability density of the time-to-failure X given that the time-to-failure Y does not exceed 300. □

Having defined the conditional probability density as the derivative of the conditional probability distribution function, let us now determine a few of its properties. First, since we showed in Exercise 4.3 that the conditional probability distribution function $F_X(x|Y \in B)$ is a monotone-increasing function of the argument x, it follows (cf. the derivation in the single random variable case) that

$$f_X(x|Y \in B) \geq 0 \tag{5.3}$$

for all x in the range sample space $\mathcal{S}_X$; that is, the conditional probability density is always nonnegative. It further follows immediately from the fundamental theorem of calculus that we get the inverse relation

$$F_X(x|Y \in B) - F_X(a|Y \in B) = \int_a^x f_X(\xi|Y \in B)\, d\xi \tag{5.4}$$

for every x in the interval $(a,b] \subset \mathcal{S}_X$, so long as the given density is continuous almost everywhere in that interval, and so long as the probability of the conditioning event is not zero. If then we let the point a recede to minus infinity and use (4.10), we get

$$F_X(x|Y \in B) = \int_{-\infty}^x f_X(\xi|Y \in B)\, d\xi \tag{5.5}$$

that is, given the conditional probability density, we may calculate the corresponding conditional-probability distribution function by integration. It further follows from (5.4) and (4.11) that

$$\int_{-\infty}^{+\infty} f_X(\xi|Y \in B)\, d\xi = 1 \tag{5.6}$$

that is, that the area under the graph of the conditional-probability density must be unity.

On setting $x = b$ in (5.4) we get

$$\int_a^b f_X(\xi|Y \in B)\,d\xi = F_X(b|Y \in B) - F_X(a|Y \in B) = P[a < X \leq b|Y \in B] \tag{5.7}$$

[cf. (4.12)]; that is, the probability that the random variable X assumes a value in the interval $(a,b]$ in its range, given that $Y \in B$, may be calculated by integrating the conditional-probability density $f_X(x|Y \in B)$ over that interval. As in the case of the probability density of a single random variable, this result can be extended to show that the probability that the event $[X \in A]$ occurs, given that $Y \in B$, can be calculated by integrating the conditional probability density over the region A; that is, we have

$$P[X \in A|Y \in B] = \int_A f_X(\xi|Y \in B)\,d\xi \tag{5.8}$$

for any event A in the range sample space $\mathcal{S}_X$.

The above results are particularly useful when the given random vector has a mixed joint-probability distribution function, that is, whenever the joint-probability distribution function is neither purely discrete nor differentiable almost everywhere. As an example of such a situation, consider the following problem.

Exercise 5.2 The input to a particular communications receiver consists of a desired binary signal and an undesired noise. Measurements show that when no input noise is present, the receiver output is $+2$ when the input signal is a 1 and is -2 when the input signal is a 0. The measurements also indicate that when both the input signal and noise are present, the receiver output is normally distributed about either $+2$ or -2, according to whether the input signal is a 1 or a 0. That is, if Y is the random variable which characterizes the signal at the receiver *output*, and X is the random variable which characterizes the signal plus noise at the receiver *output*, then we have

$$f_X(x|Y = +2) = \frac{1}{\sqrt{2\pi}} \exp\left[-\frac{(x-2)^2}{2}\right]$$

and

$$f_X(x|Y = -2) = \frac{1}{\sqrt{2\pi}} \exp\left[-\frac{(x+2)^2}{2}\right] \tag{5.9}$$

We would now like to determine the probability density of the receiver output signal plus noise, that is, of X. Assuming that the input signals 1 and 0 are equally likely:

a. Determine the joint-probability distribution function $F_{X,Y}(x,y)$ and sketch your result. (*Hint:* First determine the joint probabilities $P[X \leq x, Y = +2]$ and $P[X \leq x, Y = -2]$.)
b. Determine the probability distribution function $F_X(x)$ of X alone.
c. Determine the probability density $f_X(x)$ of X alone and sketch. □

Point conditioning

The various results derived above are all valid for any conditioning event $B \subset \mathcal{S}_Y$ so long as $P[Y \in B] > 0$. In particular, they all hold for the case in which the condition set B is a single-point set, say $B = \{y_0\}$, so long as $P[Y = y_0] > 0$. Suppose, however, that the random variables X and Y are such that their joint-probability density $f_{X,Y}$ exists and is continuous everywhere. In this case, the individual probability densities of the random variables X and Y both exist and are continuous also, and hence it follows here that

$$P[Y = y] = 0$$

for *any* $y \in \mathcal{S}_Y$. The problem before us now is how to extend our above ideas to cover this important case.

Suppose that the joint-probability density $f_{X,Y}$ of the random variables X and Y exists, is continuous, and is nonzero everywhere in $\mathcal{S}_Z$. Let A be an arbitrary event in the range space $\mathcal{S}_X$ of X and let B be the particular event

$$B \triangleq [y, y + \Delta y] \subset \mathcal{S}_Y$$

where Δy is an arbitrary small positive number. It now follows from the definition of conditional probability for random variables, (4.2), that

$$P[X \in A | y \leq Y \leq y + \Delta y] = \frac{P[X \in A, y \leq Y \leq y + \Delta y]}{P[y \leq Y \leq y + \Delta y]} \tag{5.10}$$

Since the joint-probability density $f_{X,Y}$ is assumed given here, we may obtain both probabilities on the right-hand side of (5.10) by integration, and so get the result

$$P[X \in A | y \leq Y \leq y + \Delta y] = \frac{\int_A \int_y^{y+\Delta y} f_{X,Y}(\xi,\eta)\, d\eta\, d\xi}{\int_y^{y+\Delta y} f_Y(\eta)\, d\eta} \tag{5.11}$$

If now both probability densities in (5.11) are continuous over the interval $y \leq \eta \leq y + \Delta y$, then it follows from the mean-value theorem of integral calculus[1] that there exist points y' and y'' in the interval $[y, y + \Delta y]$ such

[1] See, for example, chap. II, sec. 7 of Courant, vol. I.

that we may write

$$\int_y^{y+\Delta y} f_{X,Y}(\xi,\eta)\,d\eta = f_{X,Y}(\xi,y')\,\Delta y$$

and

$$\int_y^{y+\Delta y} f_Y(\eta)\,d\eta = f_Y(y'')\,\Delta y$$

It then further follows from these results that we may rewrite (5.11) in the form

$$P[X \in A|y \leq Y \leq y + \Delta y] = \frac{\int_A f_{X,Y}(\xi,y')\,d\xi\,\Delta y}{f_Y(y'')\,\Delta y} \tag{5.12}$$

Suppose next that we pass to the limit in (5.12) as $\Delta y \to 0$. As we do so, both y' and y'' must approach y, and hence we get the result

$$\lim_{\Delta y \to 0} P[X \in A|y \leq Y \leq y + \Delta y] = \frac{\int_A f_{X,Y}(\xi,y)\,d\xi}{f_Y(y)}$$

Now as Δy approaches zero, the interval $[y, y + \Delta y]$ reduces to the single point y, and hence the conditioning event becomes the event $Y = y$. Accordingly, we define

$$P[X \in A|Y = y] \triangleq \frac{\int_A f_{X,Y}(\xi,y)\,d\xi}{f_Y(y)} \tag{5.13}$$

whenever the joint-probability density $f_{X,Y}$ exists, is continuous, and is nonzero everywhere in $\mathcal{S}_Z$. It should be noted that this result is unambiguously defined even though the probability of the conditioning event is zero, that is, even though $P[Y = y] = 0$ for every $y \in \mathcal{S}_Y$. We thus have obtained the desired extension of our previous results.

Suppose now that A is the event that the random variable X is less than or equal to some number x:

$$A = (-\infty,x] \subset \mathcal{S}_X$$

In this case $P[X \in A|Y = y]$ becomes the conditional-probability distribution function of the random variable X subject to the hypothesis that $Y = y$; that is,

$$P[X \in A|Y = y] = F_X(x|Y = y)$$

here. A somewhat more common notation for this particular conditional-probability distribution function is $F_{X|Y}$, where

$$F_{X|Y}(x|y) \triangleq F_X(x|Y = y) \tag{5.14}$$

We shall generally use this new notation in our subsequent developments even though it is, perhaps, somewhat less explicit. It then follows, on using these results in (5.13), that

$$F_{X|Y}(x|y) = \frac{\int_{-\infty}^{x} f_{X,Y}(\xi,y)\,d\xi}{f_Y(y)} \tag{5.15}$$

If $F_{X|Y}$ is continuously differentiable with respect to x almost everywhere, we call its derivative the *conditional-probability density of X given y*, and write

$$f_{X|Y}(x|y) \triangleq \frac{dF_{X|Y}(x|y)}{dx} \tag{5.16}$$

If then we differentiate both sides of (5.15) with respect to x, we get the result that

$$f_{X|Y}(x|y) = \frac{f_{X,Y}(x,y)}{f_Y(y)} \tag{5.17}$$

that is, the conditional-probability density $f_{X|Y}$ equals the ratio of the joint-probability density $f_{X,Y}$ to the probability density f_Y. This result may also be expressed in product form to give

$$f_{X,Y}(x,y) = f_{X|Y}(x|y)f_Y(y) \tag{5.18}$$

as an expansion of the joint-probability density.

Exercise 5.3 Consider again the spacecraft-landing problem discussed in Exercise 3.6. As in that exercise, assume that the east-west component of the landing error, X, and the north-south component of the landing error, Y, have the gaussian joint-probability density

$$f_{X,Y}(x,y) = \frac{1}{2\pi}\exp\left(-\frac{x^2+y^2}{2}\right)$$

a. Determine the probability density of the north-south error component Y. [*Hint:* Remember that

$$\frac{1}{\sqrt{2\pi}}\int_{-\infty}^{+\infty} e^{-t^2/2}\,dt = 1$$

(cf. Exercise 4.5.4).]

b. Determine the conditional-probability density of the east-west error component X under the hypothesis that the north-south component of the landing error has the value y. □

Exercise 5.4 Suppose that the random variables X and Y have a gaussian joint-probability density of the form

$$f_{X,Y}(x,y) = \frac{1}{2\pi\sqrt{1-\rho^2}} \exp\left[-\frac{x^2 - 2\rho xy + y^2}{2(1-\rho^2)}\right] \tag{5.19}$$

where $|\rho| \leq 1$.

a. Show that

$$f_{X|Y}(x|y) = \frac{1}{\sqrt{2\pi(1-\rho^2)}} \exp\left[-\frac{(x-\rho y)^2}{2(1-\rho^2)}\right] \tag{5.20}$$

b. Sketch (5.20) for the case $\rho = \frac{1}{2}$ and $y = 2$.
(*Hint:* Compare with Exercise 3.3 above.) □

Having defined the conditional-probability density $f_{X|Y}$, we may now rewrite (5.13) in the more compact form

$$P[X \in A | Y = y] = \int_A f_{X|Y}(\xi|y)\, d\xi \tag{5.21}$$

If then the event A is taken to be the semi-infinite interval $(-\infty, x] \subset \mathcal{S}_X$, we get

$$F_{X|Y}(x|y) = \int_{-\infty}^{x} f_{X|Y}(\xi|y)\, d\xi \tag{5.22}$$

from which we obtain the further result

$$\int_{-\infty}^{+\infty} f_{X|Y}(\xi|y)\, d\xi = 1 \tag{5.23}$$

on setting $x = +\infty$. These results are, of course, the point-conditioning versions of the results previously obtained for more general conditioning events [cf. (5.5) and (5.6)].

Exercise 5.5 Let the random variables X and Y have the joint-probability distribution function specified in Exercise 2.2:

$$F_{X,Y}(x,y) = \begin{cases} 1 & \text{for } 1 \leq x,\ 1 \leq y \\ x & \text{for } 0 \leq x < 1,\ 1 \leq y \\ y & \text{for } 1 \leq x,\ 0 \leq y < 1 \\ xy & \text{for } 0 \leq x < 1,\ 0 \leq y < 1 \\ 0 & \text{otherwise} \end{cases}$$

a. Determine $f_{X|Y}(x|y)$ and plot for all values of x and several values of y.
b. Determine $f_{Y|X}(y|x)$ and plot for all values of y and several values of x.
c. Compare these results with those in Exercise 3.2. □

Exercise 5.6 Using the results of Exercise 5.5, and defining the event A by

$$A \triangleq (\tfrac{1}{4},\tfrac{1}{2}) \subset \mathcal{S}_X$$

calculate the conditional probability $P[X \in A | Y = y]$ for all the values of y in the interval $(0,1) \subset \mathcal{S}_Y$. □

Exercise 5.7 Suppose that the joint-probability density of the random variables X and Y exists, is continuous, and is nonzero everywhere. Show that we may then write

$$P[X \in A] = \int_{-\infty}^{+\infty} P[X \in A | Y = y] f_Y(y)\, dy \tag{5.24}$$

for any event $A \subset \mathcal{S}_X$. □

Exercise 5.8 It is often required that the operation of multiplication be performed in a control or communications system; we would like to investigate here some of the properties of this operation when the quantities being multiplied are random. Suppose (as shown in Fig. 5.1) that two quantities y and z are multiplied together to form a third quantity x. In particular, suppose that z can assume only the values $+1$ or -1, and that those values are equally likely. Now it follows from the definition of x that

$$x = \begin{cases} y & \text{when } z = +1 \\ -y & \text{when } z = -1 \end{cases}$$

and that these two cases are equally likely because the two values of z are equally likely. That is, we have here a situation in which

$$P[X = +y | Y = y] = \tfrac{1}{2}$$
$$P[X = -y | Y = y] = \tfrac{1}{2}$$

and therefore

$$P[X = x | Y = y] = 0 \qquad \text{if } x \neq \pm y$$

where X is the random variable which characterizes the possible values of the quantity x, and Y is the random variable which characterizes the

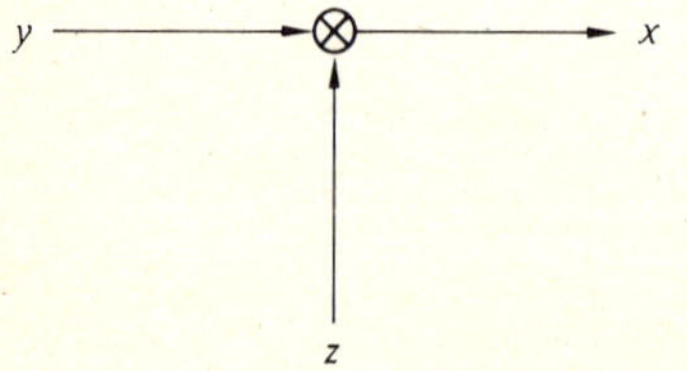

Figure 5.1 The multiplier for Exercises 5.8 and 5.9.

possible values of y. Suppose, finally, that Y is a gaussian random variable; in particular, let

$$f_Y(y) = \frac{1}{\sqrt{2\pi}} e^{-y^2/2}$$

We would now like to determine the probability density of the multiplier output.

a. Determine the conditional-probability distribution function

$$F_X(x|Y = y)$$

(Note that the result may be expressed in terms of unit-step functions.)

b. Determine the probability distribution function $F_X(x)$ of X alone. [*Hint:* Use (5.24).]

c. Using (*b*), show that

$$f_X(x) = \frac{1}{\sqrt{2\pi}} e^{-x^2/2}$$

that is, that the multiplier output is also a gaussian random variable. □

Exercise 5.9 The results of Exercise 5.8 show that, under the conditions stated there, both the multiplier input y and the multiplier output x are characterized by random variables which individually have gaussian probability densities. Nonetheless, those random variables do not have a gaussian joint-probability density. Why not? [*Hint:* Compare the result obtained in part (*a*) of Exercise 5.8 with that obtained by integrating the result in part (*a*) of Exercise 5.4.] □

Impulse densities

Problems which involve conditional-probability distribution functions of the mixed type (say, as in Exercise 5.8) may often be more easily solved through the use of impulse-function densities.

Exercise 5.10 Let X and Y be the random variables defined in Exercise 5.8. Then:

a. Determine the conditional-probability density

$$f_{X|Y}(x|y) = f_X(x|Y = y)$$

(Note that the result contains impulse functions.)

b. Using (*a*), and (5.18), determine the joint probability density $f_{X,Y}$ of the random variables X and Y.

c. Using the above results, and (3.8), determine the probability density f_X of the random variable X alone. □

6 INDEPENDENT RANDOM VARIABLES

Having just defined and discussed various conditional probability functions relating to random variables, a question automatically arises as to how to determine the degree of statistical dependence of one random variable upon another. In particular, we would like to know how to determine when random variables are "statistically independent." Now it will be recalled that our previous discussions of statistical independence were centered about the independence of events (cf. Secs. 3.7 and 3.8). The problem here then is how to apply the results of those earlier discussions to the present study of random variables.

Let us consider for the moment a pair of random variables, say X and Y, defined on some given sample space $\mathcal{S}$. As pointed out in Sec. 4.1, the ranges of those random variables, that is, $\mathcal{S}_X$ and $\mathcal{S}_Y$, respectively, may be considered to be new sample spaces which characterize new experiments related to the basic experiment defining the underlying sample space $\mathcal{S}$. With this fact in mind, it follows more or less directly from the discussion in Sec. 3.8 that the new experiments are statistically independent if and only if we have

$$P[X \in A, Y \in B] = P[X \in A]P[Y \in B] \tag{6.1}$$

for each event $A \subset \mathcal{S}_X$ and each event $B \subset \mathcal{S}_Y$. If these experiments are, in fact, statistically independent, then we shall say also that the corresponding random variables are statistically independent. That is, a pair of random variables X and Y are said to be *statistically independent* (or just *independent*) if and only if (6.1) holds true for each pair of events $A \subset \mathcal{S}_X$ and $B \subset \mathcal{S}_Y$.

While the above definition of statistically independent random variables does indeed follow logically from our earlier discussion of the statistical independence of events and experiments, it is quite awkward to use in practice. A practically more useful result can be obtained as follows: Suppose that we take A and B to be semi-infinite intervals in the range sample spaces $\mathcal{S}_X$ and $\mathcal{S}_Y$, respectively, that is, let

$$A \triangleq (-\infty, x] \subset \mathcal{S}_X$$

and (6.2)

$$B \triangleq (-\infty, y] \subset \mathcal{S}_Y$$

If now X and Y are independent random variables, then it follows [on using (6.2) in (6.1)] that

$$P[X \leq x, Y \leq y] = P[X \leq x]P[Y \leq y] \tag{6.3}$$

that is, that

$$F_{X,Y}(x,y) = F_X(x)F_Y(y) \tag{6.4}$$

for all $x \in \mathcal{S}_X$ and all $y \in \mathcal{S}_Y$. Conversely, it can also be shown that if (6.4) holds true for all $x \in \mathcal{S}_X$ and all $y \in \mathcal{S}_Y$, then (6.1) also holds true for each pair of events $A \subset \mathcal{S}_X$ and $B \subset \mathcal{S}_Y$. Thus we see that *the random variables X and Y are statistically independent if and only if their joint-probability distribution function factors into the product of their respective individual probability distribution functions.* We shall use this result as the basic definition of statistical independence for random variables from here on.

Consider, for example, the random variables X and Y with the joint-probability distribution function

$$F_{X,Y}(x,y) = \begin{cases} 1 & \text{for } 1 \leq x,\ 1 \leq y \\ x & \text{for } 0 \leq x < 1,\ 1 \leq y \\ y & \text{for } 1 \leq x,\ 0 \leq y < 1 \\ xy & \text{for } 0 \leq x < 1,\ 0 \leq y < 1 \\ 0 & \text{otherwise} \end{cases}$$

which we first introduced in Exercise 2.2. Now that exercise showed that

$$F_X(x) = \begin{cases} 1 & \text{for } 1 \leq x \\ x & \text{for } 0 \leq x < 1 \\ 0 & \text{for } x < 0 \end{cases}$$

and that

$$F_Y(y) = \begin{cases} 1 & \text{for } 1 \leq y \\ y & \text{for } 0 \leq y < 1 \\ 0 & \text{for } y < 0 \end{cases}$$

It therefore follows from these results that

$$F_{X,Y}(x,y) = F_X(x)F_Y(y)$$

for all values of x and y here, and hence that these two random variables are statistically independent. A contrasting result is shown in the following exercise.

Exercise 6.1 Let I_H and I_T be the head- and tail-indicator random variables, respectively, which we discussed in Secs. 1 and 2 of this chapter. Their joint probability distribution function is given in (2.3).

Show that I_H and I_T are *not* statistically independent random variables. □

Exercise 6.2 Let X and Y be binary random variables with the Bernoulli probability distributions

$$P[X = 1] = p = P[Y = 1]$$

and

$$P[X = 0] = q = P[Y = 0]$$

where $p + q = 1$.

a. Determine the individual probability distribution functions $F_X(x)$ and $F_Y(y)$ and plot.

b. Determine the form that the joint-probability distribution function $F_{X,Y}(x,y)$ *must* have if X and Y are to be independent random variables. Plot your result.

c. Compare the results obtained here with those obtained in Exercise 6.1. □

When X and Y are both discrete random variables, as in Exercises 6.1 and 6.2, a useful alternative equivalent definition of statistical independence becomes available. Suppose that the random variable X assumes the possible values x_i with the probabilities p_i, $i = 1, 2, 3, \ldots$, where

$$\sum_i p_i = 1$$

Suppose similarly that the random variable Y assumes the values y_j with the probabilities q_j, $j = 1, 2, 3, \ldots$, where

$$\sum_j q_j = 1$$

If now all the joint probabilities factor, that is, if

$$P[X = x_i, Y = y_j] = P[X = x_i]P[Y = y_j] = p_i q_j \tag{6.5}$$

holds for all values of i and j, then it follows directly that

$$F_{X,Y}(x,y) = F_X(x)F_Y(y) \tag{6.6}$$

for all values of $x \in \mathcal{S}_X$ and $y \in \mathcal{S}_Y$. Conversely, it can be shown (cf. Exercise 6.3) that if (6.6) holds for all $x \in \mathcal{S}_X$ and $y \in \mathcal{S}_Y$, then (6.5) holds for all values of i and j. Thus we have shown that the *discrete* random variables X and Y are statistically independent if and only if (6.5) holds for all possible pairs of values of the indices.

Exercise 6.3 Let X and Y be discrete random variables with the probability distributions defined above.

Show that if

$$F_{X,Y}(x,y) = F_X(x)F_Y(y)$$

holds for all $x \in \mathcal{S}_X$ and $y \in \mathcal{S}_Y$, then

$$P[X = x_i, Y = y_j] = P[X = x_i]P[Y = y_j]$$

for all possible pairs of values of the indices. □

If, on the other hand, the random variables X and Y are such that their joint-probability density $f_{X,Y}(x,y)$ exists for almost all x and y, then we may differentiate both sides of (6.4) and thus obtain the result

$$f_{X,Y}(x,y) = f_X(x)f_Y(y) \tag{6.7}$$

when the random variables X and Y are independent. Conversely, if (6.7) holds true for all $x \in \mathcal{S}_X$ and all $y \in \mathcal{S}_Y$, then we may integrate both sides of (6.7) and thus obtain (6.4) as a result. Thus we see that, if X and Y are random variables whose joint-probability density exists almost everywhere, then X and Y are independent random variables if and only if their joint-probability density factors into the product of their respective individual probability densities.

For example, consider the east-west and north-south landing-error-component random variables X and Y, respectively, discussed in Exercise 3.6 above; X and Y were assumed there to have the joint-probability density

$$f_{X,Y}(x,y) = \frac{1}{2\pi} \exp\left(-\frac{x^2+y^2}{2}\right) \tag{6.8}$$

for all values of x and y. Now it follows from this formula by integration (or directly from the results given in Exercise 3.3) that X and Y have the individual probability densities

$$f_X(x) = \frac{1}{\sqrt{2\pi}} e^{-x^2/2}$$

and $\qquad$ (6.9)

$$f_Y(y) = \frac{1}{\sqrt{2\pi}} e^{-y^2/2}$$

Since the product of f_X and f_Y here equals $f_{X,Y}$ as given in (6.8), it therefore follows that these random variables X and Y are statistically independent. Conversely, it also follows that if f_X and f_Y are given as in (6.9), then their joint-probability density *must* have the form given in (6.8) if X and Y are independent random variables.

Exercise 6.4 Suppose that the random variables X and Y have the two-dimensional gaussian joint-probability density introduced in Exercise 3.3 above:

$$f_{X,Y}(x,y) = \frac{1}{2\pi\sqrt{1-\rho^2}} \exp\left[-\frac{x^2 - 2\rho xy + y^2}{2(1-\rho^2)}\right] \tag{6.10}$$

As shown in Exercise 3.3, the random variables X and Y then have the individual probability densities given in (6.9).

Show that the random variables X and Y here are statistically independent if and only if the parameter ρ is zero. □

Exercise 6.5 Let X and Y be the time-to-failure random variables introduced in Exercise 3.7 above, where it was assumed that

$$f_{X,Y}(x,y) = \begin{cases} abe^{-(ax+by)} & \text{for } 0 \leq x,\ 0 \leq y \\ 0 & \text{otherwise} \end{cases}$$

and where $a = \frac{1}{200}$ and $b = \frac{1}{300}$.

Determine whether or not X and Y are independent random variables in this case. □

Exercise 6.6 The random variables X and Y have a joint-probability density $f_{X,Y}(x,y)$ which is constant over the shaded region R shown in Fig. 6.1.

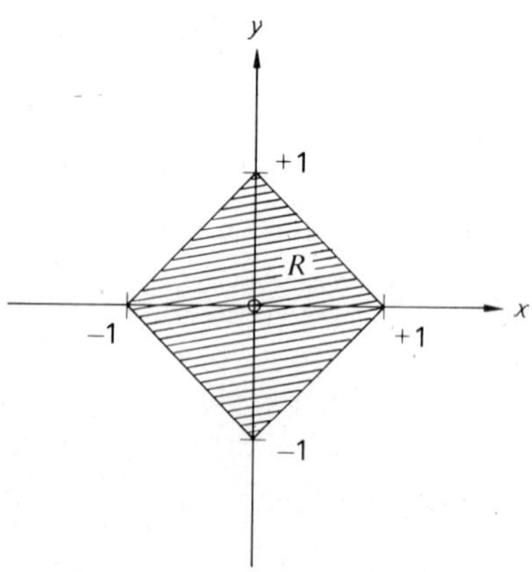

Figure 6.1 The region R in Exercise 6.6.

a. Determine a formula for the joint-probability density of X and Y.
b. Determine the individual probability densities $f_X(x)$ and $f_Y(y)$.
c. Show whether or not X and Y are independent random variables. □

Exercise 6.7 Suppose that the *independent* random variables X and Y have a joint-probability density $f_{X,Y}(x,y)$ which is constant over a bounded region R of the (x,y) plane.

a. Suppose also that the random variable X can only assume values in the x interval (a,b), while the random variable Y can only assume values in the y interval (c,d). What *must* the shape of the region R then be?
b. Suppose next that X can only assume values in either of the nonoverlapping x intervals (a_1,b_1) or (a_2,b_2), while Y can only assume values in either of the nonoverlapping y intervals (c_1,d_1) or (c_2,d_2). What *must* the shape of the region R be in this case?
c. Compare these results with the situation in Exercise 6.6. □

While the various results obtained above are for the case of a pair of random variables, they can easily be extended to cover the general case of n random variables. Thus, an extension of the argument leading to (6.4) causes us to say that the n random variables $X_1, X_2, \ldots, X_n$ are *mutually independent* if and only if

$$F_{X_1,X_2,\ldots,X_n}(x_1, x_2, \ldots, x_n) = F_{X_1}(x_1)F_{X_2}(x_2) \cdots F_{X_n}(x_n) \triangleq \prod_{i=1}^{n} F_{X_i}(x_i) \tag{6.11}$$

for all values of $x_i \in \mathcal{S}_{X_i}$, $i = 1, 2, 3, \ldots, n$. If now the n random variables X_i are all discrete random variables such that for each value of the index i the random variable X_i assumes only the possible values x_{i,k_i}, $i = 1, 2, \ldots, n$ and $k_i = 1, 2, \ldots$, then it follows [from a development parallel to that leading to the two-dimensional result (6.12)] that those random variables are mutually independent if and only if all joint probabilities factor, that is, if and only if

$$\begin{aligned} P[X_1 = x_{1,k_1}, X_2 &= x_{2,k_2}, \ldots, X_n = x_{n,k_n}] \\ &= P[X_1 = x_{1,k_1}]P[X_2 = x_{2,k_2}] \cdots P[X_n = x_{n,k_n}] \end{aligned} \tag{6.12}$$

for all sets of values of the indices $k_1, k_2, \ldots, k_n$. Finally, suppose that the n random variables $X_1, X_2, \ldots, X_n$ have an n-dimensional joint-probability density which exists almost everywhere. It then follows [from a development paralleling that leading to the two-dimensional result (6.7)] that those random variables are mutually independent if and only if

$$f_{X_1,X_2,\ldots,X_n}(x_1, x_2, \ldots, x_n) = f_{X_1}(x_1)f_{X_2}(x_2) \cdots f_{X_n}(x_n)$$
$$= \prod_{i=1}^{n} f_{X_i}(x_i) \tag{6.13}$$

for all values of $x_i \in \mathcal{S}_{X_i}$, $i = 1, 2, \ldots, n$. That is, the random variables X_i here are mutually independent if and only if their joint-probability density factors into the product of their respective individual probability densities. We shall often use these n-dimensional results in our subsequent studies of random vectors and random processes.

7 SUMMARY OF DEFINITIONS AND FORMULAS

Random vector

A finite-dimensional vector-valued function, defined on a sample space $\mathcal{S}$, whose components are random variables defined on $\mathcal{S}$, that is, the function

$$\mathbf{X}(s) = (x_1, x_2, \ldots, x_n)$$
$$= (X_1(s), X_2(s), \ldots, X_n(s)) \tag{1.2}$$

where $s \in \mathcal{S}$, is a random vector if each component X_i is a random variable defined on $\mathcal{S}$. Thus, a random vector is a finite family of random variables.

Range

$$R_{\mathbf{X}} \triangleq \{(x_1, x_2, \ldots, x_n): x_1 = X_1(s), x_2 = X_2(s), \ldots, x_n = X_n(s), s \in \mathcal{S}\} \tag{1.3}$$

Range sample space

$$S_{\mathbf{X}} = R_{\mathbf{X}}$$

Probability mapping

$$P[A \subset \mathcal{S}_{\mathbf{X}}] \triangleq P[\mathbf{X}^{-1}(A) \subset \mathcal{S}] \tag{1.4}$$

Product space

$$\mathcal{S}_{\mathbf{X}} = \mathcal{S}_{X_1} \times \mathcal{S}_{X_2} \cdots \mathcal{S}_{X_n} \tag{1.5}$$

Joint-probability distribution function

$$F_{\mathbf{Z}}(\mathbf{z}) \triangleq F_{X,Y}(x,y)$$
$$\triangleq P[X \le x, Y \le y] \tag{2.2}$$

where

$$\mathbf{Z} = (X,Y) \qquad \text{and} \qquad \mathbf{z} = (x,y)$$

Bounds on joint-probability distribution function

$$F_{X,Y}(-\infty,-\infty) = 0 \tag{2.4}$$

and

$$F_{X,Y}(+\infty,+\infty) = 1 \tag{2.5}$$

Marginal relations

$$F_{X,Y}(x,+\infty) = F_X(x) \tag{2.6}$$

and

$$F_{X,Y}(+\infty,y) = F_Y(y) \tag{2.7}$$

Joint-probability density

$$f_{\mathbf{Z}}(\mathbf{z}) \triangleq f_{X,Y}(x,y) \triangleq \frac{\partial^2 F_{X,Y}(x,y)}{\partial x\, \partial y} \tag{3.1}$$

where

$$\mathbf{Z} = (X,Y) \qquad \text{and} \qquad \mathbf{z} = (x,y)$$

Nonnegativeness

$$f_{X,Y}(x,y) \geq 0 \tag{3.3}$$

Unit area

$$\iint_{-\infty}^{+\infty} f_{X,Y}(x,y)\, dx\, dy = 1 \tag{3. 5}$$

Marginal relations

$$F_X(x) = \int_{-\infty}^{+\infty} \int_{-\infty}^{x} f_{X,Y}(\xi,\eta)\, d\xi\, d\eta \tag{3.6}$$

$$f_X(x) = \int_{-\infty}^{+\infty} f_{X,Y}(x,\eta)\, d\eta \tag{3.8}$$

and

$$F_Y(y) = \int_{-\infty}^{y} \int_{-\infty}^{+\infty} f_{X,Y}(\xi,\eta)\, d\xi\, d\eta \tag{3.7}$$

$$f_Y(y) = \int_{-\infty}^{+\infty} f_{X,Y}(\xi,y)\, d\xi \tag{3.9}$$

Two-dimensional normal (or gaussian) density

$$f_{X,Y}(x,y) = \frac{1}{2\pi\sqrt{1-\rho^2}} \exp\left[-\frac{x^2 - 2\rho xy + y^2}{2(1-\rho^2)}\right] \tag{3.10}$$

where $|\rho| \leq 1$.

Probability calculation

$$P[\mathbf{Z} \in A] = \iint\limits_A f_{X,Y}(x,y)\,dx\,dy \tag{3.12}$$

where $\mathbf{Z} = (X,Y)$ and $A \subset \mathcal{S}_{\mathbf{Z}}$.

Conditional-probability distribution function

$$F_X(x|Y \in B) \triangleq P[X \leq x|Y \in B] \tag{4.4}$$

whenever $P[Y \in B] > 0$.

Bounds on conditional-probability distribution function

$$F_X(-\infty|Y \in B) = 0 \tag{4.10}$$

and

$$F_X(+\infty|Y \in B) = 1 \tag{4.11}$$

Conditional-probability density

$$f_X(x|Y \in B) \triangleq \frac{dF_X(x|Y \in B)}{dx} \tag{5.1}$$

Nonnegativeness

$$f_X(x|Y \in B) \geq 0 \tag{5.3}$$

Unit area

$$\int_{-\infty}^{+\infty} f_X(x|Y \in B)\,dx = 1 \tag{5.6}$$

Inversion

$$F_X(x|Y \in B) = \int_{-\infty}^{x} f_X(\xi|Y \in B)\,d\xi \tag{5.5}$$

Probability calculation

$$P[X \in A|Y \in B] = \int_A f_X(\xi|Y \in B)\,d\xi \tag{5.8}$$

Point conditioning—conditional-probability distribution function

If $f_{X,Y}$ is continuous almost everywhere,

$$F_{X|Y}(x|y) \triangleq F_X(x|Y = y) \tag{5.14}$$

$$= \frac{\int_{-\infty}^{x} f_{X,Y}(\xi,y)\,d\xi}{f_Y(y)} \tag{5.15}$$

Point conditioning—conditional-probability density

$$f_{X|Y}(x|y) \triangleq \frac{dF_{X|Y}(x|y)}{dx} \tag{5.16}$$

$$= \frac{f_{X,Y}(x,y)}{f_Y(y)} \tag{5.17}$$

Probability calculation

$$P[X \in A|Y = y] = \int_A f_{X|Y}(\xi|y)\, d\xi \tag{5.21}$$

$$P[X \in A] = \int_{-\infty}^{+\infty} P[X \in A|Y = y] f_Y(y)\, dy \tag{5.24}$$

Independent random variables

The random variables X and Y are statistically independent if and only if their joint-probability distribution function factors into the product of their respective individual probability distribution functions, that is, if and only if

$$F_{X,Y}(x,y) = F_X(x)F_Y(y) \tag{6.4}$$

or (if the respective densities exist), alternatively, if and only if

$$f_{X,Y}(x,y) = f_X(x)f_Y(y) \tag{6.7}$$

REFERENCES

Courant, Richard: "Differential and Integral Calculus," Interscience Publishers, Inc., New York, vol. I, 1937; vol. II, 1936.

Cramér, Harald: "Mathematical Methods of Statistics," Princeton University Press, Princeton, N.J., 1946.

6
Functions of Random Variables

1 INTRODUCTION

Let us reconsider for the moment the kinetic energy example which we used in Chap. 4 to introduce the concept of random variable. We supposed in that example that we were studying a particle moving in a bounded one-dimensional space by measuring at some given instant of time both its position x and its velocity $v \triangleq dx/dt$, and we also supposed there that we had determined in some way or other the probability distribution P for events in the two-dimensional sample space $\mathcal{S}$ of sample points (x,v). Finally, we assumed that while we measured the position and velocity of the particle, we were also interested in the kinetic energy K of the particle, where

$$K = \tfrac{1}{2}mv^2 \tag{1.1}$$

and where m is the mass of the particle.

Now, as pointed out in Sec. 4.1, the formula in (1.1) defines, in essence, a function K which maps points in the measurement sample space $\mathcal{S}$ into points on the real line; for, given any point (x,v) in $\mathcal{S}$, we may calculate a corresponding real number using the formula

$$K(x,v) \triangleq \tfrac{1}{2}mv^2 \tag{1.2}$$

Since we may use this mapping K to induce a probability distribution on its range space R_K, we say that K is therefore a random variable. Now as we noted in Sec. 4.1, K is, of course, by no means the only random variable which can be defined on the measurement sample space $\mathcal{S}$. For example, we could also define the velocity random variable V on $\mathcal{S}$ by the formula

$$V(x,v) \triangleq v \tag{1.3}$$

for every point (x,v) in $\mathcal{S}$.

Now, as pointed out many times before, a random variable is, by definition, a function defined on a sample space which induces a probability on its range space, and hence it follows that we may consider that range space to be a new sample space. Thus, in particular, we may consider the range space R_V of the velocity random variable V to be a new sample space (say $\mathcal{S}_V$). With this fact in mind, we note that the formula given in (1.1) not only defines a function K on the measurement sample space $\mathcal{S}$ which maps points (x,v) in $\mathcal{S}$ into points k in $\mathcal{S}_K$, but it also defines a function g on the velocity sample space $\mathcal{S}_V$ which maps points v in $\mathcal{S}_V$ into points k in $\mathcal{S}_K$ by the formula

$$k = g(v) \triangleq \tfrac{1}{2}mv^2 \tag{1.4}$$

Thus it follows (on starting at a point s in the position-velocity sample space $\mathcal{S}$) that we may ultimately reach the corresponding point k in the kinetic energy sample space $\mathcal{S}_K$ *either* by going directly from $\mathcal{S}$ to $\mathcal{S}_K$ using the mapping K *or* by going first from $\mathcal{S}$ to the velocity sample space $\mathcal{S}_V$ using the mapping V and then going from $\mathcal{S}_V$ to $\mathcal{S}_K$ using the mapping g. This result is pictured in Fig. 1.1 and is expressed by the equation

$$\begin{aligned} K(x,v) &= \tfrac{1}{2}mv^2 \\ &= \tfrac{1}{2}mV^2(x,v) \\ &= g(V(x,v)) \end{aligned} \tag{1.5}$$

We say in this case that $g(V(x,v))$ is the value of the *composite function* $g(V)$ at the point (x,v), and we write

$$\begin{aligned} g(V) &= \tfrac{1}{2}mV^2 \\ &= K \end{aligned} \tag{1.6}$$

following (1.5).

In summary, we see that we may view the kinetic energy random variable which has values k in the sample space $\mathcal{S}_K$ *either* as a function K

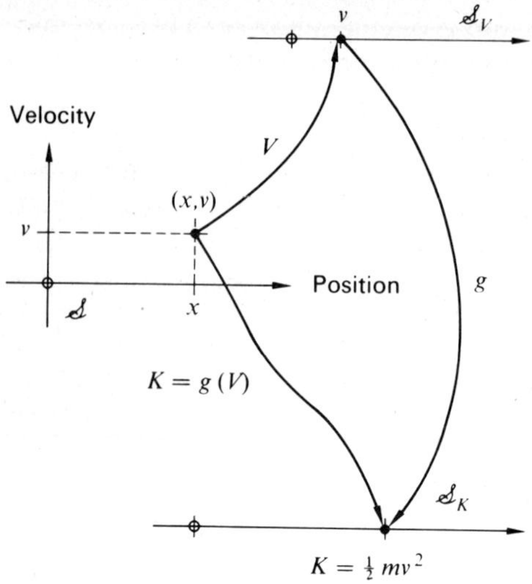

Figure 1.1 The kinetic energy mappings.

which maps directly from $\mathcal{S}$ to $\mathcal{S}_K$, *or* as a function g which maps from $\mathcal{S}_V$ to $\mathcal{S}_K$, *or* as a composite function $g(V)$ which maps from $\mathcal{S}$ to $\mathcal{S}_V$ to $\mathcal{S}_K$. Just which of these views we should take depends entirely upon the problem at hand. For example, if we are given the joint-probability distribution of the position and velocity, then we might choose to regard the kinetic energy random variable as the function K defined on $\mathcal{S}$ and determine its probability distribution function accordingly. If, on the other hand, we are given the probability distribution function of the velocity random variable V, then we might instead regard the kinetic energy random variable as the function g defined on $\mathcal{S}_V$ in order to determine its probability distribution function.

In either case, we have the problem of determining the probability distribution function of a random variable defined on some sample space when we are given the distribution of probability on that sample space. This kind of problem was discussed at length in Chap. 4 and was, in fact, solved there (at least in principle). The only reason for discussing this problem further here is that when one random variable is defined on the range sample space of another random variable, then the mapping involved becomes particularly simple (as it is then from one one-dimensional space to another) and special techniques become available.

Suppose, for example, that we are given the probability distribution function F_V of the velocity random variable V and that we wish to determine the probability distribution function F_K of the kinetic energy random variable $K = g(V)$. Since $F_K(b)$ is, by definition, the probability that the random variable K has a value less than or equal to b, let us consider the semi-infinite interval

$$B \triangleq \{k: -\infty < k \leq b\} \tag{1.7}$$

of points k in the kinetic energy sample space $\mathcal{S}_K$. Since the value $k = g(v)$ of the kinetic energy random variable K corresponding to the point v in the velocity sample space $\mathcal{S}_V$ is given by (1.4), it follows that the inverse image $g^{-1}(B)$ of B under the mapping g, that is, the event

$$g^{-1}(B) = \{v: -\infty < g(v) \leq b\}$$

may be expressed as

$$g^{-1}(B) = \left\{v: -\infty < \frac{mv^2}{2} \leq b\right\}$$

or, alternatively, as

$$g^{-1}(B) = \begin{cases} \left\{v: -\sqrt{\dfrac{2b}{m}} < v < +\sqrt{\dfrac{2b}{m}}\right\} & \text{for } b \geq 0 \\ \phi \qquad \text{for } b < 0 \end{cases} \tag{1.8}$$

Since we are dealing here with a mapping from one one-dimensional sample space ($\mathcal{S}_V$) to another ($\mathcal{S}_K$), a particularly convenient representation of the mapping may be obtained by using those sample spaces as ordinate and abscissa for a graph of the mapping. Thus, we obtain the plot shown in Fig. 1.2.

Now our usual procedure of inducing probability distributions by mappings gives

$$P[K \in B] = P[V \in g^{-1}(B)] \tag{1.9}$$

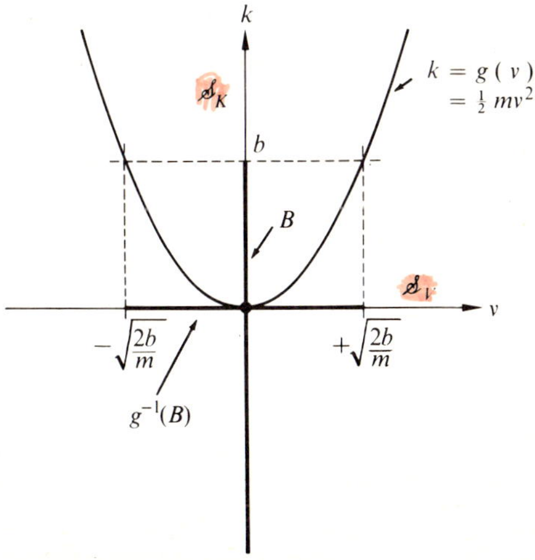

Figure 1.2 The kinetic energy mapping.

Since it follows from (1.8) that the inverse image $g^{-1}(B)$ is a closed interval in $\mathcal{S}_V$, it then follows from (4.3.15) that

$$P[V \in g^{-1}(B)] = \begin{cases} P\left[V = -\sqrt{\dfrac{2b}{m}}\right] + F_V\left(+\sqrt{\dfrac{2b}{m}}\right) - F_V\left(-\sqrt{\dfrac{2b}{m}}\right) & \text{for } 0 \le b \\ 0 & \text{otherwise} \end{cases}$$

the term $P\left[V = -\sqrt{\dfrac{2b}{m}}\right]$ vanishing if F_V is continuous at the indicated point. If next we substitute this result for the right-hand side of (1.9) and remember that the left-hand side there is simply $F_K(b)$, we finally get

$$F_K(b) = \begin{cases} P\left[V = -\sqrt{\dfrac{2b}{m}}\right] + F_V\left(+\sqrt{\dfrac{2b}{m}}\right) - F_V\left(-\sqrt{\dfrac{2b}{m}}\right) & \text{for } 0 \le b \\ 0 & \text{if } b < 0 \end{cases} \tag{1.10}$$

as a formula for the desired probability distribution function of the kinetic energy random variable which is expressed in terms of the probability distribution function of the velocity random variable.

Exercise 1.1 Suppose that the probability distribution function F_V of the velocity random variable V discussed above has a derivative f_V which exists almost everywhere.

Show that in this case the probability density f_K of the kinetic energy random variable K may then be calculated by using the formula

$$f_K(k) = \begin{cases} \dfrac{f_V\left(+\sqrt{\dfrac{2k}{m}}\right) + f_V\left(-\sqrt{\dfrac{2k}{m}}\right)}{\sqrt{2km}} & \text{for } 0 \le k \\ 0 & \text{for } k < 0 \end{cases} \tag{1.11}$$

where $m > 0$ as usual. □

Exercise 1.2 Referring to the random variables V and K in Exercise 1.1 above, let $m = 2$ so that

$$K = V^2 \tag{1.12}$$

for convenience. Suppose that the velocity random variable V has the gaussian probability density function

$$f_V(v) = \frac{1}{\sqrt{2\pi}} e^{-v^2/2} \tag{1.13}$$

a. Show that the kinetic energy random variable K has here the probability density function

$$f_K(k) = \begin{cases} \dfrac{1}{\sqrt{2\pi k}} e^{-k/2} & \text{for } 0 \leq k \\ 0 & \text{for } k < 0 \end{cases} \tag{1.14}$$

This is called a *chi-squared* probability density when written as a function of $\chi^2 = k$.

b. Plot both $f_V(v)$ and $f_K(k)$. □

A review of the above derivations shows that the obtaining of specific results was greatly facilitated by two facts: (*a*) the mapping involved was from one one-dimensional space to another, and (*b*) the probability distribution over the velocity sample space $\mathcal{S}_V$ was given in terms of the probability distribution function of V. Since both of these conditions are inherently met in any problem involving the determination of the probability distribution function of a function of a random variable, we see that the solution of such problems is quite straightforward in concept and, in fact, almost mechanical in detail. Even so, however, the problem-solving techniques used are so important that we shall spend the rest of this chapter studying them at some length while, at the same time, obtaining a number of specific results of practical importance.

2 FUNCTIONS OF RANDOM VARIABLES

While the kinetic energy example studied in the preceding section is only a particular case of the general problem of determining the probabilistic properties of a function of a random variable, it does contain within itself the essence of the solution to the general problem. Thus, suppose that we are given some arbitrary random variable, say X, and its probability distribution function, say F_X, and further that we are then asked to determine the probability distribution function of some measurable function of the given random variable, say

$$Y \triangleq g(X) \tag{2.1}$$

As in the kinetic energy example above, we view the range space $\mathcal{S}_X$ of the given random variable as a new (one-dimensional) sample space. The function Y then generates a mapping from that sample space to the range space R_Y of Y. If now Y is measurable on $\mathcal{S}_X$ [if, for every event A of interest in the range space R_Y, the probability of the inverse image $Y^{-1}(A)$ in the sample space $\mathcal{S}_X$ can in fact be determined], then Y is a random variable. Such a measurable function of a random variable is called a *Borel function*.[1] If therefore the given function g is indeed a Borel function, it follows from its mapping properties that its probability distribution function is given by the formula

$$F_Y(y) = P[\{x: g(x) \leq y, x \in \mathcal{S}_X\}] \tag{2.2}$$

If, in addition, the right-hand probability has a derivative with respect to y almost everywhere, then that derivative gives the probability density of the random variable Y and we may write

$$f_Y(y) = \frac{d}{dy} P[\{x: g(x) \leq y, x \in \mathcal{S}_X\}] \tag{2.3}$$

in such a case.

Perhaps the simplest functions of all to consider are those which are either translations or linear or both.

Exercise 2.1 Let X be a random variable with the Bernoulli probability distribution

$$P[X = 0] = p$$

and

$$P[X = 1] = q \triangleq 1 - p$$

a. Let the random variable Y be defined as a translation of the random variable X; in particular, let

$$Y \triangleq X + 1$$

Determine the probability distribution function of Y and plot your result.

b. Let the random variable Z be defined as a reflection of X followed by a translation; in particular, let

$$Z \triangleq -X + 1$$

Determine the probability distribution function of Z and plot your result. □

The general case of this type of function is considered in the next exercise.

[1] Compare with sec. 3.8 of Pfeiffer or sec. 5.3 of Loève.

Exercise 2.2 Let X be a random variable with the probability distribution function F_X. Let Y be a new random variable defined by

$$Y \triangleq aX + b \tag{2.4}$$

where a and b are real constants and a is nonzero.

Show that the probability distribution function F_Y is given by the formula

$$F_Y(y) = F_X\left(\frac{y-b}{a}\right) \tag{2.5}$$

when $a > 0$, and by

$$F_Y(y) = P\left[X = \frac{y-b}{a}\right] + 1 - F_X\left(\frac{y-b}{a}\right) \tag{2.6}$$

when $a < 0$. [*Hint:* Draw mapping diagrams using $\mathcal{S}_X$ as coordinate axes (cf. Fig. 1.2).] It should be noted that the term $P\left[X = \frac{y-b}{a}\right]$ in (2.6) is required in order to make F_Y continuous on the right as it must be. □

When the probability density of the original random variable exists, then a simplification, shown in the following exercise, results.

Exercise 2.3 Let X be a random variable with a probability density f_X which exists almost everywhere. Let Y be a new random variable defined, as in Exercise 2.2, by the formula

$$Y \triangleq aX + b \tag{2.7}$$

where a and b are real constants and a is nonzero.

a. Show that the probability density of the random variable Y is then given by

$$f_Y(y) = \frac{1}{|a|} f_X\left(\frac{y-b}{a}\right) \tag{2.8}$$

b. Suppose that X is uniformly distributed between zero and one. Determine f_Y and plot your result. (*Hint:* Remember that a may be either positive or negative.) □

The next simplest cases are those in which a new random variable is defined to be some power of an original random variable. The kinetic energy random variable discussed in Sec. 1 is an example of this case.

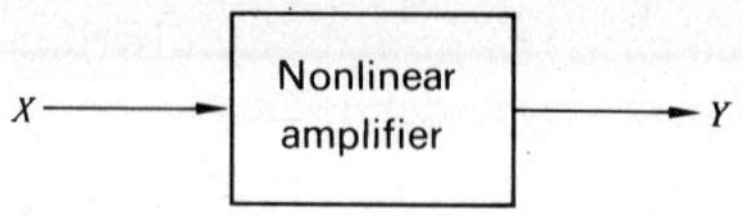

Figure 2.1

Exercise 2.4 Let the random variable X represent the possible values at a time t of the input to a nonlinear amplifier and let the random variable Y represent the corresponding values at the same time of the amplifier output, as shown schematically in Fig. 2.1. Suppose that the amplifier transfer characteristic is symmetric; in particular, suppose that X and Y are related by the transformation

$$Y \triangleq \begin{cases} X^{1/n} & \text{for } 0 \leq X \\ -|X|^{1/n} & \text{for } X < 0 \end{cases}$$

where n is a positive integer, as plotted in Fig. 2.2. It should be noted that the extreme cases of $n = 1$ and $n = +\infty$ correspond to a linear amplifier and an ideal limiter, respectively.

a. Suppose that the probability density X exists everywhere. Determine the probability density of Y.

b. Suppose that X has the gaussian probability density

$$f_X(x) = \frac{1}{\sqrt{2\pi}} e^{-x^2/2}$$

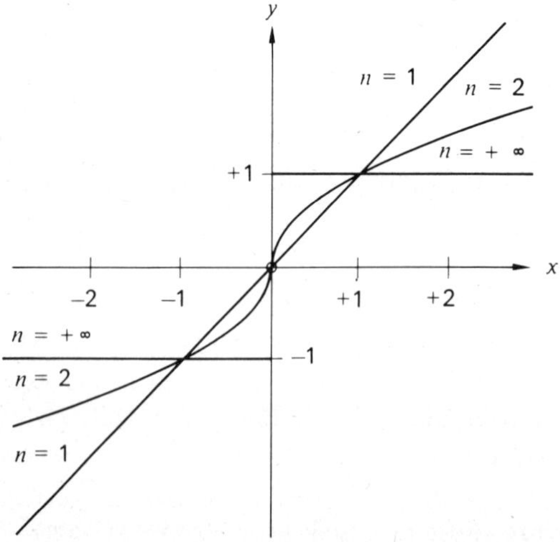

Figure 2.2 Nonlinear amplifier transfer characteristics.

Determine the resulting probability density of Y.

c. Evaluate (*b*) for the case $n = 2$ and plot your result. □

Exercise 2.5 Let us next consider the extreme cases of the nonlinear amplifier discussed in Exercise 2.4.

a. Let $n = 1$. Assuming that the probability density of the random variable X exists, determine the probability density of the output random variable X. (Stop and think!)

b. Suppose that $n = +\infty$ and that X has the gaussian probability density given above. Using the result obtained in part (*b*) of Exercise 2.4 above, show that

$$\lim_{n\to+\infty} f_Y(y) = \begin{cases} 0 & \text{for } y \neq \pm 1 \\ +\infty & \text{for } y = \pm 1 \end{cases}$$

and hence that f_Y consists of impulses in this case. Note that this result should be expected in view of the form of the transformation from X to Y in this case (cf. Fig. 2.2). □

Exercise 2.6 Let X be a gaussian random variable with the probability density

$$f_X(x) = \frac{1}{\sqrt{2\pi}} e^{-x^2/2}$$

Let Y be a new random variable defined as the cube of X:

$$Y \triangleq X^3$$

Determine the probability density of Y and plot your result. □

Next, let us consider a somewhat more complicated transformation.

Exercise 2.7 Suppose that the random variable X_t is defined by the formula (cf. Sec. 9.3)

$$X_t \triangleq \sin(\omega t + Y) \tag{2.9}$$

for all t, where ω is a constant and where Y is a uniformly distributed random variable with the probability density

$$f_Y(y) = \begin{cases} \dfrac{1}{2\pi} & \text{for } -\pi \leq y < +\pi \\ 0 & \text{otherwise} \end{cases} \tag{2.10}$$

a. Show that

$$F_{X_t}(x) = \begin{cases} 1 & \text{for } 1 \le x \\ \dfrac{1}{2} + \dfrac{1}{\pi}\sin^{-1} x & \text{for } -1 \le x < +1 \\ 0 & \text{for } x < -1 \end{cases} \tag{2.11}$$

is the probability distribution function of X_t. Plot this result.

b. Show that

$$f_{X_t}(x) = \begin{cases} 0 & \text{for } 1 < x \\ \dfrac{1}{\pi\sqrt{1 - x^2}} & \text{for } -1 < x < +1 \\ 0 & \text{for } x < -1 \end{cases} \tag{2.12}$$

is the probability density of X_t. Plot this result. □

Finally, we would like to consider a problem which is essentially the inverse of those studied above and which often arises in simulation studies.[1] Suppose that we would like to obtain a random variable with a specified probability distribution and that we have, instead, a random variable with a different probability distribution. How can we transform the undesired distribution into the desired one? Now we have seen above that transforming one random variable into another also changes the probability distribution; our problem therefore is to find the particular transformation of random variables which produces the desired transformation of probability distributions.

Exercise 2.8 We would like the random variable Y to have some given continuous probability distribution function, say F_Y, and we are given instead a random variable X which is uniformly distributed between zero and one. We would like, therefore, to determine a transformation, say g, from X to Y, that is,

$$Y = g(X) \tag{2.13}$$

which will result in Y having the desired probability distribution function F_Y.

Show that the desired transformation is

$$X = g^{-1}(Y) = F_Y(Y) \tag{2.14}$$

[1] See, for example, the review article by Chambers, or sec. 26.8, Methods of Generating Random Numbers and their Applications, of NBS publication AMS55.

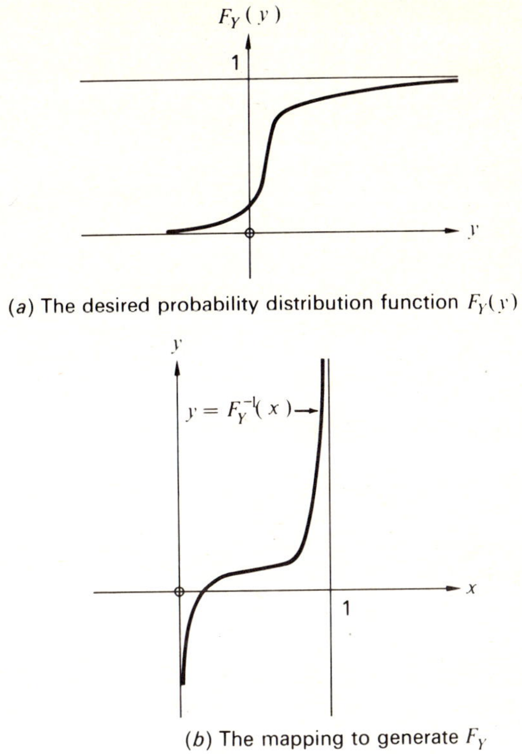

Figure 2.3 Generation of a specified probability distribution function.

that is,

$$Y = g(X) = F_Y^{-1}(X) \tag{2.15}$$

where F_Y^{-1} is the inverse of F_Y and where $0 \leq x < 1$ (cf. Fig. 2.3). □

Exercise 2.9 We would like the random variable Y to be uniformly distributed between zero and one, and we are given, instead, a random variable X with the exponential probability density

$$f_X(x) = \begin{cases} e^{-x} & \text{for } 0 \leq x \\ 0 & \text{for } x < 0 \end{cases}$$

a. Determine a monotonically increasing function

$$Y = g(X)$$

which produces the desired probability density for Y.

b. Determine a monotonically decreasing function which produces the desired result. □

3 FUNCTIONS OF RANDOM VECTORS

Now just as we are often concerned with functions of a single random variable, so also are we often interested in functions of multiple random variables, that is, of random vectors. An added complication occurs in the random vector case, however, for at times we are concerned there with *scalar* functions of random vectors while at other times we are concerned with *vector* functions of random vectors. Nonetheless, the techniques used in determining the probabilistic properties of functions of random vectors are, in either case, directly analogous to those used in the single variable case, as we shall show below.

Sums of random variables

For a specific example of the multivariate case, consider the following problem: Let X and Y be random variables with the joint-probability density function $f_{X,Y}$. Suppose that a new random variable Z is defined to be the sum of X and Y:

$$Z \triangleq X + Y \tag{3.1}$$

for example, X might be a signal and Y might be a noise. We would now like to determine the probability distribution function and density of the sum random variable Z. Paralleling the single variable case, let us consider the joint sample space $\mathcal{S}_X \times \mathcal{S}_Y$ of the random variables X and Y. The function Z is defined upon this joint sample space and hence maps points in this product space into points in its range space $\mathcal{S}_Z$. If we are to determine the probability distribution function of Z, we must determine the probabilities of the sets $(-\infty,z] \subset \mathcal{S}_Z$, for all $z \in \mathcal{S}_Z$. Now it follows from the definition of Z that the probability of the event $[Z \leq z]$ in $\mathcal{S}_Z$ is given by the probability of the event $[X + Y \leq z]$ in $\mathcal{S}_X \times \mathcal{S}_Y$. Our problem then is to evaluate the latter probability using the given joint-probability density of X and Y.

Suppose, for the moment, that we have the specific case in which X and Y are independent random variables with the uniform probability densities

$$f_X(x) = \begin{cases} 1 & \text{for } 0 \leq x < 1 \\ 0 & \text{otherwise} \end{cases} \tag{3.2}$$

and

$$f_Y(y) = \begin{cases} \frac{1}{2} & \text{for } 0 \leq y < 2 \\ 0 & \text{otherwise} \end{cases} \tag{3.3}$$

respectively. Their joint-probability density is then

$$f_{X,Y}(x,y) = \begin{cases} \frac{1}{2} & \text{for } 0 \leq x < 1, 0 \leq y < 2 \\ 0 & \text{otherwise} \end{cases} \tag{3.4}$$

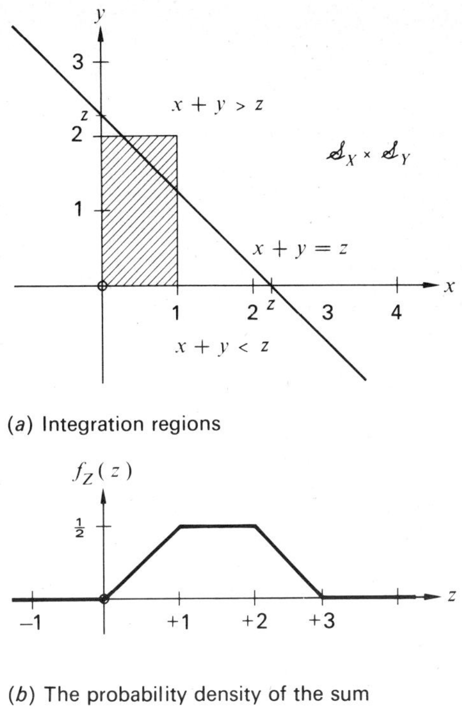

Figure 3.1 Sum of two independent, uniformly distributed random variables.

That is, this joint density has the value $\frac{1}{2}$ in the shaded region shown in Fig. 3.1*a* and is zero elsewhere.

Now the region of the joint sample space $\mathcal{S}_X \times \mathcal{S}_Y$ corresponding to the event

$$[Z \le z] = [X + Y \le z] \tag{3.5}$$

is the set of points which lie on and to the left of the line

$$z = x + y \tag{3.6}$$

plotted in that space. Since the joint density $f_{X,Y}$ here is zero outside the shaded rectangle in Fig. 3.1*a*, we see that if z is such that that rectangle lies entirely to the left of the line (3.6); that is, if

$$z \ge 3$$

then it follows that

$$P[Z \le z] = 1$$

Similarly, if z is such that the rectangle lies entirely to the right of the line (3.6); that is, if

$$z \le 0$$

then it follows that

$$P[Z \leq z] = 0$$

For intermediate values of z, the line (3.6) intersects the rectangle and we then have to calculate the probability given by that portion of the rectangle which lies to the left of the intersecting line. Thus we get the result

$$F_Z(z) = \begin{cases} 1 & \text{for } 3 \leq z \\ 1 - \dfrac{(3-z)^2}{4} & \text{for } 2 \leq z < 3 \\ \dfrac{1}{4} + \dfrac{z-1}{2} & \text{for } 1 \leq z < 2 \\ \dfrac{z^2}{4} & \text{for } 0 \leq z < 1 \\ 0 & \text{for } z < 0 \end{cases} \tag{3.7}$$

On taking the derivative of this result with respect to z, we get

$$f_Z(z) = \begin{cases} 0 & \text{for } 3 \leq z \\ \dfrac{3-z}{2} & \text{for } 2 \leq z < 3 \\ \dfrac{1}{2} & \text{for } 1 \leq z < 2 \\ \dfrac{z}{2} & \text{for } 0 \leq z < 1 \\ 0 & \text{for } z < 0 \end{cases} \tag{3.8}$$

as the probability density of the sum random variable Z. This result is plotted in Fig. 3.1*b* and is consistent, of course, with the fact that the sum cannot be less than zero or greater than three [cf. (3.2) and (3.3)].

While the specific result just derived was easy enough to obtain, it is also rather easy to derive a general formula which holds for any joint-probability density. Just as in the specific case above, we may evaluate the probability of the event $[Z \leq z]$ by integrating the joint-probability density of X and Y over that portion of the $\mathcal{S}_X \times \mathcal{S}_Y$ sample space which lies to the left of the line (3.6). Thus we get

$$\begin{aligned} F_Z(z) &= P[X + Y \leq z] \\ &= \int_{-\infty}^{+\infty} \left[\int_{-\infty}^{z-x} f_{X,Y}(x,y)\, dy \right] dx \end{aligned} \tag{3.9}$$

for the probability distribution function of the sum random variable Z. We may next obtain the probability density of Z by differentiating that result with respect to z:

$$f_Z(z) = \frac{d}{dz}\left\{\int_{-\infty}^{+\infty}\left[\int_{-\infty}^{z-x} f_{X,Y}(x,y)\,dy\right]dx\right\}$$

$$= \int_{-\infty}^{+\infty}\left[\frac{d}{dz}\int_{-\infty}^{z-x} f_{X,Y}(x,y)\,dy\right]dx$$

$$= \int_{-\infty}^{+\infty} f_{X,Y}(x,\,z-x)\,dx \tag{3.10}$$

This is the desired formula for the probability density function of the sum of two random variables. In those cases in which the random variables X and Y are statistically independent, this result reduces to the form

$$f_Z(z) = \int_{-\infty}^{+\infty} f_X(x)f_Y(z-x)\,dx \tag{3.11}$$

the integral on the right-hand side here being known as a *convolution integral*.

In view of the very great importance of the convolution integral, both in theory and in practice, we shall next work out a few specific examples of its use. Sometimes the functional forms of the probability densities f_X and f_Y are such that the integral in (3.11) above is not too difficult to evaluate analytically. For example, suppose that X and Y are independent gaussian random variables with the probability densities

$$f_X(x) = \frac{1}{\sqrt{2\pi}}e^{-x^2/2} \tag{3.12}$$

and

$$f_Y(y) = \frac{1}{\sqrt{2\pi}}e^{-y^2/2} \tag{3.13}$$

respectively. The probability density of their sum Z is then given from (3.11) by the integral

$$f_Z(z) = \frac{1}{2\pi}\int_{-\infty}^{+\infty} e^{-x^2/2}e^{-(z-x)^2/2}\,dx$$

$$= \frac{1}{2\pi}\int_{-\infty}^{+\infty} \exp\left[-\tfrac{1}{2}(z^2 - 2zx + 2x^2)\right]dx$$

In order to evaluate this integral, let us first rearrange the argument of the exponential so that it contains a perfect square in x; thus we get

$$f_Z(z) = \frac{1}{2\pi}\int_{-\infty}^{+\infty} \exp\left\{-\frac{1}{2}\left[\frac{z^2}{2} + \left(\sqrt{2}x - \frac{z}{\sqrt{2}}\right)^2\right]\right\}dx$$

$$= \frac{1}{\sqrt{2\pi}}e^{-z^2/4}\,\frac{1}{\sqrt{2\pi}}\int_{-\infty}^{+\infty}\exp\left[-\tfrac{1}{2}\left(\sqrt{2}x - \frac{z}{\sqrt{2}}\right)^2\right]dx$$

The last integral here may then be evaluated by using a change of variables (say $y = \sqrt{2}x - z/\sqrt{2}$) to give

$$f_Z(z) = \frac{1}{\sqrt{2\pi}} e^{-z^2/4} \frac{1}{\sqrt{2}} \left(\frac{1}{\sqrt{2\pi}} \int_{-\infty}^{+\infty} e^{-y^2/2}\, dy \right)$$

which, since the term in parentheses here evaluates out to unity, finally gives

$$f_Z(z) = \frac{1}{\sqrt{2\pi}\sqrt{2}} e^{-z^2/4} \tag{3.14}$$

as a formula for the probability density of the sum of two independent gaussian random variables whose individual probability densities are as given in (3.12) and (3.13). This result is plotted in Fig. 3.2. As we shall see later, f_Z as given in (3.14) is also a form of a gaussian probability density.

Unfortunately, it seldom works out in practice that the evaluation of the convolution integral can be handled as directly as in the preceding gaussian example and recourse must often be made to essentially graphical techniques. As an example of the use of such techniques, let us consider next the case of the sum of two independent random variables X and Y in which X is uniformly distributed between zero and one:

$$f_X(x) = \begin{cases} 1 & \text{for } 0 \le x < 1 \\ 0 & \text{otherwise} \end{cases} \tag{3.15}$$

and in which Y has the triangular probability density

$$f_Y(y) = \begin{cases} 2y & \text{for } 0 \le y < 1 \\ 0 & \text{otherwise} \end{cases} \tag{3.16}$$

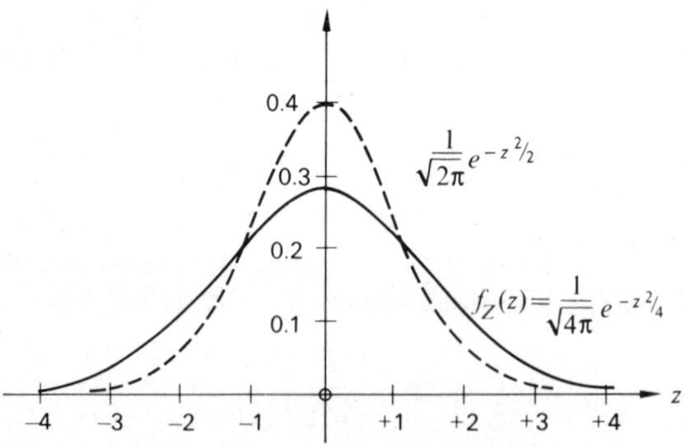

Figure 3.2 The probability density of the sum of two independent gaussian random variables.

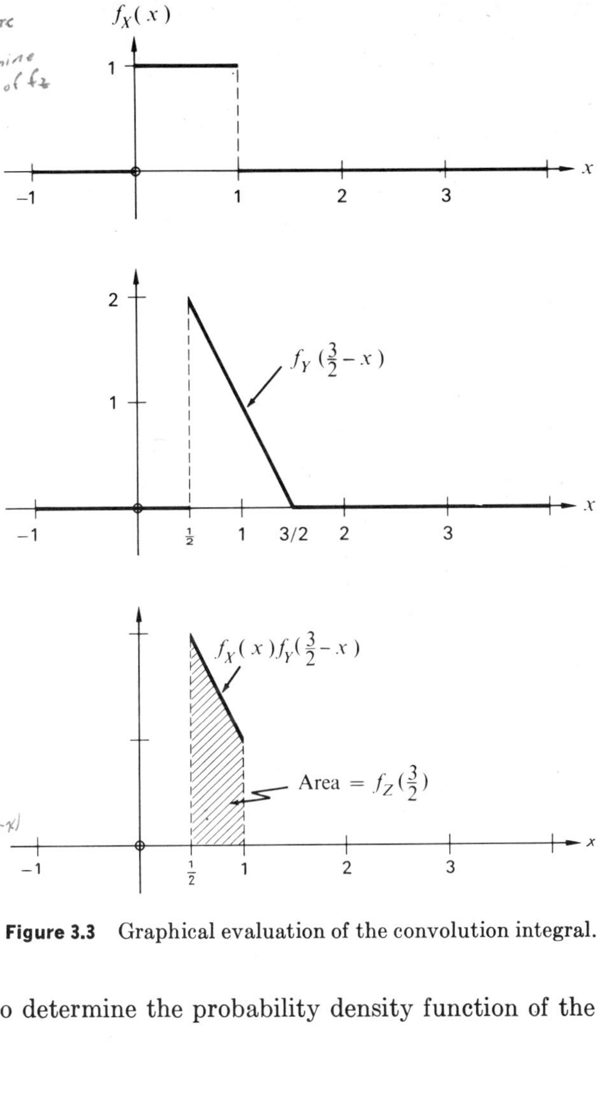

Figure 3.3 Graphical evaluation of the convolution integral.

We would now like to determine the probability density function of the sum

$$Z \triangleq X + Y$$

Suppose first that we wish to evaluate $f_Z(z)$ at the point $z = \frac{3}{2}$. The convolution integral equation (3.4) then becomes

$$f_Z(z) = \int_{-\infty}^{+\infty} f_X(x) f_Y(\tfrac{3}{2} - x)\, dx \tag{3.17}$$

The density $f_X(x)$ here is given directly by (3.15) above, while we may obtain $f_Y(\frac{3}{2} - x)$ from (3.16) by noting that changing the argument of f_Y from y to $\frac{3}{2} - x$ corresponds to reflecting f_Y about the origin (because of the $-x$) and translating it to the right by $1\frac{1}{2}$ units (for $y = 0$ corre-

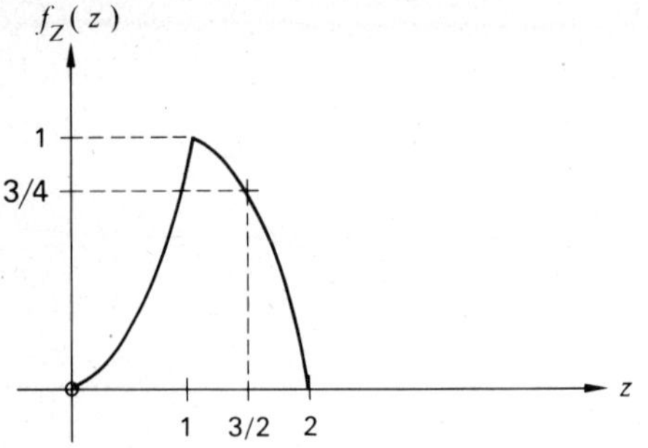

Figure 3.4 The probability density of the sum of a uniformly distributed random variable and a triangularly distributed random variable.

sponds to $\frac{3}{2} - x = 0$, or to $x = \frac{3}{2}$). These two functions and their product are sketched in Fig. 3.3. Since the integral in (3.17) is simply the area under the product function, we then get

$$\begin{aligned} f_Z(\tfrac{3}{2}) &= (1 - \tfrac{1}{2})1 + \tfrac{1}{2}(1 - \tfrac{1}{2})^2 \\ &= \tfrac{3}{4} \end{aligned}$$

(cf. Fig. 3.4). On continuing this process for other values of z, we finally get

$$f_Z(z) = \begin{cases} 2z - z^2 & \text{for } 1 \leq z < 2 \\ z^2 & \text{for } 0 \leq z < 1 \\ 0 & \text{otherwise} \end{cases}$$

as a formula for the probability density of the sum of the two random variables X and Y which have the probability densities given in (3.15) and (3.16), respectively. This result is plotted in Fig. 3.4.

Exercise 3.1 The random variables X and Y are statistically independent and have the uniform probability densities

$$f_X(x) = \begin{cases} 1 & \text{for } 0 \leq x < 1 \\ 0 & \text{otherwise} \end{cases}$$

and

$$f_Y(y) = \begin{cases} \frac{1}{2} & \text{for } 0 \leq y < 2 \\ 0 & \text{otherwise} \end{cases}$$

respectively. Let Z be the sum random variable

$$Z \triangleq X + Y$$

Determine the probability density of the sum random variable by using the convolution integral method. Plot your result and compare with (3.16). □

Much of probability and random process theory deals with sums of independent random variables. With this fact in mind, we now wish to derive a few more results concerning such sums.

Exercise 3.2 Let X, Y, and Z be independent random variables with the probability densities f_X, f_Y, and f_Z, respectively. Let the random variable W be defined as the sum of the random variables X and Y:

$$W \triangleq X + Y$$

Show that W and Z are independent random variables. □

The restriction in Exercise 3.2 to random variables which have probability densities was done purely for the sake of manipulative convenience; the final result holds just as well for the case of independent random variables with general probability distribution functions.

Exercise 3.3 Let the random variables X_1, X_2, and X_3 be statistically independent and let each have a uniform probability density, in particular, let

$$f_{X_i}(x_i) = \begin{cases} 1 & \text{for } 0 \leq x_i \leq 1 \\ 0 & \text{otherwise} \end{cases}$$

where $i = 1, 2, 3$. Let Y be the sum of these random variables:

$$Y \triangleq X_1 + X_2 + X_3$$

Determine the probability density f_Y of the sum random variable Y. (*Hint:* Determine the probability density of $W = X_1 + X_2$, then use the result of Exercise 3.2 and determine the probability density of $W + X_3$.) □

Exercise 3.4 Let the random variables $X_1, X_2, \ldots, X_n$ be independent gaussian random variables with the probability densities

$$f_{X_i}(x_i) = \frac{e^{-x_i^2/2}}{\sqrt{2\pi}}$$

for $i = 1, 2, \ldots, n$, respectively. Let the random variable Y be the sum of those random variables:

$$Y \triangleq \sum_{i=1}^{n} X_i$$

Determine the probability density f_Y of the sum random variable Y. (*Hint:* Using the result in (3.14) above for the case $n = 2$, determine f_Y for the case $n = 3$ by making use of the result of Exercise 3.2. The general result may then be determined by using induction.) □

Other scalar functions

The basic techniques used above to determine the probability density of a sum of independent random variables may equally well be applied to the determination of the statistical properties of other scalar functions of random vectors. One such function of interest is the product. For example, let x be the measured value of the width and let y be the measured value of the height of the cross section of a rectangular rod; the product xy is then the corresponding value of the cross-sectional area of the rod. If the random variable X gives the distribution of measured values x and the random variable Y gives the distribution of measured values y, then the random variable $Z = XY$ gives the distribution of the corresponding values of the cross-sectional area. With this example in mind, let us next study a few problems involving products of random variables.

Exercise 3.5 Let X and Y be independent random variables with the uniform probability densities

$$f_X(x) = \begin{cases} \dfrac{1}{A} & \text{for } 0 \leq x \leq A \\ 0 & \text{otherwise} \end{cases}$$

and

$$f_Y(y) = \begin{cases} \dfrac{1}{B} & \text{for } 0 \leq y \leq B \\ 0 & \text{otherwise} \end{cases}$$

respectively. Let Z be the product random variable:

$$Z \triangleq XY$$

a. Determine the probability distribution function F_Z of the product random variable Z.

b. Determine the probability density f_Z of the product random variable Z.

c. Plot your results for the case $A = 1$, $B = 2$. (*Hint:* Sketch the appropriate regions in $\mathcal{S}_X \times \mathcal{S}_Y$ as we did in Fig. 3.1*a* for the case of the sum of two random variables.) □

Exercise 3.6 Let X and Y be independent random variables with the probability densities f_X and f_Y, respectively. Let the new random variable Z be defined as the product of X and Y.

$$Z \triangleq XY \tag{3.18}$$

Show that the probability density of the product random variable is given by

$$f_Z(z) = \int_{-\infty}^{+\infty} \frac{1}{|x|} f_X(x) f_Y\left(\frac{z}{x}\right) dx \tag{3.19}$$

(*Hint:* Sketch the appropriate regions in $\mathcal{S}_X \times \mathcal{S}_Y$, remembering that the product may be either positive or negative.) □

The following two exercises concerning products of random variables are rather more difficult in that their solution requires some skill in manipulating and evaluating odd-looking integrals. The techniques to be used are essentially the same as those used in solving Exercises 3.5 and 3.6, however, and the results are important.

Exercise 3.7 Let X and Y be independent random variables with the probability densities

$$f_X(x) = \begin{cases} \dfrac{1}{\pi\sqrt{1 - x^2}} & \text{for } -1 < x < +1 \\ 0 & \text{elsewhere} \end{cases} \tag{3.20}$$

and

$$f_Y(y) = \begin{cases} y\, e^{-y^2/2} & \text{for } 0 \leq y \\ 0 & \text{for } y < 0 \end{cases} \tag{3.21}$$

respectively. Let Z be the product random variable XY.

Show that Z has the probability density

$$f_Z(z) = \frac{1}{\sqrt{2\pi}} e^{-z^2/2} \tag{3.22}$$

that is, that Z is a gaussian random variable. □

A moment's reflection upon the results of Exercises 2.7 and 3.7 shows that together they prove that: *If the random variable Z_t is such that*

$$Z_t \triangleq Y \sin(\omega t + \Phi) \tag{3.23}$$

for all t, where Y has a Rayleigh probability density function as in (3.21), *where ω is a constant, where Φ is uniformly distributed over $-\pi$ to $+\pi$, and where Y and Φ are independent random variables, then Z_t is a gaussian random variable with the probability density given in* (3.22).

Exercise 3.8 Let X and Y be independent random variables with the probability densities

$$f_X(x) = \frac{1}{\sqrt{2\pi}} e^{-x^2/2} \tag{3.24}$$

and

$$f_Y(y) = \begin{cases} ye^{-y^2/2} & \text{for } 0 \le y \\ 0 & \text{otherwise} \end{cases} \tag{3.25}$$

respectively. Again let Z be the product random variable XY.

Show that Z has the probability density

$$f_Z(z) = \tfrac{1}{2}e^{-|z|} \tag{3.26}$$

that is, that Z is an exponential random variable. □

Closely related to the problem of finding the probability density of a product of two random variables is that of finding the probability density of their quotient.

Exercise 3.9 Let X and Y be independent normal random variables; in particular, let

$$f_X(x) = \frac{1}{\sqrt{2\pi}} e^{-x^2/2}$$

and (3.27)

$$f_Y(y) = \frac{1}{\sqrt{2\pi}} e^{-y^2/2}$$

Suppose that the random variable Z is defined to be the quotient of Y and X; that is, let

$$Z \triangleq \frac{Y}{X} \tag{3.28}$$

Show that in this case

$$f_Z(z) = \frac{1}{\pi(1 + z^2)} \tag{3.29}$$

That is, *the quotient of two independent, identically distributed, normal random variables has a Cauchy probability density.* □

Vector functions

The only functions of random vectors which we have studied so far in this section have been scalar-valued real functions. These functions are, of course, by no means the only functions of interest. Some interesting and useful simple results concerning vector-valued functions of random vectors are as follows.

Exercise 3.10 Let $\mathbf{X}$ be a two-dimensional random vector (X_1,X_2) whose components X_1 and X_2 are independent random variables. Let

Y be the two-dimensional random vector (Y_1, Y_2) whose components are the random variables

$$Y_1 \triangleq aX_1 + b \qquad \text{and} \qquad Y_2 \triangleq cX_2 + d$$

respectively, where a, b, c, and d are real constants and where a and c are positive.

Show that the random variables Y_1 and Y_2 are statistically independent. □

Exercise 3.10 is a special case of the following result.

Exercise 3.11 Let **X** be a two-dimensional random vector (X_1, X_2) whose components X_1 and X_2 are independent random variables. Let **Y** be the two-dimensional random vector (Y_1, Y_2) whose components are the random variables

$$Y_1 \triangleq g_1(X_1) \qquad \text{and} \qquad Y_2 \triangleq g_2(X_2)$$

respectively; where g_1 and g_2 are Borel functions defined on $\mathcal{S}_{X_1}$ and $\mathcal{S}_{X_2}$, respectively (cf. Sec. 2 above).

Show that the random variables Y_1 and Y_2 are statistically independent. □

4 ONE-TO-ONE TRANSFORMATIONS

The problem of determining the probabilistic properties of a function of a random variable becomes particularly simple when the transformation of concern is one-to-one and continuously differentiable, and when the probability distribution function of the original random variable is differentiable almost everywhere. For example, let us first consider the simplest possible case, that in which a new random variable Y is defined as a monotone-increasing differentiable function g of a random variable X:

$$Y \triangleq g(X) \tag{4.1}$$

This is the situation pictured in Fig. 4.1. In this case we have

$$\begin{aligned} F_Y(y) &= P[Y \leq y] \\ &= P[X \leq h(y)] \end{aligned}$$

where h is the inverse transformation

$$h(y) \triangleq g^{-1}(y) \tag{4.2}$$

which exists for all y since g is a one-to-one transformation by hypothesis. If now we assume that the original random variable X has a probability

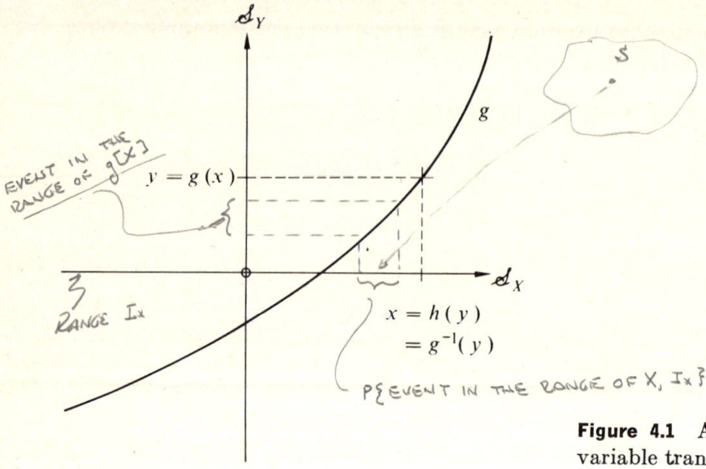

Figure 4.1 A one-to-one single-variable transformation.

density f_X, we may write

$$F_Y(y) = \int_{-\infty}^{h(y)} f_X(x)\,dx$$

On differentiating both sides of this result with respect to y (using the chain rule of differential calculus), we then get

$$f_Y(y) = f_X(h(y))\,\frac{dh}{dy} \tag{4.3}$$

as a formula for the probability density of the new random variable Y. If, on the other hand, g had been defined as a differentiable monotone-*decreasing* function with an inverse h, we would have obtained

$$F_Y(y) = \int_{h(y)}^{+\infty} f_X(x)\,dx$$

which then would have led to the result

$$f_Y(y) = -f_X(h(y))\,\frac{dh}{dy}$$

Since in this case dh/dy is negative [it is positive in (4.3)], we can express both this result and that in (4.3) by the single formula

$$f_Y(y) = f_X(h(y))\left|\frac{dh}{dy}\right| \tag{4.4}$$

This then is the desired general formula for the probability density of a one-to-one differentiable function of a random variable.

Let us now apply this technique to the determination of the probability density of the random variable Y defined in terms of the random variable X by the transformation

$$Y \triangleq aX + b \tag{4.5}$$

where a and b are real constants, where a is nonzero, and where the probability density f_X of X exists almost everywhere (cf. Exercise 2.3 above). Since a is nonzero here by hypothesis, the inverse transformation h is readily obtained from (4.5) and is

$$X = h(Y) = \frac{Y - b}{a} \tag{4.6}$$

On using this inverse transformation in (4.4) we then get

$$f_Y(y) = f_X\left(\frac{y - b}{a}\right)\frac{1}{|a|} \tag{4.7}$$

in this case [cf. Eq. (2.8)].

Exercise 4.1 Let X be a gaussian random variable with the probability density

$$f_X(x) = \frac{1}{\sqrt{2\pi}} e^{-x^2/2}$$

Let Y be a new random variable defined as the cube of X:

$$Y \triangleq X^3$$

Determine the probability density of Y and plot your result. □

The two-dimensional case

The technique for solving the two-dimensional problem is essentially that used above in the one-dimensional case; only the details are more complicated. Let **X** be a two-dimensional random vector whose components X_1 and X_2 have the joint-probability density f_{X_1,X_2}. Let **Y** be the two-dimensional random vector defined by a one-to-one continuously differentiable vector transformation **g** of the random vector **X**; that is, let

$$\begin{aligned}\mathbf{Y} \triangleq \mathbf{g}(\mathbf{X}) &= (g_1(X_1,X_2),g_2(X_1,X_2)) \\ &= (Y_1,Y_2)\end{aligned} \tag{4.8}$$

Since **g** is a one-to-one transformation by hypothesis, its inverse-transformation **h** exists and we may therefore express **X** as a function of **Y** by the formula

$$\begin{aligned}\mathbf{X} &= \mathbf{g}^{-1}(\mathbf{Y}) \triangleq \mathbf{h}(\mathbf{Y}) \\ &= (h_1(Y_1,Y_2),h_2(Y_1,Y_2)) \\ &= (X_1,X_2)\end{aligned} \tag{4.9}$$

We would now like to determine the joint-probability density of the component random variables Y_1 and Y_2.

To this end, let us consider a small rectangle A in the new sample space $\mathcal{S}_Y$ and its inverse image under the mapping $\mathbf{g}$, that is, the region

$$\mathbf{h}(A) \triangleq \mathbf{g}^{-1}(A) \tag{4.10}$$

in the original sample space $\mathcal{S}_X$. This is the situation shown schematically in Fig. 4.2.

Now it follows from (4.8) that a point $(x_1,x_2) \in \mathcal{S}_X$ is mapped into a point $(y_1,y_2) \in \mathcal{S}_Y$ by the transformation

$$y_1 = g_1(x_1,x_2)$$

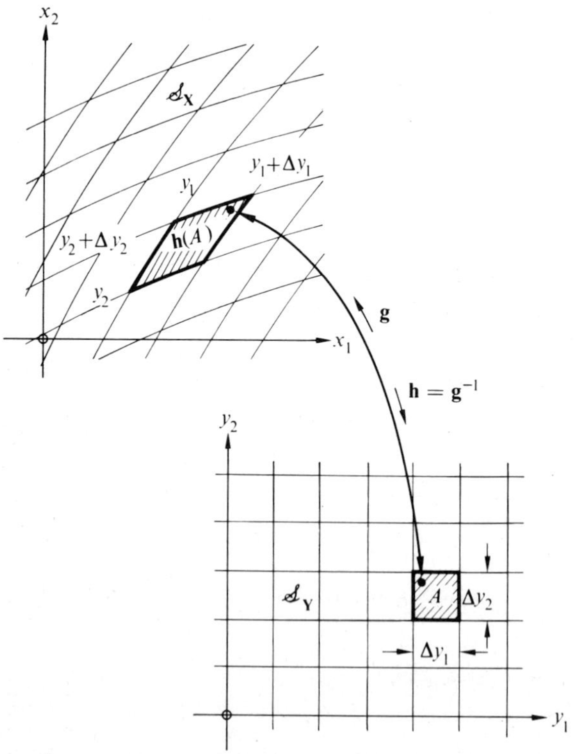

Figure 4.2 The two-dimensional mapping of incremental areas.

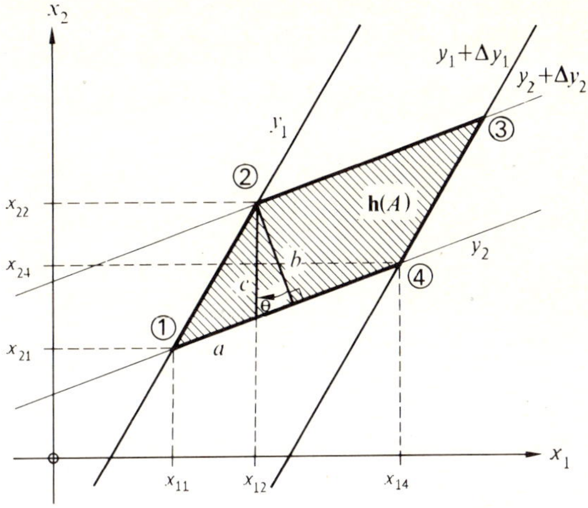

Figure 4.3 The incremental region $\mathbf{h}(A)$.

and (4.11)

$$y_2 = g_2(x_1,x_2)$$

while it follows from (4.9) that the point $(y_1,y_2) \in \mathcal{S}_Y$ is mapped back into $(x_1,x_2) \in \mathcal{S}_X$ by the inverse transformation

$$x_1 = h_1(y_1,y_2)$$

and (4.12)

$$x_2 = h_2(y_1,y_2)$$

since $\mathbf{h}$ is defined to be the inverse of $\mathbf{g}$. As y_1 and y_2 vary, Eqs. (4.12) describe contours in $\mathcal{S}_X$ corresponding to the rectangular grid formed by y_1 constant and y_2 constant in $\mathcal{S}_Y$; these are the curved contours in $\mathcal{S}_X$ shown in Fig. 4.2. Since $\mathbf{h}(A)$ is a mapping of A, it follows that

$$P[\mathbf{X} \in \mathbf{h}(A)] = P[\mathbf{Y} \in A]$$

If now the rectangle $A \subset \mathcal{S}_Y$ is small enough, we may write with a reasonable degree of accuracy

$$f_{Y_1,Y_2}(y_1,y_2)\,\Delta y_1\,\Delta y_2 = f_{X_1,X_2}(h_1(y_1,y_2),h_2(y_1,y_2))\,[\text{area of } \mathbf{h}(A)] \quad (4.13)$$

Our remaining problem then is to determine the area of the inverse-image $\mathbf{h}(A)$.

Let us, for the moment, expand the picture of the incremental region $\mathbf{h}(A)$ in the sample space $\mathcal{S}_X$ as we have done in Fig. 4.3. As $\mathbf{h}(A)$ is essentially a parallelogram, its area is equal to the product of its base a times its altitude b:

$$\text{Area } \mathbf{h}(A) = ab$$

Now examination of Fig. 4.3 shows that

$$b = c \cos \theta = c \, \frac{x_{14} - x_{11}}{a}$$

Hence, we may write

$$\text{Area } \mathbf{h}(A) = c(x_{14} - x_{11})$$

The length c may now be evaluated in terms of the coordinates of the corners of $\mathbf{h}(A)$ to give

$$\begin{aligned} c &= (x_{22} - x_{21}) - (x_{24} - x_{21}) \frac{x_{12} - x_{11}}{x_{14} - x_{11}} \\ &= \frac{(x_{22} - x_{21})(x_{14} - x_{11}) - (x_{24} - x_{21})(x_{12} - x_{11})}{x_{14} - x_{11}} \end{aligned}$$

It then follows that we get

$$\begin{aligned} \text{Area } \mathbf{h}(A) &= (x_{22} - x_{21})(x_{14} - x_{11}) - (x_{24} - x_{21})(x_{12} - x_{11}) \\ &= \begin{vmatrix} x_{14} - x_{11} & x_{12} - x_{11} \\ x_{24} - x_{21} & x_{22} - x_{21} \end{vmatrix} \end{aligned} \tag{4.14}$$

where we have expressed the result in determinant form for later convenience.

Let us next express the various terms in this result in terms of the corresponding changes in y_1 and y_2. Thus, on referring again to Fig. 4.3, we see that

$$\begin{aligned} x_{14} - x_{11} &= h_1(y_1 + \Delta y_1, y_2) - h_1(y_1, y_2) \\ &\approx \frac{\partial h_1}{\partial y_1} \Delta y_1 \end{aligned}$$

the latter result following from our assumption about the smallness of Δy_1. In an entirely similar manner, we find that we get

$$x_{12} - x_{11} \approx \frac{\partial h_1}{\partial y_2} \Delta y_2$$

$$x_{24} - x_{21} \approx \frac{\partial h_2}{\partial y_1} \Delta y_1$$

and

$$x_{22} - x_{21} \approx \frac{\partial h_2}{\partial y_2} \Delta y_2$$

On substituting these various results into (4.14) and then substituting that result back into (4.13), we finally find that the joint-probability density of the component random variables Y_1 and Y_2 is given by the formula

$$f_{Y_1,Y_2}(y_1,y_2) = f_{X_1,X_2}(h_1(y_1,y_2),h_2(y_1,y_2)) \left| \frac{\partial(h_1,h_2)}{\partial(y_1,y_2)} \right| \tag{4.15}$$

where we have used the notation

$$\frac{\partial(h_1,h_2)}{\partial(y_1,y_2)} \triangleq \begin{vmatrix} \dfrac{\partial h_1}{\partial y_1} & \dfrac{\partial h_1}{\partial y_2} \\ \dfrac{\partial h_2}{\partial y_1} & \dfrac{\partial h_2}{\partial y_2} \end{vmatrix} \tag{4.16}$$

for the *Jacobian*[1] of the transformation **h**. The magnitude of the Jacobian is taken in (4.15) for the same reason that we had to use the magnitude of the derivative in (4.14): to keep this scale factor positive. Equation (4.15) is the desired two-dimensional result.

A particularly important one-to-one transformation is the rotational transformation:

$$Y_1 \triangleq X_1 \cos\theta - X_2 \sin\theta$$

and (4.17)

$$Y_2 \triangleq X_1 \sin\theta + X_2 \cos\theta$$

in which θ is the specified angle of rotation, where X_1 and X_2 are the components of the two-dimensional random vector **X**, and where Y_1 and Y_2 are the components of the random vector **Y** which results from the rotational transformation of **X**.

Exercise 4.2

a. Determine the inverse-transformation **h(Y)** corresponding to the rotational transformation defined by (4.17), where

$$\mathbf{X} = \mathbf{h}(\mathbf{Y})$$

b. Determine the Jacobian of this inverse transformation. □

Exercise 4.3 Let the random variables X_1 and X_2 in (4.17) be independent gaussian random variables with the probability densities

$$f_{X_1}(x_1) = \frac{1}{\sqrt{2\pi}\,\sigma_1} e^{-x_1^2/2\sigma_1^2}$$

[1] The symbol $\partial(h_1,h_2)/\partial(y_1,y_2)$ used above for the Jacobian of the transformation **h** was deliberately chosen over the more common, and more compact, symbol J simply because it is more explicit. The added explicitness is often helpful in calculations.

and

$$f_{X_2}(x_2) = \frac{1}{\sqrt{2\pi}\,\sigma_2} e^{-x_2^2/2\sigma_2^2}$$

respectively, where σ_1 and σ_2 are positive real constants such that $\sigma_1 \neq \sigma_2$.

Determine the joint-probability density of the component random variables Y_1 and Y_2. (Note that the result is called a *bivariate gaussian* probability density.) □

Exercise 4.4 Let **X** be a two-dimensional random vector whose components X_1 and X_2 are independent gaussian random variables with the probability densities

$$f_{X_1}(x_1) = \frac{1}{\sqrt{2\pi}} e^{-x_1^2/2} \qquad \text{and} \qquad f_{X_2}(x_2) = \frac{1}{\sqrt{2\pi}} e^{-x_2^2/2}$$

respectively. Let **Y** be a two-dimensional random vector whose components are defined by the transformation

$$Y_1 \triangleq \sqrt{X_1^2 + X_2^2} \qquad \text{and} \qquad Y_2 \triangleq \tan^{-1}\frac{X_2}{X_1}$$

a. Determine the inverse transformation $\mathbf{h} = (h_1,h_2)$, where

$$X_1 = h_1(Y_1,Y_2) \qquad \text{and} \qquad X_2 = h_2(Y_1,Y_2)$$

(*Hint:* Draw a figure relating X_1, X_2, Y_1, and Y_2.)

b. Determine the Jacobian of the transformation **h**.

c. Determine the joint-probability density of the random variables Y_1 and Y_2. □

Exercise 4.5 Let **X** and **Y** be the two-dimensional random vectors defined in Exercise 4.4.

a. Show that Y_1 has the Rayleigh probability density

$$f_{Y_1}(y_1) = \begin{cases} y_1\, e^{-y_1^2/2} & \text{for } 0 \le y_1 \\ 0 & \text{otherwise} \end{cases}$$

b. Show that Y_2 has the uniform probability density

$$f_{Y_2}(y_2) = \begin{cases} \dfrac{1}{2\pi} & \text{for } 0 \le y_2 \le 2\pi \\ 0 & \text{otherwise} \end{cases}$$

c. Show that Y_1 and Y_2 are independent random variables. □

Once we have obtained the formula given in (4.15) for the probability density of the result of a one-to-one transformation of a two-dimensional random vector, a number of the general results previously obtained in Sec. 3 become easier to derive.

Exercise 4.6 Let $\mathbf{X}$ be a two-dimensional random vector whose components X_1 and X_2 are independent random variables with the probability densities f_{X_1} and f_{X_2}, respectively. Let $\mathbf{Y}$ be a two-dimensional random vector whose components are defined by the transformation

$$Y_1 \triangleq g_1(X_1) \qquad \text{and} \qquad Y_2 \triangleq g_2(X_2)$$

where g_1 and g_2 are differentiable one-to-one transformations.

Show, using the Jacobian, that the joint-probability density f_{Y_1,Y_2} factors, that is, that Y_1 and Y_2 are independent random variables. (Note that this exercise gives a different proof of a special subcase of the result in Exercise 3.10.) □

Exercise 4.7 Let $\mathbf{X}$ be a two-dimensional random vector whose components X_1 and X_2 are independent random variables with the probability densities f_{X_1} and f_{X_2}, respectively. Let $\mathbf{Y}$ be a two-dimensional random vector whose components are defined by the transformation

$$Y_1 \triangleq X_1 \qquad \text{and} \qquad Y_2 \triangleq X_1X_2$$

a. Find the Jacobian of the inverse transformation $\mathbf{h}$, where

$$\mathbf{X} = \mathbf{h}(\mathbf{Y})$$

b. Using (*a*), determine the joint-probability density of the component random variables Y_1 and Y_2.

c. Using (*b*), determine the probability density of the product $Y_2 = X_1X_2$.

(Note that this exercise provides an alternative method of deriving the product result in Exercise 3.6.) □

The n-dimensional case

Our above one- and two-dimensional results may now be extended by induction to the case of the general n-dimensional transformation. Let $\mathbf{X}$ be an n-dimensional random vector whose components $X_1, X_2, \ldots, X_n$ have the joint-probability density $f_{\mathbf{X}}$. Let $\mathbf{g}$ be a one-to-one continuously differentiable vector transformation which maps $\mathbf{X}$ into a new n-dimensional random vector $\mathbf{Y}$:

$$\mathbf{Y} \triangleq \mathbf{g}(\mathbf{X}) \tag{4.18}$$

If now we denote the inverse transformation to **g** by **h** and write

$$\mathbf{X} = \mathbf{h}(\mathbf{Y}) \tag{4.19}$$

then it can be shown[1] that

$$f_{\mathbf{Y}}(\mathbf{y}) = f_{\mathbf{X}}(\mathbf{h}(\mathbf{y})) \left| \frac{\partial(\mathbf{h})}{\partial(\mathbf{y})} \right| \tag{4.20}$$

is a formula for the joint-probability density of the n-dimensional random vector Y, where

$$\frac{\partial(\mathbf{h})}{\partial(\mathbf{y})} \triangleq \begin{vmatrix} \frac{\partial h_1}{\partial y_1} & \frac{\partial h_1}{\partial y_2} & \cdots & \frac{\partial h_1}{\partial y_n} \\ \frac{\partial h_2}{\partial y_1} & \frac{\partial h_2}{\partial y_2} & \cdots & \frac{\partial h_2}{\partial y_n} \\ \cdots & \cdots & \cdots & \cdots \\ \frac{\partial h_n}{\partial y_1} & \frac{\partial h_n}{\partial y_2} & \cdots & \frac{\partial h_n}{\partial y_n} \end{vmatrix} \tag{4.21}$$

is the Jacobian of the transformation **h**.

5 SUMMARY

Suppose that we are given a random variable X with the probability distribution function F_X and that we are asked to determine the probability distribution function of some measurable function of X, say

$$Y \triangleq g(X) \tag{2.1}$$

Then it follows that

$$F_Y(y) = P[\{x: g(x) \leq y,\ x \in \mathcal{S}_X\}] \tag{2.2}$$

and, if the derivative exists,

$$f_Y(y) = \frac{d}{dy} P[\{x: g(x) \leq y,\ x \in \mathcal{S}_X\}] \tag{2.3}$$

If X has a probability density f_X almost everywhere and if the transformation g is one-to-one, then

$$f_Y(y) = f_X(h(y)) \left| \frac{dh}{dy} \right| \tag{4.4}$$

where

$$h(y) \triangleq g^{-1}(y) \tag{4.2}$$

[1] See, for example, sec. 10-9, Change of Variables in Multiple Integrals, pp. 270ff, of Apostol.

that is, h is the inverse to the transformation g. Similar formulas are obtained in the case of random vectors.

Two special results of particular importance are as follows:

a. If the random variables X and Y are statistically independent and if their densities f_X and f_Y, respectively, exist almost everywhere, then the probability density of their *sum*

$$Z \triangleq X + Y \tag{3.1}$$

is given by the formula

$$f_Z(z) = \int_{-\infty}^{+\infty} f_X(x) f_Y(z - x)\, dx \tag{3.2}$$

that is, by a convolution integral.

b. If again the random variables X and Y are statistically independent and if their probability densities f_X and f_Y, respectively, exist almost everywhere, then the probability density of their product

$$Z \triangleq XY \tag{3.18}$$

is given by the formula

$$f_Z(z) = \int_{-\infty}^{+\infty} \frac{1}{|x|} f_X(x) f_Y\left(\frac{z}{x}\right) dx \tag{3.19}$$

REFERENCES

Apostol, Tom M.: "Mathematical Analysis," Addison-Wesley Publishing Company, Inc., Reading, Mass., 1957.

Chambers, R. P.: Random-number generation on digital computers, *IEEE Spectrum*, pp. 48–56, February, 1967.

Loève, Michel: "Probability Theory," 3d ed., D. Van Nostrand Company, Inc., Princeton, N.J., 1963.

National Bureau of Standards, "Handbook of Mathematical Functions," Applied Mathematics Series, AMS 55, U.S. Government Printing Office, 1964.

Pfeiffer, Paul E.: "Concepts of Probability Theory," McGraw-Hill Book Company, New York, 1965.

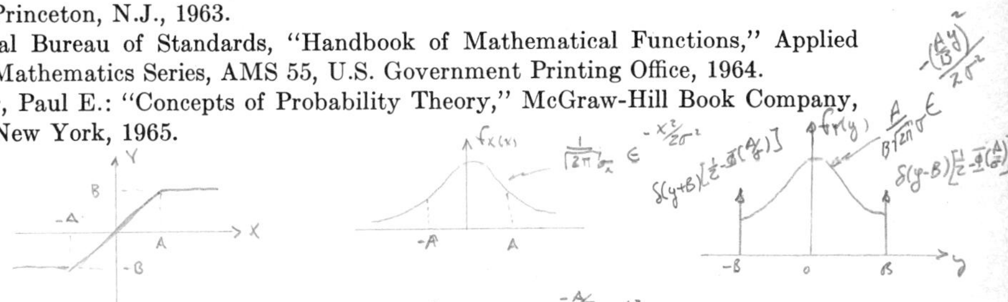

7
Statistical Averages

1 DISCRETE RANDOM VARIABLES

We introduced our subject of probability and random processes in Chap. 1 by starting with a discussion of empirical averages. We pointed out there that it was often convenient to express such averages in terms of distributions of relative frequencies and, with this in mind, then introduced probability as a model for relative frequency. Having discussed probability and its various ramifications at length in Chaps. 2 through 6, let us now return again to our introductory topic of averages.

Suppose for the moment that we are concerned with an experiment whose possible outcomes x_k, $k = 1, 2, 3, \ldots$, are described by the discrete random variable X which has the probability distribution $P[X = x_k]$. Suppose that we perform that experiment n times under essentially identical conditions and that the outcome x_k occurs n_k times in those n performances. As pointed out in Sec. 1.3, the *empirical average*

$\bar{x}$ of that specific set of data may then be calculated by using the formula

$$\bar{x} = \sum_k x_k \frac{n_k}{n} \tag{1.1}$$

where the summation is taken over all possible values of the index k [cf. (1.3.1)]. Keeping in mind the fact that *probability* is a model for *relative frequency*, we now define the *statistical average* $E[X]$ of the discrete random variable X by the formula

$$E[X] \triangleq \sum_k x_k P[X = x_k] \tag{1.2}$$

thus paralleling the definition of the empirical average [cf. (1.5.1)]. This statistical average is also often called the *mean*, or the *expectation* (whence the symbol E), or the *expected value* of the random variable X.

Finally, suppose that we repeat a number of times the procedure of performing n times the experiment whose outcomes are characterized by the random variable X (and calculating the corresponding empirical average), while increasing n with each successive sequence of experiments. As n becomes larger and larger, we would expect that the resultant values of the empirical average should become closer and closer to the value of the statistical average. While we are not yet in a position to study this convergence problem, we shall show in the next chapter that empirical averages do indeed converge to statistical averages under a wide variety of conditions.

Having defined the statistical average of a discrete random variable, let us now calculate that average for a few random variables of interest. First, let us consider the trivial (but nonetheless important) case in which the random variable X is a constant, say

$$X = a \tag{1.3}$$

where a is some real number. In this case X assumes the single value a with probability one and hence it follows from (1.2) that its expectation is simply that value. Thus we see that

$$E[a] = a \tag{1.4}$$

For a second simple example, let us consider the case of a Bernoulli random variable. That is, let us consider the random variable X which is the event indicator which has the value unity when an event of interest[1] occurs and has the value zero otherwise. Suppose that

$$P[X = 1] = p$$

[1] For example, the occurrence of a head in a flip of a coin, or the occurrence of a pulse in a communications link.

and that (1.5)

$$P[X = 0] = 1 - p$$

If now we use this distribution in (1.2), we get

$$E[X] = 1 \times p + 0 \times (1 - p)$$
$$= p \tag{1.6}$$

as the statistical average of the Bernoulli random variable X. It is interesting to note that it follows from this result that since X can assume here only the values zero and one, (1.6) shows that whenever $0 < p < 1$ the expected value is never achievable and hence that in this case, at least, that name is an inappropriate one for the statistical average.

Before we proceed to calculate further expected values it is perhaps worthwhile to reflect for a moment upon the mathematical nature of the expectation. First, we note that the expectation of a random variable is a number which helps characterize that particular random variable, for example, "the random variable X has a zero mean value." We shall therefore want to know the expected values of a number of the more common random variables. On the other hand, we also note that the expectation has the basic property that it operates on a function (in particular, on a function defined upon a sample space, that is, on a random variable) and produces a real number; such a mathematical *operator* is commonly called a *functional*. Now whenever a particular functional is as widely used as is the expectation, it is highly desirable to determine and attempt to understand its general properties, for example, whether it is linear or not, etc. We shall concern ourselves with both of these aspects of the statistical average in the remainder of this chapter, for both are important.

Exercise 1.1 Suppose that we wish to determine the average number of data blocks stored in a buffer storage unit. The unit under study has a maximum storage capacity of n data blocks, and it has been determined that the unit is just as likely to have any one number of blocks in storage as any other number. Determine the statistical average of the number of data blocks in storage. □

While the determination of the statistical average of a discrete random variable is simple in concept, it is often messy in detail. For example, let Y_n be a binomial random variable[1] and suppose that we wish to determine its mean value. That is, suppose that Y_n is a discrete random variable with the possible values $0, 1, 2, \ldots, n$ and with

[1] See, for example, Sec. 9.2, Binomial Process.

the probabilities

$$P[Y_n = k] = \binom{n}{k} p^k q^{n-k}$$
$$= \frac{n!}{k!(n-k)!} p^k q^{n-k} \qquad (1.7)$$

for $k = 0, 1, 2, \ldots, n$, respectively, where $q \triangleq 1 - p$. It then follows from (1.2) that

$$E[Y_n] = \sum_{k=0}^{n} k \frac{n!}{k!(n-k)!} p^k q^{n-k}$$

Now whenever a sum like this is to be evaluated (unless the answer is recognized immediately), it is probably useful to expand the sum and see if a familiar result emerges. Thus:

$$\begin{aligned} E[Y_n] &= 0 \frac{n!}{0!(n-0)!} p^0 q^{n-0} + 1 \frac{n!}{1!(n-1)!} pq^{n-1} \\ &\quad + 2 \frac{n!}{2!(n-2)!} p^2 q^{n-2} + \cdots \\ &\quad + (n-1) \frac{n!}{(n-1)!1!} p^{n-1} q + n \frac{n!}{n!0!} p^n \\ &= npq^{n-1} + n(n-1)p^2 q^{n-2} + \cdots + (n-1)np^{n-1}q + np^n \end{aligned}$$

We note that each term here has the common factor np and hence that we may write

$$E[Y_n] = np[q^{n-1} + (n-1)pq^{n-2} + \cdots + (n-1)p^{n-2}q + p^{n-1}]$$

A moment's reflection upon the bracketed term in this last equation should (if you remember the binomial expansion formula) lead to the result

$$\begin{aligned} q^{n-1} + (n-1)pq^{n-2} + \cdots + (n-1)p^{n-2}q + p^{n-1} &= (p+q)^{n-1} \\ &= 1 \end{aligned}$$

the last equality following from the fact that $q = 1 - p$. On combining these various results, we finally see that the expectation of the binomial random variable is

$$E[Y_n] = np \qquad (1.8)$$

and hence that the mean value increases linearly with the parameter n.

In the preceding examples (and in Exercise 1.1), the random variable under discussion had only a finite number of possible values. A case in which the random variable of interest has a countably infinite number of possible values occurs in the next exercise.

Exercise 1.2 Many physical phenomena, for example, the emission of electrons by a heated cathode, are describable in terms of a counting random variable N_t which can assume any nonnegative integer value and which has the Poisson probability distribution

Poisson

$$P[N_t = k] = \frac{e^{-\lambda t}(\lambda t)^k}{k!} \tag{1.9}$$

for all t and for $k = 0, 1, 2, \ldots$, where λ is a finite positive real constant. We shall study such random variables in detail later in Chap. 13.

Show that

$$E[N_t] = \lambda t \tag{1.10}$$

for all t. (*Hint:* Recall the power-series expansion of the exponential function.) That is, *the expected value of the Poisson counting random variable at time t is just t times the value of the parameter λ which characterizes its probability distribution.* □

As our final exercise in this section, we wish to derive one of the basic properties of the statistical average.

Exercise 1.3 Suppose that the discrete random variable X is nonnegative:

$$X(s) \geq 0 \tag{1.11}$$

for every s in the sample space upon which X is defined. Show that

$$E[X] \geq 0 \tag{1.12}$$

in this case. We will use this result in a number of our subsequent derivations. □

2 EXISTENCE (DISCRETE CASE)

Whenever the discrete random variable under consideration has a countably infinite number of possible values, for example, as in Exercise 1.2 above, the formula for the statistical average of that random variable becomes an infinite series:

$$E[X] = \sum_{k=1}^{\infty} x_k P[X = x_k] \tag{2.1}$$

Now whenever one derives a result which contains an infinite series, one would like to be assured that the same answer is obtained no matter how the series is summed. In particular, one would hope that the same answer would be obtained regardless of the order in which the terms of the series were added in calculating the sum. Unfortunately, however, this is not always the case as we shall now show.

Suppose, for example, that the discrete random variable X has the possible values

$$x_k = (-1)^k k \qquad k = 1, 2, 3, \ldots \tag{2.2}$$

with the probabilities

$$P[X = x_k] = \frac{6}{\pi^2 k^2} \tag{2.3}$$

respectively. (The factor $6/\pi^2$ is required to make the sum of the probabilities add up to unity.) The expectation of this random variable is then, using (2.1),

$$\begin{aligned} E[X] &= \sum_{k=1}^{\infty} [(-1)^k k] \frac{6}{\pi^2 k^2} = \frac{6}{\pi^2} \sum_{k=1}^{\infty} \frac{(-1)^k}{k} \\ &= \frac{6}{\pi^2}\left(-1 + \tfrac{1}{2} - \tfrac{1}{3} + \tfrac{1}{4} - \tfrac{1}{5} + \cdots\right) \end{aligned} \tag{2.4}$$

Let us now attempt to sum the series in (2.4). If the reader is familiar with power-series expansions in general, he may recall that one such expansion for the natural logarithm is[1]

$$\log(1 + x) = x - \frac{x^2}{2} + \frac{x^3}{3} - \frac{x^4}{4} + \cdots \tag{2.5}$$

If we set $x = 1$ in this expression, we get

$$\log 2 = 1 - \tfrac{1}{2} + \tfrac{1}{3} - \tfrac{1}{4} + \cdots \tag{2.6}$$

and hence, on using this result in (2.4), we obtain

$$E[X] = -\frac{6}{\pi^2} \log 2 \tag{2.7}$$

as the expectation of our given random variable.

Suppose, however, that instead of using the series expansion of the logarithm, we were to attempt to sum the series in (2.4) by first adding the positive terms and then the negative terms; that is, suppose that we were

[1] Compare with sec. 9.8, vol. I, of Apostol.

to write

$$\sum_{k=1}^{\infty} \frac{(-1)^k}{k} = (\tfrac{1}{2} + \tfrac{1}{4} + \tfrac{1}{6} + \cdots) - (1 + \tfrac{1}{3} + \tfrac{1}{5} + \cdots) \tag{2.8}$$

thus expressing our result as a difference of two series, each of which has only positive terms. The first series on the right-hand side of (2.8) is simply one-half times the *harmonic series*

$$1 + \tfrac{1}{2} + \tfrac{1}{3} + \tfrac{1}{4} + \cdots$$

which is known to be divergent.[1] The second series on the right is term-by-term greater than the first and hence is also divergent. This particular rearrangement of terms therefore leads to the result

$$\sum_{k=1}^{\infty} \frac{(-1)^k}{k} = \infty - \infty \tag{2.9}$$

which is indeterminate! It is, in fact, not too difficult to show that other rearrangements of the terms in (2.4) can be obtained which lead to other different answers—a most unfortunate result!

One classic way of avoiding such difficulties is to make use of the fact that it can be shown that if the infinite series

$$S = \sum_{k=1}^{\infty} a_k \tag{2.10}$$

is *absolutely convergent*, that is, if

$$\sum_{k=1}^{\infty} |a_k| < +\infty \tag{2.11}$$

then *any* rearrangement of the terms in S converges to the same value.[2] With this fact in mind, it is common to say that the statistical average of a discrete random variable X exists only if the summation in the formula

$$E[X] = \sum_k x_k P[X = x_k] \tag{2.12}$$

is absolutely convergent; that is, $E[X]$ exists only if

$$\sum_k |x_k P[X = x_k]| = \sum_k |x_k| P[X = x_k] < +\infty \tag{2.13}$$

[1] See sec. 9.1, vol. I, of Apostol, or sec. 26 of Bartle.
[2] See, for example, sec. 26 of Bartle.

On returning to the series in (2.4) above, we note that

$$\frac{6}{\pi^2}\sum_{k=1}^{\infty}\left|\frac{(-1)^k}{k}\right| = \frac{6}{\pi^2}\sum_{k=1}^{\infty}\frac{1}{k} \tag{2.14}$$

Since the series on the right-hand side here is the divergent harmonic series, the series expression for the expectation is not absolutely convergent in this case and we say, accordingly, that the expectation does not exist here.

3 FUNCTIONS OF DISCRETE RANDOM VARIABLES

Suppose next that we wish to determine the statistical average of some function of a discrete random variable X, say of

$$Y = g(X) \tag{3.1}$$

rather than of X itself; for example, suppose that we wish to determine the statistical average of the square of X. Now it follows from the discussion in Sec. 6.1 that if g is a measurable function of the random variable X, then $Y = g(X)$ is also a random variable, and, in fact, is a discrete random variable since X is discrete. The probability distribution of Y may therefore be obtained in the usual way by employing the mapping properties of g. Thus it follows from the results of Sec. 1 of this chapter that the statistical average of Y may be calculated using the formula

$$E[Y] = \sum_j y_j P[Y = y_j] \tag{3.2}$$

where the y_j are the various possible values of the new random variable Y.

For example, suppose that X is a discrete random variable with the three possible values $+2$, 0, and -2 and that X has the probability distribution

$$\begin{aligned} P[X = +2] &= \tfrac{1}{4} \\ P[X = 0] &= \tfrac{1}{2} \\ \text{and} \qquad P[X = -2] &= \tfrac{1}{4} \end{aligned} \tag{3.3}$$

and suppose further that we wish to obtain the statistical average of the square of X:

$$Y = g(X) = X^2 \tag{3.4}$$

Now it follows from (3.4) that Y is a discrete random variable with the two possible values, four and zero $[(+2)^2 = 4,\ 0^2 = 0,\ (-2)^2 = 4$; cf. Fig. 3.1]. The probability distribution of Y may now be determined

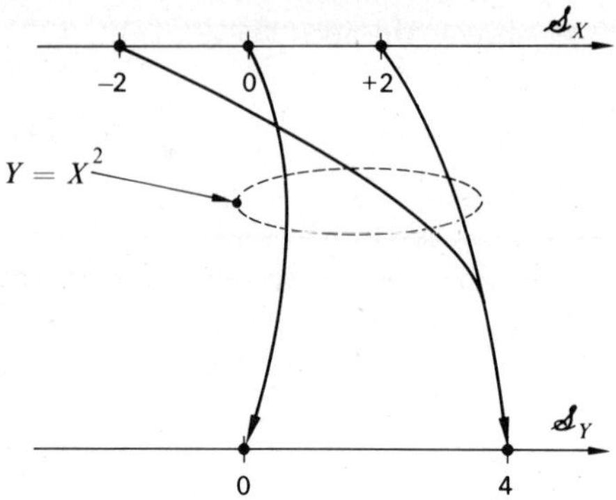

Figure 3.1 The mapping $Y = X^2$.

using the mapping (3.4) and the distribution (3.3). Thus we get

$$\begin{aligned} P[Y = 4] &= P[(X = +2) \cup (X = -2)] \\ &= P[X = +2] + P[X = -2] \\ &= \tfrac{1}{2} \end{aligned}$$

since the events $(X = +2)$ and $(X = -2)$ are mutually exclusive. Similarly,

$$\begin{aligned} P[Y = 0] &= P[X = 0] \\ &= \tfrac{1}{2} \end{aligned}$$

It therefore follows that the statistical average of Y is given by

$$\begin{aligned} E[Y] &= 0 \times P[Y = 0] + 4 \times P[Y = 4] \\ &= 2 \end{aligned} \tag{3.5}$$

While the derivation leading to (3.5) is not hard to follow, one might well suspect that if g had been a more complicated function of X, or if X had had many more possible values, then the steps required to first get the probability distribution of Y and then calculate the statistical average of Y using (3.2) would have been quite a bit more tedious to perform. With this fact in mind, a question arises as to whether or not results could be obtained which might be simpler to use in calculating expectations. That the answer is yes may be seen as follows.

A moment's reflection upon our derivation leading to (3.5) shows that we can rewrite that result in terms of the possible values of X and the probabilities of those values. Thus it follows that we may write

$$\begin{aligned} E[Y] &= 0P[Y = 0] + 4P[Y = 4] \\ &= 0P[X = 0] + 4(P[X = +2] + P[X = -2]) \\ &= (+2)^2P[X = +2] + 0^2P[X = 0] \\ &\qquad + (-2)^2P[X = -2] \end{aligned} \tag{3.6}$$

and so obtain a formula which does not explicitly involve either the possible values of Y or the probability distribution of Y. The generalization of this result is not too difficult:

Suppose that X is a discrete random variable with the possible values x_k and the corresponding probabilities $P[X = x_k]$. Suppose further that g is a measurable function defined on the sample space $\mathcal{S}_X$. It then follows that $Y = g(X)$ is a discrete random variable having, say, the possible values y_j with the corresponding probabilities $P[Y = y_j]$ given by the formula

$$P[Y = y_j] = P[X \in g^{-1}(\{y_j\})] \tag{3.7}$$

where the inverse image $g^{-1}(\{y_j\})$ is the set of points x in $\mathcal{S}_X$ such that

$$g(x) = y_j \qquad \text{that is} \qquad g^{-1}(\{y_j\}) = \{x\colon g(x) = y_j \text{ and } x \in \mathcal{S}_X\}$$

Now since X is discrete, the probabilities on the right-hand side of (3.7) may be calculated by summing up the probabilities of all the possible values x_k which fall into $g^{-1}(\{y_j\})$. Thus

$$P[X \in g^{-1}(\{y_j\})] = \sum_{x_k \in g^{-1}(\{y_j\})} P[X = x_k]$$

We may then calculate $E[Y]$ by substituting these results in (3.2) to get

$$\begin{aligned} E[Y] &= \sum_j y_j P[X \in g^{-1}(\{y_j\})] \\ &= \sum_j y_j \Big[\sum_{x_k \in g^{-1}(\{y_j\})} P[X = x_k] \Big] \end{aligned} \tag{3.8}$$

Now just as we did previously when we went from (3.5) to (3.6), we may express each of the terms y_j in (3.8) in terms of the corresponding values x_k of X which map into y_j; that is, we may write

$$g(x_k) = y_j$$

for each of the $x_k \in g^{-1}(\{y_j\})$. If we do so, we may then rewrite the resulting sum as a sum over all possible values of the x-index k. Thus we finally obtain the formula

$$E[Y] = E[g(X)] = \sum_k g(x_k) P[X = x_k] \tag{3.9}$$

for the expectation of a measurable function of a discrete random variable. This is our desired generalization of (3.8).

Exercise 3.1 Let X be a discrete random variable with the possible values $\{-1,0,+1,+2\}$, each with the probability $\frac{1}{4}$. Let g be the function defined by

$$g(x) \triangleq x^3 - x$$

for each $x \in \mathcal{S}_X$.

a. Determine the possible values of the random variable $Y = g(X)$.
b. Determine the probability distribution of Y.
c. Calculate $E[Y]$ using (3.2).
d. Calculate $E[Y]$ using (3.9). □

Having obtained a general formula for the expectation of a measurable function of a discrete random variable, we are now in a position to determine some of the general properties of the expectation: Suppose that A is a set in the range sample space $\mathcal{S}_X$ of the discrete random variable X, and that $I_A(X)$ is the set-indicator function of that event:

$$I_A(x) = \begin{cases} 1 & \text{if } x \in A \\ 0 & \text{otherwise} \end{cases} \tag{3.10}$$

The possible values x_k of X may now be split into two disjoint sets: those in A and those not in A. For each of the x_k in A, $I_A(x_k)$ is unity; while for each of the x_k not in A, $I_A(x_k)$ is zero. It therefore follows from (3.9) that

$$\begin{aligned} E[I_A(X)] &= \sum_{x_k \in A} 1P[X = x_k] + \sum_{x_k \notin A} 0P[X = x_k] \\ &= \sum_{x_k \in A} P[X = x_k] \end{aligned}$$

Since the various values x_k are disjoint, the right-most sum here is simply the probability that the random variable X has a value falling in the set A:

$$E[I_A(X)] = P[X \in A] \tag{3.11}$$

That is, *the expected value of the set indicator of the event A is simply equal to the probability of that event.*

Exercise 3.2 Let X be a discrete random variable with the possible values x_k and the corresponding probabilities $P[X = x_k]$. Suppose that

$$Y \triangleq aX \tag{3.12}$$

where a is a real constant. Show that the expectation of Y is just a times the expectation of X, that is,

$$E[aX] = aE[X] \tag{3.13}$$

when the expectation of X exists. □

Exercise 3.3 Let X be a discrete random variable with the possible values x_k and the corresponding probabilities $P[X = x_k]$. Show that

$$|E[X]| \leq E[|X|] \tag{3.14}$$

if the expectation of X exists. (*Hint:* Use the *triangle inequality*[1]

$$\left| \sum_k a_k \right| \leq \sum_k |a_k| \tag{3.15}$$

where the a_k are arbitrary real numbers.) □

4 FUNCTIONS OF DISCRETE RANDOM VECTORS

Next let us extend our concept of statistical average to cover the case of random vectors. We pointed out earlier (in Sec. 5.1) that a random vector $\mathbf{X}$ is a vector-valued function defined on a sample space, say $\mathcal{S}$, such that its components are random variables; that is, the components X_i, $i = 1, 2, \ldots, n$, of the random vector

$$\mathbf{X} \triangleq (X_1, X_2, \ldots, X_n) \tag{4.1}$$

are random variables by definition. With this in mind, we define the *statistical average of the random vector* $\mathbf{X}$ to be the vector whose components are the expectations of the respective components of $\mathbf{X}$:

$$E[\mathbf{X}] \triangleq (E[X_1], E[X_2], \ldots, E[X_n]) \tag{4.2}$$

This definition will be used regardless of whether or not the components of $\mathbf{X}$ are discrete random variables. If those components are discrete, then the averages in (4.2) can be calculated as in the preceding sections.

Suppose now that we wish to determine the statistical average of some function of a random vector. Here two distinct possibilities arise: (*a*) We may be concerned with a *scalar-valued function* of the given random vector, or (*b*) we may be concerned instead with a *vector-valued function* of that random vector. Suppose first that we are concerned with the scalar-valued function

$$Y \triangleq g(\mathbf{X})$$

of the random vector $\mathbf{X}$. One such example might be the sum of the components of $\mathbf{X}$:

$$Y = \sum_{i=1}^{n} X_i$$

If now the function g is measurable, then the result Y is an ordinary

[1] See, for example, sec. 1.21, vol. I, of Apostol, or sec. 5 of Bartle.

random variable and we may calculate its statistical average in the usual way. Thus if $\mathbf{X}$ is a discrete random vector, that is, if the components of $\mathbf{X}$ are all discrete random variables, then the result Y will have only a discrete set of possible values (say y_j, $j = 1, 2, 3, \ldots$) and hence will itself be a discrete random variable. It then follows from the results of Sec. 1 above that

$$E[Y] = \sum_j y_j P[Y = y_j] \tag{4.3}$$

where the probabilities $P[Y = y_j]$ are obtained from the joint-probability distribution of $\mathbf{X}$ through the mapping g. While (4.3) does indeed give the desired result, the calculation of the probabilities of the y_j may well be tedious to carry out and (as in the preceding section) we would like to obtain a formula for $E[Y]$ which is easier to use in calculations (even though it might look more complicated).

Suppose, for example, that the component X_i of the random vector $\mathbf{X}$ has the discrete possible values x_{k_i}. A more or less direct paraphrase of the derivation leading to the formula in (3.9) then shows that the statistical average of Y may be calculated by using the formula

$$\begin{aligned} E[Y] &= E[g(\mathbf{X})] \\ &= \sum_{k_1} \sum_{k_2} \cdots \sum_{k_n} g(x_{k_1}, x_{k_2}, \ldots, x_{k_n}) \\ &\qquad P[X_1 = x_{k_1}, X_2 = x_{k_2}, \ldots, X_n = x_{k_n}] \end{aligned} \tag{4.4}$$

where the summation is taken over all possible values of the components. We will evaluate a few specific cases of such averages somewhat later.

Suppose, on the other hand, that we wish to determine the statistical average of some vector-valued function of the random vector $\mathbf{X}$, say

$$\mathbf{Y} \triangleq \mathbf{g}(\mathbf{X})$$

that is, we have in this case the vector-valued result

$$\begin{aligned} \mathbf{Y} &= (Y_1, Y_2, \ldots, Y_m) \\ &\triangleq (g_1(\mathbf{X}), g_2(\mathbf{X}), \ldots, g_m(\mathbf{X})) \end{aligned} \tag{4.5}$$

where the components $g_j(\mathbf{X})$ are, by definition, scalar-valued functions of the random vector $\mathbf{X}$. Since it now follows from (4.2) that we may write

$$E[\mathbf{Y}] = (E[g_1(\mathbf{X})], E[g_2(\mathbf{X})], \ldots, E[g_m(\mathbf{X})]) \tag{4.6}$$

we see that the problem of determining the expectation of a vector-valued function of a random vector reduces to that of finding the expectations of a set of scalar-valued functions of the given random vector. The latter expectations can then be determined as previously discussed.

Let us now consider a specific example: Suppose that **X** is a two-dimensional discrete random vector; that is, let

$$\mathbf{X} = (X_1, X_2)$$

where X_1 has the discrete possible values x_{1i} and X_2 has the discrete possible values x_{2j}. Suppose further that we wish to determine the expected value of a weighted sum of X_1 and X_2:

$$Y \triangleq a_1X_1 + a_2X_2$$

where a_1 and a_2 are real constants. It then follows from (4.4) that

$$E[Y] = \sum_i \sum_j (a_1x_{1i} + a_2x_{2j})P[X_1 = x_{1i},\ X_2 = x_{2j}]$$

Simple expansion of this result gives

$$E[Y] = a_1 \sum_i \sum_j x_{1i}P[X_1 = x_{1i},\ X_2 = x_{2j}] \\ + a_2 \sum_i \sum_j x_{2j}P[X_1 = x_{1i},\ X_2 = x_{2j}] \tag{4.7}$$

If we sum the first term on the right-hand side of (4.7) over the index j and then sum the second term over the index i, it follows from the properties of the joint-probability that we get

$$E[Y] = a_1 \sum_i x_{1i}P[X_1 = x_{1i}] + a_2 \sum_j x_{2j}P[X_2 = x_{2j}]$$

The first term on the right here is simply a_1 times the expected value of X_1, while the second term is a_2 times the expected value of X_2. Thus we have derived the result that

$$E[a_1X_1 + a_2X_2] = a_1E[X_1] + a_2E[X_2] \tag{4.8}$$

that is, *the expected value of a weighted sum of a pair of discrete random variables equals the weighted sum of the expected values of those random variables*. Any operator such as the expectation operator E which satisfies an equation of the form of that in (4.8) is said to be a *linear functional*.[1]

Exercise 4.1 Let **X** be an n-dimensional discrete random vector with the components $X_1, X_2, \ldots, X_n$. Show that

$$E\left[\sum_{i=1}^{n} a_iX_i\right] = \sum_{i=1}^{n} a_iE[X_i] \tag{4.9}$$

where the a_i are arbitrary real constants. [*Hint:* Use induction (cf. Sec. 3.2).] □

[1] Compare with Bartle, p. 289.

Exercise 4.2 Consider the binomial random variable Y_n, $n = 1, 2, 3, \ldots$, discussed earlier in Sec. 1. We shall show later[1] that the sum of n independent Bernoulli random variables is such a binomial random variable; that is, we may write

$$Y_n = \sum_{i=1}^{n} X_i$$

where the random variables X_i each have the same probability distribution

$$P[X_i = 1] = p \qquad \text{and} \qquad P[X_i = 0] = 1 - p$$

Using (4.9), show that

$$E[Y_n] = np$$

It should be noted that the method used here to obtain this result is much simpler than that used in Sec. 1 of this chapter to get the same answer. □

We can also use the linearity result in (4.8) to obtain other general properties of the expectation.

Exercise 4.3 Suppose that the two-dimensional discrete random vector

$$\mathbf{X} = (X_1, X_2)$$

defined on a sample space $\mathcal{S}$ is such that $X_1 \leq X_2$; that is,

$$X_1(s) \leq X_2(s)$$

for each $s \in \mathcal{S}$. Show that it then follows that

$$E[X_1] \leq E[X_2]$$

(*Hint:* Use Exercise 1.3 above.) □

Sums of random variables are, of course, by no means the only functions of random vectors which are of interest. We have already seen in Sec. 6.3, for example, that one is often interested in products of random variables. Suppose, in particular, that X is a two-dimensional discrete random vector with the components X_1 and X_2, and that we wish to determine the expected value of the product of X_1 and X_2:

$$Y \triangleq X_1 X_2$$

Now it follows from (4.4) that

$$E[Y] = \sum_i \sum_j x_{1i} x_{2j} P[X_1 = x_{1i}, X_2 = x_{2j}]$$

[1] See sec. 9.2.

where the x_{1i} are the possible values of X_1 and where the x_{2j} are the possible values of X_2. Unfortunately, we cannot now proceed further, in general, unless we are able to specify the actual form of the joint-probability distribution of X_1 and X_2. A simplification does result, however, in the particular case in which the components X_1 and X_2 are independent random variables.

Exercise 4.4 Let X_1 and X_2 be the components of the two-dimensional discrete random variable **X**. Show that

$$E[X_1X_2] = E[X_1]E[X_2] \tag{4.10}$$

if X_1 and X_2 are independent random variables. □

It should be noted, however, that the converse result to that in Exercise 4.4 is not in general true; that is, specifying that two random variables are such that the expectation of their product factors is not enough in general to guarantee that those random variables are independent. We will return to this point later in Sec. 9 of this chapter.

A generalization of the result in Exercise 4.4 follows.

Exercise 4.5 Let **X** be an n-dimensional discrete random vector with the components $X_1, X_2, \ldots, X_n$, defined on a sample space $\mathcal{S}$. Suppose that these components are mutually independent random variables. Show that it then follows that

$$E[g_1(X_1)g_2(X_2) \cdots g_n(X_n)] = E[g_1(X_1)]E[g_2(X_2)] \cdots E[g_n(X_n)] \tag{4.11}$$

when each of the single-valued functions g_i, $i = 1, 2, \ldots, n$, is measurable on the sample space $\mathcal{S}$. □

5 EXTENSION TO THE CONTINUOUS CASE*

Our discussion of statistical averages has, to this point, been restricted to the case of discrete random variables. Now we have already seen in our earlier chapters that it is necessary to deal with continuous (and, sometimes, mixed) random variables in many theoretical and practical problems of interest. We would, accordingly, like to extend our development of the statistical average to cover these other cases. The basic procedure which we shall follow in making this extension is to first approximate by a discrete random variable the random variable whose expecta-

* This section is more mathematical than most. The concepts involved are important, however, and the reader should try to grasp them even though he may not later be able to reconstruct the details of the development by himself.

tion we wish to calculate, then apply our previously derived expectation results to the discrete approximation, and, finally, refine the results thus obtained by making our approximations better and better. Since we wish to focus our attention here upon the basic concepts involved, we will concentrate our effort on the simplest cases and will only sketch the further extension to the more general cases. Good rigorous treatments of the general theory may be found, for example, in chap. 4 and 11 of Royden and in chap. 4 of Bartle's book on integration.

The basic problem before us here is, given an arbitrary random variable X defined on some sample space $\mathcal{S}$, we would like to determine the expectation of some measurable function, say g, of X. That is, we would like to determine the expectation of the random variable

$$Y \triangleq g(X)$$

given that we know the probability properties of X.† In order to simplify matters, we will first assume that both the given random variable X and the measurable function g are bounded, say, as shown in Fig. 5.1.

Let us now approximate the new random variable Y with a step

† If in fact we can do so, we, of course, also immediately solve the problem of how to determine the expectation of X itself.

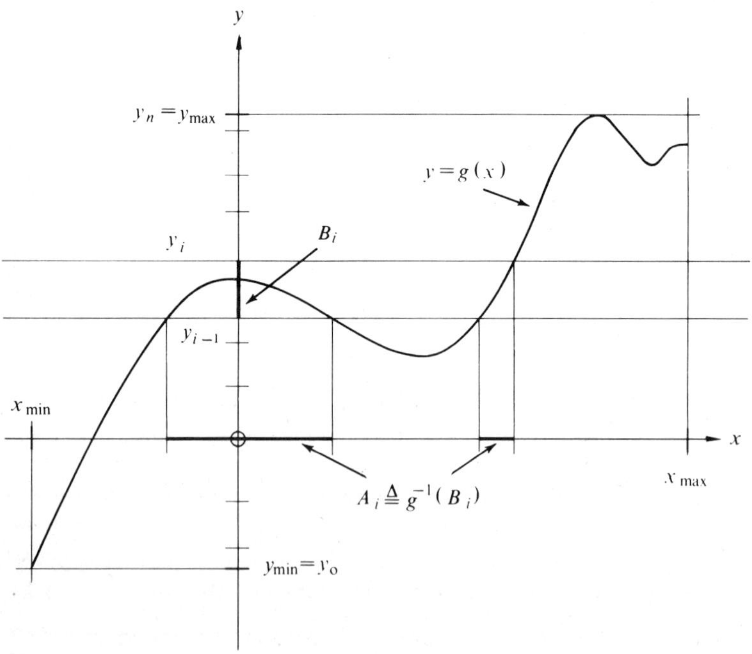

Figure 5.1 A continuous measurable function.

function by using the following procedure: Since the mapping g is assumed here to be bounded, there exist numbers $y_{\min}$ and $y_{\max}$ such that

$$y_{\min} \leq g(x) \leq y_{\max}$$

for every $x \in \mathcal{S}_X$. The interval $[y_{\min},y_{\max}]$ then bounds the range of the mapping g and hence bounds the range sample space $\mathcal{S}_Y$ of the random variable Y. Suppose next that we partition the interval $[y_{\min},y_{\max}]$ into n contiguous intervals B_i, $i = 1, 2, \ldots, n$, by choosing $n - 1$ separation points arbitrarily. Having chosen these points, let us then number them consecutively; that is, let us denote the boundary points of the partition intervals by y_i, where

$$y_0 = y_{\min} < y_1 < y_2 < \cdots < y_{n-1} < y_n = y_{\max}$$

(cf. Fig. 5.1). The intervals

$$B_1 \triangleq [y_0,y_1]$$

and (5.1)

$$B_i \triangleq (y_{i-1},y_i] \qquad \text{for } i = 2, 3, \ldots, n$$

are then disjoint and completely cover the interval $[y_{\min},y_{\max}]$; that is, they partition the range sample space $\mathcal{S}_Y$. Since to each point in $\mathcal{S}_X$ there corresponds one and only one point in $\mathcal{S}_Y$ via the mapping g, the inverse images A_i of the intervals B_i, that is, the sets

$$A_i \triangleq g^{-1}(B_i) \tag{5.2}$$

are therefore disjoint.[1] The assumed partitioning of the range sample space $\mathcal{S}_Y$ thus induces via g^{-1} a partitioning of the sample space $\mathcal{S}_X$.

Now it follows from the definition of the inverse images A_i that, for every x in a given A_i, the point

$$y = g(x)$$

falls in the corresponding interval B_i; that is,

$$y_{i-1} \leq g(x) \leq y_i \tag{5.3}$$

for each x in A_i. If our partitioning of the range of g is fine enough, that is, if the points y_{i-1} and y_i are close together, then we may form a reasonable approximation to the random variable Y throughout the interval A_i by choosing for the approximation any value which falls in

[1] For, suppose that two of the inverse images, say A_j and A_k do overlap, and consider any point x in their intersection. Since the point x is thus in the inverse images of both of the disjoint intervals B_j and B_k, it must map into two separate points in $\mathcal{S}_Y$ (one in B_j and one in B_k). This, however, contradicts the definition of g as a "function" and hence is not possible. It therefore follows that the inverse images A_j and A_k must, in fact, be disjoint.

the interval B_i or on its boundary. With this in mind, let us first approximate Y by the step function Y_n^+ which assumes the upper boundary value y_i for each x in A_i; that is, let us set

$$Y_n^+(x) \triangleq \sum_{i=1}^{n} y_i I_{A_i}(x) \tag{5.4}$$

for each x in $\mathcal{S}_X$, where I_{A_i} is the set-indicator function of the set A_i:

$$I_{A_i}(x) \triangleq \begin{cases} 1 & \text{if } x \in A_i \\ 0 & \text{if not} \end{cases}$$

Since by this definition Y_n^+ is always greater than (or at least equal to) the random variable Y, we say that Y_n^+ *approximates* Y *from above* and, accordingly, it is often called an *upper approximating function*. A typical example of such an upper approximating function is shown in Fig. 5.2.

A moment's reflection upon the construction of the approximation Y_n^+ shows that it assumes only n values throughout its range and hence that it is a discrete random variable. The expectation of Y_n^+ may therefore be calculated by using (1.2) and is

$$E[Y_n^+] = \sum_{i=1}^{n} y_i P[Y_n^+ = y_i] \tag{5.5}$$

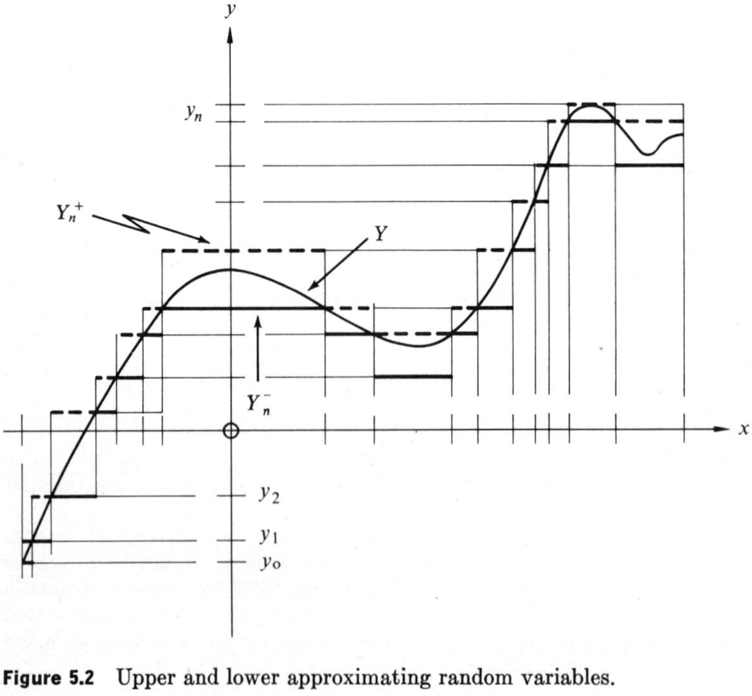

Figure 5.2 Upper and lower approximating random variables.

Now it follows from the definition of Y_n^+, and from the definition of the sets A_i and B_i, that

$$\begin{aligned} P[Y_n^+ = y_i] &= P[Y \in B_i] \\ &= P[X \in A_i] \end{aligned}$$

and therefore that (5.5) may be rewritten as

$$E[Y_n^+] = \sum_{i=1}^{n} y_i P[X \in A_i] \tag{5.6}$$

Thus we have obtained a formula for the expectation of the upper approximating function Y_n^+ which is expressed in terms of the probability properties of the original random variable X.

The upper approximating function Y_n^+ is, of course, by no means the only step-function approximation that one can make for the random variable Y. Consider, for example, the step function Y_n^-, which assumes the lower boundary value y_{i-1} for each x in A_i; that is, let us set

$$Y_n^-(x) \triangleq \sum_{i=1}^{n} y_{i-1} I_{A_i}(x) \tag{5.7}$$

for each x in $\mathcal{S}_X$, where I_{A_i} is again the set-indicator function of the set A_i. Since by this definition Y_n^- is always less than (or at most equal to) the random variable Y, we say that Y_n^- *approximates* Y *from below* and, accordingly, Y_n^- is often called a *lower approximating function*. A typical example of such a lower approximating function is also shown in Fig. 5.2. By paralleling the development leading to (5.6), it is easy to obtain

$$E[Y_n^-] = \sum_{i=1}^{n} y_{i-1} P[X \in A_i] \tag{5.8}$$

as an expression for the expectation of Y_n^-.

Suppose next that we make the partitioning of the range of g finer and finer. That is, suppose that we add more and more partition points y_i so as to make the lengths of the partition intervals B_i become shorter and shorter. It is clear that if we do so, then the approximation values y_i and y_{i-1} both become closer and closer to the values of $g(x)$ in each partition interval for which $P[X \in A_i]$ is not zero [cf. (5.3)], and hence that both the upper approximating function Y_n^+ and the lower approximating function Y_n^- approach the random variable Y more and more closely. If now the expectation of Y_n^+ approaches that of Y_n^- as we make the partitioning finer and finer, then it would seem reasonable to define the expectation of the random variable Y to be the common limiting value of those approximating expectations. That $E[Y_n^+]$ and $E[Y_n^-]$ do indeed approach each other may be seen as follows:

Consider the difference between $E[Y_n^+]$ and $E[Y_n^-]$ for a given partition. We have, from (5.6) and (5.8),

$$E[Y_n^+] - E[Y_n^-] = \sum_{i=1}^{n} y_i P[X \in A_i] - \sum_{i=1}^{n} y_{i-1} P[X \in A_i]$$
$$= \sum_{i=1}^{n} (y_i - y_{i-1}) P[X \in A_i] \tag{5.9}$$

Now it follows from the definition of y_i and y_{i-1} that $y_{i-1} < y_i$, and hence it further follows that each term in the sum in the right-most side of (5.9) is nonnegative. Thus

$$E[Y_n^+] - E[Y_n^-] \geq 0 \tag{5.10}$$

Now, given a particular partition, let us denote the maximum value of the lengths of the intervals B_i, $i = 1, 2, \ldots, n$, by $(\Delta y)_n$:

$$(\Delta y)_n \triangleq \max_i (y_i - y_{i-1})$$

It then follows from this definition that

$$y_i - y_{i-1} \leq (\Delta y)_n$$

for any value of i. On using this fact in (5.9), we obtain

$$E[Y_n^+] - E[Y_n^-] \leq (\Delta y)_n \sum_{i=1}^{n} P[X \in A_i] = (\Delta y)_n$$

When combined with (5.10), this gives the result that

$$0 \leq E[Y_n^+] - E[Y_n^-] \leq (\Delta y)_n \tag{5.11}$$

We previously pointed out that making the partitioning finer and finer implies making the lengths of the partition intervals all become smaller and smaller:

$$\lim_{n \to \infty} (\Delta y)_n = 0 \tag{5.12}$$

If therefore we use this result in (5.11), we see that the right-most term there becomes zero as n increases without limit and hence that

$$\lim_{n \to \infty} E[Y_n^+] = \lim_{n \to \infty} E[Y_n^-] \tag{5.13}$$

that is, that the limit of the expectation of the upper approximating function equals the limit of the expectation of the lower approximating function when we increase the number of partition points without limit in such a way that the lengths of the partition intervals all approach zero.

With this result in mind, we now define the expectation of the random variable $Y = g(X)$ to be that common limiting value; that is,

$$E[g(X)] \triangleq \lim_{n\to\infty} \sum_{i=1}^{n} y_i P[X \in A_i] \tag{5.14}$$

where it is understood that the limiting process is to be carried out in such a way that (5.12) is satisfied.

We assumed in our above derivation that both the random variable X and the measurable function g were bounded. These boundedness restrictions can, in fact, be removed so long as absolute convergence is obtained (cf. chap. II of Royden). The reader should be aware of the fact that since the limiting sum in (5.14) has all the properties that one usually associates with the ordinary Riemann integral, it is common in the mathematical probability literature[1] to call that limit an integral also and to denote it by the symbol $\int g\,dP$; that is,

$$\begin{aligned}\int g\,dP &\triangleq \lim_{n\to\infty} \sum_{i=1}^{n} y_i P[X \in A_i] \\ &= E[g(X)]\end{aligned} \tag{5.15}$$

Such an integral is often called a *Lebesgue-Stieltjes integral* after the French mathematician Henri Lebesgue (1875–1941) and the Dutch astronomer and mathematician Thomas Joannes Stieltjes (1856–1899), both of whom contributed to its development.

Although the derivation leading to (5.14) was carried out specifically for the single random variable case, the ideas and techniques used there can be employed to extend our results to cover the determination of the expectation of a measurable function of a general n-dimensional random vector.

Density case

While the result stated in (5.14) above is indeed useful in theoretical studies, it is, in fact, not very useful when one has to make calculations in practical problems. If, however, the random variable X has a probability density f_X, then it can be shown that (5.14) leads to the formula

$$E[g(X)] = \int_{-\infty}^{+\infty} g(x) f_X(x)\,dx \tag{5.16}$$

whenever that integral converges absolutely (cf. Sec. 7).

The derivation of (5.16) proceeds roughly as follows: Suppose first that X is a bounded random variable, that it has a continuous probability density f_X, and that we wish to determine the expectation of the random

[1] Compare with chap. 7 of Cramér or chap. II of Krickeberg.

variable

$$Y \triangleq g(X)$$

where here g is a continuous measurable function. Since these assumptions on X and g are more restrictive than those made earlier in this section, we can use directly the results obtained there. Thus, if Y_n^+ is an upper approximating function defined as in (5.4), then

$$E[Y_n^+] = \sum_{i=1}^{n} y_i P[X \in A_i] \tag{5.17}$$

is still a valid expression for its expectation. Since here we have assumed that the probability density of X exists and is continuous, we may write [using (4.5.9)]

$$P[X \in A_i] = \int_{A_i} f_X(x)\, dx \tag{5.18}$$

Suppose, for the moment, that the inverse image A_i is an interval with the endpoints x_{i-1} and x_i, where $x_{i-1} < x_i$. In this case (5.18) becomes

$$P[X \in A_i] = \int_{x_{i-1}}^{x_i} f_X(x)\, dx$$

Since f_X is continuous here by hypothesis, it follows from the mean-value theorem of calculus that we may write

$$\int_{x_{i-1}}^{x_i} f_X(x)\, dx = f_X(x_i')\, \Delta x_i \tag{5.19}$$

where x_i' is some point in the interval bounded by x_{i-1} and x_i, and where we have introduced the symbol

$$\Delta x_i \triangleq x_i - x_{i-1}$$

Now while it is certainly true (by construction) that the sets B_i are intervals, it is also true that the inverse images A_i need not be intervals (cf. Fig. 5.1). It does follow from the fact that the function g is continuous, however, that the A_i are either intervals or are unions of disjoint intervals. In either case, we may use (5.19) so long as we apply it separately to each of the subintervals which makes up an A_i. If we do so, and then substitute the results back into (5.17), we can obtain an expression of the form

$$E[Y_n^+] = \sum_j y_j f_X(x_j')\, \Delta x_j \tag{5.20}$$

Suppose finally that we refine our partitioning by increasing the number of partition intervals in $\mathcal{S}_Y$ without limit (while making certain that the lengths of all intervals tend to zero). If we do so, the summation in (5.20) tends to an ordinary integral, and thus we get the result

$$E[g(X)] = \int_{x_{\min}}^{x_{\max}} g(x)f_X(x)\,dx \tag{5.21}$$

for the expectation of a continuous measurable function g of a bounded random variable X which has a continuous probability density f_X.

If the random variable X is not bounded, for example, if its sample space $\mathcal{S}_X$ is the entire real line, then (5.21) may be extended to give

$$E[g(X)] = \int_{-\infty}^{+\infty} g(x)f_X(x)\,dx \tag{5.22}$$

whenever the integral there converges absolutely; that is, the expectation $E[g(X)]$ is said to exist only when

$$E[|g(X)|] = \int_{-\infty}^{+\infty} |g(X)|f_X(x)\,dx$$

is finite.

A further extension of the above results to cover the case of functions of random vectors may be made by paralleling more or less directly the development previously given in the discrete case. Thus, if the random variable Y is a scalar measurable function of the n-dimensional random vector $\mathbf{X}$, that is, if

$$\begin{aligned} Y &\triangleq g(\mathbf{X}) \\ &= g(X_1, X_2, \ldots, X_n) \end{aligned}$$

and if the joint-probability density of the random vector $\mathbf{X}$ exists and is continuous, then we have

$$\begin{aligned} E[g(\mathbf{X})] &= \int g(\mathbf{x})f_{\mathbf{X}}(\mathbf{x})\,d\mathbf{x} \\ &\triangleq \int_{-\infty}^{+\infty} \cdots \int_{-\infty}^{+\infty}\!\!\int g(x_1, x_2, \ldots, x_n)f_{X_1,X_2,\ldots,X_n} \\ &\qquad\qquad (x_1, x_2, \ldots, x_n)\,dx_1\,dx_2 \cdots dx_n \end{aligned} \tag{5.23}$$

whenever that multiple integral converges absolutely. If, on the other hand, we are concerned with a vector-valued function of the n-dimensional random vector $\mathbf{X}$, say with

$$\begin{aligned} \mathbf{Y} &= \mathbf{g}(\mathbf{X}) \\ &= (g_1(\mathbf{X}), g_2(\mathbf{X}), \ldots, g_m(\mathbf{X})) \end{aligned}$$

where the component functions $g_1, g_2, \ldots, g_m$ are scalar functions of $\mathbf{X}$, then it follows from (4.2) that we have

$$E[\mathbf{g}(\mathbf{X})] = (E[g_1(\mathbf{X})], E[g_2(\mathbf{X})], \ldots, E[g_m(\mathbf{X})]) \tag{5.24}$$

If then the joint-probability density of $\mathbf{X}$ exists and is continuous, the component expectations $E[g_i(\mathbf{X})]$, $i = 1, 2, \ldots, m$, may be calculated using (5.23).

We shall apply these various results in a number of specific cases in the following section and shall discuss the problem of absolute convergence in the continuous case in Sec. 7.

6 CONTINUOUS-CASE EXAMPLES AND EXERCISES

Once we have derived the general result

$$E[g(X)] = \lim_{n\to\infty} \sum_{i=1}^{n} y_i P[X \in A_i] \tag{6.1}$$

then we may show that many of the discrete random variable results obtained in Secs. 1 to 4 also hold, in fact, whether the random variable X is discrete, or has a probability density, or is a mixed random variable.

Consider, for example, the particularly simple case

$$g(X) \triangleq aX$$

in which a is a real constant. It is clear in this case that the approximating points y_i may be expressed as ax_i and hence that (6.1) becomes here

$$\begin{aligned} E[aX] &= \lim_{n\to\infty} \sum_{i=1}^{n} ax_i P[X \in A_i] \\ &= a\left(\lim_{n\to\infty} \sum_{i=1}^{n} x_i P[X \in A_i]\right) \\ &= aE[X] \end{aligned} \tag{6.2}$$

Thus we see that the result previously obtained in Exercise 3.2 for discrete random variables also holds true in general.

In a similar manner, one can also show that the following statements are valid in general.

Set indicators

$$E[I_A(X)] = P[X \in A] \tag{6.3}$$

Magnitudes

$$|E[X]| \leq E[|X|] \tag{6.4}$$

Linearity

$$E[a_1X_1 + a_2X_2] = a_1E[X_1] + a_2E[X_2] \tag{6.5}$$

Order

If $X_1 \leq X_2$, then

$$E[X_1] \leq E[X_2] \tag{6.6}$$

Factoring

If X_1 and X_2 are independent random variables, then

$$E[g_1(X_1)g_2(X_2)] = E[g_1(X_1)]E[g_2(X_2)] \tag{6.7}$$

when g_1 and g_2 are measurable functions.

Exercise 6.1 We have shown so far in this chapter that if we first define the expectation of a discrete random variable as in (1.2), then we can extend that definition to cover more general random variables, and, finally, show that expectation has the properties stated above in (6.3) to (6.7). It is sometimes convenient in more abstract treatments of the field, however, to invert this procedure and, instead, to define expectation to be a linear functional which operates on random variables [that is, to require (6.5)] and which has the set-indicator property (6.3). If this is done, then it follows that (1.2) becomes a derived formula and not a definition.[1] Thus:

Let X be a random variable with a range sample space $\mathcal{S}_X$. Suppose that $\mathcal{S}_X$ is partitioned by the sets A_i, $i = 1, 2, \ldots, n$, and that we define the random variable Y to have a constant value on each A_i; in particular, suppose that Y has the value y_i for all points $x \in A_i$. Show that if we define expectation as an operator which satisfies (6.3) and (6.5), then it follows that the expectation of the discrete random variable Y is given by the formula

$$E[Y] = \sum_{i=1}^{n} y_i P[Y = y_i]$$

that is, that the discrete expectation formula may be obtained as a derived result under the stated hypotheses. [*Hint:* cf. (5.4).] □

Density case

For a simple example of the use of the formula

$$E[g(X)] = \int_{-\infty}^{+\infty} g(x)f_X(x)\,dx \tag{6.8}$$

consider the problem of determining the expectation of the uniformly dis-

[1] Compare with sec. 2.3 of Neveu.

tributed random variable X which has the probability density

$$f_X(x) = \begin{cases} \dfrac{1}{b-a} & \text{for } a \leq x \leq b \\ 0 & \text{otherwise} \end{cases} \tag{6.9}$$

On substituting this density in (6.8), we get

$$\begin{aligned} E[X] &= \frac{1}{b-a} \int_a^b x\,dx = \frac{1}{b-a} \frac{x^2}{2} \Big|_a^b \\ &= \frac{a+b}{2} \end{aligned} \tag{6.10}$$

Thus we see that the statistical average of a uniformly distributed random variable has a value which falls halfway between the maximum and minimum values of that random variable as shown, for example, in Fig. 6.1.

As a second example, let us determine the average time to failure of a class of light bulbs whose time to failure T has an exponential probability density of the form

$$f_T(t) = \begin{cases} ae^{-at} & \text{for } 0 \leq t \\ 0 & \text{for } t < 0 \end{cases} \tag{6.11}$$

where a is a positive number. On using this density in (6.8), we get

$$\begin{aligned} E[T] &= a \int_0^{+\infty} te^{-at}\,dt \\ &= \frac{1}{a} \int_0^{+\infty} ue^{-u}\,du \end{aligned}$$

This last integral may now be evaluated either by looking it up in a table of integrals (cf. formula 861.1 of Dwight) or by using integration by parts.

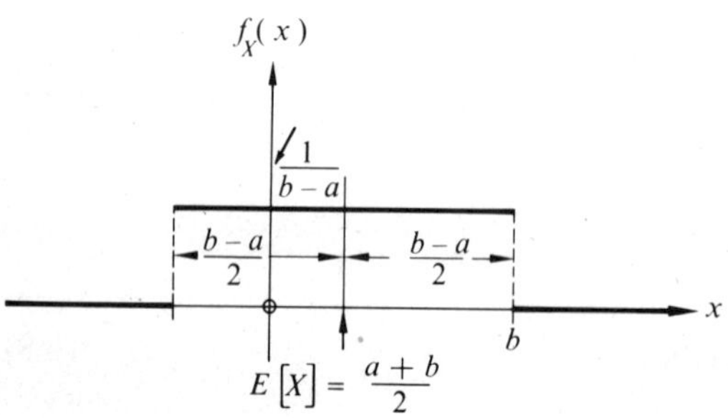

Figure 6.1 Location of the mean of a uniformly distributed random variable.

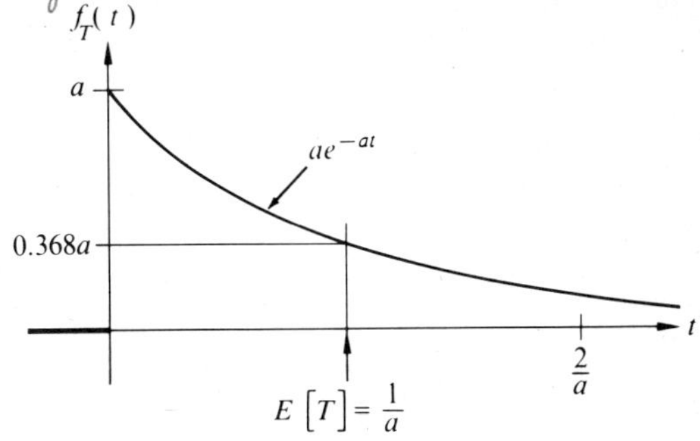

Figure 6.2 Location of the mean of an exponentially distributed random variable.

On applying this latter technique, we get

$$\int_0^{+\infty} ue^{-u}\,du = -ue^{-u}\Big|_0^{+\infty} + \int_0^{+\infty} e^{-u}\,du = 1$$

Thus it follows that

$$E[T] = \frac{1}{a} \tag{6.12}$$

that is, that the expected time to failure is simply equal to the reciprocal of the parameter a which appears in the exponential density given in (6.11). This situation is shown in Fig. 6.2.

A review of the above examples emphasizes the fact that the problem of finding the statistical average of a measurable function of a random variable which has a probability density is essentially just a problem in calculus.

Exercise 6.2 Let X be a continuous random variable with the normal probability density

$$f_X(x) = \frac{1}{\sqrt{2\pi}\,\sigma} \exp\left[-\frac{(x-m)^2}{2\sigma^2}\right] \tag{6.13}$$

where σ and m are real numbers and σ is positive. Show in this case that

$$E[X] = m \tag{6.14}$$

and hence that *the mean value of the normal random variable corresponds to the location of the peak of its probability density*, as indicated in Fig. 6.3. [*Hint:* Consider the substitution $y = x - m$ in the integral formula for $E[X]$. Also, remember that the integral over all x of (6.13) is unity.] □

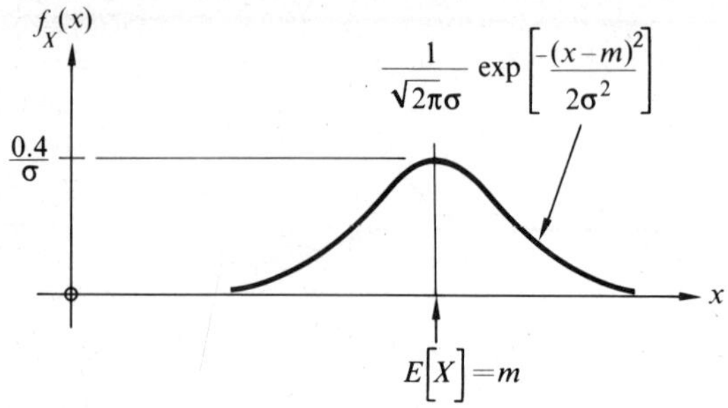

Figure 6.3 Location of the mean of a normal random variable.

Exercise 6.3 One is often interested in the expected deviation of a random variable from its mean value. While the most common measure of deviation is probably the standard deviation (or square root of the variance), which we shall meet and study in Sec. 9, another useful measure is the expected value of the magnitude of the deviation, that is, $E[|X - E[X]|]$.

Let X be a normal random variable with the probability density given in (6.13). Show that, in this case,

$$E[|X - m|] = \sqrt{\frac{2}{\pi}}\,\sigma \approx 0.798\sigma \tag{6.15}$$

that is, that *the expected value of the magnitude of the deviation of a normal random variable from its mean is about 80 percent of the standard deviation* (of the constant σ). □

Exercise 6.4 We pointed out earlier (cf. Sec. 4.5) that if the rectangular components of a spacecraft-landing error are independent, identically distributed, normal random variables, then the radius R of the landing point from the aiming point has a Rayleigh probability density of the form

$$f_R(r) = \begin{cases} \frac{r}{b} e^{-r^2/2b} & \text{for } 0 \le r \\ 0 & \text{otherwise} \end{cases} \tag{6.16}$$

where $b > 0$. Show that in this case

$$E[R] = \sqrt{\frac{b\pi}{2}} \approx 1.253\sqrt{b} \tag{6.17}$$

Thus we see that the expected value of the landing-point radius from the

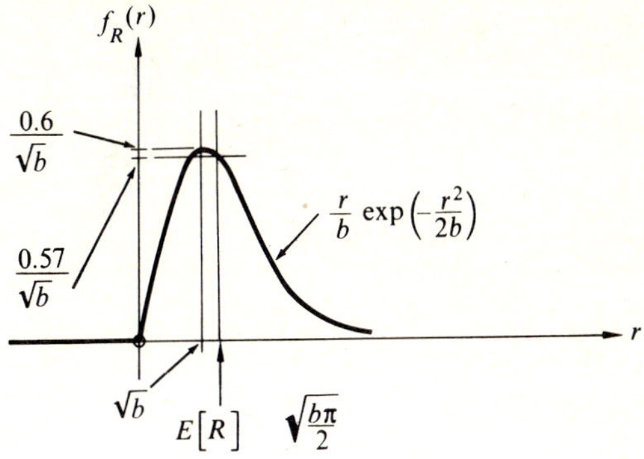

Figure 6.4 Location of the mean of a Rayleigh random variable.

aiming point is about 25 percent farther out than is the distance $\sqrt{b}$ which corresponds to the peak of the probability density (cf. Fig. 6.4). □

7 EXISTENCE (CONTINUOUS CASE)

Just as in the discrete case, so in the continuous case do we have a convergence problem. In particular, we have mentioned several times before that the expectation

$$E[g(X)] = \int_{-\infty}^{+\infty} g(x)f_X(x)\,dx \tag{7.1}$$

is said to exist only when that integral is absolutely convergent, that is, only when

$$E[|g(x)|] = \int_{-\infty}^{+\infty} |g(x)|f_X(x)\,dx$$

is finite. We would now like to investigate some aspects of this problem.

One of the basic problems here is that of dealing with infinite integrals, that is, with integrals such that either, or both, of the limits of integration are infinite. Suppose, for example, that we wish to integrate some function h from minus infinity to plus infinity. The standard definition of such an infinite integral is

$$\int_{-\infty}^{+\infty} h(x)\,dx \triangleq \lim_{\substack{a\to-\infty \\ b\to+\infty}} \int_a^b h(x)\,dx \tag{7.2}$$

and the limiting value is said to exist only if it is independent of the way a and b approach infinity.[1]

As an example of an expectation problem in which such a difficulty arises, suppose that we wish to determine the expectation of a random

[1] Compare with sec. 9.28, vol. I, of Apostol, or sec. 25 of Bartle.

variable X which has the Cauchy probability density

$$f_X(x) = \frac{1}{\pi(1 + x^2)}$$

If we use this density in (7.1) we get

$$E[X] = \frac{1}{\pi}\int_{-\infty}^{+\infty} \frac{x}{1 + x^2}\,dx$$

Now (7.2) implies that if the infinite integral here is to exist, then the limit in

$$\int_{-\infty}^{+\infty} \frac{x}{1 + x^2}\,dx = \lim_{\substack{a\to-\infty \\ b\to+\infty}} \int_a^b \frac{x}{1 + x^2}\,dx \tag{7.3}$$

must be independent of the way in which a and b approach infinity. That this condition is, in fact, not met here may be seen in the following exercise.

Exercise 7.1 Consider the integral

$$J(b) \triangleq \int_0^b \frac{x}{1 + x^2}\,dx \tag{7.4}$$

and show that

$$\lim_{b\to+\infty} J(b) = +\infty$$

[*Hint:* Consider the change of variable $y = 1 + x^2$ in (7.4).] □

With the result of Exercise 7.1 in mind, it follows that if we let a and b in (7.3) go to infinity separately, we get

$$\begin{aligned}\lim_{\substack{a\to-\infty \\ b\to+\infty}} \int_a^b \frac{x}{1 + x^2}\,dx &= \lim_{\substack{a\to-\infty \\ b\to+\infty}} \left(\int_0^b \frac{x}{1 + x^2}\,dx + \int_a^0 \frac{x}{1 + x^2}\,dx\right) \\ &= \lim_{b\to+\infty} \int_0^b \frac{x}{1 + x^2}\,dx - \lim_{a\to-\infty} \int_0^{|a|} \frac{x}{1 + x^2}\,dx \\ &= \infty - \infty\end{aligned}$$

an indeterminate result. Suppose, on the other hand, that we let a and b go to infinity together. In particular, let us take $a = b$ so as to get

$$\int_{-\infty}^{+\infty} \frac{x}{1 + x^2}\,dx = \lim_{b\to\infty} \int_{-b}^{+b} \frac{x}{1 + x^2}\,dx \tag{7.5}$$

The right-hand side here is commonly called the *Cauchy principal value* of the infinite integral on the left. Now since the integrand $x/(1 + x^2)$

is an odd function of x, it follows that the integral on the right-hand side of (7.5) is equal to zero for every finite value of b, and hence that we get

$$\lim_{b\to+\infty} \int_{-b}^{+b} \frac{x}{1+x^2}\,dx = 0$$

in this case. Thus we see that we can get quite different results according to just how we let the limits of integration in (7.3) recede to infinity. This is a most unsatisfactory result, and it should therefore seem reasonable for us to say that the expectation of a Cauchy random variable does not, in fact, exist.

Now paralleling the infinite series case, it can be shown that if the infinite integral of the absolute value of the function h is finite, that is, if

$$\int_{-\infty}^{+\infty} |h(x)|\,dx < +\infty$$

then the limit on the right-hand side of (7.2) is independent of the way the limits of integration recede to infinity.[1] When applied to expectations, this condition says that if the expected value of the absolute value of g is finite, that is, if $E[g(X)]$ is absolutely convergent, then that expectation exists.

Exercise 7.2 Show that the formula for the expected value of a Cauchy random variable is not absolutely convergent. □

8 MOMENTS

A set of statistical averages of particular importance are the expectations of the kth powers of a random variable. These averages are called the *kth moments* of the random variable, and are denoted by $E[X^k]$, $k = 1, 2, 3, \ldots$. Thus,

$$E[X^k] = \sum_j x_j{}^k P[X = x_j] \tag{8.1}$$

when X is a discrete random variable with the possible values x_j and

$$E[X^k] = \int_{-\infty}^{+\infty} x^k f_X(x)\,dx \tag{8.2}$$

when X is a continuous random variable with the probability density function f_X. As we pointed out earlier, such expectations are said to exist only if the indicated sum (or integral) is absolutely convergent, that is, only if

$$E[|X|^k] = \sum_j |x_j|^k P[X = x_j] < +\infty \tag{8.3}$$

[1] See, for example, sec. 25 of Bartle.

in the discrete case and

$$E[|X|^k] = \int_{-\infty}^{+\infty} |x|^k f_X(x)\,dx < +\infty \tag{8.4}$$

in the continuous case. We call $E[|X|^k]$ the *kth absolute moment;* hence we say that the kth moment exists only if the kth absolute moment is finite.

A "physical" feeling for the first two moments can perhaps be gained as follows. Suppose that n point masses m_i, $i = 1, 2, \ldots, n$, are located at the points x_i along the x axis, as shown, for example, in Fig. 8.1. It then follows from the principles of mechanics that the *center of mass* $\bar{x}$ of the system is given by the formula

$$\bar{x} = \frac{\sum_{i=1}^{n} x_i m_i}{\sum_{i=1}^{n} m_i} \tag{8.5}$$

while the *moment of inertia* I of the system about the origin is

$$I = \sum_{i=1}^{n} x_i^2 m_i \tag{8.6}$$

Suppose next that we arbitrarily define the "probability" distribution of mass as the collection of relative masses

$$p_i \triangleq \frac{m_i}{M} \qquad i = 1, 2, \ldots, n$$

where M is the total mass of the system:

$$M \triangleq \sum_{i=1}^{n} m_i$$

We may then rewrite (8.5) and (8.6) in terms of the p_i and thus obtain the expressions

$$\bar{x} = \frac{\sum_{i=1}^{n} x_i p_i \not{M}}{\not{M}} = E[X]$$

and

$$I = \sum_{i=1}^{n} x_i^2 p_i M = ME[X^2]$$

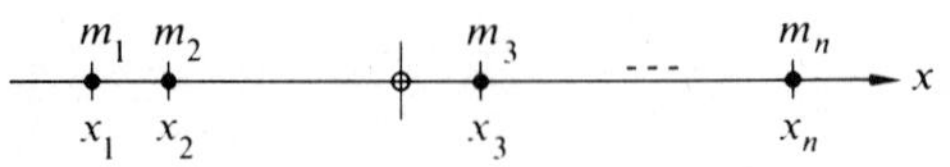

Figure 8.1 A mass distribution.

That is, the center of mass is given by the first moment (or mean value) of the distribution of mass, while the moment of inertia about the origin is given by the product of the total mass by the second moment of the distribution of mass.

One reason for the importance of the moments $E[X^k]$ lies in the fact, which we shall prove later, that if they exist for *all* values of k, and are known, then that set of moments uniquely determines the probability distribution function of the random variable X. Another reason for their importance has to do with the fact that even if all the moments do not exist, those that do help describe the nature of their attendant probability distributions. For example, reference to Figs. 6.1 to 6.4 shows that the first moment (or mean) more or less designates the center of the probability distribution. Similarly it can be shown that the square root of the second moment about the mean, that is,

$$\sigma(X) \triangleq \sqrt{E[(X - E[X])^2]} \tag{8.7}$$

provides a measure of the width of the distribution, while the third moment about the mean describes the asymmetry or skewness of the distribution.[1]

Let us now determine the moments of a few common random variables: Suppose first that we consider again a Bernoulli random variable X with the probability distribution

$$P[X = 1] = p \quad \text{and} \quad P[X = 0] = 1 - p \tag{8.8}$$

It then follows, on using this distribution in (8.1), that

$$\begin{aligned} E[X^k] &= 1^k p + 0^k(1 - p) \\ &= p \end{aligned} \tag{8.9}$$

That is, the kth moment of this Bernoulli random variable has the same value p for all values of k.

For a simple continuous example, let us consider the random variable Y which is uniformly distributed over an interval $[a,b]$. It then follows from (8.2) that we have in this case

$$\begin{aligned} E[Y^k] &= \frac{1}{b - a} \int_a^b y^k \, dy = \frac{1}{b - a} \frac{y^{k+1}}{k + 1}\bigg|_a^b \\ &= \frac{1}{k + 1} \frac{b^{k+1} - a^{k+1}}{b - a} \\ &= \frac{1}{k + 1} (b^k + b^{k-1}a + \cdots + ba^{k-1} + a^k) \end{aligned} \tag{8.10}$$

[1] Compare with secs. 15.5, 15.6, and 15.8 of Cramér.

Exercise 8.1 Consider again the time-to-failure random variable T with the exponential probability density

$$f_T(t) = \begin{cases} ae^{-at} & \text{for } 0 \leq t \\ 0 & \text{for } t < 0 \end{cases}$$

where a is a positive real number. Show that

$$E[T^k] = \frac{k!}{a^k}$$

and hence that we have in this case the result that

$$E[T^k] = k!E^k[T]$$

That is, the kth moment of the exponential time-to-failure random variable equals $k!$ times the kth power of the mean of that random variable. □

Exercise 8.2 Suppose that the random variable X has a symmetrical exponential probability density; in particular, let

$$f_X(x) = be^{-2b|x|}$$

for all x, where b is a positive real number. Determine the kth moment of X. □

Exercise 8.3 Suppose that the random variable Y has the Cauchy probability density

$$f_Y(y) = \frac{1}{\pi(1 + y^2)}$$

Show that the kth moment ($k \geq 1$) of this random variable does *not* exist when k is any even positive integer, and that it exists when k is an odd positive integer only in the sense that then the Cauchy principal values of the integrals exist and are zero. (*Hint:* See Sec. 7.) □

Having evaluated the kth moments of a few specific random variables, let us next determine some of the general properties of moments: Suppose that X is a random variable defined on a sample space $\mathcal{S}$, and suppose that the kth moment of X exists:

$$E[|X|^k] < +\infty \tag{8.11}$$

Let us consider now the jth moments, with $1 \leq j < k$. If we pick some point $s \in \mathcal{S}$ and calculate $X(s)$, it follows that either $|X(s)| \leq 1$ or $|X(s)| > 1$; we can therefore split up $\mathcal{S}$ into two regions defined by these inequalities. Consider first the region in which $|X(s)| \leq 1$ for each s.

Since the jth power of any number less than unity is itself less than unity, we have

$$|X(s)|^j \leq 1 \leq 1 + |X(s)|^k$$

throughout this region. On the other hand, consider next the region in which $|X(s)| > 1$ for each s. Here we have

$$|X(s)|^j < |X(s)|^k \leq 1 + |X(s)|^k$$

for all values of j less than k. Thus we see that we have the inequality

$$|X(s)|^j \leq 1 + |X(s)|^k$$

for any $s \in \mathcal{S}$. It therefore follows, from (6.6) and (8.11) that

$$\begin{aligned} E[|X|^j] &\leq E[1 + |X|^k] \\ &= 1 + E[|X|^k] < +\infty \end{aligned}$$

Thus it follows that *if the kth moment of a random variable exists, then so do all of the lower-order moments.*

Exercise 8.4 Let X be a random variable defined on a sample space $\mathcal{S}$. Show that if X is bounded on $\mathcal{S}$, then the kth moment of X exists for *all* values of k; that is,

$$E[|X|^k] < +\infty$$

for $k = 1, 2, 3, \ldots$. □

Exercise 8.5 Let X be a continuous random variable with a continuous probability density f_X such that the kth absolute moment of X is finite. Show that

$$\lim_{x \to -\infty} x^k F_X(x) = 0$$

and that

$$\lim_{x \to +\infty} x^k [1 - F_X(x)] = 0$$

for such a random variable, where F_X is the probability distribution function of X. □

Exercise 8.6 Let X be a random variable of the type discussed in Exercise 8.5. Using the results obtained there, show that

$$E[X] = \int_0^{+\infty} [1 - F_X(x)]\, dx - \int_{-\infty}^0 F_X(x)\, dx$$

(*Hint:* Use integration by parts.) It should be noted that this result says

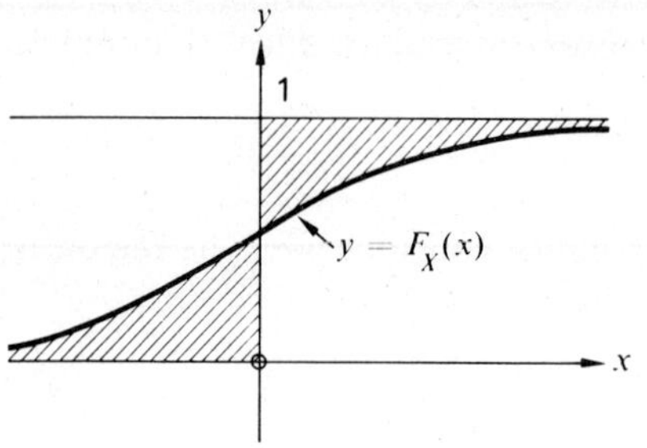

Figure 8.2 The mean as an area.

that the mean value of the random variable X is equal to the difference of the areas of the right-hand and left-hand shaded regions in Fig. 8.2. □

Central moments

It is convenient in many problems to *center* the random variables of interest, that is, to subtract off the mean values. The moments of the resulting centered random variables, the moments $E[(X - E[X])^k]$, are called the *kth central moments*.[1] A central moment of particular importance is the *variance* var (X), where

$$\operatorname{var}(X) \triangleq E[(X - E[X])^2] = \sigma^2(X) \tag{8.12}$$

The positive square root σ of the variance is often called the *standard deviation*. It follows from the definition (8.12) that the variance of a random variable gives the mean-square value of the variations about the mean of the possible values of that random variable.

Expansion of the right-hand side of (8.12) gives, on using (6.5) and (6.2),

$$\begin{aligned}\operatorname{var}(X) &= E[X^2 - 2XE[X] + E^2[X]] \\ &= E[X^2] - E^2[X]\end{aligned} \tag{8.13}$$

This formula is often more useful in calculations than is that in (8.12). For example, suppose that X is a Bernoulli random variable with the probability distribution given in (8.8) above. It then follows, from (8.9) and (8.13), that

$$\operatorname{var}(X) = p - p^2 = p(1 - p) \tag{8.14}$$

in this case.

Exercise 8.7 Let X be a discrete random variable with the Poisson probability distribution

$$P[X = k] = \frac{e^{-M}M^k}{k!} \qquad k = 0, 1, 2, \ldots$$

[1] Note that the first central moment is always zero.

where M is a positive real number. (Compare with Exercise 1.2 above.) Show that

$$\text{var}(X) = M = E[X]$$

that is, show that *the variance and mean of a Poisson random variable are equal.* □

Exercise 8.8 Let the random variable X have the normal probability density

$$f_X(x) = \frac{1}{\sqrt{2\pi}\,\sigma} \exp\left[-\frac{(x-m)^2}{2\sigma^2}\right] \tag{8.15}$$

Show that in this case

$$E[(X-m)^k] = \begin{cases} 1 \times 3 \times 5 \cdots (k-1)\sigma^k & \text{for } k \text{ even} \\ 0 & \text{for } k \text{ odd} \end{cases} \tag{8.16}$$

Since we previously showed (in Exercise 6.2 above) that in this case

$$E[X] = m$$

it then follows from (8.16) that the variance of this normal random variable is

$$\text{var}(X) = \sigma^2$$

With these results in mind, *we say that (8.15) is the probability density of a normal random variable with a mean m and a variance σ^2.* □

Finally, let us determine a few of the general properties of the variance.

Exercise 8.9 Let X be a random variable whose second moment exists, and let Y be the random variable

$$Y \triangleq aX + b$$

where a and b are real constants. Show that

$$\begin{aligned} \text{var}(Y) &= \text{var}(aX+b) \\ &= a^2 \text{var}(X) \end{aligned}$$

that is, that *multiplying a random variable by a constant multiplies its variance by the square of that constant, while adding a constant to a random variable leaves the variance unchanged.* □

Exercise 8.10 Let X be a random variable whose second moment exists, and let c be an arbitrary real number.

Show that

$$E[(X - c)^2] \geq \text{var}(X)$$

that is, show that centering a random variable about its mean gives a smaller value for the second moment $E[(X - c)^2]$ than does any other centering point c. □

Exercise 8.11 Let each of the random variables $X_1, X_2, \ldots, X_n$ have a finite second moment. Show that

$$\text{var}\left(\sum_{i=1}^{n} X_i\right) = \sum_{i=1}^{n} \text{var}(X_i)$$

when the X_i are mutually independent; in other words, *the variance of a sum of independent random variables is equal to the sum of the respective variances.* □

9 JOINT MOMENTS

Suppose now that instead of dealing with a single random variable we are, in fact, concerned with a two-dimensional random vector (with the components X and Y), defined on a sample space $\mathcal{S}$. A set of statistical averages of importance in this case are the *joint moments* $E[X^jY^k]$, where j and k may assume any positive integer values. Now

$$E[X^jY^k] = \sum_m \sum_n x_m{}^j y_n{}^k P[X = x_m, Y = y_n] \tag{9.1}$$

when X and Y are both discrete random variables [cf. (4.4)], and

$$E[X^jY^k] = \int\int_{-\infty}^{+\infty} x^j y^k f_{X,Y}(x,y)\,dx\,dy \tag{9.2}$$

when they are both continuous random variables and have a joint-probability density $f_{X,Y}$ [cf. (5.23)]. As in the single-variable case, these moments are said to exist only if the respective sums and integrals converge absolutely.

A joint moment of particular importance is the *correlation* $E[XY]$. The correlation of the centered random variables $X - E[X]$ and $Y - E[Y]$, that is, the joint moment

$$\text{cov}(X,Y) \triangleq E[(X - E[X])(Y - E[Y])] \tag{9.3}$$

is called the *covariance* of X and Y. We shall see later that the correlation and covariance have particular significance in certain estimation and sampling problems as well as in studies of linear operations on random processes.

Let us now determine a few of the general properties of the covariance.

Exercise 9.1 Let $X_1, X_2, \ldots, X_n$ all be random variables defined on a sample space $\mathcal{S}$, and let the random variable Y be defined as their sum:

$$Y \triangleq \sum_{i=1}^{n} X_i$$

Show that the variance of Y is given by the formula

$$\begin{aligned}\text{var}(Y) &= \sum_{i=1}^{n} \sum_{j=1}^{n} \text{cov}(X_i,X_j) \\ &= \sum_{i=1}^{n} \text{var}(X_i) + \sum_{\substack{i=1 \\ (i \neq j)}}^{n} \sum_{j=1}^{n} \text{cov}(X_i,X_j)\end{aligned}$$

This is the general expression for the variance of a sum of random variables (cf. Exercise 8.11). □

We shall say that the two random variables X and Y are *uncorrelated* if and only if their covariance is zero, that is, if and only if

$$\text{cov}(X,Y) = 0$$

Exercise 9.2 Let X and Y be random variables defined on the same sample space $\mathcal{S}$.

Show that X and Y are uncorrelated if and only if

$$E[XY] = E[X]E[Y]$$

(Note that both the "if" and the "only if" statements are to be proved.) □

Exercise 9.3 Let the random variables $X_1, X_2, \ldots, X_n$ all be defined on the sample space $\mathcal{S}$, and let Y be their sum.

Show that if the X_i are *pairwise uncorrelated*, that is if

$$\text{cov}(X_i,X_j) = 0$$

for all i and j such that $i \neq j$, then

$$\begin{aligned}\text{var}(Y) &\triangleq \text{var}\left(\sum_{i=1}^{n} X_i\right) \\ &= \sum_{i=1}^{n} \text{var}(X_i)\end{aligned}$$

The variance of a sum of pairwise uncorrelated random variables equals the sum of their respective variances. □

A moment's reflection on this result should remind us that we have obtained a similar result before (cf. Exercise 8.11). That such a similarity should be expected follows from our next result.

Exercise 9.4 Let X and Y be random variables defined on the same sample space $\mathcal{S}$.

Show that if X and Y are independent random variables, then they are also uncorrelated. □

That the converse statement is not necessarily true is shown by the next exercise.

Exercise 9.5 Let Z be a random variable with the uniform probability density

$$f_Z(z) = \begin{cases} \dfrac{1}{2\pi} & \text{for } 0 \leq z \leq 2\pi \\ 0 & \text{otherwise} \end{cases}$$

and let the random variables X and Y be defined by

$$X \triangleq \sin Z \qquad \text{and} \qquad Y \triangleq \cos Z$$

a. Show that X and Y are uncorrelated random variables.

b. Show that X and Y are *not* independent random variables. (*Hint:* Show that $E[X^2Y^2] \neq E[X^2]E[Y^2]$. Why does this help?) □

It follows from these last two exercises that *being uncorrelated* is a weaker (less restrictive) condition on two random variables than is *being independent*. That being uncorrelated does, however, sometimes imply statistical independence will be proved in Sec. 10.

Orthogonal random variables

As a final point in our present discussion of joint moments, it should be noted that there is another term, *orthogonal*, which is widely used to denote a relation between two random variables and which is quite similar to but *not* the same as *uncorrelated*. We say that the random variables X and Y are orthogonal if and only if it is true that

$$E[XY] = 0$$

Exercise 9.6

a. Show that if X and Y are orthogonal random variables *and* if one or both have zero means, then X and Y are uncorrelated.

b. Show that if X and Y are uncorrelated random variables *and* if one or both have zero means, then X and Y are orthogonal. □

On the other hand, consider the following results.

Exercise 9.7 Let X and Y be uncorrelated random variables with nonzero mean values.

Show that X and Y *cannot* be orthogonal random variables. □

Exercise 9.8 Let Z be a random variable such that

$$E[Z] = 0 \qquad \text{and} \qquad E[Z^2] = 1$$

In addition, let the random variables X and Y be defined by

$$X \triangleq Z + 1 \qquad \text{and} \qquad Y \triangleq Z - 1$$

a. Show that X and Y *are* orthogonal random variables.
b. Show that X and Y *are not* uncorrelated. □

In summary then, we see that if one or both of the random variables X and Y have a zero mean value, then the relations of being orthogonal and being uncorrelated are the same; on the other hand, if neither X nor Y has a zero mean, then X and Y may be either uncorrelated, or orthogonal, or neither—but not both.

10 GAUSSIAN RANDOM VECTORS

One of the most widely appearing classes of random vectors, both in theory and in practice, is the gaussian, or normal. While we will discuss this class in some considerable detail throughout our subsequent chapters, we wish to collect here a few specific results which follow most naturally from our discussion of joint moments.

The random vector **X** with components X_1 and X_2 is said to be a *two-dimensional gaussian vector*, and X_1 and X_2 are said to be *jointly gaussian random variables*, if and only if the joint probability density of X_1 and X_2 has the form

$$f_{X_1,X_2}(x_1,x_2) = \frac{1}{2\pi\sigma(X_1)\sigma(X_2)\sqrt{1-\rho^2}} \exp\left[-\frac{\left(\frac{x_1-E[X_1]}{\sigma(X_1)}\right)^2 - 2\rho\left(\frac{x_1-E[X_1]}{\sigma(X_1)}\right)\left(\frac{x_2-E[X_2]}{\sigma(X_2)}\right) + \left(\frac{x_2-E[X_2]}{\sigma(X_2)}\right)^2}{2(1-\rho^2)}\right] \tag{10.1}$$

where ρ is the *normalized covariance*

$$\rho \triangleq \frac{\text{cov}(X_1,X_2)}{\sigma(X_1)\sigma(X_2)} \tag{10.2}$$

sometimes also known as the *correlation coefficient* of X_1 and X_2.

Exercise 10.1 Show that if a pair of jointly gaussian random variables is uncorrelated, then that pair is also statistically independent. This exercise illustrates a case in which the converse to the result obtained in Exercise 9.6 above happens to be true. □

An n-dimensional random vector $\mathbf{X}$ is said to be a *gaussian random vector*, and its components $X_1, X_2, \ldots, X_n$ are called *jointly gaussian random variables*, if and only if the joint-probability density of $X_1, X_2, \ldots, X_n$ has the form

$$f_{X_1,X_2,\ldots,X_n}(x_1, x_2, \ldots, x_n)$$
$$= \frac{1}{(2\pi)^{n/2}|\mathbf{\Lambda}|^{1/2}} \exp\left[-\frac{1}{2|\mathbf{\Lambda}|}\sum_{i=1}^{n}\sum_{j=1}^{n}|\mathbf{\Lambda}|_{ij}(x_i - m_i)(x_j - m_j)\right] \quad (10.3)$$

where m_i is the mean value of X_i

$$m_i \triangleq E[X_i] \quad (10.4)$$

where $|\mathbf{\Lambda}|_{ij}$ is the cofactor of the element λ_{ij} in the determinant $|\mathbf{\Lambda}|$ of the *covariance matrix*

$$\mathbf{\Lambda} \triangleq \begin{bmatrix} \lambda_{11} & \lambda_{12} & \cdots & \lambda_{1n} \\ \lambda_{21} & \lambda_{22} & \cdots & \lambda_{2n} \\ \cdots & \cdots & \cdots & \cdots \\ \lambda_{n1} & \lambda_{n2} & \cdots & \lambda_{nn} \end{bmatrix} \quad (10.5)$$

in which the elements λ_{ij} are the covariances

$$\lambda_{ij} \triangleq \text{cov}\,(X_i,X_j) \quad (10.6)$$

and where it is assumed that the covariance matrix is nonsingular.

Exercise 10.2 Let $\mathbf{X}$ be an n-dimensional gaussian random vector with the jointly gaussian components $X_1, X_2, \ldots, X_n$.

Show that if these components are pairwise uncorrelated, that is, if

$$\text{cov}\,(X_i,X_j) = 0$$

for all $i \neq j$, then those components are also mutually independent random variables. □

11 CONDITIONAL AVERAGES

Suppose that the discrete random variables X and Y are the components of a two-dimensional random vector defined on a sample space $\mathcal{S}$. It then follows from our preceding discussions of statistical averages that we may calculate the expectation of the random variable Y either by averaging with respect to the probability distribution of Y alone or by averaging with respect to the joint-probability distribution of X and Y. Thus if the random variable X has the possible values x_i and the random variable Y has the possible values y_j, we may use either of the formulas

$$E[Y] = \sum_j y_j P[Y = y_j] \tag{11.1}$$

or

$$E[Y] = \sum_i \sum_j y_j P[X = x_i, Y = y_j] \tag{11.2}$$

Both formulas give the same result, of course.

In order to point up the possible dependence of Y on X, suppose that we express the joint probability in (11.2) in terms of the conditional-probability distribution of Y, given a value of X; that is, let us use

$$P[X = x_i, Y = y_j] = P[Y = y_j|X = x_i]P[X = x_i]$$

If we substitute this expression back into (11.2), we then get

$$E[Y] = \sum_i \left\{\sum_j y_j P[Y = y_j|X = x_i]\right\} P[X = x_i] \tag{11.3}$$

A moment's reflection upon the braced term here, that is, the term which contains the summation over the possible values of Y, should convince one that that term is a form of statistical average—an average taken with respect to a conditional probability distribution, rather than with respect to an unconditioned one. We shall accordingly call that term the *conditional expectation* of the random variable Y given that the random variable X assumes the value x_i, and we shall denote it by the symbol $E[Y|X = x_i]$:

$$E[Y|X = x_i] \triangleq \sum_j y_j P[Y = y_j|X = x_i] \tag{11.4}$$

Since the conditional probability distribution $\{P[Y = y_j|X = x_i], j = 1, 2, \ldots\}$ may in general be different for each of the possible values x_i of the random variable X, the conditional expectation $E[Y|X = x_i]$ is, in fact, a function defined on the range sample space $\mathcal{S}_X$ and hence may be considered to be a random variable. The expectation of this new random variable may then be obtained by averaging with respect to the probability distribution of X and hence is given by the right-hand side of (11.3). Thus the formula obtained by using (11.4) in (11.3), that is,

$$E[Y] = \sum_i E[Y|X = x_i]P[X = x_i] \tag{11.5}$$

says that *averaging the conditional expectation of Y (given a value of X) with respect to the random variable X yields the unconditioned average of Y.*

This technique of computing an average will be found to be quite useful in some of our subsequent developments. We shall, in addition, show later that the conditional expectation assumes importance in its own right in a study of estimation.

Exercise 11.1 Suppose that we toss a fair die[1] and a loaded coin at the same time. Let us denote the outcome of this experiment by the number of dots showing on the die if the coin shows heads and let the outcome be denoted by zero if the coin shows tails. Assuming that the number of dots showing on the die is independent of whether the coin shows heads or tails:

a. What is the conditional expectation of the outcome given that the coin shows heads?

b. What is the conditional expectation of the outcome given that the coin shows tails?

c. For what value of the probability of a head is the expected value of the outcome equal to unity? □

Our discussion of conditional expectation so far has been based on the assumption that the random variables in question were discrete random variables. Similar definitions and results can, of course, be had in other cases. Suppose, for example, that both X and Y are continuous random variables with a continuous joint-probability density $f_{X,Y}$. The conditional expectation of Y given that $X = x$ may then be defined as

$$E[Y|X = x] \triangleq \int_{-\infty}^{+\infty} y f_{Y|X}(y|x)\, dy \tag{11.6}$$

in which case we get the result

$$E[Y] = \int_{-\infty}^{+\infty} E[Y|X = x] f_X(x)\, dx \tag{11.7}$$

as the continuous analog of (11.5). If, on the other hand, we assume that X is a discrete random variable, that Y is a continuous random variable, and that the conditional probability density $f_Y(y|X = x_j)$ exists for all y and for all x_j, then it follows from our previous discussion of conditional probability density functions (Sec. 5.5) that we may define

$$E[Y|X = x_j] \triangleq \int_{-\infty}^{+\infty} y f_Y(y|X = x_j)\, dy \tag{11.8}$$

and hence that we get

$$E[Y] = \sum_j E[Y|X = x_j] P[X = x_j] \tag{11.9}$$

in this case.

[1] A *fair die* is such that the possible numbers of dots showing are all equally probable.

Having defined the conditional expectation and determined its value in a specific case, let us next determine some of its general properties.

Exercise 11.2 Let X and Y be continuous random variables with a joint-probability density $f_{X,Y}$. Let A be an event in the range space of Y and let I_A be the set-indicator function of that set.

Show that

$$E[I_A(Y)|X = x] = P[Y \in A|X = x]$$

that is, that *the conditional probability that $Y \in A$, given that $X = x$, equals the conditional expectation of the set-indicator function I_A, given that $X = x$* [cf. (6.3)]. □

Exercise 11.3 Suppose that X and Y are discrete random variables with the possible values x_i and y_j, respectively. Show that

$$E[Y|X = x_i] = E[Y]$$

for all values of i, when X and Y are statistically independent random variables. □

An example of the extension of our above results to a random vector case arises in a study of random sums of random variables. We shall study a particular case of this problem at some length in our later chapter on Poisson processes.

Exercise 11.4 Suppose that we wish to determine the expectation of the random sum

$$Z \triangleq \sum_{i=1}^{K} X_i$$

in which K is a discrete random variable having the possible values $k = 0, 1, 2, \ldots$, with probabilities $P[K = k]$ respectively, and where (given that $K = k$) the random variables $X_1, X_2, \ldots, X_k$ are independent of K and are identically distributed, discrete random variables with the possible values x_j, $j = 1, 2, \ldots$. That is,

$$P[X_i = x_j|K = k] = p_j$$

for all i, independent of K.

It now follows from our preceding discussion of the conditional average that, since (given that $K = k$) the random sum Z is a function of the random variables $X_1, X_2, \ldots,$ and X_k, we have

$$E[Z|K = k] = \sum_{j_1}\sum_{j_2} \cdots \sum_{j_k} ZP[X_1 = x_{j_1}, X_2 = x_{j_2}, \ldots, X_k = x_{j_k}|K = k]$$

a. Show that

$$E[Z|K = k] = km$$

where

$$m \triangleq \sum_j x_j p_j$$

b. Show that therefore

$$E[Z] = mE[K]$$

□

12 INEQUALITIES

From time to time in our subsequent developments we will wish to be able to calculate bounds on various probabilities and expectations. With this fact in mind, we shall next derive a few inequalities which are commonly used in probability theory. An excellent introduction to the subject of inequalities in general may be found in the short monograph by Beckenbach and Bellman, while a more complete treatment can be found in the book "Inequalities" by the same authors.

Chebyshev Inequality Suppose that the random variable X has a finite second moment; that is, let

$$E[|X|^2] < +\infty \tag{12.1}$$

A very important inequality concerning such a random variable is the *Chebyshev inequality* [named after the Russian mathematician Pafnuti L. Chebyshev (1821–1894)]:

$$P[|X - E[X]| \geq \epsilon] \leq \frac{\operatorname{var}(X)}{\epsilon^2} \tag{12.2}$$

which holds for any $\epsilon > 0$.

Proof First of all we note (using results derived in Sec. 8 above) that (12.1) implies that both the mean $E[X]$ and the variance var (X) are finite. Next, assuming that $\epsilon > 0$, the left-hand side of the Chebyshev inequality (12.2) is the probability that the random

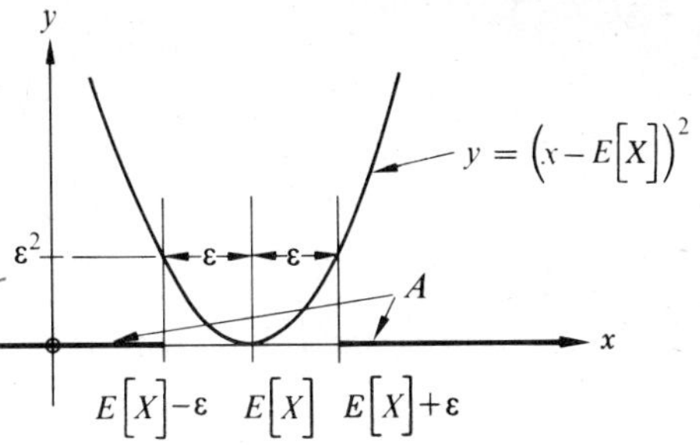

Figure 12.1 Concerning the Chebyshev inequality.

variable X has a value which falls in the region

$$A \triangleq \{x: |x - E[X]| \geq \epsilon, x \in \mathcal{S}_X\} \tag{12.3}$$

shown in Fig. 12.1.

Suppose now that we assume, for convenience, that X has a probability density f_X. We may then write

$$\text{var}(X) = \int_{-\infty}^{+\infty} (x - E[X])^2 f_X(x)\, dx \tag{12.4}$$

Since the integrand here is always nonnegative, it follows that the integral over the infinite interval $(-\infty, +\infty)$ must be greater than (or at least equal to) the integral taken over any smaller region; in particular,

$$\int_{-\infty}^{+\infty} (x - E[X])^2 f_X(x)\, dx \geq \int_A (x - E[X])^2 f_X(x)\, dx \tag{12.5}$$

Now it follows from the definition of A that

$$(x - E[X])^2 \geq \epsilon^2 \tag{12.6}$$

for every $x \in A$, and hence that

$$\int_A (x - E[X])^2 f_X(x)\, dx \geq \epsilon^2 \int_A f_X(x)\, dx \tag{12.7}$$

The integral on the right-hand side of this inequality is, however, just the probability that X assumes a value in A. Thus it follows, on combining (12.4) through (12.7), that we get the result

$$\text{var}(X) \geq \epsilon^2 P[|X - E[X]| \geq \epsilon]$$

for any $\epsilon > 0$. If next we solve for P, we finally get (12.2). While we assumed here that X had a probability density, it is not difficult to show that the same result holds in the discrete case, and hence that the Chebyshev inequality holds in general. ∎

Exercise 12.1 Perhaps the most common use of the Chebyshev inequality is in limiting arguments in theoretical studies in which the right-hand side of (12.2) can be shown to approach zero. For example:

Let X be a random variable such that its variance is zero:

$$\text{var}(X) = 0$$

Show, using the Chebyshev inequality, that

$$X = E[X]$$

with probability one; that is, show that

$$P[X = E[X]] = 1$$

(*Hint:* Pass to the limit as $\epsilon \to 0$ and use the continuity theorem of probability.) □

Exercise 12.2 While the Chebyshev inequality is indeed very useful in theoretical studies, it does not often provide in practice a very good bound on the probability $P[|X - E[X]| \geq \epsilon]$.

Suppose, for example, that the random variable X has the uniform probability density

$$f_X(x) = \begin{cases} \dfrac{1}{2\sqrt{3}} & \text{for } -\sqrt{3} \leq x \leq \sqrt{3} \\ 0 & \text{otherwise} \end{cases}$$

In this case

$$E[X] = 0 \qquad \text{and} \qquad \text{var}(X) = 1$$

a. Calculate the probability $P[|X - E[X]| \geq \epsilon]$ for this random variable and plot your result as a function of ϵ.

b. Plot the Chebyshev bound $\text{var}(X)/\epsilon^2$ for this case and compare with the result in part (*a*). □

Exercise 12.3 Generalized Chebyshev inequality The Chebyshev inequality may be generalized as follows: Let X be a random variable defined on a sample space $\mathcal{S}$ and let g be a nonnegative, even, measurable function defined on the range sample space $\mathcal{S}_X$ such that $g(x)$ is an increasing function of x for $0 \leq x \leq +\infty$ and such that $E[g(X)]$ exists. An example of such a function is shown in Fig. 12.2. Show that it follows in this case that

$$P[|X| \geq a] \leq \frac{E[g(X)]}{g(a)} \tag{12.8}$$

where a is an arbitrary positive real number. □

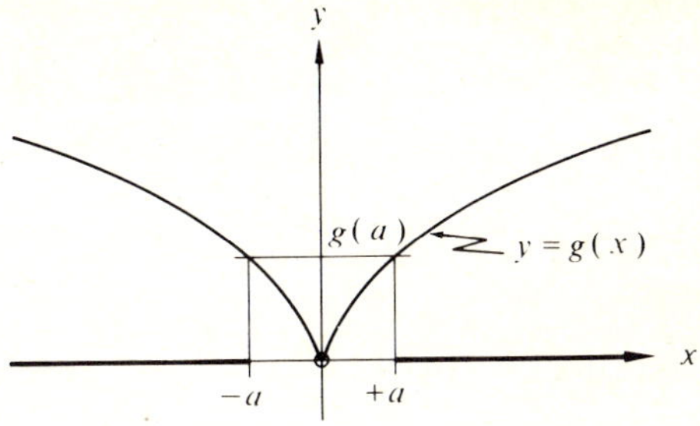

Figure 12.2 Concerning the generalized Chebyshev inequality.

CAUCHY-SCHWARZ INEQUALITY Another very important inequality is the *Cauchy-Schwarz inequality* [named after Augustin-Louis Cauchy and the German mathematician Hermann Amandus Schwarz (1843–1921)]:

$$E^2[XY] \leq E[X^2]E[Y^2] \tag{12.9}$$

in which X and Y are real random variables with finite second moments.

Proof Consider the function

$$Q(\lambda) \triangleq E[(X - \lambda Y)^2] \tag{12.10}$$

in which λ is real. Now the right-hand side of (12.10) is nonnegative no matter what may be the value of λ. Thus it follows that

$$Q(\lambda) = E[X^2] - 2\lambda E[XY] + \lambda^2 E[Y^2] \geq 0$$

such a function is said to be a *nonnegative quadratic form* in λ. A plot of Q versus λ must either not intercept the λ axis at all (say as shown in Fig. 12.3*a*) or it must touch the λ axis at one and only one point (say as shown in Fig. 12.3*b*).

Exercise 12.4

a. Find the value of λ (say λ_0) at which $Q(\lambda)$ is a minimum.
b. Using the result of (*a*), show that

$$E^2[XY] < E[X^2]E[Y^2]$$

if $Q(\lambda_0) > 0$.
c. Using the result of (*a*), show that

$$E^2[XY] = E[X^2]E[Y^2]$$

if $Q(\lambda_0) = 0$. □

On combining the results (*b*) and (*c*) of Exercise 12.4, the proof of the Cauchy-Schwarz inequality is completed. ■

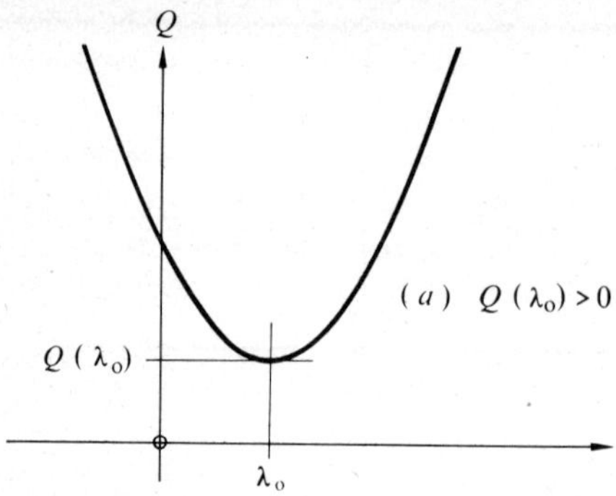

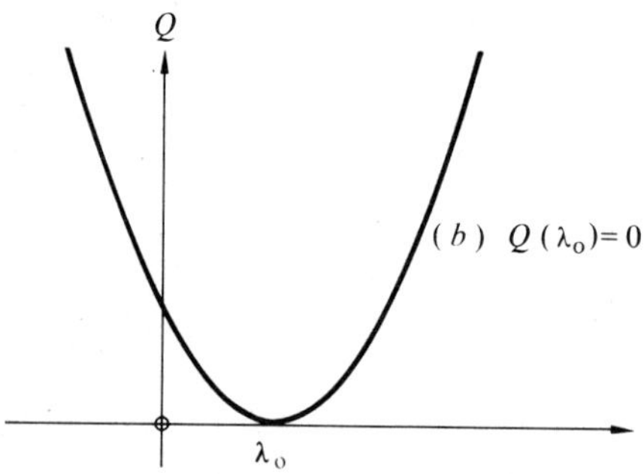

Figure 12.3 The quadratic form $Q(\lambda)$.

Exercise 12.5 Show that $Q(\lambda) = 0$, and hence that the result in part (*c*) of the preceding exercise holds if and only if

$$X = \lambda Y$$

with probability one. [*Hint:* Use (12.8) and an argument paralleling that used in proving the result in Exercise 12.1 above.] □

Exercise 12.6 As an example of the use of the Cauchy-Schwarz inequality, consider the following problem: Let X and Y be real random variables with finite second moments. Show that

$$-1 \leq \rho \leq +1$$

where

$$\rho \triangleq \frac{\text{cov}\,(X,Y)}{\sigma(X)\sigma(Y)}$$

and hence is the normalized covariance of X and Y. □

TRIANGLE INEQUALITY Suppose that X and Y are random variables defined upon a sample space $\mathcal{S}$ such that both X and Y have finite first absolute moments. It then follows from the triangle inequality for real numbers[1] that we have

$$|X(s) + Y(s)| \le |X(s)| + |Y(s)| \tag{12.11}$$

for every sample point $s \in \mathcal{S}$. It then further follows from the order inequality (6.6) that

$$E[|X + Y|] \le E[|X|] + E[|Y|] \tag{12.12}$$

This result is known as the *triangle inequality* for random variables.

Exercise 12.7 Let X and Y be random variables defined on a sample space $\mathcal{S}$ such that their rth absolute moments are finite

$$E[|X|^r] < +\infty \qquad \text{and} \qquad E[|Y|^r] < +\infty$$

where r is a positive integer. Show in this case that

$$E[|X + Y|^r] \le 2^r(E[|X|^r] + E[|Y|^r]) \tag{12.13}$$

[*Hint:* Consider separately the sets of points $s \in \mathcal{S}$ such that $|X(s)| \le |Y(s)|$ and $|X(s)| > |Y(s)|$, and apply (12.11).] □

*C_r INEQUALITY** While the coefficient 2^r in (12.13) is easy to obtain, it is by no means the best upper-bound coefficient that can be found. A better result is as follows:

Let X and Y be random variables defined on a sample space $\mathcal{S}$ with finite rth absolute moments (where now r may be any positive number). It then follows that

$$E[|X + Y|^r] \le C_r(E[|X|^r] + E[|Y|^r]) \tag{12.14}$$

where

$$C_r = \begin{cases} 1 & \text{if } 0 < r \le 1 \\ 2^{r-1} & \text{if } r > 1 \end{cases} \tag{12.15}$$

This result is known as the *C_r inequality.*

Proof For any point $s \in \mathcal{S}$, let

$$x \triangleq X(s) \qquad \text{and} \qquad y \triangleq Y(s)$$

Suppose now that we consider the function

$$\frac{|x|^r + |y|^r}{(|x| + |y|)^r} = \left(\frac{|x|}{|x| + |y|}\right)^r + \left(\frac{|y|}{|x| + |y|}\right)^r \tag{12.16}$$

[1] See, for example, sec. 1.21, vol. I, of Apostol.

* Supplementary material.

which was chosen with the form of the right-hand side of (12.14) in mind. A moment's reflection on (12.16) shows that each side is symmetric in x and y. In addition, if we denote the first term on the right-hand side by z,

$$z \triangleq \frac{|x|}{|x| + |y|}$$

then the second term on the right has the form $1 - z$, where z is bounded between zero and one, and (12.16) has the form

$$f(z) = z^r + (1 - z)^r \tag{12.17}$$

where $0 \le z \le 1$. We would now like to determine a lower bound on $f(z)$, for doing so will give us the coefficient C_r.

We note first that it follows from (12.17) that

$$f(1 - z) = f(z) \qquad \text{and that} \qquad f(0) = 1 = f(1)$$

that is, that $f(z)$ is symmetric about the point $z = \frac{1}{2}$ and has endpoint values of unity, as shown, for example, in Fig. 12.4. Since the midpoint value is

$$f(\tfrac{1}{2}) = \frac{1}{2^r} + \frac{1}{2^r} = \frac{1}{2^{r-1}}$$

we see from Fig. 12.4 that we may write

$$f(z) \ge \frac{1}{C_r}$$

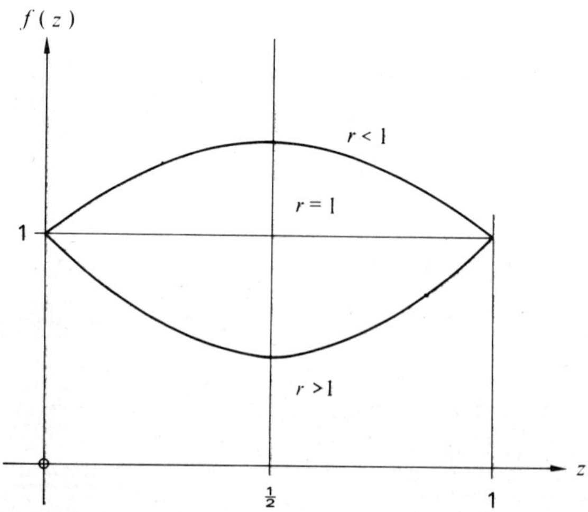

Figure 12.4 Bounds on $f(z)$.

where we have defined

$$C_r \triangleq \begin{cases} 1 & \text{if } 0 < r \leq 1 \\ 2^{r-1} & \text{if } r > 1 \end{cases}$$

Thus we have shown that

$$\frac{|x|^r + |y|^r}{(|x| + |y|)^r} \geq \frac{1}{C_r}$$

that is, that

$$(|x| + |y|)^r \leq C_r(|x|^r + |y|^r)$$

Since it follows from the triangle inequality that

$$|x + y|^r \leq (|x| + |y|)^r$$

we may then write

$$|x + y|^r \leq C_r(|x|^r + |y|^r)$$

As this result holds for all $s \in \mathcal{S}$, it finally follows from the order inequality (6.6) that we obtain the C_r inequality on taking expectations. ∎

13 SUMMARY OF DEFINITIONS AND FORMULAS

Statistical average or expectation

Let X be a discrete random variable with the possible values x_k. The statistical average of X is then given by the formula

$$E[X] \triangleq \sum_k x_k P[X = x_k] \tag{1.2}$$

Functions of random variables

Let g be a measurable function of the random variable X. Then

$$E[g(X)] = \sum_k g(x_k) P[X = x_k] \tag{3.9}$$

when X is a discrete random variable with the possible values x_k, and

$$E[g(X)] = \int_{-\infty}^{+\infty} g(x) f_X(x)\, dx \tag{5.16}$$

when X is a continuous random variable with a probability density f_X.

Random vectors

Let $\mathbf{X}$ be an n-dimensional random vector with the components X_1, $X_2, \ldots, X_n$:

$$\mathbf{X} = (X_1, X_2, \ldots, X_n)$$

The statistical average of $\mathbf{X}$ is then given by

$$E[\mathbf{X}] \triangleq (E[X_1], E[X_2], \ldots, E[X_n]) \tag{4.2}$$

If g is a measurable function on $\mathcal{S}_{\mathbf{X}}$, then

$$E[g(\mathbf{X})] = \sum_{k_1}\sum_{k_2} \cdots \sum_{k_n} g(x_{k_1}, x_{k_2}, \ldots, x_{k_n})P[X_1 = x_{k_1}, X_2 = x_{k_2}, \ldots, X_n = x_{k_n}] \tag{4.4}$$

when $\mathbf{X}$ is a discrete random vector, and

$$E[g(\mathbf{X})] = \int_{-\infty}^{+\infty} \cdots \iint_{-\infty}^{+\infty} g(x_1, x_2, \ldots, x_n) f_{X_1,X_2,\ldots,X_n}(x_1, x_2, \ldots, x_n)\, dx_1\, dx_2 \cdots dx_n \tag{5.23}$$

when $\mathbf{X}$ is a continuous random vector with a joint-probability density $f_{\mathbf{X}}$.

General properties of the statistical average

Let X be a random variable defined on a sample space $\mathcal{S}$. For any event $A \subset \mathcal{S}_X$,

$$E[I_A(X)] = P[X \in A] \qquad (3.11) \text{ and } (6.3)$$

where I_A is the set indicator of the event A.

For any real number a,

$$E[aX] = aE[X] \qquad (3.13) \text{ and } (6.2)$$

Let X_1 and X_2 be random variables defined on a sample space $\mathcal{S}$. For any real numbers a_1 and a_2:

$$E[a_1X_1 + a_2X_2] = a_1E[X_1] + a_2E[X_2] \qquad (4.8) \text{ and } (6.5)$$

that is, the expectation operator is a linear functional.

$$|E[X]| \leq E[|X|] \qquad (3.14) \text{ and } (6.4)$$

If $X(s) \geq 0$ for every $s \in \mathcal{S}$, then

$$E[X] \geq 0 \tag{1.12}$$

Let X_1 and X_2 be random variables defined on a sample space $\mathcal{S}$. If $X_1(s) \geq X_2(s)$ for every $s \in \mathcal{S}$, then

$$E[X_1] \geq E[X_2] \qquad [\text{Exercise 4.6 and } (6.6)]$$

If X_1 and X_2 are independent random variables, then

$$E[X_1X_2] = E[X_1]E[X_2] \qquad \text{(4.10) and (6.7)}$$

Moments

The kth moment of the random variable X is the expected value of the kth power of X. Thus

$$E[X^k] = \sum_j x_j{}^k P[X = x_j] \tag{8.1}$$

when X is a discrete random variable with the possible values x_j, and

$$E[X^k] = \int_{-\infty}^{+\infty} x^k f_X(x)\,dx \tag{8.2}$$

when X is a continuous random variable with a probability density f_X.

Variance

The variance var (X) of the random variable X is defined by

$$\text{var}\,(X) \triangleq E[(X - E[X])^2] \tag{8.12}$$
$$= E[X^2] - E^2[X] \tag{8.13}$$

Properties of the variance are:

a.

$$\text{var}\,(aX + b) = a^2\,\text{var}\,(X) \qquad \text{(Exercise 8.9)}$$

for any real numbers a and b.

b. For any real number c,

$$E[(X - c)^2] \geq \text{var}\,(X) \qquad \text{(Exercise 8.10)}$$

c. If $X_1, X_2, \ldots, X_n$ are mutually independent random variables (cf. Exercise 8.11), or if they are pairwise uncorrelated (cf. Exercise 9.5), then

$$\text{var}\left(\sum_{i=1}^{n} X_i\right) = \sum_{i=1}^{n} \text{var}\,(X_i)$$

Uncorrelated random variables

We say that the two random variables X and Y are uncorrelated if and only if their covariance is zero, that is, if and only if

$$\text{cov}\,(X,Y) \triangleq E[(X - E[X])(Y - E[Y])] = 0 \tag{9.3}$$

and hence if and only if

$$E[XY] = E[X]E[Y] \tag{9.12}$$

Conditional averages

Let X and Y both be discrete random variables. The conditional average of Y, given that $X = x_i$, is defined by

$$E[Y|X = x_i] \triangleq \sum_j y_j P[Y = y_j|X = x_i] \tag{11.4}$$

The basic properties of the conditional average are:

a.

$$E[Y] = \sum_i E[Y|X = x_i]P[X = x_i] \tag{11.5}$$

b. Let A be any event in $\mathcal{S}_Y$. Then

$$E[I_A(Y)|X = x] = P[Y \in A|X = x] \qquad \text{(Exercise 11.2)}$$

c. If X and Y are independent random variables, then

$$E[Y|X = x_i] = E[Y] \qquad \text{(Exercise 11.3)}$$

Averages of specific distributions

Bernoulli random variables Let X have the two possible values 0 and 1, and let

$$P[X = 1] = p \qquad \text{and} \qquad P[X = 0] = 1 - p$$

Then

$$E[X] = p \tag{1.6}$$
$$E[X^k] = p \qquad k = 1, 2, 3, \ldots \tag{8.9}$$

and

$$\text{var}\,(X) = p(1 - p) \tag{8.14}$$

Binomial random variables Let Y have the possible values 0, 1, 2, . . . , n, and let

$$P[Y = k] = \binom{n}{k} p^k q^{n-k}$$

where $q \triangleq 1 - p$. Then

$$E[Y] = np \tag{1.8}$$

and

$$\text{var}\,(Y) = npq \qquad \text{(Exercise 9.1)}$$

Poisson random variables Let X have the possible values 0, 1, 2, . . . , and let

$$P[X = k] = \frac{e^{-M}M^k}{k!}$$

where $M > 0$. Then

$$E[X] = M = \text{var}(X) \qquad \text{(Exercise 8.7)}$$

Uniform random variables Let X be uniformly distributed over the interval $[a,b]$. Then

$$E[X] = \frac{a+b}{2} \qquad (6.10)$$

and

$$E[X^k] = \frac{1}{k+1}(b^k + b^{k-1}a + \cdots + ba^{k-1} + a^k) \qquad (8.10)$$

Exponential random variables Let T have the exponential probability density

$$f_T(t) = \begin{cases} ae^{-at} & \text{for } 0 \le t \\ 0 & \text{for } t < 0 \end{cases}$$

where $a > 0$. Then

$$E[T] = \frac{1}{a} \qquad (6.12)$$

and

$$E[T^k] = \frac{k!}{a^k} \qquad \text{(Exercise 8.1)}$$

Normal random variables Let X have the normal probability density

$$f_X(x) = \frac{1}{\sqrt{2\pi}\,\sigma} \exp\left[-\frac{(x-m)^2}{2\sigma^2}\right]$$

Then

$$E[X] = m \qquad (6.14)$$

and

$$\text{var}(X) = \sigma^2 \qquad \text{(Exercise 8.8)}$$

Rayleigh random variables Let R have the Rayleigh probability density

$$f_R(r) = \begin{cases} \dfrac{r}{b} e^{-r^2/2b} & \text{for } 0 \le r \\ 0 & \text{otherwise} \end{cases}$$

Then

$$E[R] = \sqrt{\frac{b\pi}{2}} \qquad (6.17)$$

Chebyshev inequality

Suppose that the random variable X has a finite second moment. Then

$$P[|X - E[X]| \geq \epsilon] \leq \frac{\text{var }(X)}{\epsilon^2} \tag{12.2}$$

for any $\epsilon > 0$.

Generalized Chebyshev inequality

Let X be a random variable, let g be a nonnegative, even, measurable function of X such that $g(x)$ is increasing for $0 \leq x < +\infty$, and let $E[g(X)]$ exist. Then

$$P[|X| \geq a] \leq \frac{E[g(X)]}{g(a)} \tag{12.8}$$

for any $a > 0$.

Cauchy-Schwarz inequality

Let the real random variables X and Y have finite second moments. Then

$$E^2[XY] \leq E[X^2]E[Y^2] \tag{12.9}$$

C_r inequality

Let the random variables X and Y have finite rth absolute moments. Then,

$$E[|X + Y|^r] \leq C_r(E[|X|^r] + E[|Y|^r]) \tag{12.14}$$

where

$$C_r = \begin{cases} 1 & \text{if } 0 < r \leq 1 \\ 2^{r-1} & \text{if } 1 < r \end{cases} \tag{12.15}$$

REFERENCES

Apostol, Tom M.: "Calculus," Blaisdell Publishing Company, New York, vol. I, 1961; vol. II, 1962.

Bartle, Robert G.: "The Elements of Real Analysis," John Wiley & Sons, Inc., New York, 1964.

Bartle, Robert G.: "The Elements of Integration," John Wiley & Sons, Inc., New York, 1966.

Beckenbach, Edwin F., and Richard Bellman: "An Introduction to Inequalities," Random House, Inc., New York, 1961.

Beckenbach, Edwin F., and Richard Bellman: "Inequalities," 2d rev., Springer-Verlag OHG, Berlin, 1965.

Cramér, Harald: "Mathematical Methods of Statistics," Princeton University Press, Princeton, N.J., 1946.

Dwight, Herbert B.: "Tables of Integrals and Other Mathematical Data," The Macmillan Company, New York, 1947.

Krickeberg, Klaus: "Probability Theory," Addison-Wesley Publishing Company, Inc., Reading, Mass., 1965.

Loève, Michel: "Probability Theory," 3d ed., D. Van Nostrand, Inc., Princeton, N.J., 1963.

Neveu, Jacques: "Mathematical Foundations of the Calculus of Probability," Holden-Day, Inc., Publisher, San Francisco, 1965.

Royden, Halsey L.: "Real Analysis," The Macmillan Company, New York, 1963.

THE CONDITIONAL MEAN OF $h(X)$ GIVEN AN EVENT M IS:

$$E\{h(X)/M\} = \int_{-\infty}^{\infty} h(x)\, f_X(x/M)\, dx$$

$$E\{h(XY)\} = \int_{-\infty}^{\infty}\int_{-\infty}^{\infty} h(xy)\, f_{XY}(x,y)\, dx\, dy$$

LET $h(XY) = XY$ $\quad E\{XY\} = \int_{-\infty}^{\infty}\int_{-\infty}^{\infty} xy\, f_{XY}(xy)\, dx\, dy$ - CORRELATION BETWEEN X & Y

IF $h(XY) = (X-\bar{X})(Y-\bar{Y})$ $\quad E\{h(XY)\} = \overline{XY} - \bar{X}Y$ COVARIANCE

$$= \int_{-\infty}^{\infty}\int_{-\infty}^{\infty} (x-\bar{x})(y-\bar{y})\, f_{XY}(x,y)\, dx\, dy$$

IF $h(XY) = \left(\frac{X-\bar{X}}{\sigma_X}\right)\left(\frac{Y-\bar{Y}}{\sigma_Y}\right)$ $\quad E\left\{\left(\frac{X-\bar{X}}{\sigma_X}\right)\left(\frac{Y-\bar{Y}}{\sigma_Y}\right)\right\} = \rho_{XY}$ CORRELATION COEFFICIENT

FOR ZERO-MEAN GAUSSIAN R.V.

$$f_X(x) = \frac{1}{\sqrt{2\pi}\,\sigma}\, \epsilon^{-\frac{(x-m)^2}{2\sigma^2}} = \frac{1}{\sqrt{2\pi}\,\sigma}\, \epsilon^{-x^2/2\sigma^2}$$

$E[x] = \int_{-\infty}^{\infty} x\, f_x(x)\, dx$

TO WORK $\int$ LET $U = \frac{x^2}{2\sigma^2}$: $dU = \frac{2x\, dx}{2\sigma^2}$

$E[x] = \sigma/\sqrt{2\pi}$

POISSON'S THEOREM (IN TIME)

PLACE "AT RANDOM" n points IN THE INTERVAL $(0,T)$. WHAT IS THE PROBABILITY OF k OF THESE points IN THE INTERVAL (t_1, t_2)? PROB. {k points in the interval (t_1,t_2)} $= \binom{n}{k} p^k q^{n-k} = P\{k \text{ in } t_a\}$ WHERE $p = \frac{t_2 - t_1}{T} = \frac{t_a}{T}$, $q = 1-p$

IF np IS OF THE ORDER 1, THEN $\binom{n}{k} p^k q^{n-k} \approx \epsilon^{-np}(np)^k/k!$

IF $n \gg 1$ AND $p = \frac{t_a}{T} \ll 1$ and k OF THE ORDER OF $\frac{n t_a}{T} = np$

THEN $P\{k \text{ in } t_a\} \approx \epsilon^{-\frac{n t_a}{T}} \frac{\left(\frac{n t_a}{T}\right)^k}{k!}$. LETTING $\lambda = \frac{n}{T}$

THEN $P\{k \text{ in } t_a\} = \epsilon^{-\lambda t_a} \frac{(\lambda t_a)^k}{k!}$ $\quad \lambda = \frac{n}{T}$ = THE AVERAGE OCCURENCE OF POINTS IN TIME

LAW OF LARGE NUMBERS

IF $\epsilon > 0$ AND $\delta > 0$ AND n_0 IS CHOOSEN SUFFICIENTLY LARGE THEN

$P\{|k/n - p| \le \epsilon\} \ge 1 - \delta$ FOR EVERY $n > n_0$ OR $P\{|k/n - p| < \epsilon\} = 1$ FOR $n \to \infty$

$P\{|k/n - p| \le \epsilon\} \approx \mathrm{erf}\left(\frac{\epsilon \cdot n}{\sqrt{npq}}\right) - \mathrm{erf}\left(\frac{-\epsilon n}{\sqrt{npq}}\right)$ (THE ≈ IS FROM deMOIVRE LAPLACE)

8
Estimation, Sampling, and Prediction

1 INTRODUCTION

We have, up to this point, generally assumed that the probability distribution functions of the various random variables under consideration were known and that the problem before us was first to calculate some function of those probability distribution functions, for example, a mean or a variance, and then use that calculated value to predict the result which might be obtained in some specified experiment. In practice, however, we also often have to face what is, in essence, the inverse problem. That is, we often have a situation in which the probability properties of some random variable of interest are unknown a priori, and we wish to determine empirically the probability distribution function of some given random variable, or, alternatively, we might only wish to determine some simpler property, say the mean.

Consider, however, the results obtainable from a performance (or from a finite set of performances) of an experiment whose possible outcomes are characterized by the random variable of interest. Such a set of experimental results can only partly describe the probabilistic properties of the random variable in question; at best we can only hope to get an *estimate* of those properties in this way. A basic problem then in any such situation is to find a function of the measurement results which will enable us to obtain a good estimate of the particular probabilistic property of interest.

Now we have often implied in our earlier chapters that we could estimate a probability or a mean value by using sampling techniques. We shall reexamine that problem in this chapter. Having done so, we shall then study the problem of estimating, or predicting, the value of one random variable when we have observed a particular value of a second (and hopefully related) random variable.

Our goals in this chapter are simply to introduce some of the pertinent concepts in sampling and prediction theory and, further, to try to better our understanding of the relation between the mathematical world of probability and the real world of empirical data. More complete elementary treatments of this material may be found in chaps. 6 and 9 of Hoel and in chaps. 7 and 10 of Chernoff and Moses; good intermediate level discussions may be found in chaps. 7 and 8 of Mood and Graybill and in parts 3 and 4 of Freeman; and rather complete advanced treatments may be found in part III of Cramér and in chaps. 8 to 12 of Wilks.

2 THE SAMPLE MEAN

Let us consider now the specific problem of estimating the mean value of some given random variable. If our previous comments about the relation between statistical average and empirical average are to have any validity at all, then we should be able to obtain a reasonable estimate of that mean value by calculating the empirical average of some set of observed values of the random variable in question. We wish to explore this possibility in some detail in this section.

Suppose, in particular, that we are given a random variable X with a range sample space $\mathcal{S}_X$ and a probability distribution function F_X, and that we wish to estimate the mean value of X:

$$m \triangleq E[X]$$

With this in mind, suppose that we perform the basic experiment which defines the random variable X a fixed number of times in succession (say n); we shall call such a sequence of basic experiments a *combined experiment*. If now we denote by $x(i)$ the observed value of the random variable X in the ith of the n performances of the basic experiment in a given combined experiment, then the empirical average of such a set of observed values is

$$\bar{x} = \frac{1}{n} \sum_{i=1}^{n} x(i)$$

[cf. (1.2.1)]. Since we might well obtain different values for the empirical average in different performances of the *combined* experiment, those values may be viewed as sample values of a random variable characterizing the results of the combined experiment. We would like to determine the statistical properties of this random variable.

To this end, suppose that we repeat our *combined* experiment a number of times and that we focus our attention for the moment upon the results $x(i)$ of the ith basic experiment in each of the combined experiments. Now clearly those values of $x(i)$ are simply sample values of the random variable X. However, since it will be convenient in our subsequent discussion to keep separate in our minds the results of the n different basic experiments making up a combined experiment, it will also be convenient now to represent for each i ($i = 1, 2, \ldots, n$) the various observed values of $x(i)$ by a different random variable, say X_i. We shall call X_i the ith *random sample* of the random variable X. Having thus introduced the random samples X_i, we may then represent the various possible values of the empirical average $\bar{x}$ by the random variable

$$\hat{m}_n \triangleq \frac{1}{n} \sum_{i=1}^{n} X_i \tag{2.1}$$

which we shall call the *sample mean.* We have used here the symbol ($\hat{}$) to denote *estimator* and in addition, have introduced the subscript n to denote the number of random samples; *the symbol $\hat{m}_n$ thus denotes an estimator of the mean value m which is based upon n random samples.*

It follows from the definition (2.1) that the expectation of the sample mean is

$$E[\hat{m}_n] = E\left[\frac{1}{n} \sum_{i=1}^{n} X_i\right]$$
$$= \frac{1}{n} \sum_{i=1}^{n} E[X_i]$$

Since, by definition, each of the random samples X_i has the same probability distribution as does the original random variable X, it follows that

$$E[X_i] = m \qquad \text{for } i = 1, 2, \ldots, n$$

and hence that

$$E[\hat{m}_n] = m \tag{2.2}$$

Thus we see that *the expectation of the sample mean is equal to the mean value of the random variable being studied.* An estimator whose expectation is equal to the parameter being estimated is called an *unbiased estimator;* the sample mean is therefore an unbiased estimator for the mean.

While it is comforting to find that the expected value of the sample mean is indeed the desired mean value, it would also be nice to know how likely it is that a measured value of the sample mean will be close to the actual mean value being estimated. In particular, we would like to determine the probability $P[|\hat{m}_n - m| \geq \epsilon]$ that the measurement error exceeds some given bound. Now it turns out in practice, unfortunately, that we are not often able to derive an expression for the probability distribution of the sample mean (which we need in order to be able to compute the above probability). We can, however, obtain an upper bound on that probability of error by using the Chebyshev inequality (7.12.2), for it follows directly from that inequality that we may write

$$P[|\hat{m}_n - m| \geq \epsilon] \leq \frac{\text{var}\,(\hat{m}_n)}{\epsilon^2} \tag{2.3}$$

since $E[\hat{m}_n] = m$ here. If, therefore, we can calculate the variance of the sample mean, then we can also calculate an upper bound on the probability of measurement error.

Now it follows from the definition of the sample mean (2.1) that

$$\begin{aligned} \text{var}\,(\hat{m}_n) &= \text{var}\left(\frac{1}{n}\sum_{i=1}^{n} X_i\right) \\ &= \frac{1}{n^2}\,\text{var}\left(\sum_{i=1}^{n} X_i\right) \end{aligned}$$

(cf. Exercise 7.8.9). Suppose now that the results of any one of the n basic experiments which make up a combined experiment does not affect in any way the results of any other such basic experiment. In this case the n random samples X_i are pairwise independent[1] and it therefore follows (cf. Exercise 7.8.10) that

$$\begin{aligned} \text{var}\,(\hat{m}_n) &= \frac{1}{n^2}\sum_{i=1}^{n} \text{var}\,(X_i) \\ &= \frac{1}{n}\,\text{var}\,(X) \end{aligned} \tag{2.4}$$

the last equality following from the fact that all the random samples have, by definition, the same probability distribution function as does the

[1] We shall study the case of dependent random samples in the following chapter on random processes.

original random variable X. Thus we see that *the variance of the sample mean is inversely proportional to the number of random samples taken when those random samples are pairwise independent.*

We note further that it follows from this last result that we can decrease indefinitely the variance of the sample mean simply by increasing the number of independent random samples; that is, it follows directly from (2.4) that

$$\lim_{n\to\infty} \operatorname{var}(\hat{m}_n) = 0 \tag{2.5}$$

and hence that

$$\lim_{n\to\infty} E[(\hat{m}_n - m)^2] = 0 \tag{2.6}$$

whenever the random samples are pairwise independent. Because of (2.6), we say that the sample mean *converges in mean square* to the desired mean value whenever the random samples are pairwise independent. It further follows, on using (2.4) in (2.3), that

$$\lim_{n\to\infty} P[|\hat{m}_n - m| \geq \epsilon] = 0 \tag{2.7}$$

in this case, for every $\epsilon > 0$. That is, *the probability that the measurement error exceeds ϵ tends to zero for every $\epsilon > 0$ as the number of pairwise-independent random samples increases without limit.* The sample mean is said therefore to *converge in probability* to the mean value being estimated and, for that reason, the sample mean is said to be a *consistent estimator* of that mean value whenever the individual random samples are pairwise independent.

The result expressed by (2.7) is known as the *weak law of large numbers.* It says, in essence, that: *If* we measure the sample mean a large number of times for a fixed number of random samples (say n), and *if* we repeat that measurement procedure over and over again using larger and larger numbers of random samples, *then* the resultant distributions of measured values of the sample mean should become more and more concentrated about the mean value being estimated as n becomes larger and larger. That is, the probabilistic inference expressed by (2.7) is based upon the *distribution* of measured values obtained in the past and does no more than make a prediction about the *distribution* of possible results in the future. Thus, in particular, (2.7) does *not* say that if we make a single sequence of measurements upon some given random variable, then the resultant empirical averages determined from that particular single observed sequence necessarily become closer and closer to the mean value of the random variable under study as the number of random samples becomes larger and larger.

In summary, we have shown here that empirical averages calculated using experimental data can indeed be employed to give useful estimates of statistical mean value—so long as we use large enough numbers of independent observations and so long as we keep in mind the distributional implications of the weak law of large numbers.

3 RELATIVE FREQUENCY

An important special case of interest is that in which the sample mean corresponds to a relative frequency:

Suppose that we sample n times some given random variable X and that we determine for each random sample X_i ($i = 1, 2, \ldots, n$) whether or not some given event A occurs. The event A might, for example, be the event that an observed value of a random sample is positive. The set-indicator function for the event A in the ith random sample is

$$I_i \triangleq \begin{cases} 1 & \text{if } X_i \in A \\ 0 & \text{if not} \end{cases}$$

The sample mean of this set-indicator function then is

$$\hat{m}_n = \frac{1}{n} \sum_{i=1}^{n} I_i$$

Now it follows from the definition of the set-indicator function I_i that the sum here simply counts the number of times n_A that the event A occurs in the n samples:

$$\sum_{i=1}^{n} I_i = n_A$$

Thus it follows that

$$\hat{m}_n = \frac{n_A}{n} \tag{3.1}$$

that is, *the sample mean of the set-indicator function of the event A is simply the relative frequency of occurrence of that event.*

Let us now determine some of the statistical properties of the sample mean in the present case. First, we note that the set-indicator function I_i is a Bernoulli random variable with the probability distribution (for all i)

$$P[I_i = 1] = P[X_i \in A] \triangleq p$$

and

$$P[I_i = 0] = 1 - p$$

We shall suppose for the remainder of this section that the set indicators I_i are independent random variables.

Exercise 3.1 Show that

$$E\left[\frac{n_A}{n}\right] = p \qquad \text{and that} \qquad \operatorname{var}\left(\frac{n_A}{n}\right) = \frac{p(1-p)}{n}$$

that is, *the expectation of the relative frequency of occurrence of a given event is equal to the probability of occurrence of that event,* and the variance of the relative frequency varies inversely with the number of independent random samples. □

Exercise 3.2

a. Show further that the above variance achieves its maximum value when $p = \frac{1}{2}$; that is, show that

$$\operatorname{var}\left(\frac{n_A}{n}\right) \leq \frac{1}{4n}$$

whatever the value of p might be.

b. Show that it therefore follows that

$$P\left[\left|\frac{n_A}{n} - p\right| \geq \epsilon\right] \leq \frac{1}{4n\epsilon^2}$$ □

The right-hand side of the last inequality tends to zero as the number of random samples taken increases without limit (regardless of the value of p) and hence so does the left-hand side. Thus we have shown that if we sample a random variable and if the set indicators of some event A are independent then it follows that:

a. The expectation of the relative frequency of occurrence of the event A is equal to the probability of occurrence of that event, and

b. The relative frequency of occurrence of the event A converges in probability to the probability of occurrence of that event as the number of random samples increases without limit.

This result is known as the *Bernoulli theorem.*

4 RELATIVE FREQUENCY (CONTINUED)*

A review of the preceding section shows that a relative frequency obtained from independent random samples is a weighted sum of inde-

* Supplementary section. The material in this section is somewhat of a digression from the main theme of the chapter. The techniques used and the results obtained are of interest, however, in many applications.

pendent Bernoulli random variables. We shall show in Sec. 9.2 of the following chapter on random processes that such a sum has a binomial probability distribution. We now wish to make use of that result here in order to obtain more detailed properties of the relative frequency than we were able to obtain in the preceding section. First, however, we will need to develop some of the detailed properties of binomially distributed random variables.

Suppose that the random variable Y_n is defined to be the sum of n independent Bernoulli random variables, that is,

$$Y_n \triangleq \sum_{i=1}^{n} X_i$$

where here the random variables X_i are mutually independent and each has the same Bernoulli probability distribution:

$$P[X_i = 1] = p \qquad \text{and} \qquad P[X_i = 0] = q = 1 - p$$

It then follows from the results of Sec. 9.2 of the following chapter that in this case the sum random variable Y_n has the binomial probability distribution

$$\begin{aligned} P[Y_n = k] &= \binom{n}{k} p^k q^{n-k} \\ &= \frac{n!}{k!(n-k)!} p^k q^{n-k} \end{aligned} \tag{4.1}$$

where $k = 0, 1, 2, \ldots, n$. Extensive tables of the values of these probabilities are given in the National Bureau of Standards publication AMS-6.

First, we note that it follows from (4.1) that

$$\frac{P[Y_n = k]}{P[Y_n = k-1]} = \frac{(1 + n - k)p}{qk} \tag{4.2}$$

and hence that

$$\frac{P[Y_n = k]}{P[Y_n = k-1]} = 1 + \frac{(n+1)p - k}{qk} \tag{1.3}$$

The sign of the second term on the right-hand side of (4.3) changes according to whether k is smaller or larger than $(n + 1)p$. Thus we have

$$P[Y_n = k] > P[Y_n = k - 1] \qquad \text{when } k < (n+1)p \tag{4.4}$$

while, on the other hand, we have

$$P[Y_n = k] < P[Y_n = k - 1] \qquad \text{when } k > (n+1)p \tag{4.5}$$

It therefore follows that the binomial probabilities $P[Y_n = k]$ first

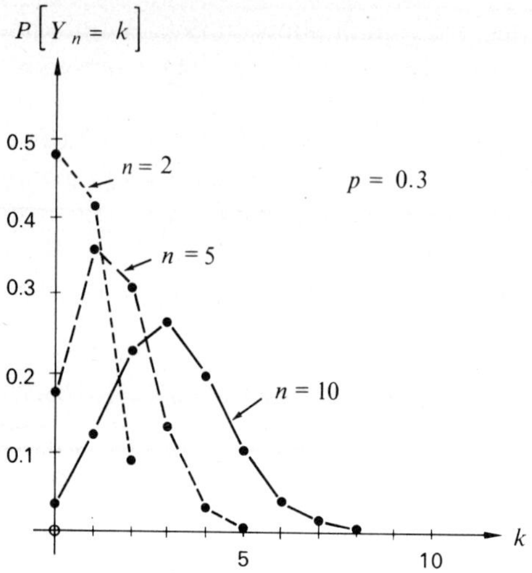

Figure 4.1 Binomial probability distributions.

increase monotonically and then decrease monotonically as k increases from zero to n, and reach a maximum near the value $k = np$. This behavior is shown graphically in Fig. 4.1 for the case $p = 0.3$. It may also be noted from that figure that, for a fixed value of p, the graph of the binomial probability distribution spreads out as n increases while its peak height decreases.

The formula for the binomial probabilities given in (4.1) becomes quite awkward to use in calculations when the value of n becomes very large. Two different approximations for these probabilities will be developed in the following sequence of exercises. The first is the normal approximation. The derivation of this approximation makes use of Stirling's formula[1] for the factorial function:

$$n! = \sqrt{2\pi}\, n^{n+(1/2)} e^{-n+(\theta/12n)}$$

where θ is a positive number such that

$$0 < \theta < 1$$

Now it follows immediately from this formula that upper and lower bounds on the factorial are

$$\sqrt{2\pi}\, n^{n+(1/2)} e^{-n} < n! < \sqrt{2\pi}\, n^{n+(1/2)} e^{-n+(1/12n)}$$

and hence (since $e^{1/12n} \to 1$ as $n \to \infty$) that

$$n! \approx \sqrt{2\pi}\, n^{n+(1/2)} e^{-n}$$

[1] Compare with formulas (3.9) in chap. 3 of Artin.

gives an approximation for the factorial which is quite valid for large values of n.

Exercise 4.1 Using the above results for the factorial, show that an upper bound for the binomial probability is given by

$$P[Y_n = k] < \left[\frac{n}{2\pi k(n-k)}\right]^{1/2} \left(\frac{np}{k}\right)^k \left(\frac{nq}{n-k}\right)^{n-k} e^{1/12n}$$

and, further, that the approximation

$$P[Y_n = k] \approx \left[\frac{n}{2\pi k(n-k)}\right]^{1/2} \left(\frac{np}{k}\right)^k \left(\frac{nq}{n-k}\right)^{n-k} \tag{4.6}$$

is therefore obtained for large n. □

We showed earlier in this section that the binomial probability distribution tended to peak around the value $k \approx np$. With this in mind, let us introduce the new variable

$$\delta \triangleq k - np$$

that is, δ measures the distance of a given value k from the location of the central peak of the distribution.

Exercise 4.2 Show that (4.6) may be rewritten in terms of the new variable δ as

$$P[Y_n = k] \approx \frac{1}{\left[2\pi npq\left(1 + \dfrac{\delta}{np}\right)\left(1 - \dfrac{\delta}{nq}\right)\right]^{1/2}} \frac{1}{\left(1 + \dfrac{\delta}{np}\right)^{np+\delta}\left(1 - \dfrac{\delta}{nq}\right)^{nq-\delta}} \tag{4.7}$$

□

We would now like to investigate the behavior of this last result as n becomes very large: First we note that, for fixed δ, and very large values of n,

$$2\pi npq\left(1 + \frac{\delta}{np}\right)\left(1 - \frac{\delta}{nq}\right) \approx 2\pi npq$$

A large n approximation for the second half of the right-hand side of (4.7) is a bit more tedious to obtain. Basically the idea is to take the logarithm of the denominator and then, realizing that δ/np and δ/nq are very small for fixed δ and large values of n, to use the power-series expansion of the natural logarithm:

$$\log (1 + x) = x - \frac{x^2}{2} + \frac{x^3}{3} - \frac{x^4}{4} + \cdots$$

which is valid for $|x| < 1$.

Exercise 4.3 Using the technique discussed above, show that the approximation

$$\left(1 + \frac{\delta}{np}\right)^{np+\delta} \left(1 - \frac{\delta}{nq}\right)^{nq+\delta} \approx \exp\left(\frac{\delta^2}{2npq}\right)$$

is obtained for fixed δ, as n becomes very large. □

On combining the various results above, we finally get the *normal approximation*

$$\begin{aligned} P[Y_n = k] &\approx \frac{1}{\sqrt{2\pi npq}} \exp\left(-\frac{\delta^2}{2npq}\right) \\ &= \frac{1}{\sqrt{2\pi npq}} \exp\left[-\frac{(k - np)^2}{2npq}\right] \end{aligned} \tag{4.8}$$

which is reasonably valid for large values of n. A detailed analysis of the error terms involved in this approximation is given in chap. 9 of Fry's book.

Unfortunately, the normal approximation derived above becomes less and less accurate when n increases and p decreases in such a way that np remains essentially constant, for example, for those cases in which n is very large and p is very small. An alternative approximation which is quite valid in this latter case may be obtained as follows.

Exercise 4.4 Let Y_n be the binomial random variable with the probability distribution given by (4.1).

a. Show that

$$P[Y_n = 0] = (1 - p)^n$$

b. On taking the logarithm of both sides of the result in (*a*), and on defining

$$\lambda \triangleq np$$

show that

$$\lim_{n \to \infty} P[Y_n = 0] = e^{-\lambda}$$

when λ is held constant. □

Exercise 4.5 Using the technique suggested in Exercise 4.4, show that

$$\lim_{n\to\infty} P[Y_n = 1] = \lambda e^{-\lambda}$$

and that

$$\lim_{n\to\infty} P[Y_n = 2] = \frac{\lambda^2}{2} e^{-\lambda}$$

when λ is held constant. □

Having obtained the results shown in Exercises 4.4 and 4.5, it then follows by induction that we may obtain the result

$$\lim_{n\to\infty} P[Y_n = k] = \frac{\lambda^k e^{-\lambda}}{k!} = \frac{(np)^k e^{-np}}{k!} \tag{4.9}$$

Thus we see that the binomial probability distribution has a *Poisson approximation* when n is very large and p is very small.

Binomial relative frequencies

Suppose now that the random variables X_i considered above are the set-indicator functions I_i introduced in the preceding section. It then follows that the random variable

$$Z_n \triangleq \frac{Y_n}{n}$$

is just the relative frequency n_A/n discussed in Sec. 3, for

$$Z_n = \frac{1}{n}\sum_{i=1}^{n} I_i = \frac{n_A}{n}$$

Now since the events $Z_n = k/n$ and $Y_n = k$ are equivalent, it follows that the probability distribution of the relative frequency is given by the binomial distribution

$$P\left[Z_n = \frac{k}{n}\right] = \binom{n}{k} p^k q^{n-k} \tag{4.10}$$

for $k = 0, 1, 2, \ldots, n$. All our above results for the properties of the binomial probabilities therefore also apply here.

A somewhat different view of the probability distribution of the relative frequencies may be obtained, however, as follows: Suppose that we replot the data in Fig. 4.1 for the present situation; this is done in Fig. 4.2. We see from this new figure that the probability distribution

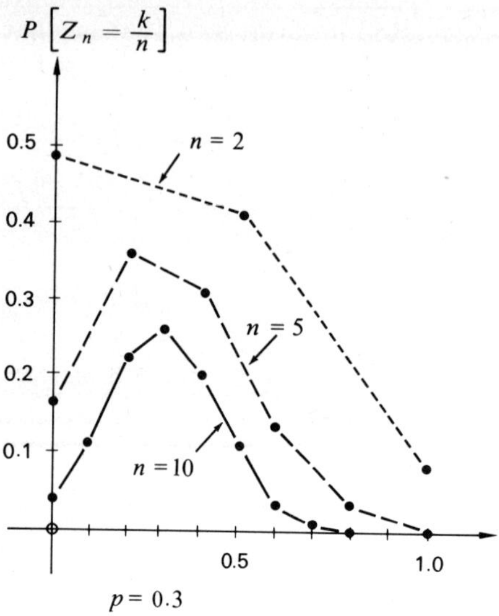

Figure 4.2 Relative-frequency probability distributions for Bernoulli random variables.

of the relative frequency becomes more and more concentrated about the value

$$\frac{k}{n} = 0.3 = p$$

as n becomes larger and larger. This result is, of course, simply a direct illustration of the Bernoulli theorem, which we derived in Sec. 3 above. It may be recalled that the derivation of the Bernoulli theorem depends upon the use of the Chebyshev inequality and upon the calculation of a bound on the variance of the relative frequency. We now wish to show that the Bernoulli theorem may also be derived directly from the properties of the binomial probability distribution.

Let us first study the behavior for large n of the probability that $Z_n > p + \epsilon$, where ϵ is some small (but fixed) positive number. It is clear from the definition of Z_n that the event $[Z_n > p + \epsilon]$ is equivalent to the event $[Y_n > n(p + \epsilon)]$, where Y_n is the binomial random variable. Thus we see that

$$P[Z_n > p + \epsilon] = P[Y_n > n(p + \epsilon)] = \sum_{k=r}^{n} P[Y_n = k] \tag{4.11}$$

where r is defined to be the next integer larger than $n(p + \epsilon)$. A bound upon this sum may be obtained by using the various results derived above.

Thus, on setting $k = r + 1$ in (4.2), we get

$$\frac{P[Y_n = r + 1]}{P[Y_n = r]} = \frac{(n - r)p}{(r + 1)q} < 1$$

the inequality following from (4.5) so long as $n\epsilon > p$. This result may be extended as in the following exercise.

Exercise 4.6 Show that

$$\frac{P[Y_n = r + j]}{P[Y_n = r]} < \left[\frac{(n - r)p}{(r + 1)q}\right]^j$$

where $j = 2, 3, 4, \ldots$. □

On using this inequality in (4.11), we then obtain the result that

$$P[Z_n > p + \epsilon] < P[Y_n = r] \sum_{j=0}^{n=r} \left[\frac{(n - r)p}{(r + 1)q}\right]^j \tag{4.12}$$

where, as pointed out above, r is the smallest integer such that

$$r > n(p + \epsilon)$$

Since the sum in (4.12) is a geometric sum, each of whose terms is less than unity, it follows that a closed-form upper bound is easily obtained.

Exercise 4.7 Show that

$$P[Z_n > p + \epsilon] < P[Y_n = r] \frac{(r + 1)q}{1 + r - (n + 1)p} \tag{4.13}$$

□

We finally are in a position to let n increase without limit.

Exercise 4.8 Show that

$$\frac{(r + 1)q}{1 + r - (n + 1)p} < \frac{[n(p + \epsilon) + 2]q}{n\epsilon + q}$$

and hence that

$$\lim_{n \to \infty} \frac{(r + 1)q}{1 + r - (n + 1)p} < c\,\frac{(p + \epsilon)q}{\epsilon}$$

when $1 < c < 2$. □

Thus it follows that the multiplying factor on the right-hand side of (4.13) is bounded for fixed ϵ, as $n \to \infty$. It can next be shown, using the results of Exercises 4.1 through 4.3, that

$$P[Y_n = r] < \frac{1}{\sqrt{2\pi npq}}$$

for all $r > n(p + \epsilon)$ and large enough n. On using these last results in (4.13), it then follows that we get

$$\lim_{n\to\infty} P[Z_n > p + \epsilon] = 0$$

In a similar manner it can be shown that

$$\lim_{n\to\infty} P[Z_n < p - \epsilon] = 0$$

and hence that

$$\lim_{n\to\infty} P[|Z_n - p| > \epsilon] = 0 \tag{4.14}$$

This result is simply the second half of the Bernoulli theorem as stated in Sec. 3.

5 MINIMUM-VARIANCE ESTIMATORS

We showed in the several preceding sections of this chapter that the sample mean is not only an intuitively obvious estimator to use for estimating the mean value of a random variable, but also that it has, in fact, desirable statistical properties in many problems of interest. Nonetheless, we now wish to reconsider the problem of estimating a mean value from a somewhat different point of view. In particular, we wish to investigate whether or not we can derive the *form* of an estimator of the mean value from a specification of the properties which we would like that estimator to have.

Suppose again that we wish to estimate the mean value m of a given random variable X by using some function of the data obtained by sampling techniques. From our studies up to this point, it seems reasonable to require that any such function $\hat{m}$ must be an *unbiased estimator* to be of interest. We shall therefore require that

$$E[\hat{m}] = m$$

If we do so, it then follows that the probability of measurement error is given by

$$P[|\hat{m} - m| \geq \epsilon] = P[|\hat{m} - E[\hat{m}]| \geq \epsilon]$$

and hence (using the Chebyshev inequality) that

$$P[|\hat{m} - m| \geq \epsilon] < \frac{\text{var}(\hat{m})}{\epsilon^2}$$

that is, the variance of the unbiased estimator determines an upper bound to the probability of measurement error. With this fact in mind, it then also seems reasonable to choose that particular unbiased estimator which has the minimum variance. We shall, for this reason, confine our attention in the remainder of this section to *minimum-variance unbiased estimators.*

Unfortunately, we still have not placed enough boundary conditions on our problem to lead us to a concrete answer. Basically, there are two approaches available to us here: We can put further restrictions upon the class of allowable estimators, or we can put further restrictions upon the class of random variables being studied. We shall consider restricting the class of allowable estimators first.

A moment's reflection upon the definition of the sample mean shows that it is a particular linear combination of the available random samples; that is, it is a particular *linear estimator.* A question therefore arises as to whether or not there is a better linear estimator of the mean than the sample mean. The following two exercises show, however, that the sample mean is a minimum-variance, unbiased, linear estimator in at least one important case.

Let $\hat{m}$ be an arbitrary linear estimator; that is, suppose that

$$\hat{m} \triangleq \sum_{i=1}^{n} a_i X_i \tag{5.1}$$

where the a_i are (as yet) arbitrary real constants, and where the X_i are random samples, as usual.

Exercise 5.1 Show that the coefficients a_i in (5.1) must satisfy the relation

$$\sum_{i=1}^{n} a_i = 1 \tag{5.2}$$

if the linear estimator $\hat{m}$ defined in (5.1) is to be an unbiased estimator of the mean m. □

Exercise 5.2 Let $\hat{m}$ be the linear estimator of m defined in (5.1). Show that if the coefficients a_i are adjusted so as to minimize the variance of $\hat{m}$ when the various samples are pairwise uncorrelated [subject to the side condition that (5.2) be satisfied], then

$$a_i = \frac{1}{n}$$

for $i = 1, 2, \ldots, n$. (*Hint:* Use Lagrange multipliers. See, for example, Apostol, vol. II, sec. 4.22.) □

It follows from the results of these two exercises that, *if the various samples are pairwise uncorrelated, then the minimum-variance, unbiased, linear sampling estimator of m has the form*

$$\hat{m} = \frac{1}{n} \sum_{i=1}^{n} X_i$$

and hence is, in fact, just the sample mean $\hat{m}_n$ defined earlier.

Suppose next that instead of bounding the problem by restricting the class of possible estimators, we consider instead restrictions on the class of random variables which we shall observe. First, however, we wish to derive a lower bound for the variance of *any* unbiased estimator.

Cramér-Rao lower bound

Since the only estimators $\hat{m}$ of interest to us are those which are unbiased, it follows that

$$E[\hat{m} - m] = 0$$

This may be expressed as

$$\int_{-\infty}^{+\infty} \cdots \int_{-\infty}^{+\infty} (\hat{m} - m) f_{\mathbf{x}}(\mathbf{x})\, d\mathbf{x} = 0 \tag{5.3}$$

where

$$f_{\mathbf{x}}(\mathbf{x}) = f_{X_1,X_2,\ldots,X_n}(x_1, x_2, \ldots, x_n)$$

is the joint-probability density of the random samples involved (which we assume exists).

If we differentiate (5.3) with respect to m (remembering that $\hat{m}$ cannot be a function of m, for we cannot use estimators of m which require an a priori knowledge of m!), we get

$$\begin{aligned} 0 &= \int_{-\infty}^{+\infty} \cdots \int_{-\infty}^{+\infty} \frac{\partial}{\partial m} [(\hat{m} - m) f_{\mathbf{x}}(\mathbf{x})]\, d\mathbf{x} \\ &= \int_{-\infty}^{+\infty} \cdots \int_{-\infty}^{+\infty} (\hat{m} - m) \frac{\partial f_{\mathbf{x}}(\mathbf{x})}{\partial m}\, d\mathbf{x} - 1 \end{aligned} \tag{5.4}$$

Now we know that

$$\frac{1}{f_{\mathbf{x}}(\mathbf{x})} \frac{\partial f_{\mathbf{x}}(\mathbf{x})}{\partial m} = \frac{\partial \log f_{\mathbf{x}}(\mathbf{x})}{\partial m}$$

If then we use this result in (5.4), we get, on rearranging terms,

$$\int_{-\infty}^{+\infty} \cdots \int_{-\infty}^{+\infty} \left[(\hat{m} - m) \frac{\partial \log f_{\mathbf{x}}(\mathbf{x})}{\partial m} \right] f_{\mathbf{x}}(\mathbf{x})\, d\mathbf{x} = 1$$

The left-hand side here is simply the expectation of the function inside the brackets:

$$E\left[(\hat{m} - m)\,\frac{\partial \log f_{\mathbf{X}}(\mathbf{x})}{\partial m}\right] = 1 \tag{5.5}$$

On squaring both sides of (5.5) and on using the Cauchy-Schwarz inequality we then get

$$1 \leq E[(\hat{m} - m)^2]E\left[\left(\frac{\partial \log f_{\mathbf{X}}(\mathbf{x})}{\partial m}\right)^2\right]$$

The first term on the right-hand side of this inequality is simply the variance of the estimator (since $\hat{m}$ is unbiased). If, therefore, we solve for that variance, we finally get the result that

$$\text{var}\,(\hat{m}) \geq \frac{1}{E\left[\left(\dfrac{\partial \log f_{\mathbf{X}}(\mathbf{x})}{\partial m}\right)^2\right]} \tag{5.6}$$

which is one form of the *Cramér-Rao* lower bound.[1] Thus we have shown that the variance of *every* unbiased estimator $\hat{m}$ of the mean m has a lower bound which is determined solely by the properties of the joint-probability density of the random samples used.

Exercise 5.3 Fill in the missing steps between (5.3) and (5.6). □

Exercise 5.4 Suppose that the random variable X under study has the probability density $f_X(x)$. Suppose further that the random samples $X_1, X_2, \ldots, X_n$ are mutually independent random variables.

Show that the Cramér-Rao lower bound becomes

$$\text{var}\,(\hat{m}) \geq \frac{1}{nE\left[\left(\dfrac{\partial \log f_X(x)}{\partial m}\right)^2\right]} \tag{5.7}$$

in the case of independent random samples. □

Having obtained the lower bound in (5.7), we are finally able to determine the form of the minimum-variance, unbiased estimator in at least one important case.

Exercise 5.5 Suppose that the random variable under study is a gaussian random variable with mean m and variance σ^2. Suppose further that the random samples are mutually independent.

[1] Compare with sec. 32.2 of Cramér.

Show that we get

$$\text{var}\,(\hat{m}) \geq \frac{\sigma^2}{n}$$

in this case. □

This last exercise shows that the variance of any unbiased estimator obtained from independent random samples of a gaussian random variable cannot be less than σ^2/n. Now the derivations in Sec. 2 show that the variance of the sample mean $\hat{m}_n$ is exactly σ^2/n in the present case. It therefore follows that *the sample mean $\hat{m}_n$ is a minimum-variance, unbiased estimator of the mean m of a gaussian random variable when the samples are independent* and, hence, that no other sampling estimator (whether linear or not) can have a lower variance in this case.

6 PREDICTION

Up to this point, we have been solely concerned with the problem of estimating the value of an unknown parameter, for example, the mean value, of some otherwise known probability distribution. We now wish to study a different estimation problem: Estimating, or predicting, the value of one random variable, given that we have observed a particular value of a second (and hopefully related) random variable. The basic problem before us here is to determine some statistical property of a pair of random variables which is useful in the given prediction problem.

Suppose, in particular, that we have a problem which involves two given variables, say X and Y. One way of getting a feeling for their statistical interrelation would be to perform the joint experiment which is characterized by X and Y a number of times and then to plot the results so obtained as points in the xy plane, that is, to plot the various observed sample points. The resultant figure, which might have one of the various forms shown in Fig. 6.1, is called a *scatter diagram*.

If the experimentally determined points are more or less evenly scattered about the xy plane (say as shown in Fig. 6.1a), then it might well be expected that the given random variables were not particularly dependent one upon the other. If, on the other hand, the observed sample points are more or less clustered about the immediate vicinity of some curve in the xy plane (say as shown in Fig. 6.1b and c), then we might expect that that curve in some way describes the interdependence of X and Y. Thus, for example, if the observed scatter diagram had a form similar to that shown in Fig. 6.1b, we would feel the random variables X and Y were essentially linearly dependent.

Suppose next that, instead of determining experimental values of

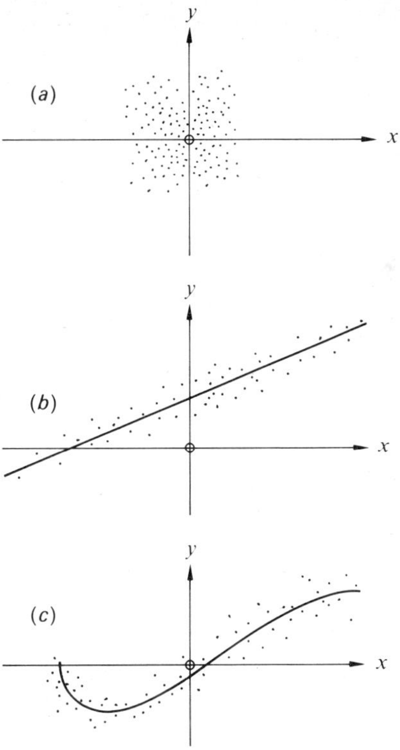

Figure 6.1 Typical scatter diagrams.

both of the random variables X and Y, we were only to measure values of X alone and that we wished to predict the corresponding values of Y. If previous measurements of both X and Y had led us to suspect that a reasonable estimate $\hat{y}$ of the value y of the random variable Y (which corresponds to a given experimental value x of the random variable X) might well be given by a straight-line function of x, then we might write

$$\hat{y} = a + bx \tag{6.1}$$

Now as x assumes different values in separate performances of the experiment characterized by the random variable X, we can use (6.1) to calculate the corresponding (different) values of the estimate $\hat{y}$. These different values of $\hat{y}$ may then be considered to be sample values of a new random variable $\hat{Y}$, defined by

$$\hat{Y} \triangleq a + bX \tag{6.2}$$

We shall call such a random variable $\hat{Y}$ an *estimator*, or *predictor*, of the values of the random variable Y. Having decided that the two random variables X and Y were linearly dependent, and having therefore decided upon a predictor $\hat{Y}$ as defined in (6.2), the problem remains to choose the values of the coefficients a and b in (6.2) in such a way that $\hat{Y}$ provides the best possible prediction of Y.

The above discussion illustrates a fundamental problem in prediction: How do you determine the specific functional form for a predictor so as to provide the best possible prediction of the random variable of interest? In order to solve such a problem one must, of course, first specify just what is meant by "best possible prediction." Unfortunately, however, this is by no means a trivial problem. The usual (and, in fact, reasonable) way out of this dilemma is to ignore the philosophical implications in "best possible" and simply specify that a predictor should have certain properties which seem important either for intuitive reasons or because they have been found useful in applications. On doing so, however, one should then be cautious about using such words as "best" or "optimum."

One convenient measure of goodness which has been found quite useful in many applications is the mean-square difference between the random variable Y and its predictor $\hat{Y}$, that is,

$$\epsilon \triangleq E[(Y - \hat{Y})^2] \tag{6.3}$$

often called the *mean-square error*.[1] With this measure of goodness, the "best" estimator is the one which minimizes the mean-square error.

Exercise 6.1 Suppose that X and Y are a pair of random variables with a joint-probability density $f_{X,Y}$, let $g(X)$ be any measurable function of the random variable X, and suppose that $g(X)$ has a finite second moment.

Show that if we use $g(X)$ as a predictor of Y,

$$\hat{Y} = g(X)$$

then it follows that the mean-square prediction error may be expressed in the form

$$E[(Y - \hat{Y})^2] = E[Y^2] - 2E[g_0(X)g(X)] + E[g^2(X)] \tag{6.4}$$

where $g_0(X)$ is the conditional expectation of Y given that $X = x$:

$$\begin{aligned} g_0(X) &\triangleq E[Y|X = x] \\ &= \int_{-\infty}^{+\infty} y f_{Y|X}(y|x)\, dy \end{aligned} \tag{6.5}$$

[cf. (7.11.6)]. □

[1] Compare with chap. 10 of Chernoff and Moses.

Exercise 6.2 Let X and Y be the random variables discussed in Exercise 6.1 and let $g(X)$ and $g_0(X)$ be defined as in that exercise.

a. Show that if we take $g_0(X)$ as a predictor for Y,

$$\hat{Y}_0 \triangleq g_0(X) = E[Y|X = x] \tag{6.6}$$

then it follows that the mean-square prediction error is in this case

$$E[(Y - \hat{Y}_0)^2] = E[Y^2] - E[g_0^2(X)] \tag{6.7}$$

b. Show that it therefore follows from (6.4) and (6.7) that

$$E[(Y - \hat{Y})^2] \geq E[(Y - \hat{Y}_0)^2] \tag{6.8}$$

for any measurable function $\hat{Y} = g(X)$ which has a finite second moment. That is, *the conditional expectation* of Y *given* X *gives the least-mean-square error of prediction of* Y. □

As an example of the determination of a conditional-expectation predictor, let us consider the following problem: Suppose that the random variable Φ is uniformly distributed over $[0,2\pi]$:

$$f_\Phi(\phi) = \begin{cases} \dfrac{1}{2\pi} & \text{for } 0 \leq \phi < 2\pi \\ 0 & \text{otherwise} \end{cases} \tag{6.9}$$

Now, given a value x of the random variable

$$X \triangleq \cos \Phi \tag{6.10}$$

(rather than given a value of the random variable Φ), we would like to determine conditional expectation predictors for both the random variable

$$Y \triangleq \sin \Phi \tag{6.11}$$

and the random variable

$$\begin{aligned} Z &\triangleq |Y| \\ &= |\sin \Phi| \end{aligned} \tag{6.12}$$

It is perhaps instructive at this point to consider the two joint sample spaces $\mathcal{S}_X \times \mathcal{S}_Y$ and $\mathcal{S}_X \times \mathcal{S}_Z$ as shown in Fig. 6.2. We note that if X assumes the value x_0 in $[-1,+1]$, then it follows that Y must assume *either* the value

$$y = +\sqrt{1 - x_0^2}$$

or the value

$$y = -\sqrt{1 - x_0^2}$$

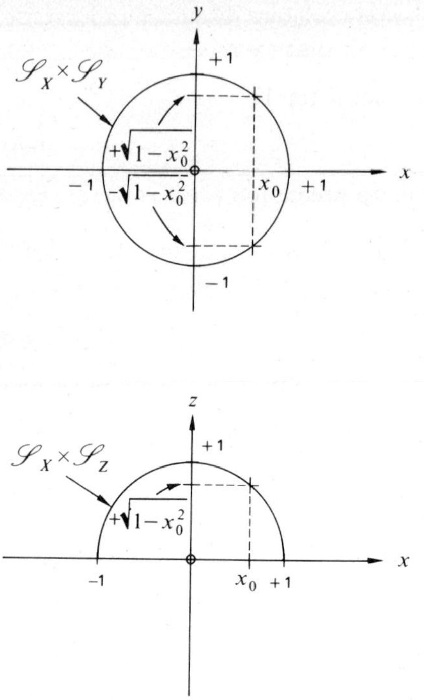

Figure 6.2 The joint sample spaces of X and Y and of X and Z.

that is, given that X assumes a value x_0 in $[-1,+1]$, then Y becomes a binary random variable. Since the random variable Φ is uniformly distributed here (by hypothesis), these two values of Y are equally likely:

$$P[Y = +\sqrt{1 - x_0{}^2}\,|X = x_0] = \tfrac{1}{2} = P[Y = -\sqrt{1 - x_0{}^2}\,|X = x_0]$$

This result holds, of course, for any value of x between -1 and $+1$. The conditional expectation of Y, given that $X = x$, is in this case

$$\begin{aligned} E[Y|X = x] &= +\sqrt{1 - x^2}\, P[Y = +\sqrt{1 - x^2}\,|X = x] \\ &\qquad - \sqrt{1 - x^2}\, P[Y = -\sqrt{1 - x^2}\,|X = x] \\ &= \frac{\sqrt{1 - x^2}}{2} - \frac{\sqrt{1 - x^2}}{2} = 0 \end{aligned}$$

Hence it follows that the conditional-expectation predictor of Y given X is zero here, that is, that

$$\begin{aligned} \hat{Y}_0 &= E[Y|X = x] \\ &= 0 \end{aligned} \tag{6.13}$$

when X and Y are defined as in (6.10) and (6.11), respectively.

We note, on the other hand, that the random variable Z can assume only the value

$$z = \sqrt{1 - x_0^2}$$

when X equals x_0 (cf. Fig. 6.2). It therefore follows that

$$P[Z = z|X = x] = \begin{cases} 1 & \text{when } z = +\sqrt{1 - x^2} \\ 0 & \text{otherwise} \end{cases}$$

and hence that

$$E[Z|X = x] = +\sqrt{1 - x^2}$$

The corresponding predictor of Z is therefore

$$\begin{aligned} \hat{Z}_0 &\triangleq E[Z|X = x] \\ &= +\sqrt{1 - x^2} \end{aligned} \tag{6.14}$$

when X and Z are defined as in (6.10) and (6.12), respectively. In this case, then, the conditional expectation predictor is a nonlinear predictor.

An example of a case in which the conditional-expectation predictor is a linear predictor is given by the following two exercises.

Exercise 6.3 Let X and Y be jointly gaussian random variables:

$$f_{X,Y}(x,y) = \frac{1}{2\pi\sigma(X)\sigma(Y)\sqrt{1 - \rho^2}} \exp\left[-\frac{\left(\dfrac{x - E[X]}{\sigma(X)}\right)^2 - 2\rho\left(\dfrac{x - E[X]}{\sigma(X)}\right)\left(\dfrac{y - E[Y]}{\sigma(Y)}\right) + \left(\dfrac{y - E[Y]}{\sigma(Y)}\right)^2}{2(1 - \rho^2)}\right] \tag{6.15}$$

where

$$\rho \triangleq \frac{\text{cov}\,(X,Y)}{\sigma(X)\sigma(Y)}$$

(cf. Sec. 7.10). Show that the conditional probability density of Y given X is

$$f_{Y|X}(y|x) = \frac{1}{\sqrt{2\pi}\,\sigma(Y)\sqrt{1 - \rho^2}} \exp\left[-\frac{\left(\dfrac{y - E[Y]}{\sigma(Y)} - \rho\,\dfrac{x - E[X]}{\sigma(X)}\right)^2}{2(1 - \rho^2)}\right] \tag{6.16}$$

in this case. (*Hint:* See Exercise 5.3.3.) □

Exercise 6.4 Let X and Y be the jointly gaussian random variables discussed in Exercise 6.3. Show that in this case

$$E[Y|X = x] = \rho \frac{\sigma(Y)}{\sigma(X)} (x - E[X]) + E[Y]$$
$$= \frac{\text{cov}(X,Y)}{\text{var}(X)} (x - E[X]) + E[Y] \tag{6.17}$$

(*Hint:* Use a technique similar to that suggested in Exercise 7.6.2.) Thus, we see that *if X and Y are jointly gaussian random variables, then it follows that the least-mean-square-error predictor of Y given that $X = x$* (that is, the conditional-expectation predictor of Y) *is a linear predictor.* □

7 LINEAR PREDICTION

Let us return now to the problem of determining the "best" *linear* predictor. In particular, suppose that, given a pair of real random variables X and Y, we wish for some reason or other to predict Y by a straight-line function of X; for example,

$$\hat{Y} \triangleq a + bX \tag{7.1}$$

Suppose further that by "best" we mean here minimum mean-square error. That is, we wish to determine the values of the coefficients a and b in (7.1) which minimize the mean-square error

$$\epsilon \triangleq E[(Y - (a + bX))^2] \tag{7.2}$$

Exercise 7.1 Show that the values of a and b which minimize (7.2) are given by the formulas

$$a = E[Y] - \frac{\text{cov}(X,Y)}{\text{var}(X)} E[X] \qquad \text{and} \qquad b = \frac{\text{cov}(X,Y)}{\text{var}(X)}$$

when both X and Y have finite second moments. □

It follows from the results of Exercise 7.1 that the least-mean-square-error linear predictor $\hat{Y}$ of the random variable Y is given by the formula

$$\hat{Y} = E[Y] + \frac{\text{cov}(X,Y)}{\text{var}(X)} (X - E[X]) \tag{7.3}$$

That is, given that the random variable X assumes a specific value x, then the least-mean-square-error linear prediction $\hat{y}$ of the value y of the ran-

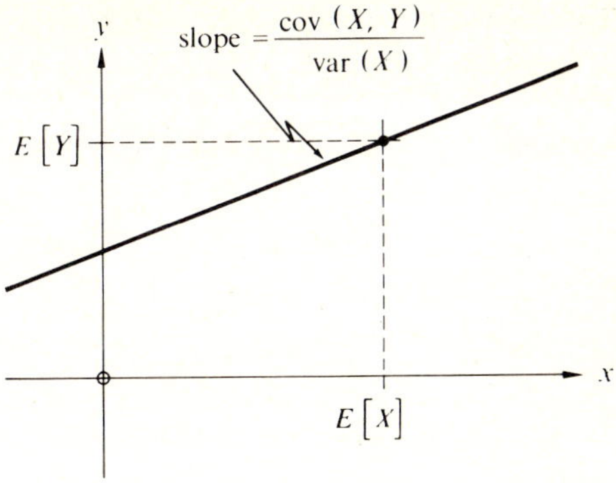

Figure 7.1 The mean-square regression line.

dom variable Y is given by the formula

$$\hat{y} \triangleq Y(x)$$
$$= E[Y] + \frac{\text{cov}\,(X,Y)}{\text{var}\,(X)}(x - E[X]) \tag{7.4}$$

which is plotted in Fig. 7.1. We see from that figure that the plot of $\hat{y}$ versus x is a straight line passing through the point $(E[X], E[Y])$ with a slope equal to $\text{cov}\,(X,Y)/\text{var}\,(X)$. This line is sometimes called the *mean-square regression line of Y on X.*

With the fact in mind that the covariance of X and Y determines the slope of the mean-square regression line, let us consider for the moment the case in which the random variables X and Y are uncorrelated, that is, the case in which

$$\text{cov}\,(X,Y) = 0$$

The slope of the regression line here is zero and hence it follows that

$$\hat{Y} = E[Y] \tag{7.5}$$

[cf. (7.4)]. That is, *if X and Y are uncorrelated random variables, then the least-mean-square-error linear prediction of Y is the mean value of Y—regardless of the observed value of X.* The same result is obtained, of course, when X and Y are independent random variables. Examples of linear prediction of random variables are given by the following two exercises.

Exercise 7.2 Suppose that the random variable Φ is uniformly distributed over $[0,2\pi]$, and let the random variables X and Y be defined by

$$X \triangleq \cos \Phi \qquad \text{and} \qquad Y \triangleq \sin \Phi$$

Determine the least-mean-square-error linear predictor for Y, given that $X = x$ and that $-1 \leq x \leq +1$. □

Exercise 7.3 Let Φ, X, and Y be defined as in Exercise 7.2, and let the random variable Z be defined by

$$\begin{aligned} Z &\triangleq |Y| \\ &= |\sin \Phi| \end{aligned}$$

Determine the least-mean-square-error linear predictor for Z, given that $X = x$ and that $-1 \leq x \leq +1$. □

We have discussed, so far, least-mean-square-error predictors in general (that is, the conditional-expectation predictors) and least-mean-square-error *linear* predictors in particular. Now we showed earlier (cf. Exercise 6.4) that the two are identical when the random variables under consideration are jointly gaussian. On the other hand, a comparison of the linear-predictor result of Exercise 7.3 for the random variables $X = \cos \Phi$ and $Z = |\sin \Phi|$ with the conditional-expectation predictor result obtained in Sec. 6 above for those same random variables shows that in this case, at least, the conditional-expectation predictor is non-linear. A feeling for when a linear predictor is most likely to be useful may, perhaps, be obtained from the following exercise.

Exercise 7.4 Suppose that the random variables X and Y have finite second moments, and consider the linear predictor

$$\hat{Y} = E[Y] + \frac{\operatorname{cov}(X,Y)}{\operatorname{var}(X)} (X - E[X])$$

derived above. Show that the prediction error for this linear predictor is given by the formula

$$E[(Y - \hat{Y})^2] = (1 - \rho^2) \operatorname{var}(Y)$$

where, as usual,

$$\rho \triangleq \frac{\operatorname{cov}(X,Y)}{\sigma(X)\sigma(Y)}$$

We see, therefore, that *the prediction error with a linear predictor becomes smaller and smaller as the random variables in question become more and more highly correlated* (that is, as ρ^2 approaches closer and closer to unity). □

8 SUMMARY OF DEFINITIONS AND FORMULAS

Sample mean

Suppose that we wish to estimate the mean value $m = E[X]$ of some given random variable X. The estimator

$$\hat{m}_n \triangleq \frac{1}{n} \sum_{i=1}^{n} X_i \tag{2.1}$$

where the X_i are random samples, is called the sample mean. This estimator has the following properties:

$$E[\hat{m}_n] = m \tag{2.2}$$

and, when the random samples X_i, $i = 1, 2, \ldots, n$, are independent random variables, a variance:

$$\text{var}\,(\hat{m}_n) = \frac{1}{n}\,\text{var}\,(X) \tag{2.4}$$

It then follows in this case that

$$\lim_{n\to\infty} P[|\hat{m}_n - m| \geq \epsilon] = 0 \tag{2.7}$$

that is, that the probability that the measurement error exceeds ϵ tends to zero for every $\epsilon > 0$ as the number of independent samples increases without limit. This result is known as the weak law of large numbers.

Relative frequency

Suppose that we sample a random variable, say X, and that we determine for each sample whether or not some given event A occurs. The random variable characterizing the relative frequency of occurrence of the event A has the statistical properties

$$E\left[\frac{n_A}{n}\right] = p \qquad \text{and} \qquad \text{var}\left(\frac{n_A}{n}\right) = \frac{p(1-p)}{n} \qquad \text{(Exercise 3.1)}$$

where

$$p \triangleq P[X \in A]$$

It then follows from these results that

$$P\left[\left|\frac{n_A}{n} - p\right| \geq \epsilon\right] \leq \frac{1}{4n\epsilon^2} \qquad \text{(Exercise 3.2)}$$

and hence that

$$\lim_{n\to\infty} P\left[\left|\frac{n_A}{n} - p\right| \geq \epsilon\right] = 0$$

for any $\epsilon > 0$. This result is known as the Bernoulli theorem.

Predictor

Let X and Y be random variables. The random variable

$$\hat{Y} = g(X)$$

is called a predictor of Y given X if, given a value of X, the corresponding value of $\hat{Y}$ is used as a prediction of the value of Y.

Mean-square error

The mean-square difference ϵ between a random variable Y and its predictor $\hat{Y}$ is called the mean-square error, that is,

$$\epsilon \triangleq E[(Y - \hat{Y})^2] \tag{6.3}$$

Least-mean-square-error predictor

The predictor $\hat{Y}_0$ of Y which minimizes the mean-square error of prediction is called the least-mean-square-error predictor. This predictor has the form

$$\hat{Y}_0 = E[Y|X = x] \tag{6.6}$$

and hence is the conditional expectation of Y given X.

Linear predictor (of Y given X)

The predictor of Y given by a straight-line function of X:

$$\hat{Y} = a + bX \tag{7.1}$$

where a and b are constants.

Least-mean-square-error linear predictor

The linear predictor of Y given X which minimizes the mean-square error is called the least-mean-square-error linear predictor. This predictor has the form

$$\hat{Y} = E[Y] + \frac{\text{cov}\,(X,Y)}{\text{var}\,(X)}(X - E[X]) \tag{7.3}$$

The mean-square prediction error given by this linear predictor is

$$E[(Y - \hat{Y})^2] = (1 - \rho^2)\,\text{var}\,(Y) \qquad \text{(Exercise 7.4)}$$

where

$$\rho \triangleq \frac{\text{cov }(X,Y)}{\sigma(X)\sigma(Y)}$$

REFERENCES

Apostol, Tom M.: "Calculus," Blaisdell Publishing Company, New York, vol. I, 1961; vol. II, 1962.

Artin, Emil: "The Gamma Function," Holt, Rinehart and Winston, Inc., New York, 1964.

Chernoff, Herman, and Lincoln E. Moses: "Elementary Decision Theory," John Wiley & Sons, Inc., New York, 1959.

Cramér, Harald: "Mathematical Methods of Statistics," Princeton University Press, Princeton, N.J., 1946.

Freeman, Harold: "Introduction to Statistical Inference," Addison-Wesley Publishing Company, Inc., Reading, Mass., 1963.

Fry, Thornton C.: "Probability and Its Engineering Uses," 2d ed., D. Van Nostrand, Inc., Princeton, N.J., 1965.

Hoel, Paul G.: "Introduction to Mathematical Statistics," 3d ed., John Wiley & Sons, Inc., New York, 1962.

Mood, Alexander M., and Franklin A. Graybill: "Introduction to the Theory of Statistics," 2d ed., McGraw-Hill Book Company, 1963.

National Bureau of Standards: "Tables of Binomial Probability Distributions," Applied Mathematics Series, AMS-6, U.S. Government Printing Office, Washington, 1950.

Wilks, Samuel S.: "Mathematical Statistics," John Wiley & Sons, Inc., New York, 1962.

9
Random Processes

1 BERNOULLI PROCESS

We saw in Chap. 4 that many probabilistic problems could be handled in terms of single random variables and we saw further in Chap. 5 that many others could be handled in terms of random vectors, that is, in terms of finite families of random variables. There are, however, many other probabilistic problems of interest which cannot be treated adequately with that machinery.

For example, consider the problem of studying an *unending* sequence of flips of a coin. In particular, let us imagine that we flip a coin at each positive-integer value of time (starting at time 1) and that we observe the result of each flip. Since a coin must show either a head or a tail on any one flip, we may describe the possible results of the particular experiment at time n by the set-indicator random variable:

$$X_n \triangleq \begin{cases} 1 & \text{if a head results from the coin flip at time } n \\ 0 & \text{if not} \end{cases} \tag{1.1}$$

As we assumed above that the sequence of coin flips was unending, it follows that we need the *infinite sequence* (or *infinite family*) of random variables

$$\{X_1, X_2, X_3, \ldots\} = \{X_n, n = 1, 2, 3, \ldots\} \tag{1.2}$$

in order to completely describe its probabilistic properties. Such an infinite family of random variables is commonly called a *random sequence* or *random process* (or *stochastic sequence* or *stochastic process*).

While we introduced the infinite sequence of random variables $\{X_n, n = 1, 2, 3, \ldots\}$ here in the context of studying an unending sequence of coin flips, it may, of course, also be used to describe any other situation in which we are concerned with whether or not some given event occurs at each of a countably infinite sequence of values of time, for example, the occurrence of zeroes or ones in a digital communications link, etc. In view of the importance of such sequences, let us now study some of their properties.

In particular, let us assume for convenience that the successive experiments which define the random variables X_n are statistically independent and, further, that the probability of occurrence of the given event of interest is independent of time. It then follows from these assumptions that the probability distribution of the random variable X_n is

$$P[X_n = 1] = p \qquad \text{and} \qquad P[X_n = 0] = q \triangleq 1 - p \tag{1.3}$$

for all values of n. Thus X_n is a Bernoulli random variable and the corresponding sequence of random variables $\{X_n, n = 1, 2, 3, \ldots\}$ is accordingly called a *Bernoulli random process* (or *Bernoulli process*, for short).

Next, let us use the assumption of independent experiments in order to determine various joint probabilities of the random variables X_n. To this end, suppose that we observe the results of performing the given basic experiment, for example, flipping a coin, a finite number of times, say n. Typical sequences (called *sample sequences*) of such experimental results are shown in Fig. 1.1.

The possible values x_1 obtainable in such a sequence at time 1 are represented by the two points 0 and 1 in the range space $\mathcal{S}_{X_1}$ of the random variable X_1; the possible pairs of values (x_1,x_2) obtainable at times 1 and 2 are represented by the 2^2 sample points (0,0), (0,1), (1,0), and (1,1) in the range space $\mathcal{S}_{X_1} \times \mathcal{S}_{X_2}$ of the two-dimensional random vector

$$\mathbf{X} = (X_1,X_2) \tag{1.4}$$

and, finally, the possible sets of n values $(x_1, x_2, \ldots, x_n)$ obtainable at

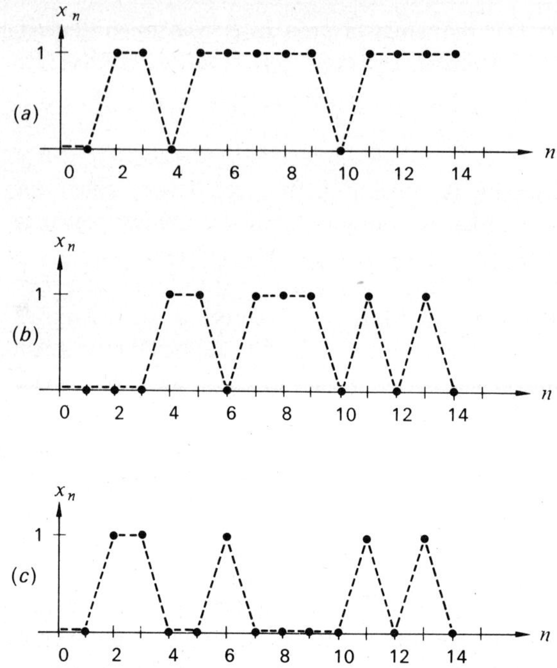

Figure 1.1 Typical Bernoulli sample sequences.

the time instants 1, 2, . . . , n are represented by the 2^n sample points (0, 0, . . . , 0), . . . , (1, 1, . . . , 1) in the range space $\mathcal{S}_{X_1} \times \mathcal{S}_{X_2} \cdots \mathcal{S}_{X_n}$ of the n-dimensional random vector

$$\mathbf{X} = (X_1, X_2, \ldots, X_n) \tag{1.5}$$

The probability distributions on these various sample spaces are particularly simple to find because of our assumption of independent experiments. Thus the joint-probability distribution on the sample space $\mathcal{S}_{X_1} \times \mathcal{S}_{X_2}$ is the set of $2^2 = 4$ probabilities

$$\begin{aligned} P[X_1 = 1, X_2 = 1] &= p^2 \\ P[X_1 = 1, X_2 = 0] &= pq \\ P[X_1 = 0, X_2 = 1] &= qp \\ P[X_1 = 0, X_2 = 0] &= q^2 \end{aligned} \tag{1.6}$$

while the joint-probability distribution on the sample space $\mathcal{S}_{X_1} \times \mathcal{S}_{X_2} \cdots \mathcal{S}_{X_n}$ is the set of 2^n probabilities

$$\begin{aligned} P[X_1 = x_1, X_2 = x_2, \ldots, X_n = x_n] \\ = P[X_1 = x_1]P[X_2 = x_2] \cdots P[X_n = x_n] \end{aligned} \tag{1.7}$$

where $x_i = 0, 1$ for $i = 1, 2, \ldots, n$.

Exercise 1.1 Determine the probabilities of occurrence of the three particular Bernoulli sample sequences shown in Fig. 1.1. □

Suppose now that we imagine a sequence of coin flips which continues without limit. The result of such an infinite sequence of flips would then be an infinite sequence of numbers, say $x_1, x_2, x_3, \ldots$. Paralleling the finite case discussed above, we may view that sequence of numbers as components of the vector

$$\mathbf{x} \triangleq (x_1, x_2, x_3, \ldots) \tag{1.8}$$

which specifies a sample point in the infinite-dimensional range space $\mathcal{S}_{X_1} \times \mathcal{S}_{X_2} \times \mathcal{S}_{X_3} \cdots$ of the random vector

$$\mathbf{X} \triangleq (X_1, X_2, X_3, \ldots) \tag{1.9}$$

The probability distribution induced on the range space of this infinite-dimensional random vector is then formed, conceptually at least, by extending the product results of the finite-dimensional case.

Averages

Since the Bernoulli process $\{X_n, n = 1, 2, 3, \ldots\}$ introduced above is simply an infinite family of ordinary Bernoulli random variables, it follows that we may calculate the statistical average of any one of its component random variables by using the techniques discussed in Chap. 7. In fact, it follows directly from results obtained in that chapter that

$$E[X_n] = p \tag{1.10}$$

[cf. (7.1.6)], and further that

$$\text{var}\,(X_n) = pq = p(1 - p) \tag{1.11}$$

[cf. (7.8.14)].

We shall, from here on, call this expectation and variance the *mean value* and *variance*, respectively, *of the Bernoulli process*. We note that it then follows from (1.10) and (1.11) that both the mean value and the variance of the Bernoulli process are constant functions of the time index n.

2 BINOMIAL PROCESS

Suppose now that at some point in a digital communications system, or in a digital data processor, we observe a Bernoulli pulse chain. That is, suppose that at the point of observation binary pulses (one denoted by the presence of a pulse and zero by the absence) can occur only at discrete values of time (called *sampling instants* here); that whether or not a pulse occurs at a given sampling time instant is independent of whether or not a

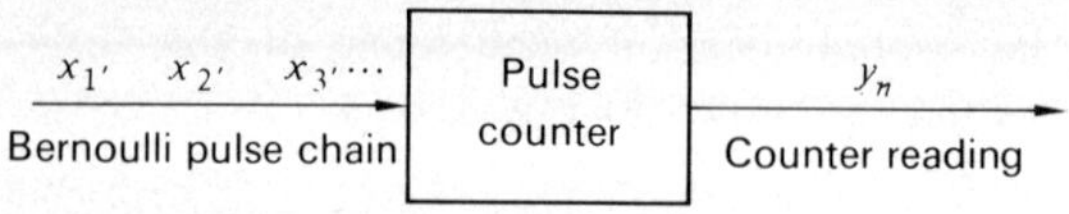

Figure 2.1 The pulse-counter system.

pulse occurs at any other sampling instant; and that the probability of occurrence of a pulse is the same for each sampling instant. Suppose, finally, that we wish to count the number of pulses which occur from sampling instant 1 through sampling instant n. Schematically, the situation is that shown in Fig. 2.1.

The problem before us now is to determine the probabilistic properties of the counter reading at time n, where $n = 1, 2, 3, \ldots$. As in the preceding section, we may describe the presence or absence of a pulse at the ith sampling instant by the set-indicator random variable

$$X_i \triangleq \begin{cases} 1 & \text{if a pulse occurs at the } i\text{th} \\ & \quad \text{sampling instant} \\ 0 & \text{if not} \end{cases} \tag{2.1}$$

It then follows from our above assumptions that this random variable has the Bernoulli probability distribution

$$P[X_i = 1] = p \qquad \text{and} \qquad P[X_i = 0] = q \triangleq 1 - p \tag{2.2}$$

and, further, that the various X_i are independent random variables. The action of the counter is to increase its reading by one each time a pulse occurs, starting with a reading of zero prior to time 1. Typical sequences of counter readings are shown in Fig. 2.2, corresponding to the three different Bernoulli sequences shown in Fig. 1.1. Now it follows from the definition of the random variables X_i, and from the description of the action of the counter, that the counter reading at time n is described by the *counting random variable*

$$Y_n \triangleq \sum_{i=1}^{n} X_i \tag{2.3}$$

Y_n is thus a discrete random variable which has the possible values $0, 1, 2, \ldots, n$.

We note that it follows from this definition that Y_n is a sum of independent random variables, each with the same probability distribution. We shall meet such sums again many times.

Let us now build up the probability distribution of the counting random variable Y_n. It follows from (2.3) that we have at time 1:

$$Y_1 = X_1 \tag{2.4}$$

and hence from (2.2) that Y_1 has the probability distribution

$$P[Y_1 = 1] = p \qquad \text{and} \qquad P[Y_1 = 0] = q \tag{2.5}$$

This situation is shown graphically by Fig. 2.3. At time 2 we have

$$Y_2 = X_1 + X_2 \tag{2.6}$$

Since X_1 might be either zero or unity, we then have the picture shown in Fig. 2.4.

A moment's reflection on this figure, and on the fact that X_1 and X_2 are independent random variables, then shows that Y_2 has the three possible values: +2, +1, and 0, with the probability distribution

$$\begin{aligned} P[Y_2 = +2] &= p^2 \\ P[Y_2 = +1] &= 2pq \\ P[Y_2 = 0] &= q^2 \end{aligned} \tag{2.7}$$

These probability values are easily obtained by following paths through Fig. 2.4 from the origin to the desired endpoint, multiplying the probabilities indicated upon each branch of a given path, and finally counting the number of distinct paths from the origin which reach the same endpoint (since distinct paths represent mutually exclusive events).

If next we continue this procedure to time n, we see that Y_n has the $n + 1$ possible values $\{0, 1, 2, \ldots, n\}$. It therefore follows that paths similar to those shown in Figs. 2.3 and 2.4 must lie between the bounds

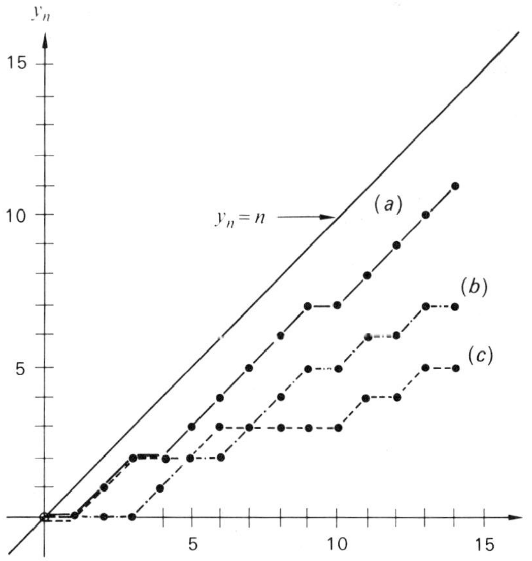

Figure 2.2 Typical counter-reading sequences.

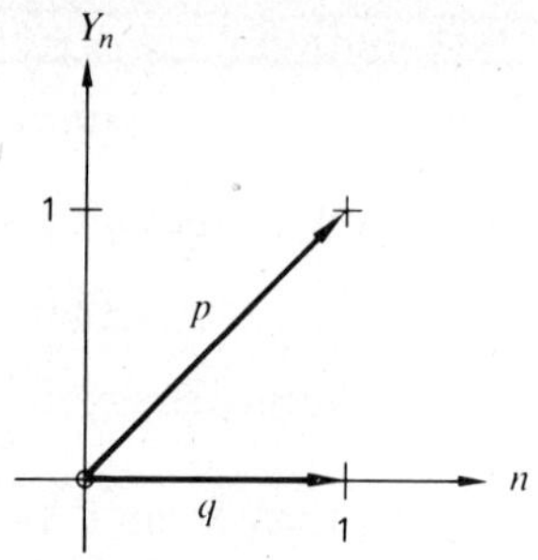

Figure 2.3 Possible paths for Y_1.

$Y_j = 0$ and $Y_j = j$ (as shown, for example, in Fig. 2.2). The probability distribution of the various possible end counts, that is, the set of probabilities $P[Y_n = k]$, $k = 0, 1, 2, \ldots, n$, may now be determined as follows: Just which value of Y_n is achieved at time n is determined by how many pulses occur in the time interval $[t_1,t_n]$. Since successive pulses are independent, the probability of occurrence of a *particular* sequence of k pulses is given simply by the product of the probabilities of occurrence of k pulses and of $n - k$ absences of pulses; that is, if we denote this probability by P^*, then we have

$$P^*[Y_n = k] = p^k q^{n-k}$$

This particular sequence of pulses is, however, not the only possible sequence having k pulses by time n; for example, there are two possible sequences each having one pulse by time 2, as shown in Fig. 2.4. The probability of obtaining any one of the various possible sequences having k pulses is equal to the total number possible of such sequences times the probability of obtaining a particular one (since the different possible paths

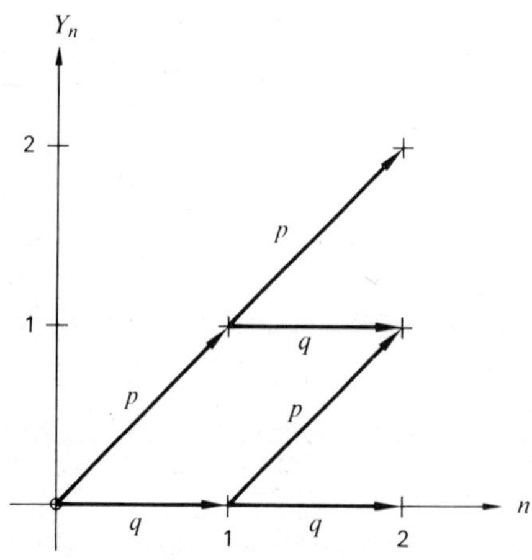

Figure 2.4 Possible paths for Y_2.

through figures such as 2.3 and 2.4 represent mutually exclusive events all having the same probability of occurrence). Since the total number of sequences which can result in $Y_n = k$ is given by the *binomial coefficient*

$$\binom{n}{k} = \frac{n!}{k!(n-k)!} \tag{2.8}$$

it follows that the probability that $Y_n = k$ is given by the formula

$$P[Y_n = k] = \binom{n}{k} p^k q^{n-k} \tag{2.9}$$

for $k = 0, 1, 2, \ldots, n$ (cf. Exercise 3.8.6). The collection of these probabilities is known as the *binomial probability distribution,* and hence the random process under discussion is known as a *binomial counting process.*

Exercise 2.1 Suppose that we represent an unending sequence of flips of a coin by a zero-one Bernoulli process as in Sec. 1 and that we count the number of heads in n flips.

Determine the probability that at least one head occurs in n flips. (Note that when $n = 2$ and $p = \frac{1}{2}$, this probability is just the probability that the head-indicator random variable I_H defined in Sec. 4.1 has the value unity.) □

It should be noted that while each counting random variable Y_n is a sum of independent random variables, the various Y_n are *not* independent random variables as shown by the following exercise.

Exercise 2.2 Let the sequence of random variables $\{Y_n, n = 1, 2, 3, \ldots\}$ be a binomial counting process as defined above.

a. Determine $P[Y_{n+1} = k | Y_n = j]$ for $k = 0, 1, 2, \ldots, n$.
b. Determine $P[Y_{n+1} = k, Y_n = j]$ and show that

$$P[Y_{n+1} = k, Y_n = j] \neq P[Y_{n+1} = k]P[Y_n = j]$$ □

It is because of this fact that the determination of the various joint-probability distributions for a binomial random process is rather more difficult than it was for the Bernoulli process. We shall return to this point at a later time when we have more powerful tools to work with.

The binomial counting process discussed above is, of course, not the only binomial random process which can be generated from a Bernoulli process. Consider, for example, the following situation. Suppose that instead of a (0,1) Bernoulli process, we have a $(-1,+1)$ Bernoulli process;

that is, if a head (pulse) occurs at time j, set $x_j = +1$, while if a tail (no pulse) occurs at time j, set $x_j = -1$, and let

$$P[X_j = +1] = p \qquad \text{and} \qquad P[X_j = -1] = q \triangleq 1 - p \tag{2.10}$$

We would now like to apply this new Bernoulli process to a counter and determine the probability distribution of the counter reading.

One way of visualizing the present situation is to imagine a particle moving in a plane as follows: Suppose that we consider the lattice of points (j,k) in the xy plane, where $j = 0, 1, 2, \ldots, n$ and $k = 0, 1, 2, \ldots, n$. Suppose further that at time j, a particle moves up 1 unit in y if $X_j = +1$ and moves down 1 unit if $X_j = -1$. The random variable Y_n describing the y coordinate of the particle at time n is then given by

$$Y_n = \sum_{j=1}^{n} X_j \tag{2.11}$$

where $Y_0 = 0$ and where X_j here is a $(-1,+1)$ Bernoulli random variable, that is, by a counting random variable. The resulting trajectory of the particle as it moves through the xy plane, where the x coordinate indicates time, is called a one-dimensional *random walk*. Random walks in various dimensions have been widely used to model a number of different physical phenomena.[1]

Exercise 2.3 Let the sequence of random variables $\{Y_n, n = 0, 1, 2, \ldots\}$ describe the one-dimensional random walk discussed above.

a. Plot all possible paths for $n = 3$ and $n = 4$, and indicate the probabilities of the various path branches (as done in Fig. 2.4 for the binomial process).

b. Determine the probability distributions for the coordinate random variables Y_3 and Y_4, respectively. □

Exercise 2.4 For the one-dimensional random walk discussed above, determine the probability distribution of the coordinate random variable Y_n. Does it matter whether n is odd or even? □

The mean value of the binomial counting process $\{Y_n, n = 1, 2, 3, \ldots\}$ may now be obtained by taking the expectation of both sides of (2.3) and using (2.2). Thus we get

$$E[Y_n] = np \tag{2.12}$$

Thus we see that the mean value of the binomial counting process increases linearly with the time index n.

[1] See, for example, Chandrasekhar's paper.

Exercise 2.5 Let $\{Y_n, n = 1, 2, 3, \ldots\}$ be a binomial counting process, where the binomial counting random variable Y_n is defined as in (2.3) above.

a. Show that the variance of the binomial counting random variable is given by the formula

$$\text{var}\,(Y_n) = npq \tag{2.13}$$

(*Hint:* cf. Exercise 7.8.11.) Thus we see that the variance of the binomial counting process increases linearly with the time index n.

b. In many problems involving counting processes, it is of interest to study the *counting difference*, or *counting increment* $Y_m - Y_n$, where $m \neq n$. Show that the variance of the counting difference is given by the formula

$$\text{var}\,(Y_m - Y_n) = |m - n|pq \tag{2.14}$$

in the case of the binomial counting process. We note that this result is directly proportional to the magnitude of the difference of the time indices. □

Exercise 2.6 Using the results of the preceding exercise, show that the covariance of the two binomial counting random variables Y_m and Y_n $(m \neq n)$ is given by the formula

$$\text{cov}\,(Y_m, Y_n) = pq \min\,(m,n) \tag{2.15}$$

where min (m,n) denotes the smaller of the indices m and n. [*Hint:* Express var $(Y_m - Y_n)$ in terms of this covariance and the variances of the individual counting random variables Y_m and Y_n.] □

3 SINE WAVE PROCESS

Both the Bernoulli process and the binomial process introduced above are characterized by the fact that the experiments which define the random variables of interest can occur only at a specified set of discrete time instants, for example, at the times 1, 2, etc. That not all random phenomena of interest fit into this pattern is shown by the following example.

An electronics instrument manufacturer produces sine wave oscillators, that is, instruments which generate outputs of the form

$$x(t) = v \sin\,(\omega t + \phi) \tag{3.1}$$

where v is the peak amplitude of the sine wave, $f \triangleq \omega/2\pi$ is the sine wave frequency, and ϕ is a phase angle measured relative to a given origin.

While it is desired that the parameters v and ω be the same for all oscillators of the same model, manufacturing tolerances on the components used to make the oscillators cause these parameters to vary from oscillator to oscillator (but not with time for a particular oscillator).[1] If, therefore, we measure the output of a particular oscillator at a specified time, say t_1, then that output will depend not only upon the time t_1 but also upon the specific values of v, ω, and ϕ which pertain to that oscillator.

This situation may be modeled probabilistically as follows: Suppose that the values of the parameters v, ω, and ϕ have been measured for each of a large number of oscillators all of the same model, and that that data have been used to determine the empirical distributions of the values of those parameters. Those empirical distributions may then be taken as the probability distributions of the random variables V, Ω, and Φ which characterize the possible values of the parameters v, ω, and ϕ, respectively. Further, the outputs at the time t_1 of the various oscillators may then be characterized by the random variable

$$X_{t_1} \triangleq V \sin (\Omega t_1 + \Phi) \tag{3.2}$$

following (3.1). Since a similar result is obtained for each value of time, we may characterize the collection of oscillator-output waveforms by a family of random variables X_t, where now the index t ranges over the set of times of interest. Thus, if we are interested in the oscillator outputs at any time t after some arbitrarily chosen time origin (called zero). then we are concerned with the family of random variables

$$\{X_{t,}\ 0 \leq t\}$$

in which the random variable X_t is related to the parameter random variables V, Ω, and Φ by the formula

$$X_t = V \sin (\Omega t + \Phi) \tag{3.3}$$

We call this family of random variables a *sine wave random process* (or *sine wave process*).

The specific sine waveform obtained from a particular oscillator, that is, the function of time determined by a particular set of values of the constants v, ω, and ϕ, is called a *sample function* (or *realization*) of this process; several typical sine wave sample functions are shown in Fig. 3.1. Suppose now that we observe the output of each of the oscillators at the single time instant t_1; that is, suppose that we determine the values $x(t_1)$ of the various sample functions (cf. Fig. 3.1). The set of values $x(t_1)$ thus obtained is then characterized by the random variable X_{t_1}.

[1] In addition, the parameter ϕ also varies from oscillator to oscillator, but neither the manufacturer nor the user generally cares that it does.

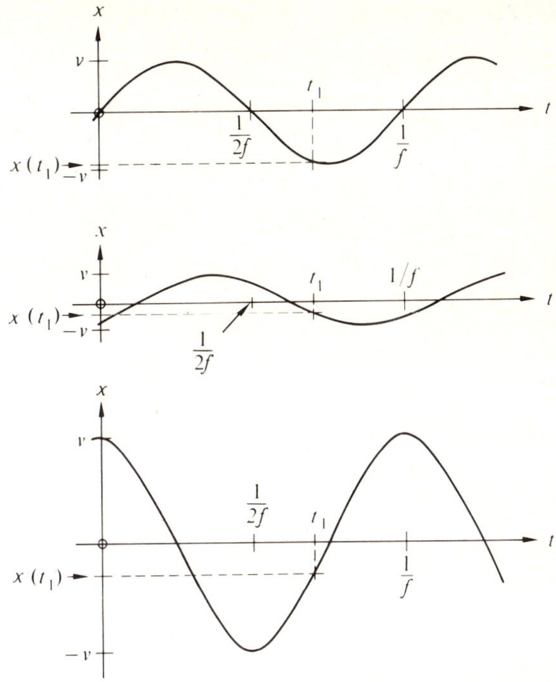

Figure 3.1 Sine wave sample functions.

It should be noted that the time parameter t which indexes the random variables X_t may assume here *any* value whatsoever throughout the interval $[0, +\infty)$ and is not restricted to just a discrete set of values.

Exercise 3.1 Suppose that all of the oscillators have, in fact, the same reference phase angle ϕ (in particular, suppose that $\phi = \pi/2$ for all oscillators) and the same frequency $f = \omega/2\pi$. That is, consider the sine wave process $\{X_t, 0 \leq t\}$, where

$$X_t = V \cos \omega t \tag{3.4}$$

for all t in $[0, +\infty)$ and where ω is a constant. Suppose also that the amplitude V is a uniformly distributed random variable, in particular:

$$f_V(v) = \begin{cases} 1 & \text{for } 0 \leq v \leq 1 \\ 0 & \text{otherwise} \end{cases} \tag{3.5}$$

a. Sketch a few sample functions of this process.

b. Determine the probability density of the random variable X_t, and plot, for

$$t = 0, \frac{\pi}{4\omega}, \frac{3\pi}{4\omega}, \frac{\pi}{\omega}$$

(*Hint:* Remember that $\cos \omega t$ is just a number for a fixed value of t.)

c. What is the probability density of X_t when $t = \pi/2\omega$?
(*Hint:* Consider the probability distribution function first.) □

Exercise 3.2 A random process closely related to the above sine wave process is the process $\{Y_t, -\infty < t < +\infty\}$ in which

$$Y_t \triangleq X \cos \omega t + Z \sin \omega t \tag{3.6}$$

for all values of t, where X and Z are random variables and where ω is a constant.

a. Sketch several typical sample functions of this random process.

Suppose that X and Z are independent gaussian random variables with the probability densities

$$f_X(x) = \frac{1}{\sqrt{2\pi}} e^{-x^2/2} \qquad \text{and} \qquad f_Z(z) = \frac{1}{\sqrt{2\pi}} e^{-z^2/2}$$

b. Calculate the probability of the event

$$A \triangleq \left[\frac{2\omega}{\pi} \int_0^{\pi/\omega} Y_t^2\, dt > C\right]$$

as a function of the value of the parameter C. (*Hint:* Determine the region in $\mathcal{S}_X \times \mathcal{S}_Z$ which corresponds to A.) □

Exercise 3.3 Let the sine wave process $\{X_t, -\infty < t < +\infty\}$ be such that

$$X_t \triangleq Y \cos \omega t$$

for all t, where ω is a constant and where Y is uniformly distributed between zero and one. Determine the statistical average $E[X_t]$ of this random process. □

Exercise 3.4 Let the sine wave process $\{X_t, -\infty < t < +\infty\}$ be such that

$$X_t \triangleq \cos(\omega t + Z)$$

for all t, where ω is a constant and where Z is uniformly distributed between zero and 2π. Determine the statistical average $E[X_t]$ of this random process. □

Exercise 3.5 Let the sine wave process $\{X_t, -\infty < t < +\infty\}$ be such that

$$X_t \triangleq Y \cos(\omega t + Z)$$

for all t, where ω is a constant, where Y is uniformly distributed between zero and one, where Z is uniformly distributed between zero and 2π, and where Y and Z are independent random variables. Determine the statistical average $E[X_t]$ of this random process. □

4 RANDOM PROCESS DESCRIPTIONS

Let us review for a moment the Bernoulli and binomial random processes introduced in the first two sections of this chapter. We pointed out there that each may be thought of as a sequence (or *indexed family*) of random variables, say, as

$$\{X_n, n = 1, 2, 3, \ldots\} = \{X_1, X_2, X_3, \ldots\} \tag{4.1}$$

Then given a particular value of the index n, say k, we have the single random variable X_k with the range space $\mathcal{S}_{X_k}$ and the attendant probability distribution function F_{X_k} defined on $\mathcal{S}_{X_k}$. Similarly, if we are given a finite set of indices, say $\{1, 2, \ldots, j\}$, we obtain a j-dimensional random vector

$$\mathbf{X} = (X_1, X_2, \ldots, X_j) \tag{4.2}$$

with the j-dimensional range space

$$\mathcal{S}_{\mathbf{X}} = \mathcal{S}_{X_1} \times \mathcal{S}_{X_2} \cdots \mathcal{S}_{X_j} \tag{4.3}$$

and the corresponding j-variate joint-probability distribution function

$$F_{\mathbf{X}} = F_{X_1, X_2, \ldots, X_j} \tag{4.4}$$

defined on $\mathcal{S}_{\mathbf{X}}$. Such finite-dimensional joint-probability functions must, of course, be *symmetric*, in the sense that identical functions are obtained under permutations of the pairs (X_i, x_i), for example,

$$F_{X_3,X_1,X_2}(x_3,x_1,x_2) = F_{X_1,X_2,X_3}(x_1,x_2,x_3)$$

In addition, these joint-probability distribution functions must be *consistent* in that, given any j-dimensional joint-probability distribution function, we must be able to obtain any lower-dimensional probability distribution function from it in the usual manner; in other words, we must have

$$F_{X_1,X_3,X_7}(x_1,x_3,x_7) = F_{X_1,X_2,X_3,X_4,X_5,X_6,X_7}(x_1,+\infty,x_3,+\infty,+\infty,+\infty,x_7)$$

Such a description of a Bernoulli or binomial random process as an indexed family of random variables with its corresponding collection of finite-dimensional joint-probability distribution functions is complete in the sense that *any* meaningful probabilistic questions concerning the process can be answered (in principle, at least) once the various probability functions are specified.

In view of these facts, it seems natural to consider (as we did in Sec. 1) the entire sequence $\{X_n, n = 1, 2, 3, \ldots\}$ to be an infinite-dimensional random vector

$$\mathbf{X} = (X_1, X_2, X_3, \ldots) \tag{4.5}$$

with the infinite-dimensional range space

$$\mathcal{S}_{\mathbf{X}} = \mathcal{S}_{X_1} \times \mathcal{S}_{X_2} \times \mathcal{S}_{X_3} \cdots \tag{4.6}$$

and a corresponding infinite-dimensional joint-probability distribution function

$$F_{\mathbf{X}} = F_{X_1,X_2,X_3,\ldots} \tag{4.7}$$

which is a conceptual extension of the finite-dimensional joint-probability distribution functions discussed previously.

Suppose now that we consider a particular sample point, say

$$s = (x_1, x_2, x_3, \ldots) \tag{4.8}$$

in the infinite-dimensional range space $\mathcal{S}_{\mathbf{X}}$. The components x_i of this vector sample point specify the particular sample values of the entire family of random variables making up the process; that is, the components of an s specify a particular *sample sequence*, or *sample function*, of the given random process. Thus it follows that we may view any such random process either as an indexed family of random variables $\{X_n, n = 1, 2, 3, \ldots\}$—where picking a value of the index n specifies a particular random variable—or as a random vector $\mathbf{X}(s)$ with an infinite-dimensional range space $\mathcal{S}_{\mathbf{X}}$ of sample points s—where picking a particular sample point s in $\mathcal{S}_{\mathbf{X}}$ specifies a particular sample function. With these alternative views of the same random process in mind, we are led to describe such a random process as a function of two variables: the index n and the sample point s; that is, we will sometimes find it convenient to use the notation

$$X(n,s) \triangleq \{X_n, n = 1, 2, 3, \ldots\} \tag{4.9}$$

when discussing such processes.

While both the Bernoulli and the binomial random processes are made up of discrete random variables defined at discrete instants of time, we have already seen in the preceding section that not all processes of interest are so formed. In particular, we saw there that the sine wave process $\{X_t, 0 \leq t\}$, where

$$X_t \triangleq V \sin (\Omega t + \Phi)$$

is made up of continuous random variables defined at any of a continuum of values of time. In spite of these differences, *we note that the random process in either case may be viewed as an indexed family of random variables.* We shall, accordingly, commonly denote a random process by the symbol[1]

$$\{X_t, t \in T\}$$

where X_t is the random variable which characterizes the possible values at time t of the various sample functions making up the process and where T denotes the set of possible values of the *index*, or *parameter*, t. If the index set T is a countably infinite set, for example, the positive integers as in the case of the Bernoulli process, we say that we are dealing with a *discrete-parameter* random process. If, on the other hand, the index set T is a continuum such as the semi-infinite interval

$$T = \{t: 0 \leq t\}$$

as in our discussion of the sine wave process, then we say that we have a *continuous-parameter* random process. In any case, if we pick a particular value of t, say $t' \in T$, then we have a single random variable $X_{t'}$, while if we pick a finite set of values of t, say $\{t_1, t_2, \ldots, t_j\}$, with all $t_i \in T$, then we have a finite-dimensional random vector

$$\mathbf{X} = (X_{t_1}, X_{t_2}, \ldots, X_{t_j})$$

If next we wish to extend our discrete-parameter concepts of countably infinite-dimensional random vectors with countably infinite-dimensional range spaces to continuous-parameter random processes, certain technical difficulties unfortunately arise. Now we can, of course, imagine an abstract sample space $\mathcal{S}$ in which each sample point s represents a particular sample function of the given random process. When it comes to

[1] We shall at times drop the notation $t \in T$ for the sake of compactness and just use the symbol $\{X_t\}$ to denote a random process, so long as there is no ambiguity as to which parameter set T is to be used.

discussing the probability distributions induced on such sample spaces, however, problems arise which are well beyond our present abilities. Nevertheless, it can be said that, for essentially all the problems of interest to us, it is quite sufficient to require that families of finite-dimensional joint-probability distribution functions exist which are symmetric and consistent.[1] In most cases then, we may handle continuous-parameter random processes in essentially the same manner that we treated the discrete-parameter random processes above.

In particular, we shall find it convenient to describe all random processes as functions of two variables: the index $t \in T$ and the sample point $s \in \mathcal{S}$, and to write on occasion

$$X(t,s) \triangleq \{X_t, t \in T\} \tag{4.10}$$

It therefore follows, as in the discrete-parameter case, that if we fix the value of t, say as t', then we get the *random variable*

$$X_{t'}(s) = X(t',s) \tag{4.11}$$

while if we fix the value of s, say as s', then we get the *sample function*

$$x_{s'}(t) = X(t,s') \tag{4.12}$$

In many problems of interest, the required finite-dimensional joint-probability distribution functions are explicitly available for all finite sets of indices. For example, having defined gaussian random vectors in Sec 7.10, we shall say that a real random process $\{X_t, t \in T\}$ is a *gaussian random process* if and only if for every n (given any arbitrary set of n time instants $t_1, t_2, \ldots, t_n$ in the index set T), the random variables $X_{t_1}, X_{t_2}, \ldots, X_{t_n}$ form the components of an n-dimensional gaussian random vector, that is, if and only if they are jointly gaussian random variables. In this case those n random variables have the joint-probability density

$$f_{X_{t_1}, X_{t_2}, X_{t_n}}(x_1, x_2, \ldots, x_n) = \frac{1}{(2\pi)^{n/2}|\mathbf{\Lambda}|^{1/2}} \exp\left[-\frac{1}{2|\mathbf{\Lambda}|}\sum_{i=1}^{n}\sum_{j=1}^{n} |\mathbf{\Lambda}|_{ij}(x_i - m_i)(x_j - m_j)\right]$$

where m_i is the mean value of X_{t_i}

$$m_i \triangleq E[X_{t_i}]$$

[1] A rather readable treatment of the mathematical problems involved here is given in secs. 3.1 to 3.4 and sec. 3.6 of Cramér and Leadbetter. A much more advanced discussion is given in chap. 3 of Neveu, while the definitive advanced treatment is given in chap. 2 of Doob.

where $|\mathbf{\Lambda}|_{ij}$ is the cofactor of the element λ_{ij} in the determinant $|\mathbf{\Lambda}|$ of the covariance matrix $\mathbf{\Lambda}$ in which the elements λ_{ij} are the covariances

$$\lambda_{ij} \triangleq \text{cov}\,(X_{t_i},X_{t_j})$$

and where it is assumed that the covariance matrix is nonsingular. While such joint-probability functions are explicitly available in some cases, this is unfortunately not always the case. Such a situation commonly arises, for example, when the random process under study is defined essentially only in terms of its sample functions, for example, as in the case of the sine wave process discussed in the preceding section.

5 STATIONARITY

Let us reconsider again the Bernoulli process $\{X_n,\ n = 1, 2, 3, \ldots\}$ which we introduced in Sec. 1 of this chapter. We showed there that the probability distribution of the random variable X_n was the Bernoulli distribution

$$P[X_n = 1] = p \qquad \text{and} \qquad P[X_n = 0] = 1 - p$$

and, further, that the joint probabilities of the random variables $\{X_1, X_2, \ldots, X_n\}$ were given by the product of the probabilities of the individual random variables $X_1, X_2, \ldots, X_n$. A moment's reflection upon these results shows that neither the individual probabilities nor any of the joint probabilities of this process are functions of time. Consider, on the other hand, the binomial counting process which we introduced in Sec. 2. We showed there that the counting random variable Y_n had the binomial probability distribution

$$P[Y_n = k] = \binom{n}{k} p^k(1 - p)^{n-k}$$

for $k = 0, 1, \ldots, n$. Since both the binomial coefficient $\binom{n}{k}$ and the term $(1 - p)^{n-k}$ are functions of the time index n, we see that the probability distribution of the binomial counting random variable most defiitely is a function of time.

Now we shall see later (when we study the problem of sampling a random process) that the distinction between random processes which exhibit a stationarity in time and those processes which do not exhibit such a stationarity is of considerable importance. We shall, in fact, find it useful to define several different kinds of stationarity.

In particular, we shall from this point on say that the random process $\{X_t, t \in T\}$ is *stationary* if and only if all of the finite-dimensional probability distribution functions are invariant under shifts of the time origin.

That is, the random process $\{X_t, t \in T\}$ is stationary if and only if for every value of n (including $n = 1$), and for every set of time instants $\{t_i \in T, i = 1, 2, \ldots, n\}$ it follows that

$$F_{X_{t_1},X_{t_2},\ldots,X_{t_n}}(x_1, x_2, \ldots, x_n) = F_{X_{t_1+\tau},X_{t_2+\tau},\ldots,X_{t_n+\tau}}(x_1, x_2, \ldots, x_n) \quad (5.1)$$

holds true for all values of $x_1, x_2, \ldots, x_n$ and all τ such that $(t_i + \tau) \in T$ for all i. If these conditions are not met, then that random process is said to be *nonstationary*.

The above conditions for stationarity are often unnecessarily restrictive in applications and, accordingly, we shall define a weaker form of stationarity later in this chapter. With this fact in mind, random processes which satisfy the conditions specified above will sometimes be said to be *stationary in the strict sense* and called *strictly stationary* random processes. Thus the Bernoulli process is stationary in the strict sense, while the binomial counting process is nonstationary.

Exercise 5.1 Consider the sine wave process $\{X_t, 0 \leq t\}$, where

$$X_t \triangleq V \cos \omega t$$

which we studied in Exercise 3.1 above.

Show whether or not this random process is stationary in the strict sense. □

Exercise 5.2 Consider the random process $\{X_t, -\infty < t < +\infty\}$, where

$$X_t \triangleq Z$$

for all t, and where Z is a gaussian random variable with the probability density

$$f_Z(z) = \frac{1}{\sqrt{2\pi}} \exp\left(-\frac{z^2}{2}\right)$$

a. Sketch several typical sample functions of this random process.
b. Determine the probability density of X_t.
c. Show whether or not this random process is stationary in the strict sense. □

6 COVARIANCE AND CORRELATION FUNCTIONS

We noted in Secs. 8.6 and 8.7 of the preceding chapter on Estimation, Sampling, and Prediction that an estimation problem of some interest is that of estimating or predicting the value of one random variable, given that we have observed a particular value of a second random variable. This problem becomes of particular importance when the random variables in question are from the same random process. Thus, for example, suppose that we are given a random process $\{X_t, t \in T\}$, that we observe at time $t_1 \in T$ the value $x(t_1)$ of a sample function of the given random process, and that we wish to predict the value of that sample function at a later time, say $t_2 \in T$. Suppose further that, purely for the sake of simplicity of implementation, we wish to use a linear predictor. It then follows [from (8.4)] that the least-mean-square-error linear prediction of $x(t_2)$, given $x(t_1)$, is given by the formula

$$\hat{x}(t_2) = E[X_{t_1}] + \frac{\text{cov}\,(X_{t_1},X_{t_2})}{\text{var}\,(X_{t_1})}\,(x(t_1) - E[X_{t_1}]) \tag{6.1}$$

We note that this prediction depends upon the mean, the variance, and the covariance of the given random process; this dependence is one of the reasons that these functions are so important in a study of random processes. Having already discussed means and variances of random processes, we wish to study in this section some of the detailed properties of covariances of random processes.

Suppose now that a given real random process has a finite mean-square value, that is,

$$E[X_t^2] < +\infty$$

for all $t \in T$. Suppose further that t_1 and t_2 are any two t values in the index set T. The covariance of the corresponding random variables X_{t_1} and X_{t_2}, respectively, will commonly be a function of those two time instants, and it will therefore often be convenient to use a notation which emphasizes that fact. With this in mind, we shall from now on generally denote the covariance of X_{t_1} and X_{t_2} by the symbol $K_X(t_1,t_2)$ and call it the *covariance function* of the random process $\{X_t, t \in T\}$:

$$K_X(t_1,t_2) \triangleq \text{cov}\,(X_{t_1},X_{t_2}) \tag{6.2}$$

It then follows from the definition of the covariance that

$$\begin{aligned} K_X(t_1,t_2) &= E[(X_{t_1} - E[X_{t_1}])(X_{t_2} - E[X_{t_2}])] \\ &= E[X_{t_1}X_{t_2}] - E[X_{t_1}]E[X_{t_2}] \\ &= R_X(t_1,t_2) - E[X_{t_1}]E[X_{t_2}] \end{aligned} \tag{6.3}$$

where we have introduced the symbol $R_X(t_1,t_2)$ for the expectation of $X_{t_1}X_{t_2}$:

$$R_X(t_1,t_2) \triangleq E[X_{t_1}X_{t_2}] \tag{6.4}$$

This correlation is called the *autocorrelation function* of the random process

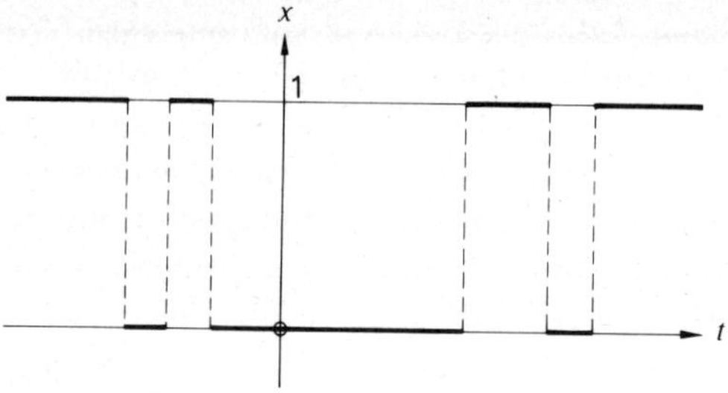

Figure 6.1 A random-telegraph sample function.

$\{X_t, t \in T\}$.† In some problems of interest, it will be more convenient to deal directly with the autocorrelation function than with the covariance function; in others, vice versa.

As a first example of the calculation of covariance and autocorrelation functions, let us consider the case of the so-called *random-telegraph process*. This is a real random process $\{X_t, \; -\infty < t < +\infty\}$ whose sample functions at any instant of time t may assume only the values zero or one. It is assumed that these two values are equally likely, that is, that

$$P[X_t = 0] = \tfrac{1}{2} = P[X_t = 1] \tag{6.5}$$

for all t; that the probability $P[k,\tau]$ that k transversals from one value to another occur in a time interval of length τ is given by the Poisson probability distribution

$$P[k,\tau] = \frac{(\lambda\tau)^k e^{-\lambda\tau}}{k!} \tag{6.6}$$

for $k = 0, 1, 2, \ldots$ (where λ is the average number of transversals per unit time); and, finally, that the occurrence of k transversals in an interval of length τ is statistically independent of the value assumed by any particular sample function at the start of the given interval. A typical sample function of such a process might be as shown in Fig. 6.1.

Exercise 6.1 For the above random-telegraph process:

a. Show that

$$E[X_t] = \tfrac{1}{2}$$

for all t.

† We denote the autocorrelation function by R_X in conformity with the bulk of the available literature. The unfortunate confusion which might arise from using the same symbol to denote both an autocorrelation function and a range space should be avoidable by referring to the context in which the symbol occurs.

b. Show that

$$R_X(t_1,t_2) = P[X_{t_1} = 1, X_{t_2} = 1]$$

for all t_1 and t_2.

c. Show further that

$$R_X(t_1,t_2) = \tfrac{1}{2}P\,[k \text{ even}]$$

where $P\,[k$ even$]$ is the probability that the number of transversals which occur in an interval of duration $|t_1 - t_2|$ is even. □

Exercise 6.2 Using (6.6) and the results of Exercise 6.1:

a. Show that

$$P\,[k \text{ even}] = \tfrac{1}{2} + \tfrac{1}{2}\exp(-2\lambda|t_1 - t_2|)$$

and hence that

$$R_X(t_1,t_2) = \tfrac{1}{4} + \tfrac{1}{4}\exp(-2\lambda|t_1 - t_2|)$$

(*Hint:* Recall the power-series expansion of an exponential function.)

b. Show further that

$$K_X(t_1,t_2) = \tfrac{1}{4}\exp(-2\lambda|t_1 - t_2|)$$

The covariance function of the random-telegraph process is shown in Fig. 6.2. □

Suppose next that we wish to estimate the value of a sample function of a random-telegraph process at a time t_2, given that we know its value at a time t_1. It then follows, on using the results of Exercises 6.1 and 6.2 in (6.1), that we get

$$\hat{x}(t_2) = \exp(-2\lambda|t_2 - t_1|)[x(t_1) - \tfrac{1}{2}] + \tfrac{1}{2} \tag{6.7}$$

If now t_1 and t_2 are so close together that

$$|t_2 - t_1| \ll \frac{1}{2\lambda}$$

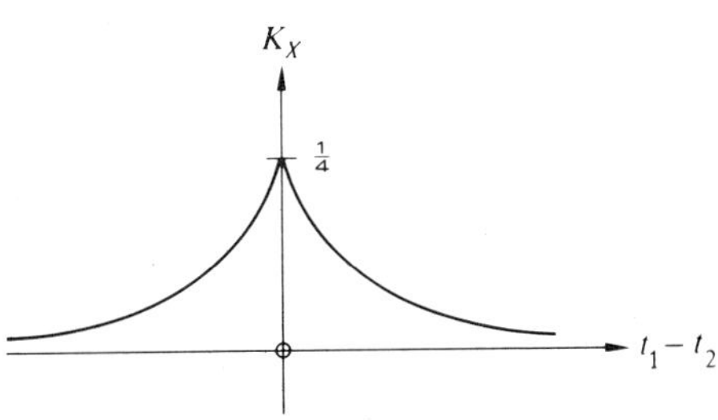

Figure 6.2 Covariance function of the random-telegraph process.

then the exponential in (6.7) becomes nearly equal to unity and we get

$$\hat{x}(t_2) \approx x(t_1)$$

If, on the other hand, t_1 and t_2 are so far apart that

$$|t_2 - t_1| \gg \frac{1}{2\lambda}$$

then the exponential in (6.7) practically vanishes and we get

$$\hat{x}(t_2) \approx \tfrac{1}{2}$$

Since, by definition, a random-telegraph wave must assume either the value zero or the value unity at every instant of time, this last result is not very useful. Thus it follows that our ability to use a linear estimator in order to predict the value of a random-telegraph sample function at one instant of time (given that we know its value at another instant) is strongly dependent upon the time separation of the two given instants. This conclusion agrees, of course, with the related general result obtained earlier in Exercise 8.7.4.

Further examples of the calculation of autocorrelation functions and covariance functions are given by the following exercises.

Exercise 6.3 Let the random process $\{X_t, \; -\infty < t < +\infty\}$ be such that

$$X_t \triangleq Y \cos t$$

for all t, where Y is a uniformly distributed random variable with the probability density

$$f_Y(y) = \begin{cases} 1 & \text{for } 0 \le y \le 1, \\ 0 & \text{otherwise} \end{cases}$$

a. Show that

$$R_X(t_1,t_2) = \tfrac{1}{3} \cos t_1 \cos t_2$$

b. Show that

$$K_X(t_1,t_2) = \frac{1}{12} \cos t_1 \cos t_2$$

(cf. Exercise 3.3). □

Exercise 6.4 Let the random process $\{X_t, \; -\infty < t < +\infty\}$ be such that

$$X_t \triangleq \sin(\omega t + Y)$$

for all t, where ω is a constant and where Y is a uniformly distributed random variable with the probability density

$$f_Y(y) = \begin{cases} \dfrac{1}{2\pi} & \text{for } 0 \le y \le 2\pi \\ 0 & \text{otherwise} \end{cases}$$

Show that in this case,

$$\begin{aligned} R_X(t_1,t_2) &= K_X(t_1,t_2) \\ &= \tfrac{1}{2} \cos \omega(t_1 - t_2) \end{aligned}$$

for all t_1 and t_2. □

An examination of the results in Exercise 6.3 shows that the autocorrelation function and the covariance function of the $\{Y \cos t\}$ process both depend upon the time instants t_1 and t_2 separately. The results in Exercises 6.2 and 6.4 show, on the other hand, that the autocorrelation functions and the covariance functions of the random-telegraph and $\{\sin (\omega t + Y)\}$ processes depend only upon the time difference $t_1 - t_2$ and not upon the time instants t_1 and t_2 separately. We shall study random processes of this latter type in more detail in Sec. 7 of this chapter.

Multiple random processes

A covariance may, of course, relate random variables drawn from two different random processes as well as from a single random process. For example, let $\{X_t, t \in T\}$ and $\{Y_t, t \in T\}$ be a pair of real random processes, each with a finite mean-square value:

$$E[X_t^2] < +\infty \qquad \text{and} \qquad E[Y_t^2] < +\infty$$

for all $t \in T$. Suppose that t_1 and t_2 are any two t values in the common index set T. Paralleling the single process case, the covariance of the random variables X_{t_1} and Y_{t_2} will, in general, be a function of the time instants t_1 and t_2. We shall, accordingly, denote that covariance by the symbol $K_{XY}(t_1,t_2)$:

$$K_{XY}(t_1,t_2) \triangleq \operatorname{cov}(X_{t_1}, Y_{t_2}) \tag{6.8}$$

(note that the first subscript on K goes with the first time instant and the second subscript goes with the second time instant). It then follows from the definition of the covariance that

$$\begin{aligned} K_{XY}(t_1,t_2) &= E[(X_{t_1} - E[X_{t_1}])(Y_{t_2} - E[Y_{t_2}])] \\ &= E[X_{t_1}Y_{t_2}] - E[X_{t_1}]E[Y_{t_2}] \\ &= R_{XY}(t_1,t_2) - E[X_{t_1}]E[Y_{t_2}] \end{aligned} \tag{6.9}$$

where we have introduced the symbol $R_{XY}(t_1,t_2)$ for the expectation of $X_{t_1}Y_{t_2}$:

$$R_{XY}(t_1,t_2) \triangleq E[X_{t_1}Y_{t_2}] \tag{6.10}$$

This expectation is called the *cross-correlation function* of the given pair of random processes. Just which of the functions K_{XY} and R_{XY} is the most useful depends entirely upon the problem at hand.

7 STATIONARITY (CONTINUED)

We noted in the study of the random-telegraph process made in the preceding section that that process had a constant mean value and that its autocorrelation function depended only upon the difference of the two time instants involved and not on those instants separately. Similar results are obtained, of course, for any (strictly) stationary random process since, by the definition of stationarity, all the probability distribution functions of such a process are invariant under translations of the time origin.

Consider, for example, the stationary real random process $\{X_t, -\infty < t < +\infty\}$ whose probability density $f_{X_t}(x)$ exists for all t and all x. Then if t' is an arbitrarily chosen time instant, it follows from the assumption of stationarity that

$$f_{X_t}(x) = f_{X_{t'}}(x) \tag{7.1}$$

for all values of x and for all values of t. It therefore also follows that, for $k = 1, 2, 3, \ldots,$

$$\begin{aligned} E[X_t{}^k] &= \int_{-\infty}^{+\infty} x^k f_{X_t}(x)\,dx = \int_{-\infty}^{+\infty} x^k f_{X_{t'}}(x)\,dx \\ &= E[X_{t'}{}^k] \end{aligned} \tag{7.2}$$

for all values of t. Since the same result would be obtained for a stationary random process in which X_t is a discrete random variable, we see that *the kth moment of any stationary random process is a constant function of time.*

Exercise 7.1 Let $\{X_t, -\infty < t < +\infty\}$ be a strictly stationary real random process whose second moment is finite, that is, a process such that

$$E[X_t{}^2] < +\infty$$

for all t. Suppose also that the joint-probability density $f_{X_t,X_{t'}}$ exists for all times t and t'.

Show that then

$$R_X(t_1 + \delta, t_2 + \delta) = R_X(t_1,t_2) \tag{7.3}$$

holds for any value of δ, where t_1 and t_2 are arbitrarily chosen instants of time. That is, *the autocorrelation function of a strictly stationary random process is unchanged when we shift both time instants by the same amount*, that is, is invariant under a shift of the time origin. □

It is often convenient to rewrite (7.3) in terms of the time difference

$$\tau \triangleq t_1 - t_2 \tag{7.4}$$

If then we define

$$t \triangleq t_1 + \delta$$

an equivalent statement to that in (7.3) is

$$R_X(t, t - \tau) = R_X(t_1, t_1 - \tau) \tag{7.5}$$

which holds for any value of t, where t_1 is any arbitrarily chosen time instant. We note that in this case the autocorrelation function is a function only of the time difference τ (and not of t_1 or t_2 separately). For example, we see from the results of Exercise 6.2 above that the autocorrelation function of the random telegraph process may be expressed as

$$R_X(t_1, t_1 - \tau) = \tfrac{1}{4} + \tfrac{1}{4} \exp\left(-2\lambda|\tau|\right)$$

which is a function of τ, but not of t_1. Since in such cases the autocorrelation function is a function of a single variable, it is often convenient to introduce the notation

$$R_X(\tau) \triangleq R_X(t, t - \tau) \tag{7.6}$$

when dealing with stationary random processes.

The restriction to stationary cases gives the autocorrelation function certain special properties as shown by the results of the following exercises.

Exercise 7.2 Let $\{X_t, -\infty < t < +\infty\}$ be a stationary real random process. Show that in this case

$$R_X(-\tau) = R_X(\tau)$$

That is, *the autocorrelation function of a stationary real random process is an even function of its argument.* □

Exercise 7.3 Let $\{X_t, -\infty < t < +\infty\}$ be a stationary, real random process. Show that

$$|R_X(\tau)| \leq E[X_t^2] = R_X(0)$$

for all t. (*Hint:* Use the Cauchy-Schwarz inequality.) Thus, we see that *the autocorrelation function $R_X(\tau)$ of a stationary, real random process has its maximum magnitude at the origin in τ.* □

Exercise 7.4 Let $\{X_t, -\infty < t < +\infty\}$ be a stationary, real random process. Show that *if the autocorrelation function $R_X(\tau)$ of the given process is continuous at $\tau = 0$, then it is also continuous at every other value of τ.* (*Hint:* Consider $E[(X_{\tau+h} - X_\tau)X_0]$ and use the Cauchy-Schwarz inequality.) □

The net effect of the three preceding exercises is to limit severely the class of functions which are admissible as autocorrelation functions of stationary, real random processes; e.g., functions of the type shown in Fig. 7.1*a* are ruled out by the requirement of evenness (Exercise 7.2), functions of the type shown in Fig. 7.1*b* are ruled out by the requirement that the maximum magnitude must occur at the origin (Exercise 7.3), and functions of the type shown in Fig. 7.1*c* are ruled out by the requirement that the autocorrelation function of a stationary real random process must be continuous everywhere if it is continuous at the origin (Exercise 7.4).

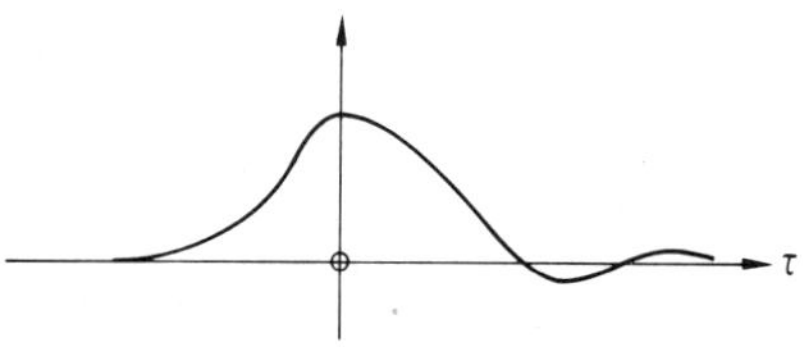

(*a*) Not an even function of τ

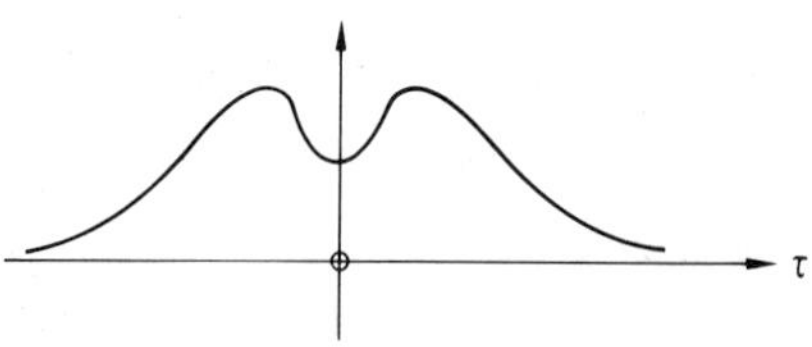

(*b*) Maximum not at origin

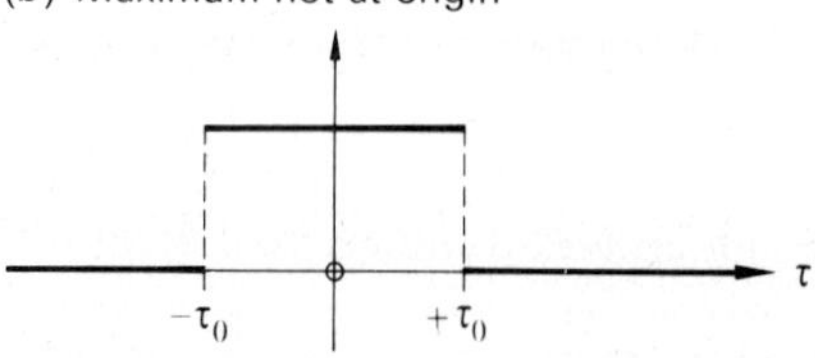

(*c*) Continuous at origin but discontinuous at $\pm\tau_0$

Figure 7.1 Functions which are inadmissible as autocorrelation functions of stationary, real random processes.

Wide-sense stationarity

As we shall see in our subsequent developments, there are many problems of interest in which a given random process, say $\{X_t, -\infty < t < +\infty\}$, is not stationary in the strict sense, but which, nonetheless, is such that given any arbitrary time instant, say t_1, the equations

$$E[X_t] = E[X_{t_1}] \tag{7.7}$$

and

$$R_X(t, t - \tau) = R_X(t_1, t_1 - \tau) \tag{7.8}$$

do hold true for all values of t. A random process which satisfies Eqs. (7.7) *and* (7.8) for all values of t is said to be *stationary in the wide sense*, or *stationary to the second order*. We shall use the notation

$$R_X(\tau) \triangleq R_X(t, t - \tau) \tag{7.9}$$

with such processes just as we did in the strictly stationary case.

A moment's reflection upon the results of Exercises 7.2 to 7.4 should show that those results really require only that the process be real and that the conditions expressed by (7.7) and (7.8) hold true, that is, that the given process be real and stationary in the wide sense and not that it necessarily be stationary in the strict sense. Thus it follows that the autocorrelation function $R_X(\tau)$ of any wide-sense stationary, real random process is an even function of its argument τ, that it has its maximum magnitude at the origin in τ, and that if it is continuous at the origin in τ, then it must be continuous everywhere.

It should be noted further that the derivations leading to (7.2) and (7.3) show that *if a random process is stationary in the strict sense, then it must also be stationary in the wide sense.* That the converse does not necessarily hold true, however, is shown by the following exercise.

Exercise 7.5 Let the random process $\{Z_t, -\infty < t < +\infty\}$ be such that

$$Z_t \triangleq Y \cos t + X \sin t$$

for all values of t, where X and Y are independent binary random variables each of which assumes the values -1 and $+2$ with the probabilities $\frac{2}{3}$ and $\frac{1}{3}$, respectively.

a. Calculate $E[Z_t]$ and $R_Z(t_1,t_2)$.
b. Show that this process is stationary in the wide sense but *not* in the strict sense. (*Hint:* Consider $E[Z_t^3]$.) □

There is, nonetheless, a very important case in which wide-sense stationarity does imply strict-sense stationarity.

Exercise 7.6 Let $\{X_t, -\infty < t < +\infty\}$ be a real gaussian random process (as defined in Sec. 4 above).

Show that if this process is stationary in the wide sense, then it is also stationary in the strict sense. □

Exercise 7.7 Let $\{X_t, -\infty < t < +\infty\}$ be a wide-sense, stationary, real random process.

a. Show that if

$$R_X(T) = R_X(0)$$

for some $T > 0$, then it follows with probability one that

$$X_{t+T} = X_t$$

for all t. (*Hint:* Consider $E[|X_{t+T} - X_t|^2]$ and use an argument paralleling that used to solve Exercise 7.12.1.)

b. Show further that the correlation function of such a process is periodic with period T; that is, show that

$$R_X(\tau + T) = R_X(\tau)$$

for all τ. A wide-sense stationary random process whose autocorrelation function is periodic is said to be a *periodic random process.* □

Exercise 7.8 Let $\{X_t, -\infty < t < +\infty\}$ be a wide-sense, stationary, real random process whose autocorrelation function is periodic with a fundamental period T. Show in this case that the least-mean-square-error linear estimate of the value of a sample function at a time which is an integral number of periods away from an observation instant is equal to the observed value. That is, show that

$$\hat{x}(t + kT) = x(t)$$

for any t, where k is any integer. □

Covariance stationarity

It is quite common in many problems to deal with random processes which have zero mean values or, if the mean values are not zero, to deal with processes which are centered about their means. In such cases the correlation and covariance functions become the same, and, for this reason, it is often common to define a second-order stationarity in terms of the covariance function. A random process $\{X_t, -\infty < t < +\infty\}$ is said to be *covariance stationary* if, given an arbitrary value of t, say t_1, the covariance function K_X of the process satisfies the equation

$$K_X(t, t - \tau) = K_X(t_1, t_1 - \tau) \tag{7.10}$$

for all values of t. It should be noted that while the concepts of wide sense stationarity and covariance stationarity are similar, they are not quite the same, as shown by the following exercises.

Exercise 7.9 Let $\{X_t, \ -\infty < t < +\infty\}$ be a real random process. Show that if this process is stationary in the wide sense, then it must also be covariance stationary. □

Exercise 7.10 Let the random process $\{Z_t, \ -\infty < t < +\infty\}$ be such that

$$Z_t \triangleq (Y + 1) \cos t + X \sin t$$

for all t, where X and Y are independent real random variables for which

$$E[X] = 0 = E[Y] \qquad \text{and} \qquad E[X^2] = 1 = E[Y^2]$$

a. Show that the given process is covariance stationary.
b. Show that the given random process is *not* stationary in either the wide sense or the strict sense. □

Multiple random processes

Let us next consider a pair of real random processes $\{X_t, \ -\infty < t < +\infty\}$ and $\{Y_t, \ -\infty < t < +\infty\}$ which are *jointly stationary* in the strict sense, that is, stationary in the sense that all their joint-probability distribution functions are invariant under translations of the time origin. It then follows, for arbitrary t_1, that

$$E[X_t] = E[X_{t_1}] \qquad \text{and} \qquad E[Y_t] = E[Y_{t_1}] \tag{7.11}$$

for all t, and that

$$R_X(t, t - \tau) = R_X(t_1, t_1 - \tau) \qquad \text{and} \qquad R_Y(t, t - \tau) = R_Y(t_1, t_1 - \tau) \tag{7.12}$$

for all values of t. It further follows from the assumption of joint stationarity in the strict sense that the equations

$$R_{XY}(t, t - \tau) = R_{XY}(t_1, t_1 - \tau)$$

and

$$R_{YX}(t, t - \tau) = R_{YX}(t_1, t_1 - \tau) \tag{7.13}$$

also hold true for arbitrary t_1 and all t. Since these cross-correlation functions are here functions only of the time difference τ (and not of t or t_1 separately) it is convenient to introduce the notation

$$R_{XY}(\tau) \triangleq R_{XY}(t, t - \tau) \qquad \text{and} \qquad R_{YX}(\tau) \triangleq R_{YX}(t, t - \tau) \tag{7.14}$$

Exercise 7.11 Let the random process $\{X_t, -\infty < t < +\infty\}$ and $\{Y_t, -\infty < t < +\infty\}$ be jointly stationary in the strict sense. Show that

$$R_{YX}(-\tau) = R_{XY}(\tau)$$

in this case. □

Suppose next that a pair of random processes $\{X_t, -\infty < t < +\infty\}$ and $\{Y_t, -\infty < t < +\infty\}$ are such that Eqs. (7.11) through (7.13) all hold true for all values of t and any value of t_1. Such a pair of processes is said to be *jointly stationary in the wide sense.* Note that we have shown above that *if a pair of random processes is jointly stationary in the strict sense then that pair is also jointly stationary in the wide sense.* The converse statement is, however, not necessarily true. Further, it is quite possible for a pair of random processes to be *individually* stationary in the wide sense and yet not be *jointly* stationary in the wide sense, as shown by the following example.

Exercise 7.12 Let the two random process $\{U_t, -\infty < t < +\infty\}$ and $\{V_t, -\infty < t < +\infty\}$ be such that

$$U_t \triangleq X \cos t + Y \sin t \qquad \text{and} \qquad V_t \triangleq Y \cos t + X \sin t$$

for all t, where X and Y are independent real random variables for which

$$E[X] = 0 = E[Y] \qquad \text{and} \qquad E[X^2] = 1 = E[Y^2]$$

a. Show that the U_t and V_t processes are individually stationary in the wide sense.

b. Show that those processes are *not* jointly stationary in the wide sense. □

8 SAMPLING A RANDOM PROCESS

We have, by now, developed enough tools to be able to study the problem of estimating the mean value of a random process by using sampling techniques.[1] Suppose, therefore, that we are given a random process $\{X_t, t \in T\}$ and that we wish to estimate its mean value $E[X_t]$. Now since the random variable X_t specified by the particular value of the index t is an ordinary random variable, we can directly apply our earlier discussion of this problem as given in Sec. 8.2. Different performances of the basic experiment discussed in that section correspond to different sample functions of the given random process in the present case. Thus

[1] It should be realized that we can generally obtain better estimators by *not* restricting ourselves to using sampling techniques (cf. appendix B of Davenport, Johnson, and Middleton, or sec. 3, chap. 4, of Van Trees).

if we choose a particular value of $t \in T$, say t_1, and observe the values $x(t_1)$ of n different sample functions at that time, then we can form a sample mean from those n sample-function values and use it to obtain an estimate of the mean value $E[X_t]$.

Now, in general, the mean value $E[X_t]$ is a function of the time index t. Suppose, however, that the given random process is stationary, at least in the wide sense. In this case, it follows from the discussions in Secs. 5 and 7 of this chapter that the mean value of the given process has the same value for all values of the index t. That is, given the wide-sense stationary random process $\{X_t, -\infty < t < +\infty\}$, then it follows that there exists a finite number m such that

$$E[X_t] = m \tag{8.1}$$

for all values of the index t. With this fact in mind, we are led to an alternative method of sampling.

Suppose that we observe a *particular* sample function of the given process, say $x(t)$, and that we observe its values $x(t_i)$ at n different time instants $t_1, t_2, \ldots, t_n$, as shown, for example, in Fig. 8.1. The empirical average of this particular set of sample function values then is

$$\bar{x} = \frac{1}{n} \sum_{i=1}^{n} x(t_i) \tag{8.2}$$

Suppose now that we were to repeat this sampling procedure on other *different* sample functions of the given process while using the *same* set of time instants. The various measured values $x(t_i)$ obtained from the different sample functions at the single time instant t_i are all sample values of the random variable which corresponds to that time instant; that is, the various $x(t_i)$ are sample values of the random variable

$$X_i \triangleq X_{t_i} \tag{8.3}$$

The various *sets* of sample values lead [when substituted in (8.2)] to

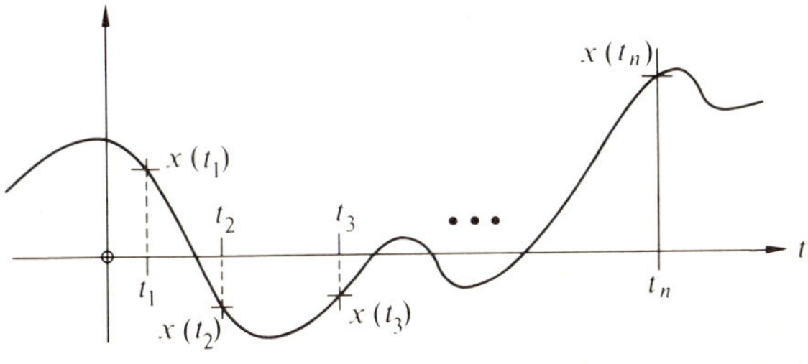

Figure 8.1 Sampling a sample function.

different values of the arithmetic average x; these latter values may then be thought of as different sample values of the sample mean

$$\hat{m}_n \triangleq \frac{1}{n} \sum_{i=1}^{n} X_i \tag{8.4}$$

The expected value of this sample mean is

$$\begin{aligned} E[\hat{m}_n] &= E\left[\frac{1}{n} \sum_{i=1}^{n} X_i\right] \\ &= \frac{1}{n} \sum_{i=1}^{n} E[X_i] \end{aligned}$$

Since we are dealing here with a wide-sense stationary random process, it follows that the expected values of all of the X_i are the same, and hence that we get the result

$$E[\hat{m}_n] = m \tag{8.5}$$

That is, *the expected value of a sample mean obtained by sampling a wide-sense stationary random process along a sample function in time is equal to the constant mean value of that random process.*

As in our earlier study of the sample mean, the question of how good an estimate of the mean value the sample mean gives is most easily answered by studying the properties of the variance of the sample mean. Now it follows from the definition of the sample mean, and from the results of Exercise 7.9.3, that

$$\begin{aligned} \text{var}\,(\hat{m}_n) &= \frac{1}{n^2} \sum_{i=1}^{n} \sum_{j=1}^{n} \text{cov}\,(X_i,X_j) \\ &= \frac{1}{n^2} \sum_{i=1}^{n} \sum_{j=1}^{n} K_X(t_i - t_j) \end{aligned} \tag{8.6}$$

where K_X is the covariance function of the random process being studied. In order to proceed further with calculations, however, we need to be able to specify the actual functional form of the covariance function K_X. We can, nonetheless, obtain some limiting results as follows.

Suppose first that the sample random variables X_i are all *highly correlated;* in particular, suppose that

$$\text{cov}\,(X_i,X_j) = \text{var}\,(X_i) \triangleq \sigma^2 \tag{8.7}$$

for all values of i and j. This would more or less be the situation, for example, if the duration T_0 of the total interval over which the individual samples were taken was so short that

$$K_X(t_i - t_j) \approx K_X(0)$$

for all values of i and j (say as shown in Fig. 8.2). It then follows from (8.6) that

$$\text{var}(\hat{m}_n) = \sigma^2 \tag{8.8}$$

in this case. That is, *the variance of the sample mean equals the variance of the process being studied when the samples taken are all highly correlated.* Since this result is independent of the number of sampling instants used, it follows that a measurement taken at a single sampling instant provides just as good (or bad) an estimate of the desired mean value as can be obtained by using *any* greater number of sampling instants—whenever highly correlated samples are obtained. *Whether or not a good estimate can be obtained at all in such a case therefore depends strictly upon just how concentrated the probability distribution of an individual sample is about its mean value.*

Suppose, on the other hand, that the individual samples making up the sample mean are *pairwise uncorrelated*; that is, suppose that

$$\text{cov}(X_i,X_j) = \begin{cases} \sigma^2 & \text{if } i = j \\ 0 & \text{if } i \neq j \end{cases} \tag{8.9}$$

This situation would be obtained, for example, if the covariance function of the random process being studied was as shown in Fig. 8.2 and if the minimum separation between adjacent sampling instants was greater than the value T_1 indicated in that figure. It then follows, on using (8.9) in (8.6), that we get the result

$$\text{var}(\hat{m}_n) = \frac{\sigma^2}{n} \tag{8.10}$$

whenever the individual samples are pairwise uncorrelated; that is, *the variance of the sample mean is inversely proportional to the number of samples taken when the samples are pairwise uncorrelated.*

We noted that it follows from (8.10)—as in the independent sample

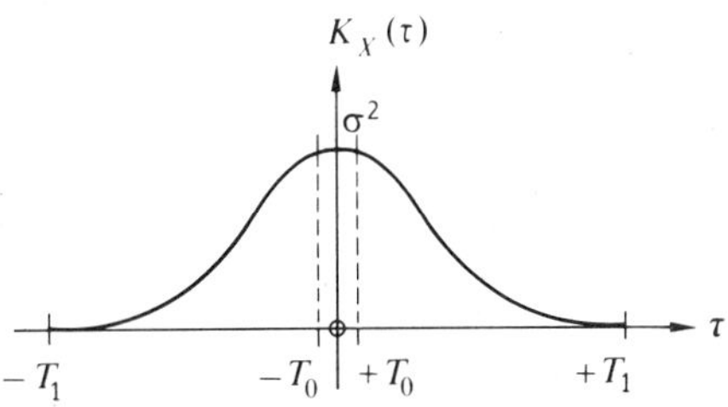

Figure 8.2 A typical covariance function.

case studied in Sec. 8.2—that we can decrease indefinitely the variance of the sample mean here simply by increasing the number of samples taken. That is, it follows from (8.10) that

$$\lim_{n\to\infty} \text{var}\,(\hat{m}_n) = 0 \tag{8.11}$$

whenever the samples are pairwise uncorrelated. If now we apply this result to the Chebyshev inequality, we obtain the further result that

$$\lim_{n\to\infty} P[|\hat{m}_n - m| \geq \epsilon] = 0 \tag{8.12}$$

Thus we see that *the sample mean converges in probability to the mean value of the given wide-sense stationary random process as the number of uncorrelated samples increases without limit;* that is, the weak law of large numbers holds in this case. Such results are obtained, of course, only if the duration of the total time interval over which measurements are made increases at least as fast as does the number of samples taken since, as shown by Fig. 8.2, a minimum spacing must be maintained between samples in order to keep them uncorrelated.

9 PERIODIC SAMPLING

A sampling case of considerable practical importance is that in which a wide-sense stationary random process is sampled periodically. Suppose, for example, that we take n samples of the wide-sense stationary random processes $\{X_t, -\infty < t < +\infty\}$ during an observation interval which starts at $t = 0$ and lasts until $t = T$, and suppose, further, that the n sampling instants are spaced equally throughout that interval (the last occurring at $t = T$). If now we define

$$\Delta t \triangleq \frac{T}{n}$$

(that is, Δt is the intersample period) it then follows that the sampling instants may be expressed in terms of Δt as

$$t_i = i\,\Delta t \qquad i = 1, 2, \ldots, n$$

Since the given process is stationary in the wide sense by hypothesis, it follows that

$$\begin{aligned} \text{cov}\,(X_i,X_j) &= K_X(t_i - t_j) \\ &= K_X[(i - j)\,\Delta t] \end{aligned}$$

and hence, on using this result in our previously derived expression (8.6) for the variance of the sample mean, that we get

$$\operatorname{var}(\hat{m}_n) = \frac{1}{n^2} \sum_{i=1}^{n} \sum_{j=1}^{n} K_X[(i-j)\,\Delta t] \tag{9.1}$$

as a formula for the variance of the sample mean obtained by periodic sampling of a wide-sense stationary random process.

The double summation in (9.1) may be reduced to a single summation as follows: Consider the covariance matrix of the random variables X_i and X_j. It follows from the previously derived properties of the covariance function (cf. Secs. 6 and 7) that this matrix has the form

$$\mathbf{K}_X = \begin{bmatrix} K_X(0) & K_X(\Delta t) & \cdots & K_X[(n-1)\,\Delta t] \\ K_X(\Delta t) & K_X(0) & \cdots & K_X[(n-2)\,\Delta t] \\ \cdots & \cdots & \cdots & \cdots \\ K_X[(n-1)\,\Delta t] & K_X[(n-2)\,\Delta t] & \cdots & K_X(0) \end{bmatrix} \tag{9.2}$$

in the present case. The summation in (9.1) is over all elements of this matrix. Inspection of (9.2) shows that the n terms along the principal diagonal are all equal to $K_X(0)$. Further inspection shows that there are $n - k$ terms, all equal to $K_X(k\,\Delta t)$, along both of the two diagonals which are displaced k positions from the principal diagonal, one to the right and one to the left. With these facts in mind we may then evaluate the double sum in (9.1) by summing along the diagonals of the covariance matrix (instead of along the rows or columns) to get

$$\operatorname{var}(\hat{m}_n) = \frac{nK_X(0)}{n^2} + \frac{2}{n^2} \sum_{k=1}^{n-1} (n-k)K_X(k\,\Delta t) \tag{9.3}$$

Now since

$$K_X(0) = \operatorname{var}(X_t) \triangleq \sigma^2$$

(9.3) may be rewritten as

$$\operatorname{var}(\hat{m}_n) = \frac{\sigma^2}{n} + \frac{2}{n} \sum_{k=1}^{n-1} \left(1 - \frac{k}{n}\right) K_X(k\,\Delta t) \tag{9.4}$$

The first term on the right-hand side of (9.4) is that obtained previously for the case of uncorrelated samples; the remaining terms account for the intersample correlations.

Let us next investigate the behavior of the variance of the sample mean as we let the number n of samples in an observation interval increase without limit while we hold fixed the duration T of the overall observation interval.

Exercise 9.1 Show that (9.4) gives

$$\lim_{n\to\infty} \operatorname{var}(\hat{m}_n) = \frac{2}{T}\int_0^T \left(1 - \frac{\tau}{T}\right) K_X(\tau)\, d\tau \tag{9.5}$$

if we pass to the limit $n \to \infty$ while keeping T fixed (so that $\Delta t \to 0$). □

Thus, we see that *when estimating the mean value of a wide-sense stationary random process by periodic sampling over a fixed duration observation interval, the variance of the sample mean does not, in general, vanish as the number of samples increases without limit.* The reason is, of course, that as the number of samples becomes larger and larger, adjacent samples become closer and closer together and hence become more and more highly correlated. The integral in (9.5) expresses the effect of this increase in adjacent sample correlation.

Exercise 9.2 Let $\{X_t, -\infty < t < +\infty\}$ be a wide-sense stationary random process with a mean m and the covariance function

$$K_X(\tau) = \begin{cases} \sigma^2\left(1 - \dfrac{|\tau|}{\tau_0}\right) & \text{for } 0 \le |\tau| \le \tau_0 \\ 0 & \text{for } \tau_0 < |\tau| \end{cases}$$

Suppose that the sample mean of this random process is obtained by periodically sampling over an observation interval of duration $T > \tau_0$.

Show that an upper bound on the variance of the sample mean is given by the inequality

$$\operatorname{var}(\hat{m}_n) < \sigma^2\left(\frac{1}{n} + \frac{\tau_0}{T}\right)$$

in the present case. [*Hint:* View the sum in (9.4) as the area of a step-function approximation to an integral.] □

Our discussion above focused upon the behavior of the variance of the sample mean as we increased the number of samples while keeping fixed the duration of the overall observation interval. Let us now consider what happens to the sample mean itself under these conditions. It follows from the definition of the sample mean that we may write

$$\hat{m}_n = \frac{1}{n}\sum_{i=1}^{n} X_i = \frac{1}{T}\sum_{i=1}^{n} X_{i\Delta t}\,\Delta t$$

in the present case of periodic sampling. If we let $n \to \infty$ (while forcing $\Delta t \to 0$ by keeping T fixed), it seems reasonable to assume that the sample

mean should approach, in some sense, the random variable

$$\hat{m}_T \triangleq \frac{1}{T} \int_0^T X_t \, dt \tag{9.6}$$

called the *sample time average.* A moment's reflection on (9.6) should convince one that this sample time average is a random variable whose probability distribution function describes the distribution of possible values taken on by the empirical time average

$$\frac{1}{T} \int_0^T x(t) \, dt$$

as we evaluate the latter for the various sample functions $x(t)$ which make up the given random process. We shall study further the properties of the sample time average in Chap. 10.

10 SUMMARY OF DEFINITIONS AND FORMULAS

Random process

An indexed family of random variables,

$$\{X_t, t \in T\}$$

where T denotes the set of possible values of the index t. If T is a countably infinite set, then the process is called a discrete-parameter random process; if T is a continuum, then the process is called a continuous-parameter random process.

Bernoulli process

A random process $\{X_n, n = 1, 2, 3, \ldots\}$ in which the random variables X_n are Bernoulli random variables, for example, where

$$P[X_n = 1] = p \qquad \text{and} \qquad P[X_n = 0] = 1 - p \tag{1.3}$$

and where the X_n are statistically independent random variables.

The mean value of the Bernoulli process is

$$E[X_n] = p \tag{1.10}$$

and the variance of the process is

$$\text{var}\,(X_n) = pq = p(1 - p) \tag{1.11}$$

Binomial counting process

A random process $\{Y_n, n = 1, 2, 3, \ldots\}$ in which the counting random variable Y_n is defined to be a sum of independent Bernoulli random variables:

$$Y_n \triangleq \sum_{i=1}^{n} X_i \tag{2.3}$$

If the X_i can assume only the values $+1$ or 0, then

$$P[Y_n = k] = \binom{n}{k} p^k(1-p)^{n-k} \tag{2.9}$$

for $k = 0, 1, 2, \ldots, n$.

The mean value of the binomial counting process is

$$E[Y_n] = np \tag{2.12}$$

and the variance is

$$\operatorname{var}(Y_n) = npq \tag{2.13}$$

The covariance of the two different counting random variables Y_m and Y_n is

$$\operatorname{cov}(Y_m,Y_n) = pq \min(m,n) \tag{2.15}$$

while the variance of the counting difference, or counting increment $Y_m - Y_n$, is

$$\operatorname{var}(Y_m - Y_n) = |m - n|\, pq \tag{2.14}$$

Sine wave process

A random process $\{X_t, t \in T\}$, where the index set T is a continuum, and where

$$X_t \triangleq V \sin(\Omega t + \Phi) \tag{3.2}$$

for all values of t in T. V, Ω, and Φ here are random variables.

Stationarity (strict sense)

A random process $\{X_T, t \in T\}$ is stationary (in the strict sense) if and only if all of the finite-dimensional probability distribution functions are invariant under shifts of the time origin.

Covariance function

Given the random process $\{X_t, t \in T\}$, the function

$$K_X(t_1,t_2) \triangleq \operatorname{cov}(X_{t_1},X_{t_2}) \tag{6.2}$$

where $t_1 \in T$ and $t_2 \in T$, is called the covariance function of the given random process.

Autocorrelation function

Given the random process $\{X_t,\ t \in T\}$, the function

$$R_X(t_1,t_2) \triangleq E[X_{t_1}X_{t_2}] \tag{6.4}$$

where t_1 and t_2 are in T, is called the autocorrelation function of the given random process. It then follows that

$$K_X(t_1,t_2) = R_X(t_1,t_2) - E[X_{t_1}]E[X_{t_2}] \tag{6.3}$$

Multiple random processes

Given the pair of random processes $\{X_t,\ t \in T\}$ and $\{Y_t,\ t \in T\}$, we define the (cross) covariance function

$$K_{XY}(t_1,t_2) \triangleq \operatorname{cov}(X_{t_1},Y_{t_2}) \tag{6.8}$$

and the cross-correlation function

$$R_{XY}(t_1,t_2) \triangleq E[X_{t_1}Y_{t_2}] \tag{6.10}$$

where t_1 and t_2 are in T in both cases.

Stationary random processes

Let $\{X_t,\ -\infty < t < +\infty\}$ be a strictly stationary real random process. It then follows that

$$E[X_t] = E[X_{t_1}] \qquad \text{[see (7.2)]}$$

and

$$R_X(t,\ t-\tau) = R_X(t_1,\ t_1-\tau) \tag{7.5}$$

both hold for any value of t, where t_1 is an arbitrarily chosen time instant. We generally write in this case

$$E[X_t] = E[X]$$

and

$$R_X(t,\ t-\tau) = R_X(\tau) \tag{7.6}$$

for all t. It then follows that

$$R_X(-\tau) = R_X(\tau) \qquad \text{(Exercise 7.2)}$$

and

$$|R_X(\tau)| \le R_X(0) \qquad \text{(Exercise 7.3)}$$

Stationarity (wide sense)

Let $\{X_t,\ -\infty < t < +\infty\}$ be a real random process such that

$$E[X_t] = E[X_{t_1}]$$

and

$$R_X(t,\ t-\tau) = R_X(t_1,\ t_1-\tau)$$

both hold for some t_1 and all t, the given random process is then said to be stationary in the wide sense.

Multiple stationary random processes

Let the random processes $\{X_t, -\infty < t < +\infty\}$ and $\{Y_t, -\infty < t < +\infty\}$ be jointly stationary in the strict sense. Then it follows that the relations

$$E[X_t] = E[X_{t_1}] \qquad E[Y_t] = E[Y_{t_1}] \tag{7.11}$$

and

$$R_X(t, t-\tau) = R_X(t_1, t_1-\tau) \qquad R_Y(t, t-\tau) = R_Y(t_1, t_1-\tau) \tag{7.12}$$

and

$$R_{XY}(t, t-\tau) = R_{XY}(t_1, t_1-\tau) \qquad R_{YX}(t, t-\tau) = R_{YX}(t_1, t_1-\tau) \tag{7.13}$$

all hold for some t_1 and all t.

Conversely if the random processes $\{X_t, -\infty < t < +\infty\}$ and $\{Y_t, -\infty < t < +\infty\}$ are such that (7.11), (7.12), and (7.13) all hold, then those processes are said to be jointly stationary in the wide sense.

Sample mean

Consider the wide-sense stationary random process $\{X_t, -\infty < t < +\infty\}$ whose second moment is finite. Suppose that we sample that process at the n time instants $t_1, t_2, \ldots t_n$. The estimator

$$\hat{m}_n \triangleq \frac{1}{n}\sum_{i=1}^{n} X_i \tag{8.4}$$

where

$$X_i \triangleq X_{t_i} \tag{8.3}$$

is called the sample mean. This estimator has the properties:

$$E[\hat{m}_n] = m \tag{8.5}$$

where m is the mean value of the given process, and

$$\text{var}\,(\hat{m}_n) = \frac{1}{n^2}\sum_{i=1}^{n}\sum_{j=1}^{n} K_X(t_i,t_j) \tag{8.6}$$

Special cases are:

$$\text{var}\,(\hat{m}_n) = \sigma^2 \tag{8.8}$$

(where σ^2 is the variance of the process) when the samples X_i are all highly correlated, and

$$\text{var}\,(\hat{m}_n) = \frac{\sigma^2}{n} \tag{8.10}$$

when the samples X_i are pairwise uncorrelated. It then follows in the latter case that

$$\lim_{n \to \infty} P[|\hat{m}_n - m| > \epsilon] = 0 \tag{8.12}$$

that is, that the weak law of large numbers holds in this case.

Periodic sampling

Let the wide-sense stationary random process $\{X_t, \ -\infty < t < +\infty\}$ be sampled periodically throughout the interval $0 \le t \le T$ in such a way that there are n sampling instants equally spaced throughout that interval (the last at $t = T$). The variance of the sample mean is given in this case by the formula

$$\text{var}\,(\hat{m}_n) = \frac{\sigma^2}{n} + \frac{2}{n} \sum_{k=1}^{n-1} \left(1 - \frac{k}{n}\right) K_X(k\ \Delta t) \tag{9.4}$$

where $\Delta t = T/n$. It therefore follows that

$$\lim_{n \to \infty} \text{var}\,(\hat{m}_n) = \frac{2}{T} \int_0^T \left(1 - \frac{\tau}{T}\right) K_X(\tau)\, d\tau \tag{9.5}$$

if we pass to the limit $n \to \infty$ while keeping T fixed.

REFERENCES

Cramér, Harald, and M. R. Leadbetter: "Stationary and Related Stochastic Processes," John Wiley & Sons, Inc., New York, 1967.

Chandrasekhar, Subrahmanyan: Stochastic Problems in Physics and Astronomy, *Rev. Mod. Phys.*, vol. 15, no. 1, pp. 1–89, 1943 (reprinted in Nelson Wax: "Selected Papers on Noise and Stochastic Processes," pp. 3–91, Dover Publications, Inc., New York, 1954).

Davenport, Wilbur B., Jr., Richard A. Johnson, and David Middleton: Statistical Errors in Measurements on Random Time Functions, *J. Appl. Phys.*, vol. 23, no. 4, pp. 377–388, 1952.

Doob, Joseph: "Stochastic Processes," John Wiley & Sons, Inc., New York, 1953.

Neveu, Jacques: "Mathematical Foundations of the Calculus of Probability," Holden-Day, Inc., Publisher, San Francisco, 1965.

Van Trees, Harry L., Jr.: "Detection, Estimation, and Modulation Theory," pt. I, John Wiley & Sons, Inc., New York, 1968.

10
Linear Transformations

1 TWO-DIMENSIONAL VECTORS

In the preceding chapter, we first introduced some of the simplest random processes, then gave a general definition of a random process, and finally discussed some of the basic properties of random processes. In this chapter, we wish to introduce the problem of operating upon random processes; in particular, we wish to study the effects of linear transformations on random processes. With the characterization of a random process as an infinite-dimensional random vector in mind, we shall study first, however, the effects of linear transformations on finite-dimensional random vectors, and then we shall extend our results to the random process case.

One example of a linear transformation which we have already met in the two-dimensional random vector case is the rotational transformation introduced in Sec. 6.4. In this case, a two-dimensional random vector

$$\mathbf{Y} = (Y_1, Y_2)$$

is generated from the two-dimensional random vector

$$\mathbf{X} = (X_1, X_2)$$

by the rotational transformation

$$Y_1 = X_1 \cos\theta - X_2 \sin\theta \qquad \text{and} \qquad Y_2 = X_1 \sin\theta + X_2 \cos\theta \tag{1.1}$$

in which θ is the angle of rotation [cf. (6.4.17)]. That this transformation is indeed a linear transformation may be seen as follows: Suppose that the random vector **X** is a weighted sum of two other two-dimensional random vectors; in particular, let

$$\mathbf{X} = a\mathbf{W} + b\mathbf{Z}$$

where a and b are arbitrary real numbers. That is,

$$\begin{aligned} \mathbf{X} &= (X_1, X_2) \\ &= (aW_1 + bZ_1,\ aW_2 + bZ_2) \end{aligned}$$

where

$$\mathbf{W} = (W_1, W_2) \qquad \text{and} \qquad \mathbf{Z} = (Z_1, Z_2)$$

Rotation of this vector **X** [using (1.1)] produces the vector **Y** with the components

$$\begin{aligned} Y_1 &= (aW_1 + bZ_1)\cos\theta - (aW_2 + bZ_2)\sin\theta \\ &= a(W_1 \cos\theta - W_2 \sin\theta) + b(Z_1 \cos\theta - Z_2 \sin\theta) \end{aligned} \tag{1.2}$$

and

$$\begin{aligned} Y_2 &= (aW_1 + bZ_1)\sin\theta + (aW_2 + bZ_2)\cos\theta \\ &= a(W_1 \sin\theta + W_2 \cos\theta) + b(Z_1 \sin\theta + Z_2 \cos\theta) \end{aligned} \tag{1.3}$$

If now we denote the rotational transformation by the vector function **g** and write

$$\mathbf{Y} = \mathbf{g}(\mathbf{X}) \tag{1.4}$$

then it follows from (1.2) and (1.3) that **g** satisfies the equation

$$\mathbf{g}(a\mathbf{W} + b\mathbf{Z}) = a\mathbf{g}(\mathbf{W}) + b\mathbf{g}(\mathbf{Z}) \tag{1.5}$$

Any transformation **g** which satisfies (1.5) is, by definition, a *linear transformation*. Thus we see that the rotational transformation (1.1) is a linear transformation.

Exercise 1.1 Fill in the missing steps to show that the rotational transformation (1.1) satisfies (1.5). □

Exercise 1.2 Suppose that the two-dimensional random vector $\mathbf{Y} = (Y_1, Y_2)$ is generated from the two-dimensional random vector $\mathbf{X} = (X_1, X_2)$ by the transformation

$$Y_1 = X_1^2 \qquad \text{and} \qquad Y_2 = X_2^2$$

Show that this transformation is not a linear transformation. □

The most general linear transformation $\mathbf{g}$ which takes a two-dimensional real random vector $\mathbf{X}$ into a real random vector $\mathbf{Y}$ [as in (1.4)] is that defined by the equations

$$\begin{aligned} Y_1 &= g_{11}X_1 + g_{12}X_2 \\ Y_2 &= g_{21}X_1 + g_{22}X_2 \end{aligned} \tag{1.6}$$

where the coefficients g_{ij} are arbitrary real numbers.[1] We showed earlier (in Sec. 6.4) that if $\mathbf{g}$ has an inverse, say

$$\mathbf{h} = (h_1, h_2)$$

and if $\mathbf{X}$ has a joint-probability density f_{X_1,X_2}, then the joint-probability density of $\mathbf{Y}$ is given by the formula

$$f_{Y_1,Y_2}(y_1,y_2) = f_{X_1,X_2}(h_1(y_1,y_2),h_2(y_1,y_2)) \left| \frac{\partial(h_1,h_2)}{\partial(y_1,y_2)} \right| \tag{1.7}$$

Exercise 1.3 What condition must be satisfied by the coefficients g_{ij} in (1.6) in order that the transformation $\mathbf{g}$ have an inverse? □

Now in some applications, the transformation $\mathbf{g}$ may not have an inverse and then (1.7) cannot be used. In others, we may not know the joint-probability density of $\mathbf{X}$ and again (1.7) cannot be used. Even in such cases, however, if we are given the moments of $\mathbf{X}$, we can readily calculate those of $\mathbf{Y}$. Thus, for example, if we are given the mean values $E[X_1]$ and $E[X_2]$, we obtain

$$E[Y_1] = g_{11}E[X_1] + g_{12}E[X_2]$$

and

$$E[Y_2] = g_{21}E[X_1] + g_{22}E[X_2] \tag{1.8}$$

as formulas for the mean values of Y_1 and Y_2.

Exercise 1.4 Suppose that the means and variances of X_1 and X_2 are known and that the covariance of X_1 and X_2 is also known. Derive

[1] Compare with secs. 2.5 and 2.6 of Kreider et al., or chap. 3 of Hoffman and Kunze. Note that if $\mathbf{X}$ and $\mathbf{Y}$ are complex random vectors, then the coefficients g_{ij} may be complex numbers.

formulas for the covariance of Y_1 and Y_2, and for the variances of Y_1 and Y_2. □

Exercise 1.5 Let $\mathbf{X}$ be a two-dimensional random vector whose components X_1 and X_2 have zero means, a covariance $\operatorname{cov}(X_1,X_2)$, and variances σ_1^2 and σ_2^2 respectively. Suppose that the random vector $\mathbf{Y} = (Y_1,Y_2)$ is generated from $\mathbf{X}$ by the rotational transformation defined by (1.1).

a. Show that the components Y_1 and Y_2 are uncorrelated if the angle of rotation θ is chosen such that

$$\tan 2\theta = \frac{2\operatorname{cov}(X_1,X_2)}{\sigma_2^2 - \sigma_1^2}$$

whenever $\sigma_1^2 \neq \sigma_2^2$.

b. Show that if $\sigma_1^2 = \sigma_2^2$, then Y_1 and Y_2 are uncorrelated if $\theta = \pi/4$.

c. Show that if $\sigma_1^2 = \sigma_2^2$ and if X_1 and X_2 are uncorrelated, then Y_1 and Y_2 are uncorrelated for *any* angle of rotation.

Thus we see that any pair of random variables with zero means can be transformed into a pair of uncorrelated random variables simply by rotation. □

2 n-DIMENSIONAL VECTORS

The results just obtained in the two-dimensional case may now be extended more or less directly to the n-dimensional case. For example, suppose that the m-dimensional real random vector

$$\mathbf{Y} = (Y_1, Y_2, \ldots, Y_m)$$

is generated from the n-dimensional real random vector (where $n \geq m$)

$$\mathbf{X} = (X_1, X_2, \ldots, X_n)$$

by a linear transformation $\mathbf{g}$; that is, suppose that

$$\mathbf{Y} = \mathbf{g}(\mathbf{X}) \tag{2.1}$$

where $\mathbf{g}$ satisfies the equation

$$\mathbf{g}(a\mathbf{W} + b\mathbf{Z}) = a\mathbf{g}(\mathbf{W}) + b\mathbf{g}(\mathbf{Z})$$

where a and b are arbitrary real constants and $\mathbf{W}$ and $\mathbf{Z}$ are arbitrary real random vectors. Paralleling the two-dimensional case [cf. (1.6)], the most general form of $\mathbf{g}$ here is that defined by the system of linear equations

$$\begin{aligned}
Y_1 &= g_{11}X_1 + g_{12}X_2 + \cdots + g_{1n}X_n \\
Y_2 &= g_{21}X_1 + g_{22}X_2 + \cdots + g_{2n}X_n \\
&\cdots\cdots\cdots\cdots\cdots\cdots\cdots\cdots \\
Y_m &= g_{m1}X_1 + g_{m2}X_2 + \cdots + g_{mn}X_n
\end{aligned} \tag{2.2}$$

where the coefficients g_{ij} are arbitrary real numbers.[1] A more compact form of (2.2) is

$$Y_i = \sum_{j=1}^{n} g_{ij}X_j \qquad i = 1, 2, \ldots, m \tag{2.3}$$

We see then that the sample mean $\hat{m}_n$ is an example of a linear transformation of an n-dimensional random vector, for [cf. (8.2.1)]

$$\begin{aligned}\hat{m}_n &\triangleq \frac{1}{n}\sum_{i=1}^{n} X_i \\ &= \frac{1}{n}X_1 + \frac{1}{n}X_2 + \cdots + \frac{1}{n}X_n\end{aligned}$$

As in the two-dimensional case, sometimes we wish to determine the joint-probability density of the random vector **Y** and sometimes we need only to determine the means, variances, and covariances of the components of **Y**. We note that in order to use the results of Sec. 6.4 in calculating the joint-probability density of **Y**, the transformation **g** must be one to one; in particular, we must have $m = n$ and the determinant of the coefficients g_{ij} must not be zero.[2] These restrictions need not be satisfied, of course, if we only wish to determine moments of the components of **Y**. Thus, for example, it follows immediately from (2.3), and from the linearity of the expectation, that the mean of the ith component of **Y** is given by

$$E[Y_i] = \sum_{j=1}^{n} g_{ij}E[X_j] \tag{2.4}$$

whether **g** has an inverse or not. In order to calculate the mean $E[Y_i]$, then, we only need to know the means of the components of the random vector **X**.

Exercise 2.1 Let **X** be an n dimensional real random vector and suppose that the random vector **Y** is generated from **X** by the linear transformation (2.2).

a. Show that

$$\operatorname{cov}(Y_i, Y_k) = \sum_{j=1}^{n}\sum_{r=1}^{n} g_{ij}g_{kr} \operatorname{cov}(X_j, X_r)$$

b. Determine the variance of Y_i. □

[1] Note that again if **X** and **Y** are complex random vectors, then the coefficients g_{ij} may be complex numbers.

[2] Compare with sec. 1.4 of Hildebrand or sec. 5.4 of Hoffman and Kunze.

It is perhaps worth pointing out at this time that not only are the above n-dimensional results important in their own right but also that they are direct analogs of the results which we shall obtain later in this chapter in the case of linear transformations of random processes.

3 MATRIX FORMULATION*

If the number of components of the original random vector $\mathbf{X}$ and of the resultant random vector $\mathbf{Y}$ (that is, n and m, respectively) are very large, the manipulations required in carrying out any detailed calculation can become quite unwieldy and it then becomes relatively easy to lose sight of where one is headed. With this fact in mind, matrices are often introduced at this point in order to make the notation more compact. We wish, accordingly, to derive here some of the n-dimensional results in matrix form.

To this end, let us represent the random vectors $\mathbf{X}$ and $\mathbf{Y}$ by the column matrices formed from their components:

$$\mathbf{X} = \begin{bmatrix} X_1 \\ X_2 \\ \cdot \\ \cdot \\ \cdot \\ X_n \end{bmatrix} \qquad \mathbf{Y} = \begin{bmatrix} Y_1 \\ Y_2 \\ \cdot \\ \cdot \\ \cdot \\ Y_m \end{bmatrix}$$

If now we define the *transformation matrix* $\mathbf{g}$ to be the $m \times n$ matrix formed of the coefficients g_{ij} in Eqs. (2.2),

$$\mathbf{g} = \begin{bmatrix} g_{11} & g_{12} & \cdots & g_{1n} \\ g_{21} & g_{22} & \cdots & g_{2n} \\ \cdots & \cdots & \cdots & \cdots \\ g_{m1} & g_{m2} & \cdots & g_{mn} \end{bmatrix}$$

then it follows from the usual rule for matrix multiplication that we get

$$\begin{bmatrix} g_{11} & g_{12} & \cdots & g_{1n} \\ g_{21} & g_{22} & \cdots & g_{2n} \\ \cdots & \cdots & \cdots & \cdots \\ g_{m1} & g_{m2} & \cdots & g_{mn} \end{bmatrix} \begin{bmatrix} X_1 \\ X_2 \\ \cdot \\ \cdot \\ \cdot \\ X_n \end{bmatrix} = \begin{bmatrix} Y_1 \\ Y_2 \\ \cdot \\ \cdot \\ \cdot \\ Y_m \end{bmatrix}$$

* Supplementary section. A prior knowledge of elementary matrix manipulations is needed here. The results of this section will be used only in the subsequent chapter on the gaussian process.

where

$$Y_i = \sum_{j=1}^{n} g_{ij}X_j \qquad i = 1, 2, \ldots, m$$

Thus we see that Eqs. (2.2) may be written in the more compact matrix form as

$$\mathbf{Y} = \mathbf{gX} \tag{3.1}$$

We may now derive a matrix formula for the expectation of the resultant vector **Y** in exactly the same way. Thus, suppose that we use column matrix representations of the expectations of the random vectors **X** and **Y**; that is, suppose that we write

$$E[\mathbf{X}] = \begin{bmatrix} E[X_1] \\ E[X_2] \\ \cdot \\ \cdot \\ \cdot \\ E[X_n] \end{bmatrix} \qquad E[\mathbf{Y}] = \begin{bmatrix} E[Y_1] \\ E[Y_2] \\ \cdot \\ \cdot \\ \cdot \\ E[Y_m] \end{bmatrix}$$

It then follows that we obtain

$$E[\mathbf{Y}] = \mathbf{g}E[\mathbf{X}] \tag{3.2}$$

as the matrix version of (2.4).

Finally, let us now derive the matrix relation between the covariances of the random vector **X** and the covariances of the random vector **Y**. By definition, the covariance of the j and r components of **X** is

$$\operatorname{cov}(X_j,X_r) = E[(X_j - E[X_j])(X_r - E[X_r])]$$

Since we will have to deal with centered random variables (for example, $X_j - E[X_j]$) in our following manipulations of covariances, it is convenient at this time to introduce the new random variables

$$\xi_j \triangleq X - E[X_j] \qquad \text{and} \qquad \eta_i \triangleq Y_i - E[Y_i]$$

If therefore we denote the covariance of X_j and X_r by γ_{jr} and the covariance of Y_i and Y_k by λ_{ik},

$$\gamma_{jr} \triangleq \operatorname{cov}(X_j,X_r) \qquad \text{and} \qquad \lambda_{ik} \triangleq \operatorname{cov}(Y_i,Y_k)$$

then it follows that we may also write

$$\gamma_{jr} = E[\xi_j\xi_r] \qquad \text{and} \qquad \lambda_{ik} = E[\eta_i\eta_k] \tag{3.3}$$

The *covariance matrices* of the random vectors $\mathbf{X}$ and $\mathbf{Y}$ are then defined by the formulas

$$\mathbf{\Gamma} \triangleq \begin{bmatrix} \gamma_{11} & \gamma_{12} & \cdots & \gamma_{1n} \\ \gamma_{21} & \gamma_{22} & \cdots & \gamma_{2n} \\ \cdots & \cdots & \cdots & \cdots \\ \gamma_{n1} & \gamma_{n2} & \cdots & \gamma_{nn} \end{bmatrix} \tag{3.4}$$

and

$$\mathbf{\Lambda} \triangleq \begin{bmatrix} \lambda_{11} & \lambda_{12} & \cdots & \lambda_{1m} \\ \lambda_{21} & \lambda_{22} & \cdots & \lambda_{2m} \\ \cdots & \cdots & \cdots & \cdots \\ \lambda_{m1} & \lambda_{m2} & \cdots & \lambda_{mm} \end{bmatrix} \tag{3.5}$$

Suppose next that we represent the centered random vector

$$\mathbf{n} \triangleq \mathbf{Y} - E[\mathbf{Y}] \tag{3.6}$$

in the usual manner by the column matrix of its components; that is,

$$\mathbf{n} = \begin{bmatrix} \eta_1 \\ \eta_2 \\ \cdot \\ \cdot \\ \cdot \\ \eta_m \end{bmatrix} = \begin{bmatrix} Y_1 - E[Y_1] \\ Y_2 - E[Y_2] \\ \cdot \\ \cdot \\ \cdot \\ Y_m - E[Y_m] \end{bmatrix}$$

The *transpose* $\mathbf{n}^T$ of this vector is defined to be the row matrix formed of the components of $\mathbf{n}$:

$$\mathbf{n}^T \triangleq [\eta_1 \quad \eta_2 \quad \cdots \quad \eta_m]$$

If now we consider the matrix product $\mathbf{n}\mathbf{n}^T$, we get

$$\mathbf{n}\mathbf{n}^T = \begin{bmatrix} \eta_1 \\ \eta_2 \\ \cdot \\ \cdot \\ \cdot \\ \eta_m \end{bmatrix} [\eta_1 \quad \eta_2 \quad \cdots \quad \eta_m]$$

$$= \begin{bmatrix} \eta_1\eta_1 & \eta_1\eta_2 & \cdots & \eta_1\eta_m \\ \eta_2\eta_1 & \eta_2\eta_2 & \cdots & \eta_2\eta_m \\ \cdots & \cdots & \cdots & \cdots \\ \eta_m\eta_1 & \eta_m\eta_2 & \cdots & \eta_m\eta_m \end{bmatrix}$$

Paralleling the definition of the expectation of a random vector as the vector whose components are the expectations of the respective random components, *we define the expectation of a matrix to be the matrix of the expectations of the respective elements of the original matrix.* Thus, for example,

$$E[\mathbf{n}\mathbf{n}^T] = \begin{bmatrix} E[\eta_1\eta_1] & E[\eta_1\eta_2] & \cdots & E[\eta_1\eta_m] \\ E[\eta_2\eta_1] & E[\eta_2\eta_2] & \cdots & E[\eta_2\eta_m] \\ \cdots & \cdots & \cdots & \cdots \\ E[\eta_m\eta_1] & E[\eta_m\eta_2] & \cdots & E[\eta_m\eta_m] \end{bmatrix}$$

It therefore follows from this result, and from (3.3) and (3.5), that we may write

$$\mathbf{\Lambda} = E[\mathbf{n}\mathbf{n}^T] \tag{3.7}$$

We may now use this result to determine our desired relation between $\mathbf{\Lambda}$ and $\mathbf{\Gamma}$.

First of all, we note that it follows on using (3.1) and (3.2) in (3.6) that we get

$$\mathbf{n} = \mathbf{g}\boldsymbol{\xi}$$

where $\boldsymbol{\xi}$ is the centered random variable based on $\mathbf{X}$; that is,

$$\boldsymbol{\xi} \triangleq \mathbf{X} - E[\mathbf{X}]$$

It therefore follows that

$$\begin{aligned} \mathbf{n}\mathbf{n}^T &= \mathbf{g}\boldsymbol{\xi}(\mathbf{g}\boldsymbol{\xi})^T \\ &= \mathbf{g}\boldsymbol{\xi}\boldsymbol{\xi}^T\mathbf{g}^T \end{aligned}$$

the last equality following from the matrix rule for the transpose of a product of matrices. On taking the expectation of both sides of this last result, we then obtain the formula

$$E[\mathbf{n}\mathbf{n}^T] = \mathbf{g}E[\boldsymbol{\xi}\boldsymbol{\xi}^T]\mathbf{g}^T \tag{3.8}$$

Now, paralleling the steps leading to (3.7), it is easy to show that

$$\mathbf{\Gamma} = E[\boldsymbol{\xi}\boldsymbol{\xi}^T] \tag{3.9}$$

and hence, using (3.7) and (3.9) in (3.8), it follows that we finally obtain

$$\mathbf{\Lambda} = \mathbf{g}\mathbf{\Gamma}\mathbf{g}^T \tag{3.10}$$

as our desired matrix relation between the covariances of the random vector $\mathbf{Y}$ and those of the random vector $\mathbf{X}$. We shall find this result to be quite useful in our subsequent study of the gaussian process.

Exercise 3.1 Using the matrix definitions of $\mathbf{g}$, $\mathbf{\Gamma}$, and $\mathbf{\Lambda}$, verify (3.10) by applying the usual rule for matrix multiplication and comparing your result with that of Exercise 2.1. □

4 TIME AVERAGES

Having studied some of the effects of linear transformations on random vectors, we would next like to extend our study to the case of linear transformations operating on random processes; in particular, we wish to obtain formulas for various moments of the results of such transformations.

As our first example, let us reconsider the problem of estimating the mean of a wide-sense stationary random process. Now we pointed out in Sec. 9.9 that if we increased the number of samples without limit, while holding fixed the duration of the total observation interval, then the sample mean converged to a time average over the observation interval [cf. Eq. (9.9.6)]. Let us now study directly the statistical properties of such "short-time" averages (rather than calculating them as limits of sampling results as done in Sec. 9.9).

Suppose that a sample function $x(t)$ of the real random process $\{X_t, \ -\infty < t < +\infty\}$ forms the input to a device which measures a running time average, as shown schematically in Fig. 4.1. The output of such a device is

$$y(t) = \frac{1}{T} \int_{t-T}^{t} x(t')\, dt' \tag{4.1}$$

In other words, the output at time t is the time average of the input taken over the preceding time interval of duration T. Such a device is sometimes called a *short-time averager.*

Suppose next that we repeat this experiment with a number of different averagers (each with a different input sample function taken from the given input process) and that we observe the output of each averager at the same time t. The resultant set of averager-output values may then be thought of as different sample values of a new random variable Y_t, defined by the formula

$$Y_t \triangleq \frac{1}{T} \int_{t-T}^{t} X_{t'}\, dt' \tag{4.2}$$

If now we repeat this entire procedure for a number of different values of the time index t, we get correspondingly a set of random variables indexed by those time instants. That is, we may represent the output of the averager by the random process $\{Y_t, \ -\infty < t < +\infty\}$ where Y_t is

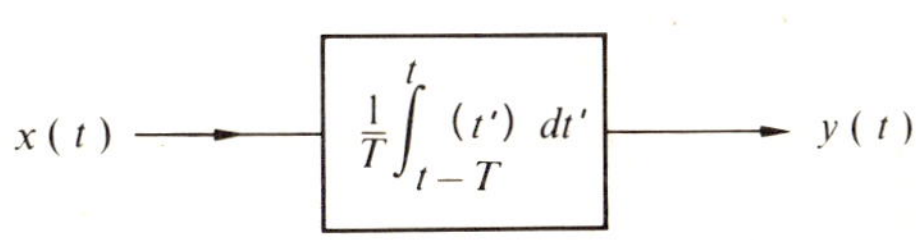

Figure 4.1 The short-time averager.

defined by (4.2). We next wish to determine some of the statistical properties of this output random process.

Exercise 4.1 Suppose that the random process $\{X_t, -\infty < t < +\infty\}$ is such that

$$X_t \triangleq Z \cos t$$

for all t, where the random variable Z is uniformly distributed over [0,1]. Suppose further that the X_t process is the input to a short-time averager.

a. Show that the averager-output random process $\{Y_t, -\infty < t < +\infty\}$ is such that

$$Y_t = Z\left(\frac{\sin\frac{T}{2}}{\frac{T}{2}}\right)\cos\left(t - \frac{T}{2}\right)$$

b. Show that the mean value and the autocorrelation function of the output processes are then given by

$$E[Y_t] = \frac{1}{2}\frac{\sin\frac{T}{2}}{\frac{T}{2}}\cos\left(t - \frac{T}{2}\right)$$

and

$$R_Y(t_1,t_2) = \frac{1}{3}\left(\frac{\sin\frac{T}{2}}{\frac{T}{2}}\right)^2\cos\left(t_1 - \frac{T}{2}\right)\cos\left(t_2 - \frac{T}{2}\right)$$

respectively (cf. Exercise 9.6.3). □

Exercise 4.2 Suppose that the random process $\{X_t, -\infty < t < +\infty\}$ is such that

$$X_t \triangleq \sin(\omega t + Z)$$

for all t, where ω is a constant and where the random variable Z is uniformly distributed over $(0,2\pi]$. Suppose further that the X_t process is the input to a short-time averager.

a. Show that in this case the averager-output process $\{Y_t, -\infty < t < +\infty\}$ is such that

$$Y_t = \frac{\sin\left(\frac{\omega T}{2}\right)}{\frac{\omega T}{2}} \sin\left(\omega t - \frac{\omega T}{2} + Z\right)$$

b. Show that the mean value and the autocorrelation function of the output random process are then given by

$$E[Y_t] = 0$$

and

$$R_Y(t_1,t_2) = \frac{1}{2}\left(\frac{\sin\frac{\omega T}{2}}{\frac{\omega T}{2}}\right)^2 \cos\left[\omega(t_1 - t_2)\right]$$

respectively (cf. Exercise 9.6.4). □

Let us now look a bit more carefully at the derivation of the above results. It follows from (4.2) that we have

$$E[Y_t] = E\left[\frac{1}{T}\int_{t-T}^{t} X_{t'}\,dt'\right]$$

We see then that the evaluation of the output mean requires, in essence, the evaluation of an iterated integral: first over time and then over a sample space (that is, the expectation integral). In both of the above exercises we were actually able to carry through the evaluation of the time integral first and then take the expectation of the result. In other cases, however, it often proves more convenient to be able to interchange the order of these integrations. Now it can be shown that such an interchange of order of integration may in fact be made under a reasonable variety of conditions[1] and hence that we may usually write

$$E[Y_t] = \frac{1}{T}\int_{t-T}^{t} E[X_{t'}]\,dt' \tag{4.3}$$

Similarly, it can also be shown that while in general

$$R_Y(t_1,t_2) = E\left[\frac{1}{T^2}\int_{t_2-T}^{t_2}\int_{t_1-T}^{t_1} X_{t'}X_{t''}\,dt'\,dt''\right]$$

[1] The pertinent mathematical theorem is called the Fubini theorem after the Italian mathematician Guido Ghiron Fubini (1879–1943) (cf. sec. 4, chap. 12, of Royden).

we may usually interchange the orders of time integration and expectation and so obtain the formula

$$R_Y(t_1,t_2) = \frac{1}{T^2}\int_{t_2-T}^{t_2}\int_{t_1-T}^{t_1} R_X(t',t'')\,dt'\,dt'' \tag{4.4}$$

Exercise 4.3 Show that we may rewrite (4.4) in the somewhat more convenient form

$$R_Y(t_1,t_2) = \frac{1}{T^2}\iint_0^T R_X(\tau_1 + t_1 - T,\, \tau_2 + t_2 - T)\,d\tau_1\,d\tau_2$$

by making appropriate changes of variables. □

Exercise 4.4 Suppose that the input random process $\{X_t, -\infty < t < +\infty\}$ to a short-time averager is stationary in the wide sense with a mean m_X and an autocorrelation function R_X.

a. Show that in this case the output mean is

$$E[Y_t] = m_X$$

and hence that it is a constant function of time.

b. Show that the output autocorrelation function is

$$R_Y(t_1,t_2) = \frac{1}{T^2}\iint_0^T R_X(t_1 - t_2 + \tau_1 - \tau_2)\,d\tau_1\,d\tau_2$$

in this case, and hence that it is a function of the time difference $t_1 - t_2$ rather than of t_1 and t_2 separately. □

An examination of the results of Exercise 4.4 shows that *if the input to a short-time averager is a wide-sense stationary random process, then so is its output.*

Estimation of a mean

Since the mean value of the averager output equals the mean value of the input random process when that input is stationary in the wide sense, a question arises as to whether or not we may use the averager output at some given time t as an estimate of the mean of the input process. With this question in mind, let us now investigate the variance of the averager output.

Exercise 4.5* Suppose that the input $\{X_t, -\infty < t < +\infty\}$ to a short-time averager is stationary in the wide sense with a mean m_X and a covariance function K_X.

* Supplementary exercise, but note the result expressed by Eq. (4.5).

a. Show that the variance of the averager output is

$$\operatorname{var}(Y_t) = \frac{1}{T^2} \iint\limits_0^T K_X(\tau_1 - \tau_2)\, d\tau_1\, d\tau_2$$

in this case and hence is a constant function of t.

b. Show that we may rewrite the result in (*a*) in the form

$$\operatorname{var}(Y_t) = \frac{2}{T} \int_0^T \left(1 - \frac{\tau}{T}\right) K_X(\tau)\, d\tau \tag{4.5}$$

(*Hint:* Make use of the symmetry properties of K_X, paralleling the arguments used in Sec. 9.9 when evaluating the variance of the sample mean.) □

It should be noted that the result in (4.5) has exactly the same form as does the result derived previously [cf. Eq. (9.9.5)] for the limiting value of the variance of the sample mean obtained as we let the number of samples increase without limit (while keeping fixed the duration of the total observation interval).

Exercise 4.6 Using the formula derived in Exercise 4.5 for the variance of the output of a short-time averager in response to a wide-sense stationary random process, show that if the covariance function of the input random process is absolutely integrable,

$$\int_{-\infty}^{+\infty} |K_X(\tau)|\, d\tau = C < +\infty$$

where C is some positive real constant, then it follows that

$$\operatorname{var}(Y_t) < \frac{C}{T}$$

and hence that

$$\lim_{T \to \infty} \operatorname{var}(Y_t) = 0$$

in this case. □

We see from the results of Exercise 4.6 that the variance of the averager output tends to zero as the duration of the averaging interval increases without limit. That is, *the output of the short-time averager converges in mean square (and hence in probability) to the mean of the input random process as the duration of the integration interval increases without limit.* The short-time averager thus provides a useful[1] estimator of the input mean whenever the process being studied has an absolutely integrable covariance function.

Exercise 4.7 Suppose that the input $\{X_t, -\infty < t < +\infty\}$ to a short-time averager is a constant random process; in particular, suppose that

$$X_t = Z$$

for all t, where Z is a gaussian random variable with zero mean and unit variance.

Show that the variance of the averager output is

$$\text{var}(Y_t) = 1$$

and hence that

$$\lim_{T\to\infty} \text{var}(Y_t) = 1 \neq 0$$

in this case. Note that the covariance function of the input process is *not* absolutely integrable here. □

Exercise 4.8 Suppose that the input $\{X_t, -\infty < t < +\infty\}$ to a short-time averager is a wide-sense stationary random process with a mean m_X and a covariance function

$$K_X(\tau) = \begin{cases} \sigma_X{}^2\left(1 - \dfrac{|\tau|}{\tau_0}\right) & \text{for } 0 \le |\tau| \le \tau_0 \\ 0 & \text{for } \tau_0 \le |\tau| \end{cases}$$

a. Show that the variance of the averager output is given by

$$\text{var}(Y_t) = \begin{cases} \sigma_X{}^2\left(1 - \dfrac{1}{3}\dfrac{T}{\tau_0}\right) & \text{for } 0 \le T \le \tau_0 \\ \sigma_X{}^2\dfrac{\tau_0}{T}\left(1 - \dfrac{1}{3}\dfrac{\tau_0}{T}\right) & \text{for } \tau_0 \le T \end{cases}$$

b. Let $\alpha \triangleq T/\tau_0$ and plot $\text{var}(Y_t)/\sigma_X{}^2$ versus α using the result of part (*a*). Plot also the upper bound

$$\frac{1}{\alpha} = \frac{\tau_0}{T}$$

for comparison. □

[1] Note that we have said "useful" here and not "best." Compare with appendix B of Davenport, Johnson, and Middleton or chap. 4 of Van Trees.

Exercise 4.9 Suppose that the input $\{X_t, -\infty < t < +\infty\}$ to a short-time averager is a random-telegraph process as discussed in Sec. 9.6. In particular, suppose that the input process is stationary in the wide sense with a mean value of $\frac{1}{2}$ and a covariance function

$$K_X(\tau) = \tfrac{1}{4}e^{-2\lambda|\tau|}$$

a. Show that in this case the variance of the averager output is, for $0 \leq T$,

$$\text{var}\,(Y_t) = \frac{1}{4\lambda T}\left(1 - \frac{1}{2\lambda T} + \frac{1}{2\lambda T}e^{-2\lambda T}\right)$$

b. Let $\alpha \triangleq 2\lambda T$ and plot var $(Y_t)/\sigma_X{}^2$ versus α using the result in (*a*). Plot also the upper bound

$$\frac{2}{\alpha} = \frac{1}{\lambda T}$$

for comparison. □

5 WEIGHTING FUNCTIONS

Let us return for the moment to the defining equation for the output of the short-time averager, that is, to the equation

$$y(t) = \frac{1}{T}\int_{t-T}^{t} x(t')\,dt' \tag{5.1}$$

which expresses the averager output as an integral of the input, and in which the integration is taken from a point T seconds in the past and proceeds up to the present. The situation here then is that pictured in Fig. 5.1.

Suppose now that we introduce the change of variables

$$\tau = t - t'$$

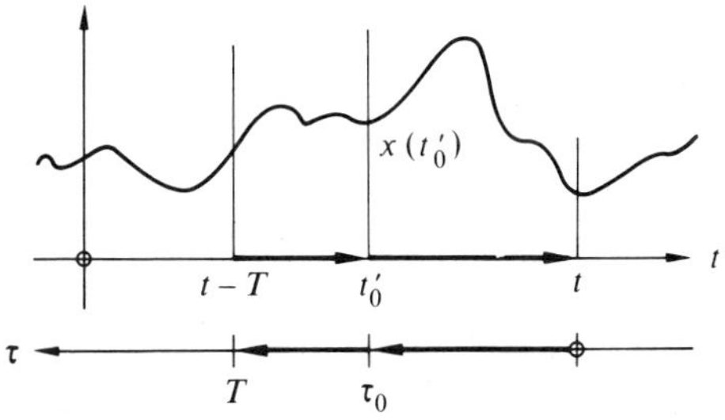

Figure 5.1 Time relations for the evaluation of the short-time averaging integral.

We may then rewrite (5.1) in the form

$$\begin{aligned} y(t) &= -\frac{1}{T}\int_T^0 x(t-\tau)\,d\tau \\ &= \frac{1}{T}\int_0^T x(t-\tau)\,d\tau \end{aligned} \tag{5.2}$$

where the integration here is taken along the τ axis as pictured in Fig. 5.1, that is, where now we integrate back in time from the time of observation t to a point T seconds previously. Suppose finally that we introduce a new function h which expresses the operation of the short-time averager and which is defined by the formula

$$h(t) \triangleq \begin{cases} \dfrac{1}{T} & \text{for } 0 \le t \le T \\ 0 & \text{otherwise} \end{cases} \tag{5.3}$$

Then using (5.3), we may rewrite (5.2) in the form

$$y(t) = \int_{-\infty}^{+\infty} h(\tau)x(t-\tau)\,d\tau \tag{5.4}$$

Thus we see that we may express the output of the short-time averager as the convolution of the input with the function h which characterizes the averager. Since that convolution is simply a weighted integral of the input, we call h the *weighting function* of the short-time averager.

That many other systems of interest may also be characterized by an input-output relation of the form given in (5.4) may be seen as follows: Suppose first that the output $y(t)$ of a given system is expressed in terms of the system input $x(t)$ by a first-order, linear, ordinary differential equation with constant coefficients. In particular, suppose that

$$T\frac{dy(t)}{dt} + y(t) = x(t) \tag{5.5}$$

where T is a positive real constant.

Exercise 5.1 Suppose that the output $y(t)$ and the input $x(t)$ of a given system are related by the first-order differential equation in (5.5). Show that if both the input and the output are zero for $t < 0$, that is, if the output is zero before the input starts, then we may determine the output either by using the formula

$$y(t) = \frac{1}{T}\int_0^t x(t')\exp\left(-\frac{t-t'}{T}\right)dt'$$

or by using the formula

$$y(t) = \frac{1}{T}\int_0^t x(t - \tau)e^{-\tau/T}\, d\tau \tag{5.6}$$

where t assumes only nonnegative values in either case. □

Since we assumed that $x(t) = 0$ for $t < 0$ in this last exercise, it follows that $x(t - \tau) = 0$ for $\tau > t$ there, and hence that we may rewrite (5.6) in the form

$$y(t) = \frac{1}{T}\int_0^{+\infty} e^{-\tau/T}x(t - \tau)\, d\tau$$

If, therefore, we define a weighting function h here by the formula

$$h(t) \triangleq \begin{cases} \dfrac{1}{T}\, e^{-t/T} & \text{for } 0 \leq t \\ 0 & \text{otherwise} \end{cases} \tag{5.7}$$

then we may write

$$y(t) = \int_{-\infty}^{+\infty} h(\tau)x(t - \tau)\, d\tau \tag{5.8}$$

Thus we see that the system whose input-output relation is specified by the differential equation (5.5) may also have its output expressed as a convolution of its input with a system weighting function. It may be shown, in fact, that *the output of any system characterized by an nth order, linear, ordinary differential equation with constant coefficients may be expressed in terms of its input by a convolution integral of the form in (5.8)*; the weighting function h in this case is often called the *Green's function* of the system[1] [after the English mathematician George Green (1793–1841)].

Exercise 5.2 Suppose that the input to a system characterized by a convolution integral as in (5.8) is a unit-impulse function; that is, let

$$x(t) = \delta(t)$$

(cf. Sec. 4.6). Show that the system output in response to such an input is

$$y(t) = h(t)$$

where $h(t)$ is the system weighting function. Thus we note that *the system weighting function is simply the response of the system to a unit-impulse input function.* The system weighting function is accordingly often called the *unit-impulse response* of the system. □

[1] See, for example, secs. 4.4 and 4.5 of Kreider et al., or secs. 2.1 to 2.8 of James, Nichols, and Phillips.

Having shown that there are a number of systems of interest for which we may express the system output as a convolution of the input with a system weighting function, let us next determine a few of the general properties of such systems. First, let us determine the effect of time-shifting the system input. In particular, suppose that the system response to the input $x(t)$ is $y(t)$, and that we apply a shifted version of $x(t)$ at the input, say

$$x_0(t) \triangleq x(t - t')$$

The output $y_0(t)$ in response to this new input is

$$\begin{aligned} y_0(t) &= \int_{-\infty}^{+\infty} h(\tau)x_0(t-\tau)\,d\tau \\ &= \int_{-\infty}^{+\infty} h(\tau)x(t-t'-\tau)\,d\tau \\ &= y(t-t') \end{aligned}$$

Thus we see that *if the input $x(t)$ gives rise to the output $y(t)$, then the input $x(t - t')$ gives rise to the output $y(t - t')$.* We say therefore that such systems are *time-invariant systems.*

Next, suppose that the input to the system is a weighted sum of two other input functions; for example, suppose that

$$x(t) = a_1x_1(t) + a_2x_2(t)$$

where a_1 and a_2 are arbitrary real numbers. The corresponding system output is then

$$\begin{aligned} y(t) &= \int_{-\infty}^{+\infty} h(\tau)[a_1x_1(t-\tau) + a_2x_2(t-\tau)]\,d\tau \\ &= a_1\int_{-\infty}^{+\infty} h(\tau)x_1(t-\tau)\,d\tau + a_2\int_{-\infty}^{+\infty} h(\tau)x_2(t-\tau)\,d\tau \\ &= a_1y_1(t) + a_2y_2(t) \end{aligned}$$

where $y_1(t)$ is the system response to $x_1(t)$ and $y_2(t)$ is the response to $x_2(t)$. Thus it follows that if $y_i(t)$ is the system response to the input $x_i(t)$, then the system response to the input $a_1x_1(t) + a_2x_2(t)$ is $a_1y_1(t) + a_2y_2(t)$; that is, *the system is a linear system* [cf. the corresponding result in the vector case, (1.7)].

In summary, we see that *if a system is such that its input $x(t)$ and its output $y(t)$ are related by a convolution integral, that is, if we can write*

$$y(t) = \int_{-\infty}^{+\infty} h(\tau)x(t-\tau)\,d\tau \tag{5.9}$$

then that system is a time-invariant linear system. We shall make specific use of this fact in the next chapter.

It is important to note at this time that it is often necessary to place certain restrictions upon the allowable forms of the system weighting function. For example, if it is desired that the system act in "real" time, that is, if the system is to generate an output continuously as the input unfolds in time, then we require that the system not give an output before an input is applied. Recalling the derivation of the convolution integral relation in the differential equation case (cf. Exercise 5.1), we see that this is equivalent to requiring that the weighting function be zero for negative values of its argument:

$$h(t) = 0 \qquad \text{for } t < 0 \tag{5.10}$$

A system whose weighting function satisfies (5.10) is said to be a *realizable* (or *physically realizable*) system.

A second requirement commonly placed upon such a system is that it be *stable* in the sense that every bounded input must give rise to a bounded output.

Exercise 5.3 Consider a time-invariant linear system whose input-output relation is expressible in the convolution integral form (5.9).

a. Show that if the system weighting function is absolutely integrable, that is, if

$$\int_{-\infty}^{+\infty} |h(t)|\, dt < +\infty$$

then that system is stable in the sense that every bounded input gives a bounded output.

b. Show that if the system is stable in the above sense, then its weighting function must be absolutely integrable. (*Hint:* Consider the output at time zero in response to an input derived from the time reverse of the system weighting function.)

It follows from (*a*) and (*b*) that a time-invariant linear system is stable (in the sense that every bounded input gives rise to a bounded output) if and only if the system weighting function is absolutely integrable. □

Exercise 5.4 Consider the time-invariant linear system shown in Fig. 5.2.

a. Determine the system weighting function.
b. Determine whether or not the system is stable. □

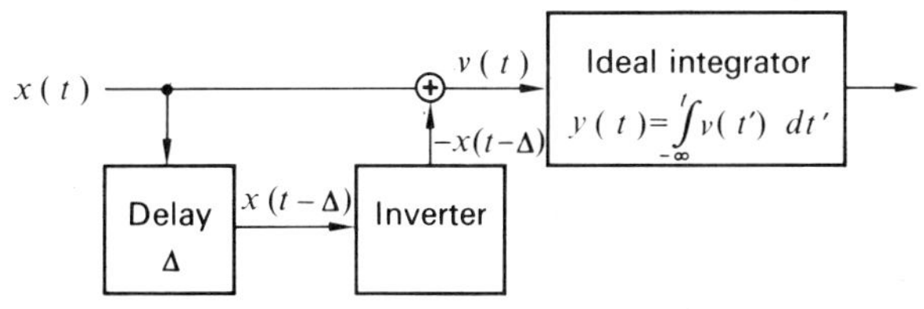

Figure 5.2 An integrating system.

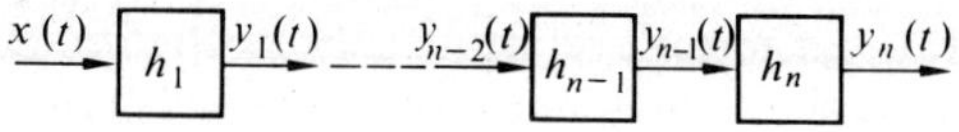

Figure 5.3 A cascade of n time-invariant, stable linear systems.

Exercise 5.5 Show that if the input $x(t)$ and the output $y(t)$ of a time-invariant linear system are related by the convolution integral

$$y(t) = \int_{-\infty}^{+\infty} h(\tau)x(t - \tau)\, d\tau$$

then we also may write

$$y(t) = \int_{-\infty}^{+\infty} h(t - \tau)x(\tau)\, d\tau \tag{5.11}$$

□

Exercise 5.6 Suppose that $x(t)$ is the input to a cascade of time-invariant, stable, linear systems (as shown in Fig. 5.3) and suppose that the weighting function of the ith system is denoted by h_i.

Show that the output of the nth system may be expressed as the n-fold convolution of the input with the n respective weighting functions. That is, show that

$$y_n(t) = \iint\limits_{-\infty}^{+\infty} \cdots \int\limits_{-\infty}^{+\infty} h_n(t - \tau_n)h_{n-1}(\tau_n - \tau_{n-1}) \cdots h_1(\tau_2 - \tau_1)x(\tau_1)\, d\tau_1 \cdots d\tau_{n-1}\, d\tau_n$$

[*Hint:* Use induction on (5.11).] □

6 OUTPUT MOMENTS

Suppose that the input to a time-invariant linear system with a weighting function h is a sample function $x(t)$ of the random process $\{X_t, -\infty < t < +\infty\}$. The system output in response to the input sample function is then the time function

$$y(t) = \int_{-\infty}^{+\infty} h(\tau)x(t - \tau)\, d\tau \tag{6.1}$$

Similarly, other sample functions of the input process will generate other output functions where the relation between the output and the input in each case again has the convolution integral form given in (6.1). This collection of output functions may then be thought of as the sample functions of the output random process $\{Y_t, -\infty < t < +\infty\}$, where

$$Y_t \triangleq \int_{-\infty}^{+\infty} h(\tau)X_{t-\tau}\, d\tau \tag{6.2}$$

This integral formula is the direct analog of the summation formula for the result of a linear transformation acting on a random vector [cf. Eq. (2.3)].

Having obtained (6.2) as the relation between the input and output random processes for a time-invariant linear system, we would now like to determine the probabilistic properties of the system output when we know those of its input. Unfortunately, however, it turns out that one can almost never calculate the probability distribution functions of the system-output random process unless the system input is a gaussian random process.[1] It is, nonetheless, a relatively straightforward problem to determine the various moments of the system-output random process, as we shall now show.

Let us consider first the mean value of the output random process. It follows from (6.2) that

$$E[Y_t] = E\left[\int_{-\infty}^{+\infty} h(\tau)X_{t-\tau}\,d\tau\right]$$

Assuming, as in the case of the short-time averager, that we may interchange the operations of time integration and statistical averaging, we get

$$E[Y_t] = \int_{-\infty}^{+\infty} h(\tau)E[X_{t-\tau}]\,d\tau \tag{6.3}$$

[cf. Eq. (2.4)]. In order to proceed further we generally need to be able to express both the input mean value and the system weighting function as explicit functions of time. Certain special results can easily be obtained, however, without this detailed knowledge.

Exercise 6.1 Let the random process $\{X_t, -\infty < t < +\infty\}$ be the input to a time-invariant linear system with a weighting function h.

Show that if the mean value of the input process is bounded for all t and if the system is stable, then the output mean value is also bounded for all t (cf. Exercise 5.3). □

Exercise 6.2 Let the input random process $\{X_t, -\infty < t < +\infty\}$ to a time-invariant linear system with a weighting function h have a constant mean value; in particular, let

$$E[X_t] = m_X \qquad \text{for all } t$$

Show that

$$E[Y_t] = m_X \int_{-\infty}^{+\infty} h(\tau)\,d\tau$$

[1] It can be shown (cf. sec. 8.4 of Davenport and Root) that if the input to a stable, linear, time-invariant system is a gaussian random process, then so is its output; the nongaussian input case is nowhere near as simple (cf. sec. 9.5 of Davenport and Root).

that is, that the output mean value is also a constant function of time in this case. □

Exercise 6.3 Let the random process $\{X_t, \; -\infty < t < +\infty\}$ be the input to a time-invariant linear system whose input $x(t)$ and output $y(t)$ are related by the first-order linear differential equation

$$T\frac{dy(t)}{dt} + y(t) = x(t)$$

where T is a positive real constant.

Show that if the input mean value is a constant function of time m_X, then the output mean value is equal to that input mean value:

$$E[Y_t] = m_X$$

for all t in this case [cf. Eqs. (5.7) and (5.8)]. □

Next, let us consider the autocorrelation function of the output of a time-invariant, stable linear system. Now it follows from (6.2) that

$$\begin{aligned} R_Y(t_1,t_2) &= E[Y_{t_1}Y_{t_2}] \\ &= E\left[\iint\limits_{-\infty}^{+\infty} h(\tau_1)h(\tau_2)X_{t_1-\tau_1}X_{t_2-\tau_2}\, d\tau_1\, d\tau_2\right] \end{aligned}$$

Hence, on interchanging the orders of integration and expectation, we get the result

$$R_Y(t_1,t_2) = \iint\limits_{-\infty}^{+\infty} h(\tau_1)h(\tau_2)R_X(t_1 - \tau_1, t_2 - \tau_2)\, d\tau_1\, d\tau_2 \tag{6.4}$$

In order to proceed further, we would like to have both the system weighting function and the input autocorrelation function given as explicit functions of their arguments. We can obtain a result of considerable importance, however, when the input random process is stationary in the wide sense. In that case we may write

$$R_X(t_1 - \tau_1, t_2 - \tau_2) = R_X(t_1 - \tau_1 - t_2 + \tau_2)$$

and hence (6.4) becomes in this case

$$R_Y(t_1,t_2) = \iint\limits_{-\infty}^{+\infty} h(\tau_1)h(\tau_2)R_X[(t_1 - t_2) - \tau_1 + \tau_2]\, d\tau_1\, d\tau_2 \tag{6.5}$$

It should be noted that this result is a function only of the time difference $t_1 - t_2$ and not of t_1 and t_2 separately. On combining this result with that of Exercise 6.2, we see that *if the input to a time-invariant, stable linear system is a wide-sense stationary random process then the output of that system is also a wide-sense stationary random process.*

Exercise 6.4 Let K_X and K_Y be the covariance functions of the input and output random processes, respectively, for a time-invariant, stable linear system.

Show that we have

$$K_Y(t_1,t_2) = \iint\limits_{-\infty}^{+\infty} h(\tau_1)h(\tau_2)K_X(t_1 - \tau_1, t_2 - \tau_2)\, d\tau_1\, d\tau_2$$

in general, and that we get

$$K_Y(\tau) = \iint\limits_{-\infty}^{+\infty} h(\tau_1)h(\tau_2)K_X(\tau - \tau_1 + \tau_2)\, d\tau_1\, d\tau_2$$

when the input random process is stationary in the wide sense (cf. Exercise 2.1). □

Not only does the weighting-function characterization of a time-invariant, stable linear system enable us to determine useful formulas for the system-output mean and autocorrelation and covariance functions, but also it enables us to determine similarly the cross-correlation function of the system output and input. Thus, for example:

Exercise 6.5 Let h be the weighting function of a time-invariant, stable linear system. Let the system input $\{X_t, -\infty < t < +\infty\}$ be a wide-sense stationary random process.

Show that the crosscorrelation function of the system output and input is given by the formula

$$R_{YX}(\tau) = \int_{-\infty}^{+\infty} h(t')R_X(\tau - t')\, dt'$$

in this case. □

System correlation functions

A review of the various results obtained above shows that the calculation of the output autocorrelation function requires the evaluation of a double integral. It is often convenient to convert this double integral in such a way that we end up with the iteration of two single integrals, one of which is completely determined by the system weighting function. The following exercises illustrate this technique.

Exercise 6.6 Let $\{X_t, -\infty < t < +\infty\}$ and $\{Y_t, -\infty < t < +\infty\}$ be the input and output random processes, respectively, of a time-invariant, stable linear system with a system weighting function h and suppose that the input process is stationary in the wide sense. Let us now introduce a new function

$$R_h(\tau) \triangleq \int_{-\infty}^{+\infty} h(t)h(t-\tau)\,dt \tag{6.6}$$

which we shall call the *system correlation function.*

a. Show that we may express the output autocorrelation function as a convolution of the system correlation function and the autocorrelation function of the input process; in particular, show that

$$R_Y(\tau) = \int_{-\infty}^{+\infty} R_h(t')R_X(\tau - t')\,dt'$$

b. Show that it therefore follows that the variance of the system output is given by the formula

$$\text{var}\,(Y_t) = \int_{-\infty}^{+\infty} R_h(t')K_X(t')\,dt' \tag{6.7}$$

for all t in this case. □

Exercise 6.7 Show that the system correlation function defined in (6.6) is an even function of its argument:

$$R_h(-\tau) = R_h(\tau)$$

□

Exercise 6.8 Let a wide-sense stationary random process $\{X_t, -\infty < t < +\infty\}$ with mean m_X and covariance function K_X be the input to a short-time averager (as discussed in Secs. 4 and 5).

a. Determine the system correlation function of the short-time averager.
b. Using the results of Exercise 6.6*a* and Eq. (6.7) above, rederive the result given in (*b*) of Exercise 4.5. □

Exercise 6.9 Let h be the weighting function of a system whose input $x(t)$ and output $y(t)$ are related by the first-order, linear differential equation

$$T\frac{dy(t)}{dt} + y(t) = x(t)$$

where T is a positive real constant.

Determine the system correlation function of this system. □

7 SUMMARY OF DEFINITIONS AND FORMULAS

n-dimensional case

Suppose that the m-dimensional real random vector

$$\mathbf{Y} = (Y_1, Y_2, \ldots, Y_m)$$

is generated from the n-dimensional real random vector

$$\mathbf{X} = (X_1, X_2, \ldots, X_n)$$

by the transformation $\mathbf{g}$, that is,

$$\mathbf{Y} = \mathbf{g}(\mathbf{X}) \tag{2.1}$$

We say that $\mathbf{g}$ is a linear transformation if and only if it satisfies the relation

$$\mathbf{g}(a\mathbf{W} + b\mathbf{Z}) = a\mathbf{g}(\mathbf{W}) + b\mathbf{g}(\mathbf{Z})$$

where a and b are arbitrary real constants and where $\mathbf{W}$ and $\mathbf{Z}$ are arbitrary real random vectors.

The general n-dimensional form of $\mathbf{g}$ is that defined by the system of linear equations

$$\begin{aligned} Y_1 &= g_{11}X_1 + g_{12}X_2 + \cdots + g_{1n}X_n \\ Y_2 &= g_{21}X_1 + g_{22}X_2 + \cdots + g_{2n}X_n \\ &\cdots\cdots\cdots\cdots\cdots\cdots \\ Y_m &= g_{m1}X_1 + g_{m2}X_2 + \cdots + g_{mn}X_n \end{aligned} \tag{2.2}$$

in other words, by

$$Y_i = \sum_{j=1}^{n} g_{ij}X_j \qquad i = 1, 2, \ldots, m \tag{2.3}$$

where the coefficients g_{ij} are arbitrary real numbers.

The mean of the ith component of $\mathbf{Y}$ is then

$$E[Y_i] = \sum_{j=1}^{n} g_{ij}E[X_j] \tag{2.4}$$

The covariance of Y_i and Y_k is

$$\operatorname{cov}(Y_i, Y_k) = \sum_{j=1}^{n} \sum_{r=1}^{n} g_{ij}g_{kr} \operatorname{cov}(X_j, X_r)$$

(cf. Exercise 2.1).

Matrix formulation

If we represent the random vectors **X** and **Y** by column matrices,

$$\mathbf{X} = \begin{bmatrix} X_1 \\ X_2 \\ \cdot \\ \cdot \\ \cdot \\ X_n \end{bmatrix} \qquad \mathbf{Y} = \begin{bmatrix} Y_1 \\ Y_2 \\ \cdot \\ \cdot \\ \cdot \\ Y_m \end{bmatrix}$$

and if we define the transformation matrix **g** to be the $m \times n$ matrix formed of the coefficients g_{ij} in (2.2),

$$\mathbf{g} = \begin{bmatrix} g_{11} & g_{12} & \cdots & g_{1n} \\ g_{21} & g_{22} & \cdots & g_{2n} \\ \cdot & \cdot & \cdots & \cdot \\ g_{m1} & g_{m2} & \cdots & g_{mn} \end{bmatrix}$$

then we may write

$$\mathbf{Y} = \mathbf{gX} \tag{3.1}$$

as the matrix form of (2.2). If we similarly represent $E[\mathbf{X}]$ and $E[\mathbf{Y}]$ by column matrices, we get

$$E[\mathbf{Y}] = \mathbf{g}E[\mathbf{X}] \tag{3.2}$$

as the matrix version of (2.4). Finally if we define the covariance matrices

$$\mathbf{\Gamma} \triangleq \begin{bmatrix} \gamma_{11} & \gamma_{12} & \cdots & \gamma_{1n} \\ \gamma_{21} & \gamma_{22} & \cdots & \gamma_{2n} \\ \cdot & \cdot & \cdots & \cdot \\ \gamma_{n1} & \gamma_{n2} & \cdots & \gamma_{nn} \end{bmatrix} \tag{3.4}$$

and

$$\mathbf{\Lambda} \triangleq \begin{bmatrix} \lambda_{11} & \lambda_{12} & \cdots & \lambda_{1m} \\ \lambda_{21} & \lambda_{22} & \cdots & \lambda_{2m} \\ \cdot & \cdot & \cdots & \cdot \\ \lambda_{m1} & \lambda_{m2} & \cdots & \lambda_{mm} \end{bmatrix} \tag{3.5}$$

where

$$\gamma_{jr} \triangleq \operatorname{cov}(X_j, X_r) \qquad \text{and} \qquad \lambda_{ik} \triangleq \operatorname{cov}(Y_i, Y_k)$$

then it follows that we get

$$\mathbf{\Lambda} = \mathbf{g\Gamma g}^T \tag{3.10}$$

as the matrix relation between the covariance matrix of **Y** and that of **X**, where $\mathbf{g}^T$ is the transpose of the transformation matrix **g**.

Time averages

Suppose that the random process $\{Y_t, -\infty < t < +\infty\}$ is defined in terms of the random process $\{X_t, -\infty < t < +\infty\}$ by the relation

$$Y_t \triangleq \frac{1}{T}\int_{t-T}^{t} X_{t'}\,dt' \tag{4.2}$$

we then say that Y_t is formed by taking a short-time average of the $\{X_t\}$ process.

The mean of the output process is

$$E[Y_t] = \frac{1}{T}\int_{t-T}^{t} E[X_{t'}]\,dt' \tag{4.3}$$

and the output autocorrelation function is

$$R_Y(t_1,t_2) = \frac{1}{T^2}\iint_0^T R_X(\tau_1 + t_1 - T, \tau_2 + t_2 - T)\,d\tau_1\,d\tau_2$$

(cf. Exercise 4.3). If the input process is stationary in the wide sense, we then get

$$E[Y_t] = m_X \triangleq E[X_t]$$

and

$$R_Y(t_1,t_2) = \frac{1}{T^2}\iint_0^T R_X(t_1 - t_2 + \tau_1 - \tau_2)\,d\tau_1\,d\tau_2$$

(c.f. Exercise 4.4). The variance of the output is then

$$\text{var}(Y_t) = \frac{2}{T}\int_0^T \left(1 - \frac{\tau}{T}\right) K_X(\tau)\,d\tau \tag{4.5}$$

if $\{X_t\}$ is wide-sense stationary. It further follows that

$$\text{var}(Y_t) < \frac{C}{T}$$

where C is some positive real constant, if K_X is absolutely integrable (cf. Exercise 4.6).

Weighting functions

Many linear systems are such that their input $x(t)$ and output $y(t)$ are related by a convolution integral:

$$y(t) = \int_{-\infty}^{+\infty} h(\tau)x(t-\tau)\,d\tau \tag{5.9}$$

In such a case we call $h(t)$ the system weighting function and it follows that the system is time invariant; that is, if $y(t)$ is the system output in response to an input $x(t)$, then $y(t - t')$ is the system output in response to the time-shifted input $x(t - t')$.

Output moments

Suppose that $\{Y_t, -\infty < t < +\infty\}$ is the random process at the output of a linear system characterized by a system function h and that $\{X_t, -\infty < t < +\infty\}$ is the system input process. We write in this case

$$Y_t = \int_{-\infty}^{+\infty} h(\tau) X_{t-\tau}\, d\tau \tag{6.2}$$

The mean of the output process is then

$$E[Y_t] = \int_{-\infty}^{+\infty} h(\tau) E[X_{t-\tau}]\, d\tau \tag{6.3}$$

which becomes

$$E[Y_t] = m_X \int_{-\infty}^{+\infty} h(\tau)\, d\tau$$

when $\{X_t\}$ is wide-sense stationary with a mean m_X (cf. Exercise 6.2). The covariance function of the output process is

$$K_Y(t_1,t_2) = \iint_{-\infty}^{+\infty} h(\tau_1)h(\tau_2)K_X(t_1 - \tau_1,\, t_2 - \tau_2)\, d\tau_1\, d\tau_2$$

in general, and becomes

$$K_Y(\tau) = \iint_{-\infty}^{+\infty} h(\tau_1)h(\tau_2)K_X(\tau - \tau_1 + \tau_2)\, d\tau_1\, d\tau_2$$

when $\{X_t\}$ is stationary in the wide sense (cf. Exercise 6.4).

The cross-correlation function of the system output and input is given by the formula

$$R_{YX}(\tau) = \int_{-\infty}^{+\infty} h(t')R_X(\tau - t')\, dt'$$

when the input is wide-sense stationary (cf. Exercise 6.5).

When the system-input process is stationary in the wide sense, it is often convenient to introduce the system correlation function

$$R_h(\tau) \triangleq \int_{-\infty}^{+\infty} h(t)h(t - \tau)\, dt \tag{6.6}$$

We may then calculate the output autocorrelation function using the formula

$$R_Y(\tau) = \int_{-\infty}^{+\infty} R_h(t')R_X(\tau - t')\, dt'$$

(cf. Exercise 6.6) and the output variance using the formula

$$\text{var}\,(Y_t) = \int_{-\infty}^{+\infty} R_h(t')K_X(t')\,dt' \tag{6.7}$$

REFERENCES

Davenport, Wilbur B., Jr., Richard A. Johnson, and David Middleton: Statistical Errors in Measurements on Random Time Functions, *J. Appl. Phys.*, vol. 23, no. 4, pp. 377–388, 1952.

Davenport, Wilbur B., Jr., and William L. Root: "Random Signals and Noise," McGraw-Hill Book Company, New York, 1958.

Hildebrand, Francis B.: "Methods of Applied Mathematics," Prentice-Hall, Inc., Englewood Cliffs, N.J., 1952.

Hoffman, Kenneth, and Ray Kunze: "Linear Algebra," Prentice-Hall, Inc., Englewood Cliffs, N.J., 1961.

James, Hubert M., Nathaniel B. Nichols, and Ralph S. Phillips: "Theory of Servomechanisms," McGraw-Hill Book Company, New York, 1947.

Kreider, Donald L., Robert G. Kuller, Donald R. Ostberg, and Fred W. Perkins: "An Introduction to Linear Analysis," Addison-Wesley Publishing Company, Inc., Reading, Mass., 1966.

Royden, Halsey L.: "Real Analysis," The Macmillan Company, New York, 1963.

Van Trees, Harry L., Jr.: "Detection, Estimation, and Modulation Theory," pt. 1, John Wiley & Sons, Inc., New York, 1968.

11 Spectral Analysis

1 INTRODUCTION

The preceding chapter was concerned with what might be called the *time-domain analysis* of linear transformations; that is, we expressed all our transformations there as operations in time (for example, as an integral of an input over some given time interval). We shall in this chapter, on the other hand, be concerned with the *frequency-domain analysis* of linear systems. In particular, we shall first show that any stable, linear, time-invariant system may be completely characterized by its response to sinusoids of different frequencies. With this fact in mind, we shall next study the problem of representing system inputs and outputs as "sums" of sinusoids (this is the so-called problem of *spectral analysis*). Finally, we shall combine the results of these studies to determine the response of stable, linear, time-invariant systems to random inputs.

The material in this chapter rests heavily upon the theory of Fourier series and integrals [named after the French mathematician Jean Baptiste Joseph Fourier (1768–1830)]. While we shall not be able to give more than a heuristic treatment of these topics here, we do hope that the reader will be able to obtain some intuitive feeling for the concepts involved. An excellent introduction to Fourier theory (including a discussion of the heat-flow problem which led Fourier into his studies) may be found in Seeley's monograph. A good intermediate level treatment of Fourier series may be found in chaps. 9 and 10 of Kreider et al., while similar level treatments of the Fourier integral may be found in the books by Bracewell and Papoulis. Definitive advanced treatments of Fourier integrals are given in Bochner's monograph and in Titchmarsh's book.

2 SINE WAVE IN, SINE WAVE OUT*

We pointed out in Sec. 10.5 that many systems of interest to us were such that the system output could be calculated by convolving the input with the system weighting function. We further showed there that such systems were both linear and time invariant and, in addition, we obtained a condition upon the system weighting function which would ensure that the system was stable in the sense that every bounded input would give rise to a bounded output. Now as we shall show below, *every stable, linear, time-invariant system has the property that a sinusoidal input always gives rise to a sinusoidal output of the same frequency.* Further, we shall also show that essentially any waveform of interest to us can be represented by a weighted sum of sinusoids. It is the combination of these two facts which makes it possible to develop a frequency-domain analysis of such systems and thus enables us to introduce a whole new and powerful set of analytic tools.

In order to emphasize the fact that our result is, in fact, a property of *every* stable, linear, time-invariant system (and not just of those for which we choose to express the input-output relation in terms of a convolution integral), let us use for the moment the abstract operator notation

$$y(t) = \mathcal{O}[x(t)] \tag{2.1}$$

where $x(t)$ is the system input, $y(t)$ is the system output, and $\mathcal{O}$ denotes the operator acting on $x(t)$ to produce $y(t)$, for example, $\mathcal{O}$ might represent the operation of integration or differentiation or convolution. The situation then is that shown schematically by Fig. 2.1.

Now by *linear* operator we mean an operator $\mathcal{O}$ such that for every possible pair of inputs, say $x_1(t)$ and $x_2(t)$, and every pair of numbers a_1

* This section may be skipped by the reader who is already familiar with the frequency analysis of linear systems.

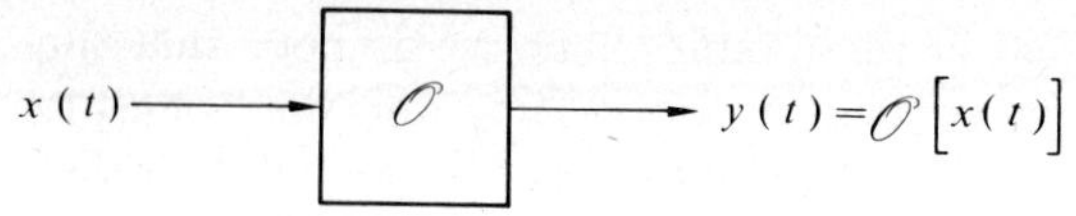

Figure 2.1 The system schema.

and a_2, $\mathcal{O}$ must satisfy the usual linearity condition:[1]

$$\mathcal{O}[a_1x_1(t) + a_2x_2(t)] = a_1\mathcal{O}[x_1(t)] + a_2\mathcal{O}[x_2(t)] \tag{2.2}$$

by *time-invariant* operator, we mean an operator $\mathcal{O}$ such that if

$$y(t) \triangleq \mathcal{O}[x(t)] \tag{2.3}$$

then it follows that

$$\mathcal{O}[x(t + t')] = y(t + t') \tag{2.4}$$

that is, if $y(t)$ is the output response to the input $x(t)$, then the response to $x(t)$ shifted in time by an amount t' is simply $y(t)$ shifted in time by the same amount.

Exercise 2.1 Suppose that a system acts as a *differentiator;* that is, suppose that a system operator $\mathcal{O}$ is defined by the equation

$$y(t) = \mathcal{O}[x(t)] \triangleq \frac{dx(t)}{dt}$$

where $x(t)$ is the system input and $y(t)$ is the system output.

Show that this operator is both linear and time invariant. □

Exercise 2.2 A common element of many communications systems is the *modulator* sketched in Fig. 2.2. This system is described by the

[1] While our discussion in the preceding chapter was limited mainly to real linear transformation of real vectors or functions, we shall find it convenient in this chapter to extend our discussion to cover the complex case. Thus the coefficients a_1 and a_2 and the functions x_1 and x_2 in (2.2) may be either real or complex.

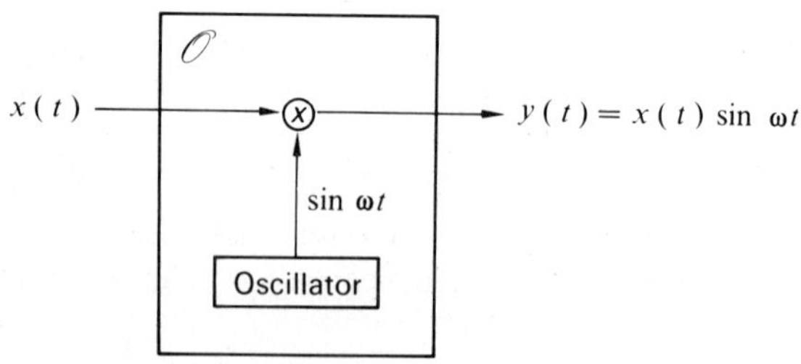

Figure 2.2 A modulator.

operator equation

$$y(t) = \mathcal{O}[x(t)] \triangleq x(t) \sin \omega t$$

where ω is a constant.

Show that while this modulator is a linear operator, it is *not* a time-invariant operator. □

Exercise 2.3 Another widely used system element is the *squarer*, that is, the element whose operator is defined by the equation

$$y(t) = \mathcal{O}[x(t)] \triangleq x^2(t)$$

Show that while the squarer is a time-invariant operator, it is *not* a linear operator. □

Our goal now is to show that if an operator $\mathcal{O}$ is in fact a stable, linear, time-variant operator, that is, if $\mathcal{O}$ satisfies (2.2) through (2.4), then a sinusoidal input produces a sinusoidal output. Since any real sinusoid can be represented by a complex exponential and since it turns out to be generally easier to manipulate complex exponentials than real sinusoids, we shall find it convenient to consider, for the moment, input functions of the form

$$\begin{aligned} \xi(t) &\triangleq Ae^{i\omega t} \\ &= |A| \cos (\omega t + \phi) + i|A| \sin (\omega t + \phi) \end{aligned}$$

where $\omega = 2\pi f$, $i = \sqrt{-1}$, and where A is a complex number (though possibly a function of the frequency f) with magnitude $|A|$ and phase ϕ.

Exercise 2.4 The reason for our interest in complex functions here comes in part from the following result: Suppose that $\xi(t)$ is an arbitrary complex function of time:

$$\xi(t) = \text{Re}\,[\xi(t)] + i\,\text{Im}\,[\xi(t)]$$

where Re denotes *real part of*, where Im denotes *imaginary part of*, and where Re $[\xi(t)]$ and Im $[\xi(t)]$ are both real functions of time. Suppose further that $\xi(t)$ is the input to a stable linear system described by the operator $\mathcal{O}$.

Show that

$$\mathcal{O}[\text{Re}\,[\xi(t)]] = \text{Re}\,[\mathcal{O}[\xi(t)]]$$

That is, the response of a stable linear system to a real function of time may be calculated by first taking that real input to be the real part of a complex function of time, and then determining the real part of the system response to that complex input. □

Suppose that we denote the response of a stable, linear, time-invariant system $\mathcal{O}$ to the complex input

$$\xi(t) = e^{i\omega t}$$

by $\eta(t)$; that is, let us write

$$\eta(t) \triangleq \mathcal{O}[e^{i\omega t}]$$

If now we apply the input $e^{i\omega(t+t')}$, then the corresponding system response is

$$\mathcal{O}[e^{i\omega(t+t')}] = \eta(t + t') \tag{2.5}$$

since the system is time invariant by hypothesis. Suppose, for the moment, that we regard t' as the time variable and t as a parameter. It then follows in this case that

$$\begin{aligned} \mathcal{O}[e^{i\omega(t+t')}] &= \mathcal{O}[e^{i\omega t}e^{i\omega t'}] \\ &= e^{i\omega t}\mathcal{O}[e^{i\omega t'}] \\ &= e^{i\omega t}\eta(t') \end{aligned} \tag{2.6}$$

since the system is assumed to be linear. On combining (2.5) and (2.6), we get

$$\eta(t + t') = e^{i\omega t}\eta(t')$$

If therefore we set $t' = 0$, we obtain the result

$$\eta(t) = \eta(0)e^{i\omega t} \tag{2.7}$$

which holds for all values of the parameter t.

This last result essentially completes the proof of our assertion that a sinusoid input implies a sinusoidal output. For, suppose that we apply the sinusoidal input

$$x(t) = |A| \cos (\omega t + \phi) = \text{Re } [Ae^{i\omega t}]$$

to the system. It then follows from (2.7), and from the result of Exercise 2.4, that if we set

$$|\eta(0)|e^{i\theta} = \eta(0)$$

then it follows that we get the sinusoidal response

$$\begin{aligned} y(t) &= \text{Re } [\mathcal{O}[Ae^{i\omega t}]] \\ &= \text{Re } [A\eta(0)e^{i\omega t}] \\ &= |A|\, |\eta(0)| \cos (\omega t + \phi + \theta) \end{aligned}$$

thus confirming our assertion.

A moment's reflection upon the above derivation shows that our results may be viewed in two somewhat different ways: On the one hand, if we are told that the system input is the complex function

$$\xi(t) = Ae^{i\omega t} \tag{2.8}$$

and if we are given the complex number $\eta(0)$ which characterizes the system, then we can calculate the corresponding system output by using the formula

$$\eta(t) = A\eta(0)e^{i\omega t} \tag{2.9}$$

On the other hand, if we are not given $\eta(0)$ (but rather wish to calculate it), then we can apply a complex input as in (2.8) and measure the corresponding complex output, say

$$\eta(t) = Be^{i\omega t} \tag{2.10}$$

where B is a complex number. On comparing (2.9) and (2.10), we see that we can then solve for $\eta(0)$ and thus get the result

$$\eta(0) = \frac{B}{A} \tag{2.11}$$

Suppose now that in order to emphasize the possible frequency dependence of the various complex quantities above, we rewrite (2.11) in the form

$$H(f) \triangleq \eta(0) = \frac{B(f)}{A(f)} \tag{2.12}$$

where we have introduced the symbol $H(f)$; we shall call $H(f)$ the *system function* of the given linear system. It then follows from our preceding development that *the system function $H(f)$ completely characterizes the behavior of the given system at the frequency f and, further, that it may be determined by taking the ratio of the complex amplitudes of the output and input when the input is sinusoidal.*

For example, suppose that the system under consideration is a differentiator (cf. Exercise 2.1). That is, suppose that

$$y(t) = \mathcal{O}[x(t)] = \frac{dx(t)}{dt}$$

If now we apply the complex sinusoidal input

$$x(t) = Ae^{i\omega t} = Ae^{i2\pi ft}$$

then it follows that the system output is

$$\begin{aligned} y(t) &= \frac{d(Ae^{i2\pi ft})}{dt} \\ &= i2\pi f A e^{i2\pi ft} \end{aligned}$$

It therefore follows from (2.12) that we get

$$H(f) = i2\pi f \tag{2.13}$$

as a formula for the system function of a differentiator.

Exercise 2.5 Suppose that a given stable, linear, time-invariant system is characterized by a system weighting function $h(t)$ and hence that its input and output are related by the convolution integral

$$y(t) = \int_{-\infty}^{+\infty} h(\tau)x(t-\tau)\,d\tau$$

Show that for such a system, the system function H is given by the formula

$$H(f) = \int_{-\infty}^{+\infty} h(\tau)e^{-i2\pi f\tau}\,d\tau \tag{2.14}$$

We say in this case that the system function is the *Fourier transform* of the system weighting function (Fourier transforms will be discussed further in the next section). □

Exercise 2.6 Using (2.14), show that the system function which corresponds to the short-time averager system weighting function

$$h(t) = \begin{cases} \dfrac{1}{T} & \text{for } 0 \le t < T \\ 0 & \text{otherwise} \end{cases}$$

is given by the formula

$$H(f) = \frac{\sin \pi f T}{\pi f T}\, e^{-i\pi f T}$$

Determine the magnitude and phase functions of H as functions of frequency, and plot. □

Exercise 2.7 Using (2.14), show that the system function which corresponds to the system weighting function

$$h(t) = \begin{cases} \dfrac{1}{T}\, e^{-t/T} & \text{for } 0 \le t \\ 0 & \text{for } t < 0 \end{cases}$$

(cf. Exercise 10.5.1), is given by the formula

$$H(f) = \frac{1}{1 + i2\pi f T}$$

Determine the magnitude and phase of H as functions of frequency, and plot. □

3 FOURIER ANALYSIS*

We showed in the preceding section that if we apply the complex sinusoid

$$\xi(t) = A(f)e^{i2\pi ft} \tag{3.1}$$

to the input of a stable, linear, time-invariant system, then the complex output $\eta(t)$ of that system is given by the formula

$$\eta(t) = A(f)H(f)e^{i2\pi ft} \tag{3.2}$$

where $H(f)$ is the system function of the given system evaluated at the input frequency f. If therefore we apply a sum of such complex sinusoids to the system input, for example, if we have

$$\xi(t) = \sum_{k=1}^{n} A(f_k)e^{i2\pi f_k t} \tag{3.3}$$

where the input frequencies f_k are arbitrarily chosen, then it follows from the linearity of the system that the corresponding output is equal to the sum of the outputs due to the respective inputs:

$$\eta(t) = \sum_{k=1}^{n} A(f_k)H(f_k)e^{i2\pi f_k t} \tag{3.4}$$

Suppose next that the input frequencies f_k are all harmonically related; in particular, suppose that

$$f_k = \frac{k}{T} \qquad k = 1, 2, \ldots, n$$

where T is the period of the lowest input-frequency component. It then follows that

$$\xi(t) = \sum_{k=1}^{n} A\left(\frac{k}{T}\right) e^{i2\pi kt/T}$$

and (3.5)

$$\eta(t) = \sum_{k=1}^{n} A\left(\frac{k}{T}\right) H\left(\frac{k}{T}\right) e^{i2\pi kt/T}$$

* This section may be skipped by the reader who is already familiar with the Fourier analysis of linear systems.

in this case. Now we note that since k is an integer and since the exponential function is periodic with a period 2π, we have

$$\begin{aligned} e^{i2\pi k(t+T)/T} &= e^{i2\pi kt/T}e^{i2\pi k} \\ &= e^{i2\pi kt/T} \end{aligned}$$

It therefore follows that

$$\xi(t+T) = \xi(t) \qquad \text{and} \qquad \eta(t+T) = \eta(t)$$

for all t. That is, the input and output are both periodic with the same period when the input is a sum of harmonically related sinusoids. While we obtained this result for the case of a finite number of sinusoids at the input, the same result is obtained when the input has a countably infinite number of terms. Thus, if we have

$$\xi(t) = \sum_{k=-\infty}^{+\infty} A\left(\frac{k}{T}\right) e^{i2\pi kt/T} \tag{3.6a}$$

at the input, then we get

$$\eta(t) = \sum_{k=-\infty}^{+\infty} A\left(\frac{k}{T}\right) H\left(\frac{k}{T}\right) e^{i2\pi kt/T} \tag{3.6b}$$

at the output (so long as both series converge), and it follows that

$$\xi(t+T) = \xi(t) \qquad \text{and} \qquad \eta(t+T) = \eta(t)$$

for all t in this case also, that is, that again both the input and output are periodic with the same period.

Suppose next that instead of being given a system input expressed in the form of a sum of harmonically related sinusoids, we are told only that the input is some periodic function with period T (say, as shown in Fig. 3.1) and, further, that it is desired to obtain an expression for the system output. Now we know that if we could somehow express the system input as a sum of sinusoids, then we could also directly apply our preceding results in order to calculate the system output. That is, if

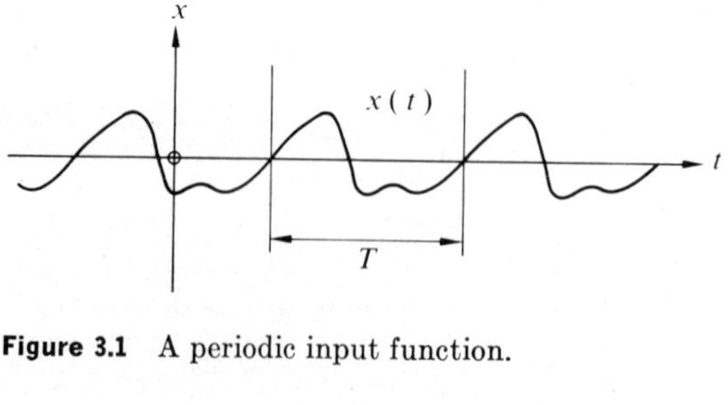

Figure 3.1 A periodic input function.

we could in fact express the periodic system input $\xi(t)$ in the form

$$\xi(t) = \sum_{k=-\infty}^{+\infty} \alpha_k e^{i2\pi kt/T} \tag{3.7}$$

then we know from our preceding results that we could write

$$\eta(t) = \sum_{k=-\infty}^{+\infty} \alpha_k H\left(\frac{k}{T}\right) e^{i2\pi kt/T}$$

as an expression for the output. The problems remaining, therefore, are how to determine the coefficients α_k in (3.7) when we are given a periodic input $\xi(t)$, and whether or not the series formed using such coefficients does indeed converge to the given input $\xi(t)$. The solution to the first of these problems depends upon the fact that the exponential $e^{i2\pi kt/T}$ is orthogonal over any time interval of length T as shown by the following exercise.

Exercise 3.1

a. Show that

$$\frac{1}{T}\int_{-T/2}^{+T/2} e^{i2\pi(k-r)t/T}\,dt = \begin{cases} 1 & \text{if } k = r \\ 0 & \text{if } k \neq r \end{cases}$$

b. Show that if the series in (3.7) does in fact converge to the system input $\xi(t)$, then it follows on using (*a*) that the coefficients α_k there may be calculated using the formula

$$\alpha_k = \frac{1}{T}\int_{-T/2}^{+T/2} \xi(t)e^{-i2\pi kt/T}\,dt$$

c. Show finally that if $\xi(t)$ is a real function of time, then

$$\alpha_{-k} = \alpha_k^*$$

where α_k^* is the complex conjugate of the coefficient α_k; that is, the coefficients α_k and α_{-k} are complex conjugates of each other when $\xi(t)$ is real. □

Coefficients calculated using the formula in Exercise 3.1*b* are called *Fourier coefficients* and the series resulting from the use of such coefficients is called a *Fourier series*. It is beyond us to prove it here, but nonetheless it can be shown that the Fourier series does in fact converge in most practical cases to the function from which its coefficients were determined (cf. chap. 10 of Kreider et al.).

In summary, it follows from our preceding development (and from the results in Exercise 3.1) that if the input $x(t)$ to a stable, linear, time-invariant system with system function H is a real periodic function with period T, then we can express that input in the form of a Fourier series, that is, as

$$x(t) = \sum_{k=-\infty}^{+\infty} \alpha_k e^{i2\pi kt/T} \tag{3.8}$$

where the Fourier coefficients α_k are given by the formula

$$\alpha_k = \frac{1}{T} \int_{-T/2}^{+T/2} x(t) e^{-i2\pi kt/T}\, dt \tag{3.9}$$

and hence the system output may be calculated using the series formula

$$y(t) = \sum_{k=-\infty}^{+\infty} \alpha_k H\left(\frac{k}{T}\right) e^{i2\pi kt/T} \tag{3.10}$$

so long as that series converges.

Finally, let us suppose that the system input $x(t)$ is neither sinusoidal nor periodic but is, instead, a transient function of time as sketched, for example, in Fig. 3.2a. Our problem now is to determine the system response to such an input using frequency-analysis techniques. One way of attacking this problem is to approximate the input $x(t)$ using the Fourier series $x_T(t)$ formed by expanding $x(t)$ over the interval of length T, which extends from $-T/2$ to $+T/2$. Now (as is the case in Fig. 3.2), the periodic function $x_T(t)$ thus obtained is probably not a very good approximation to the transient input $x(t)$ outside of the expansion interval. If, however, we let the duration of the expansion interval become larger and larger, then $x_T(t)$ approximates $x(t)$ better and better over longer and longer time intervals. With this fact in mind, let us investigate the limiting behavior of the approximating Fourier series as we let T approach

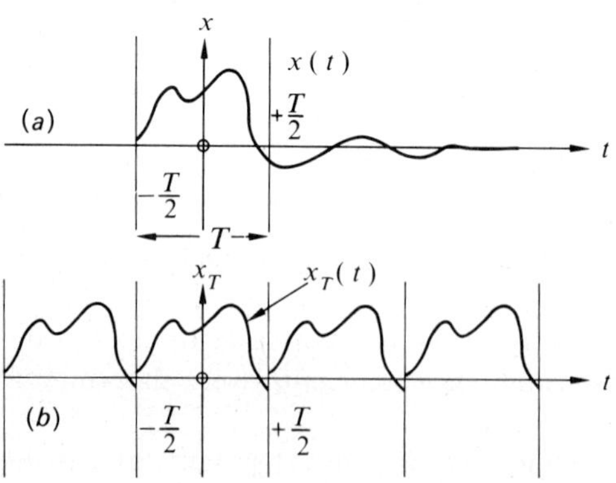

Figure 3.2 A transient input and its periodic approximation.

infinity. For convenience, let us rewrite (3.8) in the form

$$x_T(t) = \sum_{k=-\infty}^{+\infty} T\alpha_k e^{i2\pi f_k t} \frac{1}{T} \tag{3.11}$$

where we have defined $f_k \triangleq k/T$. Since it follows from this definition of f_k that the separation between adjacent values is just $1/T$, it is convenient to introduce the additional notation

$$\Delta f_k \triangleq \frac{1}{T}$$

Equation (3.11) may then be rewritten in the form

$$x_T(t) = \sum_{k=-\infty}^{+\infty} T\alpha_k e^{i2\pi f_k t}\,\Delta f_k \tag{3.12}$$

where, using (3.9),

$$T\alpha_k = \int_{-T/2}^{+T/2} x(t)e^{-i2\pi f_k t}\,dt$$

If next we let $T \to \infty$ (while making $k \to \infty$ so that $f = f_k$ remains fixed), the formula for the coefficient $T\alpha_k$ becomes

$$\begin{aligned} X(f) &\triangleq \lim_{T\to\infty} T\alpha_k \\ &= \int_{-\infty}^{+\infty} x(t)e^{-i2\pi ft}\,dt \end{aligned} \tag{3.13}$$

while the infinite series in (3.12) becomes the integral

$$\begin{aligned} x(t) &= \lim_{T\to\infty} x_T(t) \\ &= \int_{-\infty}^{+\infty} X(f)e^{i2\pi ft}\,df \end{aligned} \tag{3.14}$$

Thus we see that we have not only obtained a frequency decomposition $X(f)$ of the transient function $x(t)$, but also we have shown that if we are given such a frequency decomposition then we can recapture the original function by evaluating the integral in (3.14).

The integrals in (3.13) and (3.14) are called *Fourier integrals.* The frequency function $X(f)$ defined in (3.13) is said to be a *Fourier transform* of $x(t)$, in which case the time function $x(t)$ calculated by (3.14) is said to be an *inverse Fourier transform* of $X(f)$; together they are said to form a *Fourier transform pair.* While our above derivation of the Fourier transform pair relation between $x(t)$ and $X(f)$ is by no means rigorous, it does nonetheless indicate essentially the line of reasoning followed by Fourier in obtaining that result.[1] The validity of the transform-pair result

[1] Compare with sec. 1 of chap. 9 of Fourier's "Analytic Theory of Heat."

depends, of course, upon the existence of the respective integrals. An investigation of this existence problem is beyond us here but may be found, for example, in chap. 4 of Seeley; we shall be rather cavalier about this problem in the remainder of this text and we shall generally assume that Fourier integrals exist whenever we need them—they are much too useful to do without.

Having introduced the idea of a Fourier transform, let us now return to the problem of studying the passage of a transient function through a linear system. To this end, let us reconsider for the moment the periodic input approximation $x_T(t)$ with T finite; in particular, let us reconsider the formula

$$x_T(t) = \sum_{k=-\infty}^{+\infty} T\alpha_k e^{i2\pi f_k t} \, \Delta f_k$$

If we apply this periodic function to the input of a stable, linear, time-invariant system, it follows from (3.10) that the correponding periodic output $y_T(t)$ is given by

$$y_T(t) = \sum_{k=-\infty}^{+\infty} T\alpha_k H(f_k) e^{i2\pi f_k t} \, \Delta f_k$$

If we let $T \to \infty$, the factor $T\alpha_k$ tends to $X(f)$ as above and the output approximation $y_T(t)$ tends to the actual output $y(t)$. Thus we get in the limit the result that

$$y(t) = \int_{-\infty}^{+\infty} X(f)H(f)e^{i2\pi ft} \, df \tag{3.15}$$

that is, the output $y(t)$ is given by the Fourier transform of the product of the input Fourier transform $X(f)$ and the system function $H(f)$. If then we define

$$Y(f) \triangleq \int_{-\infty}^{+\infty} y(t)e^{-i2\pi ft} \, dt \tag{3.16}$$

it follows that $y(t)$ and $Y(f)$ thus form a Fourier transform pair, that is, that we may write

$$y(t) = \int_{-\infty}^{+\infty} Y(f)e^{i2\pi ft} \, df \tag{3.17}$$

It therefore follows from this transform-pair relation that if we take the Fourier transform of (3.15), we obtain the result that

$$Y(f) = X(f)H(f) \tag{3.18}$$

Thus we see that *the Fourier transform of the output of a stable, linear, time-invariant system in response to a transient input is equal to the Fourier transform of that input multiplied by the system function of the given system.*

Suppose now that the system output and input are also related by a convolution integral as discussed in Chap. 10:

$$y(t) = \int_{-\infty}^{+\infty} h(\tau)x(t - \tau)\, d\tau \tag{3.19}$$

This then is the time-domain equivalent to (3.18). Now we have already seen (cf. Exercise 2.5) that the system function H of a stable, linear, time-invariant system is a Fourier transform of its weighting function h; in particular,

$$H(f) = \int_{-\infty}^{+\infty} h(\tau)e^{-i2\pi f\tau}\, d\tau \tag{3.20}$$

It therefore follows that $h(t)$ and $H(f)$ form a Fourier transform pair and hence that the weighting function is an inverse Fourier transform of the system function:

$$h(\tau) = \int_{-\infty}^{+\infty} H(f)e^{i2\pi f\tau}\, df \tag{3.21}$$

With this transform-pair relationship in mind, we see on studying (3.18) and (3.19) that the Fourier transform of the convolution of the system weighting function and the system input is equal to the product of their respective Fourier transforms. This result is, in fact, a special case of the general result that *the Fourier transform of the convolution of any two transformable functions is equal to the product of their respective Fourier transforms.*

We have tried to show in this section that the output of a stable, linear, time-invariant system may be calculated by using its system function whether the system input is a sinusoid, a sum of sinusoids, a periodic function, or a transient function. It therefore follows from the results of this and the preceding chapter that we may determine the output of a stable, linear, time-invariant system in response to its input *either* by using the system weighting function $h(t)$ and making calculations in the time domain *or* by using the corresponding system function $H(f)$ and making calculations in the frequency domain. Just which technique is easier to use depends entirely upon the particular problem at hand.

Exercise 3.2 A somewhat more direct derivation of (3.21) may be made by using impulse functions as follows: Suppose that the input to a stable, linear, time-invariant system is the unit-impulse function; that is,

$$x(t) = \delta(t)$$

(cf. Sec. 4.6).

a. Show that the corresponding Fourier transform of this input is a constant function of frequency; in particular, show that

$$X(f) = 1$$

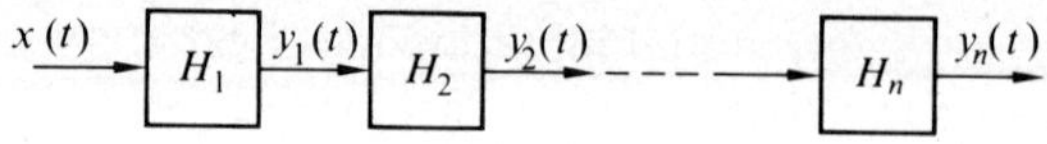

Figure 3.3 A cascade of n linear systems.

b. Show that the system output corresponding to this input is given by the formula

$$y(t) = \int_{-\infty}^{+\infty} H(f)e^{i2\pi ft}\,df$$

and hence that

$$h(t) = \int_{-\infty}^{+\infty} H(f)e^{i2\pi ft}\,df$$

that is, that $h(t)$ is the inverse transform of $H(f)$. □

Exercise 3.3 Suppose that $x(t)$ is the input to a cascade of n stable, linear, time-invariant systems and that $y_n(t)$ is the output of the cascade. Suppose further that the system function of the kth linear system is H_k, as indicated in Fig. 3.3.

Show that the Fourier transform $Y_n(f)$ of the cascade output is equal to the product of the input Fourier transform and the various system functions:

$$Y_n(f) = H_1(f)H_2(f) \cdots H_n(f)X(f)$$

On comparing this result with that in Exercise 10.5.6 *we see that we have here replaced a multiple convolution in the time domain by a multiple product in the frequency domain.* It is often much easier to determine the output of a cascade by calculating the inverse Fourier transform of $Y_n(f)$ as given above than it is by evaluating the equivalent multiple-convolution integral. □

4 SPECTRAL DENSITY

Having introduced the frequency characterization of linear systems, let us now return to the study of random processes in linear systems and reconsider that problem using frequency-domain ideas.

Suppose, in particular, that the wide-sense stationary random process $\{X_t, -\infty < t < +\infty\}$ is the input to a stable, linear, time-invariant system and that $\{Y_t, -\infty < t < +\infty\}$ is the corresponding output random process. We showed in Sec. 10.6 that the autocorrelation function of the output random process is given in this case by the formula

$$R_Y(\tau) = \iint_{-\infty}^{+\infty} h(\tau_1)h(\tau_2)R_X(\tau - \tau_1 + \tau_2)\,d\tau_1\,d\tau_2$$

where h is the weighting function of the given linear system and where R_X and R_Y are the autocorrelation functions of the input and output random processes, respectively (cf. 10.6.5). The mean-square value of the system output may now be obtained by setting $\tau = 0$ in this expression; thus we get the result

$$E[Y_t^2] = \iint_{-\infty}^{+\infty} h(\tau_1)h(\tau_2)R_X(\tau_2 - \tau_1)\, d\tau_1\, d\tau_2 \tag{4.1}$$

which is a constant function of time. We next wish to derive a frequency-domain equivalent to this result.

First, let us express the system weighting function h as the inverse Fourier transform of the system function H; in particular, let us make the substitution

$$h(\tau_1) = \int_{-\infty}^{+\infty} H(f)e^{i2\pi f\tau_1}\, df$$

Thus we get

$$\begin{aligned} E[Y_t^2] &= \iint_{-\infty}^{+\infty} \left(\int_{-\infty}^{+\infty} H(f)e^{i2\pi f\tau_1}\, df\right) h(\tau_2)R_X(\tau_2 - \tau_1)\, d\tau_1\, d\tau_2 \\ &= \int_{-\infty}^{+\infty} H(f)\, df \int_{-\infty}^{+\infty} h(\tau_2)\, d\tau_2 \int_{-\infty}^{+\infty} R_X(\tau_2 - \tau_1)e^{i2\pi f\tau_1}\, d\tau_1 \end{aligned} \tag{4.2}$$

Next, let us change variables in the last integral on the right-most side of (4.2) by using the substitution

$$\tau \triangleq \tau_2 - \tau_1$$

We may then rewrite (4.2) in the form

$$E[Y_t^2] = \int_{-\infty}^{+\infty} H(f)\, df \int_{-\infty}^{+\infty} h(\tau_2)e^{i2\pi f\tau_2}\, d\tau_2 \int_{-\infty}^{+\infty} R_X(\tau)e^{-i2\pi f\tau}\, d\tau$$

Since the middle integral here is simply the complex conjugate $H^*(f)$ of $H(f)$ [cf. Eq. (3.20)], we therefore obtain the result

$$E[Y_t^2] = \int_{-\infty}^{+\infty} |H(f)|^2\, df \int_{-\infty}^{+\infty} R_X(\tau)e^{-i2\pi f\tau}\, d\tau \tag{4.3}$$

A moment's reflection upon formula (4.3) should show that the last integral on the right-hand side is simply the Fourier transform of the autocorrelation function R_X of the input random process. If we denote that transform by S_X,

$$S_X(f) \triangleq \int_{-\infty}^{+\infty} R_X(\tau)e^{-i2\pi f\tau}\, d\tau \tag{4.4}$$

then it follows that S_X and the autocorrelation function R_X form a Fourier transform pair, and hence that we may determine that autocorrelation function from S_X by using the inverse Fourier transform relation

$$R_X(\tau) = \int_{-\infty}^{+\infty} S_X(f)e^{i2\pi f\tau}\,df \tag{4.5}$$

We shall call the function S_X the *spectral density* of the wide-sense stationary random process $\{X_t, -\infty < t < +\infty\}$.

On using (4.4) in (4.3), we finally obtain the result that

$$E[Y_t^2] = \int_{-\infty}^{+\infty} |H(f)|^2 S_X(f)\,df \tag{4.6}$$

Thus we see that *the mean-square value of the output of a stable, linear, time-invariant system in response to a wide-sense stationary random process is given by the integral over all frequencies of the spectral density of the input random process multiplied by the squared magnitude of the system function.* This is the desired frequency-domain equivalent to the time-domain formula in (4.1).

With the frequency-domain result (4.6) in mind, let us next see if we can obtain a somewhat more physical feeling for the spectral density. Suppose that the linear system under study is a narrowband filter; in particular, suppose that the filter system function H has a unit magnitude over the narrow frequency band extending from f_0 to $f_0 + \Delta f$ (where $\Delta f \ll f_0$) and is zero elsewhere:

$$|H(f)| = \begin{cases} 1 & \text{for } f_0 \le |f_0| \le f_0 + \Delta f \\ 0 & \text{elsewhere} \end{cases}$$

as shown in Fig. 4.1. If we now substitute this form of $|H(f)|$ in (4.6), we find that the mean-square value of the narrowband filter output is given by the expression

$$E[Y_t^2] = 2\int_{f_0}^{f_0+\Delta f} S_X(f)\,df$$

where we have made use of the fact that since the input random process is real, then its autocorrelation function R_X is an even function of its argument and hence so is its spectral density $S_X(f)$ [cf. Eq. (5.9)]. Assuming that the spectral density is continuous over the narrow passband of the

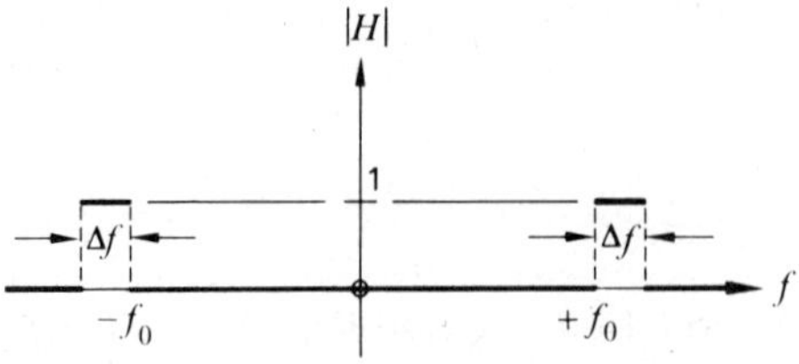

Figure 4.1 A narrowband filter system function.

filter, it then follows from the mean-value theorem of calculus that we may write

$$E[Y_t^2] = 2S_X(f_1)\,\Delta f \tag{4.7}$$

where f_1 is some frequency in the passband of the filter, that is, where $f_0 \le f_1 \le f_0 + \Delta f$. Since the output random process $\{Y_t, -\infty < t < +\infty\}$ here corresponds to that part of the input random process which falls in a narrow frequency band centered about the frequency f_1, we see that the spectral density of the input process provides (when evaluated at the frequency f_1) a measure of the "power" in the input process in that narrowband.

Suppose, for example, that the random process $\{X_t, -\infty < t < +\infty\}$ is a zero-mean version of the random-telegraph process introduced earlier in Sec. 9.6; in particular, suppose that $\{X_t\}$ is a random telegraph process which has the two possible values $+\frac{1}{2}$ and $-\frac{1}{2}$. The autocorrelation function of this process is

$$R_X(\tau) = \tfrac{1}{4}e^{-2\lambda|\tau|}$$

where λ is the average number of transversals from one possible value to another per unit time (cf. Exercise 9.6.2). Using this autocorrelation function in (4.4) then gives

$$\begin{aligned} S_X(f) &= \frac{1}{4}\int_{-\infty}^{+\infty} \exp\,(-2\lambda|\tau| - i2\pi f\tau)\,d\tau \\ &= \frac{1}{4}\int_{0}^{+\infty} e^{-2(\lambda+i\pi f)\tau}\,d\tau + \frac{1}{4}\int_{-\infty}^{0} e^{+2(\lambda-i\pi f)\tau}\,d\tau \\ &= \frac{1}{8}\left(\frac{1}{\lambda+i\pi f} + \frac{1}{\lambda - i\pi f}\right) \\ &= \frac{1}{4\lambda}\,\frac{1}{1+\left(\dfrac{\pi f}{\lambda}\right)^2} \end{aligned}$$

as a formula for the spectral density of this random-telegraph process. The Fourier transform pair R_X and S_X is plotted in Fig. 4.2 for this case. We see from that figure that the spectral density of the random-telegraph process has its maximum value at zero frequency and that it tends to zero monotonically with increasing frequency. We call the frequency at which such a spectral density has half of its zero-frequency value the *half-power* frequency of the process; the half-power frequency of the random-telegraph process is thus λ/π (that is, $1/\pi$ times the average number of transversals per unit time).

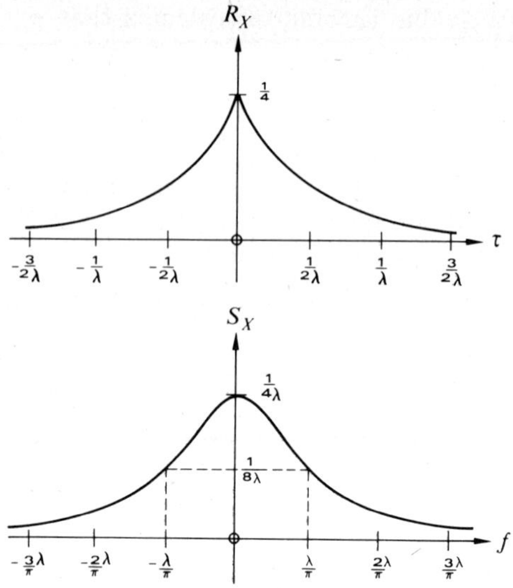

Figure 4.2 The autocorrelation function and the spectral density of the zero-mean random-telegraph process.

Exercise 4.1 Let $\{X_t, \ -\infty < t < +\infty\}$ be a wide-sense stationary random process with the autocorrelation function

$$R_X(\tau) = \sigma^2 \frac{\sin 2\pi f_0 \tau}{2\pi f_0 \tau}$$

where σ is a positive constant.

a. Determine the spectral density of this process. (*Hint:* Remember that R_X and S_X form a Fourier transform pair, and review Exercise 2.6 of this chapter.)

b. Plot $R_X(\tau)$ and $S_X(f)$ for this case. □

Exercise 4.2 Let $\{X_t, \ -\infty < t < +\infty\}$ be a wide-sense stationary random process with the autocorrelation function

$$R_X(\tau) = \begin{cases} \sigma^2\left(1 - \dfrac{|\tau|}{\tau_0}\right) & \text{for } 0 \le |\tau| \le \tau_0 \\ 0 & \text{otherwise} \end{cases}$$

where σ is a positive constant.

a. Determine the spectral density of this process. (*Hint:* Recall that the transform of a convolution is the product of the respective transforms, and review Exercise 6.3.3).

b. Plot $R_X(\tau)$ and $S_X(f)$ for this case. □

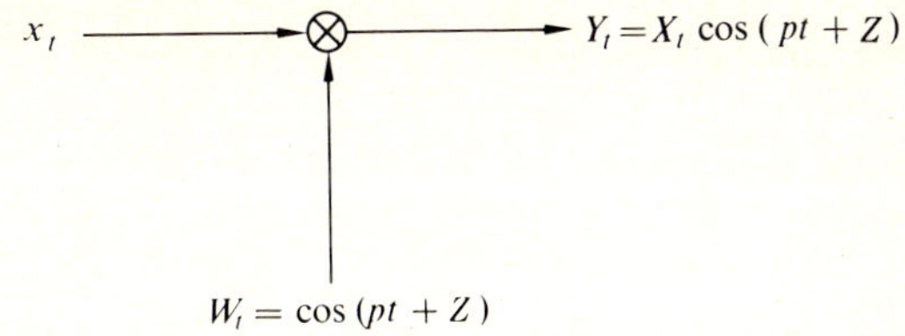

Figure 4.3 A modulator.

Exercise 4.3 Suppose that the input $\{X_t, -\infty < t < +\infty\}$ to the modulator shown in Fig. 4.3 is a zero-mean, wide-sense stationary random process whose spectral density $S_X(f)$ is zero for values of frequency greater in magnitude than some frequency f_0, say, for example, as shown in Fig. 4.4. Let the other modulator input be the sinusoidal random process $\{W_t, -\infty < t < +\infty\}$ in which

$$W_t = \cos(pt + Z)$$

where $f_p \triangleq p/2\pi \gg f_0$, where the random variable Z is uniformly distributed over $[0,2\pi]$, and where the random variables X_t and Z are statistically independent for all t. The modulator-output random process $\{Y_t, -\infty < t < +\infty\}$ then is such that

$$Y_t = X_t W_t$$
$$= X_t \cos(pt + Z)$$

a. Show that the output process is stationary in the wide sense and that its autocorrelation function is given by

$$R_Y(\tau) = \tfrac{1}{2} R_X(\tau) \cos p\tau$$

b. Show that the spectral density of the output process is

$$S_Y(f) = \tfrac{1}{4} S_X(f - f_p) + \tfrac{1}{4} S_X(f + f_p)$$

that is, the modulator output spectral density is obtained by shifting the input spectral density S_X to center on $+f_p$ ond on $-f_p$.

c. Assuming that S_X has the shape shown in Fig. 4.4, sketch S_Y as a function of frequency. □

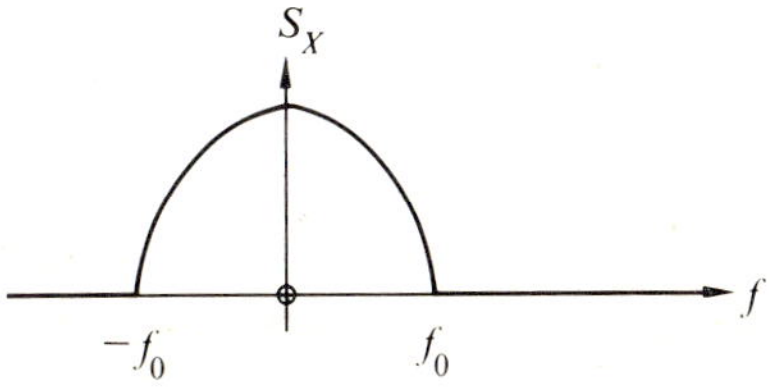

Figure 4.4 The modulator input spectral density.

Exercise 4.4 Exercise 4.3 is a particular case of the following general result: Let $\{X_t, \ -\infty < t < +\infty\}$ and $\{W_t, \ -\infty < t < +\infty\}$ be statistically independent, wide-sense stationary random processes. Let a new random process $\{Y_t, \ -\infty < t < +\infty\}$ be defined as the product of $\{X_t\}$ and $\{W_t\}$:

$$Y_t \triangleq X_t W_t$$

for all t.

a. Show that

$$\begin{aligned} R_Y(t, t - \tau) &= R_Y(\tau) \\ &= R_X(\tau)R_W(\tau) \end{aligned}$$

in this case.

b. Show that if the spectral densities S_X and S_W both exist, then

$$S_Y(f) = \int_{-\infty}^{+\infty} S_X(f')S_W(f' - f)\, df'$$

Thus we see again that the Fourier transform of a product is the convolution of the respective Fourier transforms. [Note that S_W in Exercise 4.3 exists only in the impulse-function sense (cf. Exercise 4.5*b* below).] □

Exercise 4.5 It should be noted that two of the random processes which often arise in applied systems problems have spectral densities which exist only in the impulse-function sense:

a. Let $\{X_t, \ -\infty < t < +\infty\}$ be a *constant* random process with the autocorrelation function

$$R_X(\tau) = m^2 \qquad \text{for all } \tau$$

where m is a constant. Show that the spectral density of this process is

$$S_X(f) = m^2\delta(f)$$

where δ is a unit-impulse function. (*Hint:* Remember that R_X and S_X form a Fourier transform pair, and work the problem backwards.)

b. Let $\{Y_t, \ -\infty < t < +\infty\}$ be a sinusoidal random process with the autocorrelation function

$$R_Y(\tau) = \sigma^2 \cos p\tau$$

where σ and p are positive real constants. Show that the spectral density of this process is

$$S_Y(f) = \frac{\sigma^2}{2}\left[\delta\left(f - \frac{p}{2\pi}\right) + \delta\left(f + \frac{p}{2\pi}\right)\right]$$

where again δ is a unit-impulse function.

c. Sketch both of these Fourier transform pairs. □

5 SOME GENERAL PROPERTIES OF THE SPECTRAL DENSITY

We defined in the preceding section the spectral density S_X of the wide-sense stationary random process $\{X_t, -\infty < t < +\infty\}$ to be the Fourier transform of the autocorrelation function R_X of the process; that is, we set

$$S_X(f) \triangleq \int_{-\infty}^{+\infty} R_X(\tau)e^{-i2\pi f\tau}\,d\tau \tag{5.1}$$

whenever that integral exists. We then pointed out that it therefore followed that we could also write

$$R_X(\tau) = \int_{-\infty}^{+\infty} S_X(f)e^{i2\pi f\tau}\,df \tag{5.2}$$

that is, that *the spectral density and the autocorrelation function form a Fourier transform pair.* We now wish to determine some of the general properties of the spectral density.

First of all, we note on setting $f = 0$ in (5.1) that it follows that

$$S_X(0) = \int_{-\infty}^{+\infty} R_X(\tau)\,d\tau \tag{5.3}$$

in other words, that *the zero-frequency value of the spectral density equals the total area under the graph of the autocorrelation function.* Similarly, if we set $\tau = 0$ in (5.2), it then follows that we get

$$E[X_t^2] = R_X(0) = \int_{-\infty}^{+\infty} S_X(f)\,df \tag{5.4}$$

that is, that *the mean-square value of a wide-sense stationary random process equals the total area under the graph of the spectral density.* This latter result is, of course, entirely consistent with the discussion in the preceding section in which we pointed out that the value of the spectral density at a frequency f gave a measure of the "power" in the given process within a narrow frequency band centered on f.

It should also be noted that it is an immediate consequence of Eq. (4.7) in that discussion that, since $E[Y_t^2]$ must always be nonnegative, it follows that

$$S_X(f) \geq 0 \tag{5.5}$$

for all f; *the spectral density is always nonnegative.*

Suppose next that we expand the exponential in (5.1) as $\cos 2\pi ft - i \sin 2\pi ft$ to obtain the result

$$S_X(f) = \int_{-\infty}^{+\infty} R_X(\tau) \cos (2\pi f\tau)\, d\tau - i \int_{-\infty}^{+\infty} R_X(\tau) \sin (2\pi f\tau)\, d\tau \tag{5.6}$$

Now if the random process $\{X_t, -\infty < t < +\infty\}$ is a real random process, then its autocorrelation function is an even function of its argument; that is,

$$R_X(-\tau) = R_X(\tau)$$

in this case. The imaginary term in (5.6) then vanishes and we therefore get the result that

$$S_X(f) = \int_{-\infty}^{+\infty} R_X(\tau) \cos (2\pi f\tau)\, d\tau \tag{5.7}$$

when $\{X_t\}$ is a real random process. Similarly, it follows from (5.2) that we also get the result

$$R_X(\tau) = \int_{-\infty}^{+\infty} S_X(f) \cos (2\pi f\tau)\, df \tag{5.8}$$

in this case. That is, *the spectral density and the autocorrelation function of a real wide-sense stationary random process form a Fourier cosine-transform pair.* Whether the cosine-transform-pair relation or the exponential-transform-pair relation given by (5.1) and (5.2) is the easier to use depends upon the problem at hand.

One result does follow directly from the cosine-transform formula (5.7), however: We note that the only place in which the frequency f occurs on the right-hand side of that formula is in the argument of the cosine. Since the cosine is an even function of its argument, it therefore follows that

$$S_X(-f) = S_X(f) \tag{5.9}$$

the spectral density of a real random process is an even function of frequency.

6 SPECTRAL ANALYSIS OF LINEAR SYSTEMS

Let us now return again to the study of random processes in linear systems. We pointed out earlier that if $\{X_t, -\infty < t < +\infty\}$ and $\{Y_t, -\infty < t < +\infty\}$ are the input and output random processes, respectively, of a stable, linear time-invariant system and if $\{X_t\}$ is stationary in the wide sense, then the autocorrelation function of the output random process is given by the formula

$$R_Y(\tau) = \iint_{-\infty}^{+\infty} h(\tau_1)h(\tau_2)R_X(\tau - \tau_1 + \tau_2)\, d\tau_1\, d\tau_2$$

where h is the weighting function of the given linear system. If now we take the Fourier transform of both sides of this equation, we obtain after some manipulation the result

$$S_Y(f) = |H(f)|^2 S_X(f) \tag{6.1}$$

Thus we see that *the output spectral density equals the input spectral density multiplied by the squared magnitude of the system function.*

Suppose, for example, that the linear system under study is a differentiator. We showed earlier [cf. (2.13)] that the system function of a differentiator is

$$H(f) = i2\pi f$$

It therefore follows that the spectral density S_Y of the differentiator output is

$$S_Y(f) = 4\pi^2 f^2 S_X(f) \tag{6.2}$$

where S_X is the spectral density of the differentiator input.

Exercise 6.1 Suppose that the input to a differentiator is the zero-mean random-telegraph process discussed in Sec. 4; that is, suppose that the differentiator input process $\{X_t, -\infty < t < +\infty\}$ has the autocorrelation function

$$R_X(\tau) = \tfrac{1}{4}e^{-2\lambda|\tau|}$$

and the spectral density

$$S_X(f) = \frac{1}{4\lambda}\cdot\left[\frac{1}{1 + \left(\frac{\pi f}{\lambda}\right)^2}\right]$$

a. Determine the spectral density of the differentiator output, and plot your result as a function of frequency.
b. Determine the mean-square value of the differentiator output. (Be careful!) □

Exercise 6.2 Let $\{X_t, -\infty < t < +\infty\}$ be a real wide-sense stationary random process with a spectral density S_X. Let $\{Y_t, -\infty < t < +\infty\}$ be the output process in response to the X_t process of the system shown in Fig. 6.1.

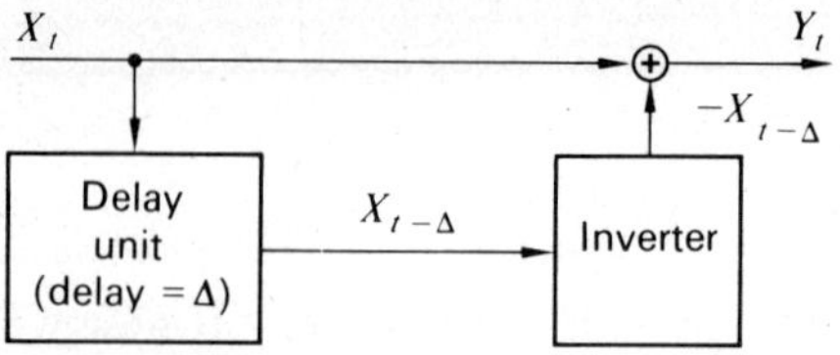

Figure 6.1 A comb filter.

a. Show that the system function of the given system has the squared-magnitude

$$|H(f)|^2 = 2(1 - \cos 2\pi\Delta f)$$

(where Δ is the time delay of the delay unit) and hence that the output spectral density is

$$S_Y(f) = 2(1 - \cos 2\pi\Delta f)S_X(f)$$

Such a system is sometimes called a *comb filter* (plot $|H(f)|^2$ versus f to see why).

b. Show that for values of frequency which are small compared to $1/\Delta$ we obtain the approximate result

$$S_Y(f) \approx (4\pi^2 f^2)\Delta^2 S_X(f)$$

On comparing this result with that in (6.2), we note that the present system acts like a differentiator for low-frequency inputs. (Why should that result not be very surprising?) □

System spectral functions

We pointed out in Sec. 10.6 that it is sometimes convenient to introduce the use of the system correlation function

$$R_h(\tau) \triangleq \int_{-\infty}^{+\infty} h(t)h(t - \tau)\, dt$$

when studying the passage of a wide-sense stationary random process through a stable, linear, time-invariant system. We also showed there that we could then calculate the autocorrelation function of the system output by using the formula

$$R_Y(\tau) = \int_{-\infty}^{+\infty} R_h(t)R_X(\tau - t)\, dt$$

We now wish to obtain a frequency-domain equivalent to this formula. To this end, let us introduce the corresponding *system spectral function* S_h defined as the Fourier transform of the system correlation function:

$$S_h(f) \triangleq \int_{-\infty}^{+\infty} R_h(\tau)e^{-i2\pi f\tau}\, d\tau \tag{6.3}$$

It then follows that we may write

$$R_h(\tau) = \int_{-\infty}^{+\infty} S_h(f)e^{i2\pi f}\, df \tag{6.4}$$

that is, that *the system correlation function and the system spectral function form a Fourier transform pair.*

Exercise 6.3 Let the wide-sense stationary random process $\{X_t, -\infty < t < +\infty\}$ and $\{Y_t, -\infty < t < +\infty\}$ be the input and output, respectively, of a stable, linear, time-invariant system with the system weighting function h and system function H.

Show that the system spectral function is related to the system function H by the formula

$$S_h(f) = |H(f)|^2$$

and hence that the system output spectral density is

$$S_Y(f) = S_h(f)S_X(f)$$

where S_X is the input spectral density. □

7 NARROWBAND FILTERING

A situation commonly arising in practical systems studies is that in which a wide-sense stationary random process is passed through a *narrowband filter*, that is, through a stable, linear, time-invariant system whose system function H has a nonzero magnitude essentially only over some region of frequency which is narrow as compared to the frequency extent of the spectral density of the filter input. If such a narrowband filter has its passband centered on zero frequency, it is called a *low-pass* filter; if its passband is centered on some nonzero frequency (which is greater than the filter "bandwidth"), it is called a *bandpass* filter. Typical narrowband filter characteristics are shown in Fig. 7.1.

Since the passband of a narrowband filter is, by design, small compared to the frequency extent of the spectral density of its input, it is often quite reasonable to assume that the input spectral density is essentially constant over the passband of the filter. That is, it is often quite reasonable to assume that

$$S_X(f) = S_0$$

for all f such that $|H(f)| > 0$, where S_0 is a positive constant. The spectral density of the filter output is then

$$S_Y(f) = |H(f)|^2S_0$$

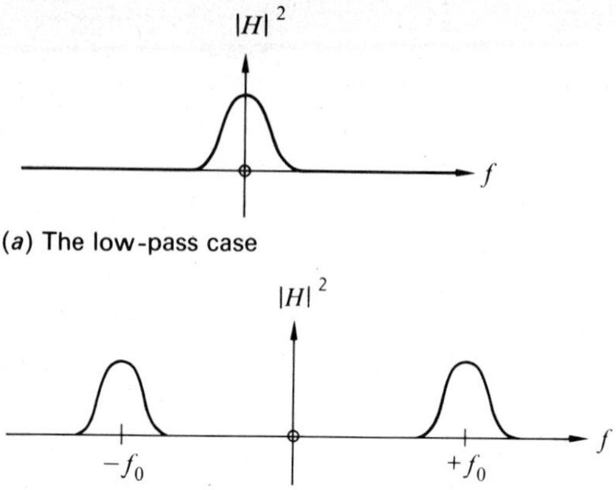

(a) The low-pass case

(b) The bandpass case

Figure 7.1 Narrowband filter characteristics.

and hence the output mean-square value is

$$E[Y_t^2] = S_0 \int_{-\infty}^{+\infty} |H(f)|^2 \, df$$

in this case.

In view of this last result, it is often convenient in narrowband cases to characterize such a filter by an *effective gain* G_{eff} defined as

$$G_{\text{eff}} \triangleq |H(f)|^2_{\max} \tag{7.1}$$

(that is, as the maximum value of the squared magnitude of the filter system function in its passband) and by an *effective bandwidth* B_{eff} defined as

$$B_{\text{eff}} \triangleq \frac{\int_0^{+\infty} |H(f)|^2 \, df}{G_{\text{eff}}} = \int_0^{+\infty} \frac{|H(f)|^2}{|H(f)|^2_{\max}} \, df \tag{7.2}$$

A review of our introductory discussion of the spectral density shows that the idealized narrowband filter used there (cf. Fig. 4.1) has an effective gain of unity and an effective bandwidth Δf.

We may now express the mean-square value of the narrowband filter output in terms of the parameters G_{eff} and B_{eff} by the formula

$$E[Y_t^2] = 2S_0 G_{\text{eff}} B_{\text{eff}} \tag{7.3}$$

when the input spectral density may be assumed to be constant over the filter passband.

Exercise 7.1 Show that it follows from the definition of the effective bandwidth that

$$B_{\text{eff}} = \frac{R_h(0)}{2G_{\text{eff}}}$$

where R_h is the system correlation function of the given narrowband filter. □

Exercise 7.2 Suppose that a short-time averager is used as a low-pass filter; that is, consider the low-pass filter weighting function

$$h(t) = \begin{cases} \dfrac{1}{T} & \text{for } 0 \leq t \leq T \\ 0 & \text{otherwise} \end{cases}$$

Determine the effective gain and the effective bandwidth of the short-time averager. □

Exercise 7.3 Another type of linear system commonly used as a low-pass filter is that characterized by the first-order, linear differential equation

$$T\frac{dy}{dt} + y(t) = x(t)$$

where $y(t)$ is the system output in response to the input $x(t)$. This low-pass filter has the system weighting function

$$h(t) = \begin{cases} \dfrac{1}{T}e^{-t/T} & \text{for } 0 \leq t \\ 0 & \text{for } t < 0 \end{cases}$$

Determine the effective gain and effective bandwidth of such a system. □

8 CROSS-SPECTRAL DENSITIES

Just as we introduced the spectral density in Sec. 4 in order to provide a measure of the frequency distribution of a single random process, so now it is convenient to also introduce cross-spectral densities in order to provide a measure of the frequency interrelationship between two random processes. In particular, suppose that $\{X_t, \; -\infty < t < +\infty\}$ and $\{Y_t, \; -\infty < t < +\infty\}$ are two *jointly* wide-sense stationary random processes with the cross-correlation functions R_{XY} and R_{YX}. We then define the *cross-spectral densities* S_{XY} and S_{YX} of that pair of random processes to be the Fourier transforms of the respective cross-correlation functions; that is, we get

$$S_{XY}(f) \triangleq \int_{-\infty}^{+\infty} R_{XY}(\tau)e^{-i2\pi f\tau}\,d\tau$$
$$S_{YX}(f) \triangleq \int_{-\infty}^{+\infty} R_{YX}(\tau)e^{-i2\pi f\tau}\,d\tau \tag{8.1}$$

whenever those integrals converge absolutely. The cross-correlation functions and the cross-spectral densities thus form Fourier transform pairs, and hence we may write

$$R_{XY}(\tau) = \int_{-\infty}^{+\infty} S_{XY}(f)e^{i2\pi f\tau}\,df$$
$$R_{YX}(\tau) = \int_{-\infty}^{+\infty} S_{YX}(f)e^{i2\pi f\tau}\,df \tag{8.2}$$

in the case of a pair of random processes which are jointly stationary in the wide sense.

Exercise 8.1 Suppose that the random processes $\{X_t, -\infty < t < +\infty\}$ and $\{Y_t, -\infty < t < +\infty\}$ are individually stationary in the wide sense. Consider the sum random process $\{Z_t, -\infty < t < +\infty\}$ where

$$Z_t \triangleq X_t + Y_t$$

for all t.

a. Show that

$$R_Z(t_1,t_2) = R_X(t_1 - t_2) + R_Y(t_1 - t_2) + R_{XY}(t_1,t_2) + R_{YX}(t_1,t_2)$$

and hence, on defining $\tau \triangleq t_1 - t_2$, that

$$R_Z(\tau) = R_X(\tau) + R_Y(\tau) + R_{XY}(\tau) + R_{YX}(\tau)$$

when the $\{X_t\}$ and $\{Y_t\}$ processes are also jointly stationary in the wide sense.

b. Show that it therefore follows that if the $\{X_t\}$ and $\{Y_t\}$ processes are jointly stationary in the wide sense, then the spectral density of their sum is given by the formula

$$S_Z(f) = S_X(f) + S_Y(f) + S_{XY}(f) + S_{YX}(f)$$

c. Finally, show that the sum spectral density is the sum of the respective spectral densities:

$$S_Z(f) = S_X(f) + S_Y(f)$$

when the wide-sense stationary random processes $\{X_t\}$ and $\{Y_t\}$ are uncorrelated.

Thus we see that *the cross-spectral densities S_{XY} and S_{YX} represent the spectral components which must be added to the individual spectral*

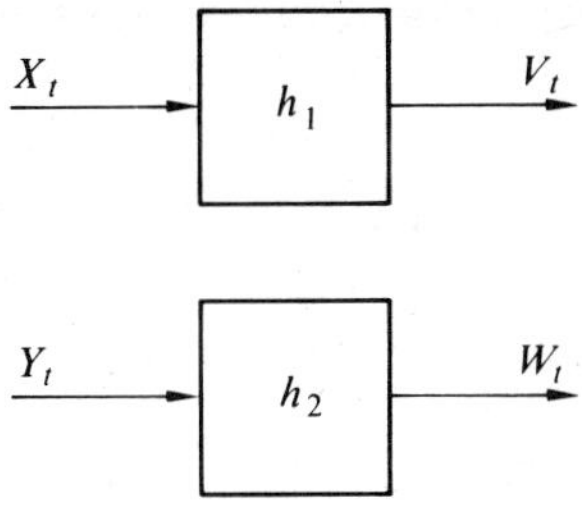

Figure 8.1 A pair of linear systems.

densities of a pair of correlated random processes in order to obtain the spectral density of their sum. □

We may perhaps get a better feeling for the cross-spectral density by considering for the moment the problem of passing two *jointly* wide-sense stationary random processes through a pair of separate, stable, linear, time-invariant systems as shown in Fig. 8.1. In particular, suppose that the random process $\{X_t, -\infty < t < +\infty\}$ is the input to system 1 and that the random process $\{Y_t, -\infty < t < +\infty\}$ is the input to system 2. The cross-correlation function of the corresponding output processes $\{V_t, -\infty < t < +\infty\}$ and $\{W_t, -\infty < t < +\infty\}$, respectively, is then

$$\begin{aligned} R_{VW}(t_1,t_2) &= E\left[\int_{-\infty}^{+\infty} h_1(\tau_1)X_{t_1-\tau_1}\,d\tau_1 \int_{-\infty}^{+\infty} h_2(\tau_2)Y_{t_2-\tau_2}\,d\tau_2\right] \\ &= \iint_{-\infty}^{+\infty} h_1(\tau_1)h_2(\tau_2)R_{XY}(t_1 - \tau_1, t_2 - \tau_2)\,d\tau_1\,d\tau_2 \end{aligned} \tag{8.3}$$

Since the input processes are jointly stationary by hypothesis, we may introduce the time difference $\tau \triangleq t_1 - t_2$ and rewrite (8.3) as

$$R_{VW}(\tau) = \iint_{-\infty}^{+\infty} h_1(\tau_1)h_2(\tau_2)R_{XY}(\tau - \tau_1 + \tau_2)\,d\tau_1\,d\tau_2$$

Exercise 8.2 On defining the cross-spectral density S_{VW} of the system outputs to be the Fourier transform of the output cross-correlation function R_{VW}, show that

$$S_{VW}(f) = H_1(f)H_2^*(f)S_{XY}(f) \tag{8.4}$$

where $H_2^*(f)$ is the complex conjugate of $H_2(f)$. □

An interesting special case of this last result is obtained when the two systems under study are narrowband filters with nonoverlapping passbands as shown, for example, in Fig. 8.2. It is convenient at this point to express the system functions of the narrowband filters in terms

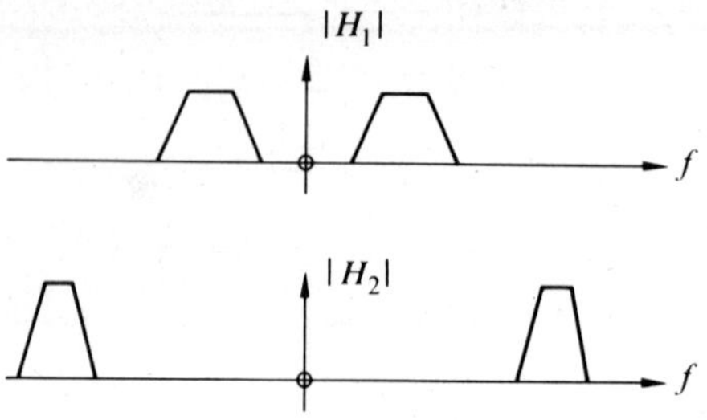

Figure 8.2 Nonoverlapping passbands

of their magnitude and phase functions:

$$H_1(f) = |H_1(f)|e^{i\theta_1(f)} \qquad \text{and} \qquad H_2(f) = |H_2(f)|e^{i\theta_2(f)}$$

If then we rewrite (8.4) in terms of the filter magnitude and phase functions, we get

$$S_{VW}(f) = |H_1(f)|\,|H_2(f)| \exp \{i[\theta_1(f) - \theta_2(f)]\}S_{XY}(f)$$

Now filter passbands are said to be nonoverlapping if and only if

$$|H_1(f)| \neq 0 \text{ implies } |H_2(f)| = 0$$

and

$$|H_2(f)| \neq 0 \text{ implies } |H_1(f)| = 0$$

With this fact in mind, we see that since our narrowband filters are assumed here to have nonoverlapping passbands, it follows in this case that

$$S_{VW}(f) = 0$$

for all values of f, and hence that

$$R_{VW}(\tau) = 0$$

for all values of τ. Thus we see that *if two jointly stationary random processes are applied as inputs to a pair of stable, linear, time-invariant filters whose passbands are nonoverlapping, then the filter output processes are uncorrelated.*

An important subcase of the present situation is that in which the two filters are both excited by the same wide-sense stationary input as shown, for example, in Fig. 8.3. It then follows from our above results that *nonoverlapping portions of the frequency spectrum of a single wide-sense stationary random process correspond to uncorrelated subprocesses.*

It must be emphasized that all of the results obtained above depend strictly upon the assumption that the pair of input random processes are

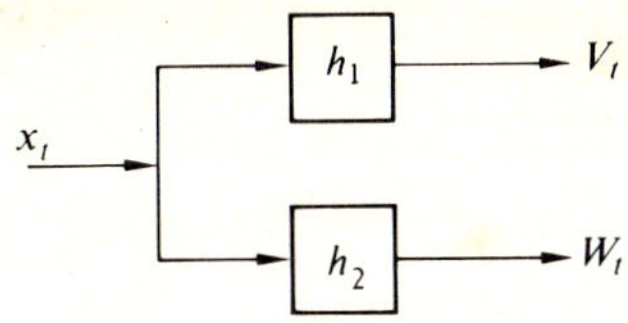

Figure 8.3 Filters with a common input.

jointly stationary, at least in the wide sense. The following example illustrates this point:

Let $\{Y_t, -\infty < t < +\infty\}$ be the narrowband, bandpass, wide-sense stationary random process defined in Exercise 4.3. That is, let Y_t be defined as

$$Y_t \triangleq X_t \cos(pt + Z)$$

for all t, where $\{X_t, -\infty < t < +\infty\}$ is a zero-mean wide-sense stationary random process whose spectral density is zero for values of frequency greater in magnitude than some frequency f_0 as shown in Fig. 4.3, where $f_p \triangleq p/2\pi \gg f_0$, where the random variable Z is uniformly distributed over $[0,2\pi]$, and where the random variables X_t and Z are statistically independent for all t. Suppose now that the $\{Y_t\}$ process is multiplied by $\cos qt$ (where q is a positive constant such that $2\pi f_0 \ll q < p/3$). The resultant process $\{V_t, -\infty < t < +\infty\}$ is then defined by

$$\begin{aligned} V_t &\triangleq Y_t \cos qt = X_t \cos(pt + Z) \cos qt \\ &= \frac{X_t}{2} \cos[(p + q)t + Z] + \frac{X_t}{2} \cos[(p - q)t + Z] \end{aligned} \tag{8.5}$$

We call the first term on the right-most side of (8.5) the *upper sideband* and write

$$U_t \triangleq \frac{X_t}{2} \cos[(p + q)t + Z]$$

and we call the second term there the *lower sideband* and write

$$L_t \triangleq \frac{X_t}{2} \cos[(p - q)t + Z]$$

Suppose finally that the $\{V_t\}$ process is applied to upper and lower sideband filters which then separate out the $\{U_t\}$ and $\{L_t\}$ processes, respectively. This is the situation pictured in Fig. 8.4.

Exercise 8.3

a. Show that the upper and lower sideband processes $\{U_t\}$ and $\{L_t\}$ are individually stationary in the wide sense.

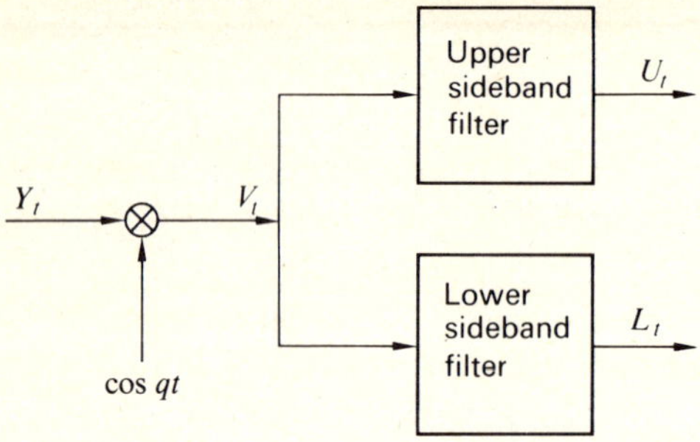

Figure 8.4 Sideband generation and separation.

b. Show that the cross-correlation function $R_{UL}(t_1,t_2)$ does *not* vanish for all t_1 and t_2 and, further, that it is *not* a function only of the time difference $t_1 - t_2$. Thus we see that not only are the upper and lower sideband processes correlated but also that they are *not* jointly stationary in the wide sense. □

Exercise 8.4 Show that the multiplier output process $\{V_t\}$ is *not* stationary in the wide sense. □

9 EPILOGUE

We have tried in this chapter to give an introduction to some of the basic aspects of the spectral analysis of wide-sense stationary random processes and to the frequency-domain analysis of the passage of such processes through stable, linear, time-invariant systems. It should be noted that even though we restricted our attention to such processes and systems, our results were confined to formulas for the mean values and the correlation functions and spectra of the system inputs and outputs. The further problem of determining the probability distribution function of the system output is quite difficult even in such simple cases[1] unless the input is a gaussian process (in which case the output is also a gaussian process as we shall show later in Chap. 14). Nonetheless, our results obtained in this chapter are adequate and useful in many practical applications.

One important special case of the spectral analysis of random processes in linear systems which occurs in many practical problems is that of the analysis of thermal agitation noise in linear electrical networks. While the techniques we have developed above are completely adequate to handle such problems, a special notation and terminology has been developed by workers in that field. A good introduction to this topic is given in chap. 2 of Bennett's book, while a somewhat more complete treatment may be found in chap. 9 of Davenport and Root.

[1] See, for example, sec. 9-5 of Davenport and Root.

Another problem area related to the subject matter of this chapter which, although quite important, has not been discussed here at all, is that of the synthesis of optimum linear systems. The basic (though rather advanced) reference for the frequency-domain approach to this problem is Wiener's book. A good intermediate level treatment of this approach is given by Lee, while a somewhat more advanced treatment is given by Yaglom. A time-domain approach to the optimum linear system area is discussed in the books by Deutsch and Van Trees.

Finally, there is the very important topic of estimating spectral densities from physical data. Now we pointed out in our introduction to the spectral density that evaluating the spectral density at some frequency f_0 gives a measure of the "power" in the process in a small frequency band centered about f_0. With this in mind, an estimate of the value of the spectral density at some frequency f_0 may be made by using a system which consists of a narrowband filter (centered on f_0) followed by a square-law device and an averaging circuit.[1] While such a system forms, in fact, the basis of many of the radiometers used in radio and radar astronomy,[2] a more practical method in many applications consists of sampling the process under study and then processing the resultant data with a digital computer. A by now classical introduction to this latter technique may be found in Blackman and Tukey's monograph, while a more recent comprehensive treatment may be found in the book by Jenkins and Watts.

10 SUMMARY OF DEFINITIONS AND FORMULAS

Fourier transforms

Let $x(t)$ be an absolutely integrable function. The Fourier transform of $x(t)$ is then

$$X(f) \triangleq \int_{-\infty}^{+\infty} x(t)e^{-i2\pi ft}\,dt \tag{3.13}$$

The corresponding inverse-transform relation is

$$x(t) = \int_{-\infty}^{+\infty} X(f)e^{i2\pi ft}\,df \tag{3.14}$$

and we say that $X(f)$ and $x(t)$ form a Fourier transform pair.

[1] See, for example, Exercise 14.3.3 of this book and pt. 4 of Davenport, Johnson, and Middleton.

[2] See, for example, chap. 10, Detectors for Radar Astronomy, by Robert Price, in Evans and Hagfors.

System functions

Let $h(t)$ be the weighting function of a stable, linear, time-invariant linear system. We then call the Fourier transform

$$H(f) \triangleq \int_{-\infty}^{+\infty} h(\tau) e^{-i2\pi f\tau}\, d\tau \tag{3.20}$$

the system function of the given linear system. The corresponding inverse transform relation is

$$h(\tau) = \int_{-\infty}^{+\infty} H(f) e^{i2\pi f\tau}\, df \tag{3.21}$$

It further follows that if $X(f)$ is the Fourier transform of the input to such a system, then the Fourier transform of the system output $y(t)$ is given by the formula

$$Y(f) = X(f)H(f) \tag{3.18}$$

This is the frequency-domain equivalent to the convolution integral formula in the time domain

$$y(t) = \int_{-\infty}^{+\infty} h(\tau) x(t - \tau)\, d\tau \tag{3.19}$$

Spectral density

Let $\{X_t, \ -\infty < t < +\infty\}$ be a wide-sense stationary random process with the autocorrelation function R_X. We call the Fourier transform S_X of R_X, that is,

$$S_X(f) \triangleq \int_{-\infty}^{+\infty} R_X(\tau) e^{-i2\pi f\tau}\, d\tau \tag{4.4}$$

the spectral density of the given process. The autocorrelation function and the spectral density thus form a Fourier transform pair, and hence we also have the inverse-transform formula

$$R_X(\tau) = \int_{-\infty}^{+\infty} S_X(f) e^{i2\pi f\tau}\, df \tag{4.5}$$

It follows from these formulas that

$$S_X(0) = \int_{-\infty}^{+\infty} R_X(\tau)\, d\tau \tag{5.3}$$

and

$$E[X_t^2] = R_X(0) = \int_{-\infty}^{+\infty} S_X(f)\, df \tag{5.4}$$

A further property of the spectral density is

$$S_X(f) \geq 0 \qquad \text{for all } f \tag{5.5}$$

that is, the spectral density is always nonnegative. Also, if $\{X_t\}$ is a real random process, then

$$S_X(-f) = S_X(f) \tag{5.6}$$

that is, the spectral density of a real random process is an even function of its argument.

Spectral analysis of linear systems

If we pass a wide-sense stationary random process $\{X_t, -\infty < t < +\infty\}$ through a stable, linear, time-invariant system with a system function H, the spectral density of the system output process $\{Y_t, -\infty < t < +\infty\}$ is then

$$S_Y(f) = |H(f)|^2 S_X(f) \tag{6.1}$$

and hence the mean-square value of the output is

$$E[Y_t^2] = \int_{-\infty}^{+\infty} |H(f)|^2 S_X(f)\, df \tag{4.6}$$

If H has the value unity over a narrowband of width Δf centered about a frequency f, then (4.6) reduces to

$$E[Y_t^2] = 2S_X(f)\,\Delta f \tag{4.7}$$

Cross-spectral density

Suppose that $\{X_t, -\infty < t < +\infty\}$ and $\{Y_t, -\infty < t < +\infty\}$ are two jointly wide-sense stationary random processes with a cross-correlation function R_{XY}. We call the Fourier transform S_{XY} of R_{XY}; that is,

$$S_{XY}(f) \triangleq \int_{-\infty}^{+\infty} R_{XY}(\tau) e^{-i2\pi f\tau}\, d\tau \tag{8.1}$$

the cross-spectral density of those processes. It then follows that

$$R_{XY}(\tau) = \int_{-\infty}^{+\infty} S_{XY}(f) e^{i2\pi f\tau}\, df \tag{8.2}$$

REFERENCES

Bennett, William R.: "Electrical Noise," McGraw-Hill Book Company, New York, 1960.

Blackman, Ralph B., and John W. Tukey: "The Measurement of Power Spectra," Dover Publications, Inc., New York, 1958.

Bochner, Salomon: "Lectures on Fourier Integrals," Ann. Math. Studies, no. 42, Princeton University Press, Princeton, N.J., 1959.

Bracewell, Ron: "The Fourier Transform and Its Applications," McGraw-Hill Book Company, New York, 1965.

Davenport, Wilbur B., Jr., Richard A. Johnson, and David Middleton: Statistical Errors in Measurements on Random Time Functions, *J. Appl. Phys.*, vol. 23, no. 4, pp. 377–388, 1952.

Davenport, Wilbur B., Jr., and William L. Root: "Random Signals and Noise," McGraw-Hill Book Company, New York, 1958.

Deutsch, Ralph: "Estimation Theory," Prentice-Hall, Inc., Englewood Cliffs, N.J., 1965.

Evans, John V., and Tor Hagfors: "Radar Astronomy," McGraw-Hill Book Company, New York, 1968.

Fourier, Jean Baptiste Joseph: "The Analytical Theory of Heat," Dover Publications, Inc., New York, 1955 [reprint of an 1878 English Translation (by Alexander Freeman) of the 1822 French original].

Jenkins, Gwilym M., and Donald G. Watts: "Spectral Analysis and Its Applications," Holden-Day, Inc., Publisher, San Francisco, 1968.

Lee, Yuk-Wing: "Statistical Theory of Communication," John Wiley & Sons, Inc., New York, 1960.

Papoulis, Athanasios: "The Fourier Integral and Its Applications," McGraw-Hill Book Company, New York, 1962.

Seeley, Robert T.: "An Introduction to Fourier Series and Integrals," W. A. Benjamin, Inc., New York, 1966.

Titchmarsh, Edward Charles: "Introduction to the Theory of Fourier Integrals," Clarendon Press, Oxford, 1948.

Van Trees, Harry L., Jr.: "Detection, Estimation, and Modulation Theory," pt. 1, John Wiley & Sons, Inc., 1968.

Wiener, Norbert: "Extrapolation, Interpolation, and Smoothing of Stationary Time Series," John Wiley & Sons, Inc., New York, 1949.

Yaglom, Akiva Moisseevich: "Stationary Random Functions," Prentice-Hall, Inc., Englewood Cliffs, N.J., 1962.

12
Sums of Independent Random Variables

1 INTRODUCTION

Much of the theory of random processes has developed out of attempts to provide mathematical models for various physical random phenomena. A more or less typical example is that of Brownian motion, named after the Scottish botanist Robert Brown (1773–1858). In 1828 Brown published his discovery that flower pollen suspended in water and viewed with a microscope exhibited persistent and highly irregular random motion. A typical path of such a pollen particle might well look as shown in Fig. 1.1. After numerous experiments with liquid suspensions of many other kinds of small particles (both animate and inanimate), Brown decided that he had discovered some primitive form of life which was present in all kinds of matter. Such a conclusion led, of course, to many other experiments and theories,[1] and it was not until 1905 that the German physicist Albert Einstein (1879–1955) published the first physically sound theory of the phenomenon. The first rigorous mathematical

[1] See, for example, sec. 1.2 of MacDonald's book.

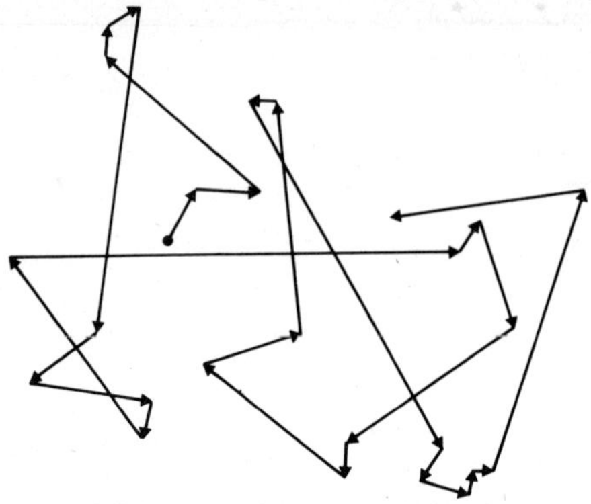

Figure 1.1 A typical pollen-particle path.

treatment of the random process describing Brownian motion was published in 1923 by the American mathematician Norbert Wiener (1894–1964); for this reason, the Brownian motion process is often called the Wiener process in the mathematics literature.

Basic to an explanation of Brownian motion is the fact that while a pollen particle seems quite small to a human being, it is in fact quite large compared to the molecules of the liquid in which it is suspended. Now due to thermal agitation, the liquid molecules are always in motion and continually collide with the pollen particles. Since in any given time interval the number and direction of liquid molecules striking a pollen particle is essentially a matter of chance, the pollen particle is subjected to a randomly varying force causing it to have the observed Brownian motion.

Let us now consider the modeling of such a Brownian motion. We shall, for convenience, consider only one-dimensional Brownian motion, for example, the projection of the particle motion onto a single space-coordinate axis. The basic idea here is that the motion of any such pollen particle in any given time interval is due to collisions which are independent of those which occur in any other nonoverlapping time interval. That is, the changes in particle position during nonoverlapping time intervals are describable by statistically independent random variables. In particular, if $y(t)$ is the y projection of the particle position at the time t, and if the n time instants $t_1, t_2, \ldots, t_n$ are ordered,

$$t_1 < t_2 < \cdots < t_n$$

then the changes in position, that is, *increments*,

$$x_k \triangleq y(t_k) - y(t_{k-1})$$

are representable by statistically independent random variables (say X_k). Thus, if we take the position of the particle at time zero to be the coordinate origin, we may represent the particle position at time t_n by the random variable

$$Y_{t_n} = \sum_{k=1}^{n} X_k \tag{1.1}$$

where the increments X_k are mutually independent random variables. The family of the sum random variables $\{Y_t, 0 \leq t\}$ is then called the *Brownian motion process.*

A moment's reflection upon (1.1) should make us realize that we are dealing here with a process which is conceptually quite similar to both the binomial counting process and the random-walk process previously introduced in Chap. 9, for, in each case, the random variable of interest is defined to be a sum of independent random variables—the only difference being in the particular probability distribution assumed for the increments and in whether the time parameter t is discrete or continuous. Such random processes (called *independent increment processes*) occur in many physical situations and are, accordingly, of considerable practical and theoretical interest.

We shall derive some of the basic moment properties of the general independent-increment process in the next section. We shall then attack the problem of determining the probability distribution function of the sum random variable, for example, of Y_{t_n} in (1.1), and, to that end, we shall find it desirable to introduce the Fourier transform of a probability density, which we shall call the *characteristic function.* Next, we shall determine some of the properties of the characteristic function and use it to determine the probability distributions of a few sum random variables. Finally, we shall study the limiting behavior of the sum random variable as the number of increments increases without limit, thus leading to the important central limit theorem. We shall then see as one conclusion of that theorem that the Brownian motion process, that is, the Wiener process, is a gaussian random process.

2 INDEPENDENT INCREMENT PROCESSES

Let us reconsider for the moment the binomial counting process introduced in Sec. 9.2. This is the random process $\{Y_n, n = 1, 2, 3, \ldots\}$ in which the sum random variable Y_n is defined as the sum of the mutually independent increment random variables $X_1, X_2, \ldots, X_n$:

$$Y_n \triangleq \sum_{i=1}^{n} X_i$$

where the increments all have the same Bernoulli probability distribution

$$P[X_i = 1] = p \qquad \text{and} \qquad P[X_i = 0] = q \triangleq 1 - p$$

for all $i = 1, 2, \ldots, n$. We showed in Sec. 9.2 that it followed from these assumptions that the sum random variable Y_n had the binomial probability distribution

$$P[Y_n = k] = \binom{n}{k} p^k q^{n-k}$$

for $k = 0, 1, 2, \ldots, n$. We then further showed that both the mean and the variance of the process increase linearly with time; in particular, we showed that

$$E[Y_n] = np$$

(cf. (9.2.12)], and that

$$\operatorname{var}(Y_n) = npq$$

[cf. (9.2.13)]. We also studied some of the other moment properties of a pair of sum variables in that section and showed, in particular, that the covariance of the pair Y_m and Y_n is given by the formula

$$\operatorname{cov}(Y_m, Y_n) = pq \min(m,n)$$

[cf. (9.2.15)], and that the variance of their difference is

$$\operatorname{var}(Y_m - Y_n) = |m - n|pq$$

[cf. (9.2.14)]. We now wish to show that these various moment properties of the binomial counting process are simply special cases of the results to be obtained for the general independent increment process.

We say that the real random process $\{Y_t, 0 \leq t\}$ is an *independent-increment process* if for every set of time instants

$$t_0 = 0 < t_1 < t_2 < \cdots < t_n$$

the increments

$$Y_{t_1} - Y_{t_0}, Y_{t_2} - Y_{t_1}, \ldots, Y_{t_n} - Y_{t_{n-1}}$$

are mutually independent random variables. If, *in addition*, we assume that the process starts at zero, that is, if we set

$$Y_0 = 0 \tag{2.1}$$

then it follows that we may write

$$Y_{t_n} = \sum_{i=1}^{n} X_i \tag{2.2}$$

where we have denoted the increment $Y_{t_i} - Y_{t_{i-1}}$ by X_i:

$$X_i \triangleq Y_{t_1} - Y_{t_{i-1}} \tag{2.3}$$

for $i = 1, 2, \ldots, n$. We note that it follows from this definition that the increments corresponding to any two *nonoverlapping* time intervals are independent random variables.

In our preceding discussion of the binomial counting process we stated that the increments there all had the same probability distribution. A moment's reflection upon that requirement shows that an equivalent assumption in the general independent increment case is the assumption that the probability distribution function of an increment is independent of a translation in time; that is, for $0 \leq t_i < t_j$,

$$F_{Y_{t_j+\tau} - Y_{t_i+\tau}}(x) = F_{Y_{t_j} - Y_{t_i}}(x) \tag{2.4}$$

for all values of x and for all time shifts τ such that $0 \leq t_i + \tau < t_j + \tau$. An independent increment process which has this property is said to have *stationary increments;* we shall restrict our remaining discussions of independent increment processes to processes which have stationary increments.

Suppose that the real random process $\{Y_t, 0 \leq t\}$, where $Y_0 \triangleq 0$, is an independent increment process with stationary increments. We now wish to derive expressions for the mean and variance of such a process. First, we note that we may always write

$$Y_{t_1+t_2} = (Y_{t_1+t_2} - Y_{t_1}) + (Y_{t_1} - Y_0)$$

since $Y_0 = 0$ here. If, therefore, we take the expectation of both sides of this equation, we get

$$E[Y_{t_1+t_2}] = E[Y_{t_1+t_2} - Y_{t_1}] + E[Y_{t_1} - Y_0]$$

Now since $Y_0 = 0$, we have

$$E[Y_{t_1} - Y_0] = E[Y_{t_1} - 0] = E[Y_{t_1}]$$

In addition, since the given process has stationary increments by hypothesis, we have

$$\begin{aligned} E[Y_{t_1+t_2} - Y_{t_1}] &= E[Y_{t_1+t_2-t_1} - Y_{t_1-t_1}] \\ &= E[Y_{t_2} - Y_0] = E[Y_{t_2}] \end{aligned}$$

On substituting these last two results back into our formula for $E[Y_{t_1+t_2}]$, we see that *the expectation of an independent increment process with stationary increments must satisfy the linear-functional equation*

$$E[Y_{t_1+t_2}] = E[Y_{t_1}] + E[Y_{t_2}] \tag{2.5}$$

It is not too difficult to show that the variance of such a process must also satisfy a linear-functional equation.

Exercise 2.1 Let $\{Y_t, 0 \leq t\}$ be an independent-increment process with stationary increments and suppose that $Y_0 \triangleq 0$. Show that

$$\text{var}\,(Y_{t_1+t_2}) = \text{var}\,(Y_{t_1}) + \text{var}\,(Y_{t_2})$$

in this case. □

We see, therefore, that both the mean and the variance of a random process with independent and stationary increments must satisfy linear functional equations of the form

$$g(t_1 + t_2) = g(t_1) + g(t_2) \tag{2.6}$$

With the binomial-counting-process mean and variance results in mind, it seems reasonable to expect that a linearly increasing function of time will satisfy (2.6). It is, in fact, quite simple to show (cf. Sec. 3 below) that

$$g(t) = at$$

(where a is some constant) is a solution of the linear functional equation; the value of the constant a can be determined by evaluating g at time $t = 1$. Thus we find that

$$g(t) = g(1)t \tag{2.7}$$

is a solution to the linear functional equation. One can show, in fact, that this is the *only* possible solution if some additional restriction is placed upon the function g [such as continuity or boundedness on any finite interval (cf. Sec. 3)].

It follows from this last result [and from (2.5) and Exercise 2.1] that the mean of an independent increment process with stationary increments must have the form

$$E[Y_t] = mt \tag{2.8}$$

where the constant m is the value of the mean at time $t = 1$; that is,

$$m \triangleq E[Y_t]\Big|_{t=1} \tag{2.9}$$

Similarly, it follows that the variance of such a process must have the form

$$\text{var}\,(Y_t) = \sigma^2 t \tag{2.10}$$

where the constant σ^2 is the value of the variance at time $t = 1$:

$$\sigma^2 \triangleq \text{var}\,(Y_t)\Big|_{t=1} \tag{2.11}$$

The results previously obtained for the binomial counting process are then obviously special cases of these general results. We note that it follows from (2.8) and (2.10) that *a random process with stationary and independent increments cannot be a stationary random process* (in either the strict or the wide sense).

Exercise 2.2 Let $\{Y_t, 0 \le t\}$ be an independent-increment process with stationary increments and suppose that $Y_0 \triangleq 0$. Let t and t' be arbitrary positive time instants. Show that

$$\operatorname{var}(Y_t - Y_{t'}) = \sigma^2|t - t'|$$

and that

$$\operatorname{cov}(Y_t, Y_{t'}) = \sigma^2 \min(t, t')$$

where

$$\sigma^2 \triangleq \operatorname{var}(Y_t)\Big|_{t=1}$$

(*Hint:* Recall your solutions of Exercises 9.2.5 and 9.2.6) □

Exercise 2.3 Let $\{Y_t, 0 \le t\}$ be an independent-increment process with stationary increments and suppose that $Y_0 \triangleq 0$. Suppose further that the increment means are zero and hence that

$$E[Y_t] = 0 \qquad \text{for all } t \ge 0$$

Consider now the result of operating upon such a random process with a short-time averager. In particular, let $\{Z_t, 0 \le t\}$ be the output random process from such an averager:

$$Z_t \triangleq \frac{1}{t}\int_0^t Y_{t'}\, dt'$$

Show that the mean of the averager output is zero,

$$E[Z_t] = 0 \qquad \text{for all } t \ge 0$$

and that the variance of the averager output is given by the formula

$$\operatorname{var}(Z_t) = \frac{\sigma^2}{3} t$$

where again

$$\sigma^2 \triangleq \operatorname{var}(Y_t)\Big|_{t=1}$$

[*Hint:* Be careful of your regions of integration when evaluating the variance of Z_t!]

Thus we see that *the variance of the short-time averager output in response to an independent-increment process increases linearly with time,* rather than decreasing inversely with time as in the case of a stationary random process input (cf. Exercise 10.4.6). □

3 LINEAR-FUNCTIONAL EQUATION

We showed in the preceding section that both the mean and the variance of a random process with stationary and independent increments satisfy linear-functional equations, that is, equations of the form

$$g(t_1 + t_2) = g(t_1) + g(t_2)$$

We also pointed out there that one solution of this equation was the linear function

$$g(t) = at$$

where a is some constant. While such an answer seems intuitively obvious, it is interesting to note that, in fact, the linear-functional equation also has solutions which are highly discontinuous and which are unbounded in every interval and, further, that the linear solution is unique only when some restriction is placed upon the possible form of g. Although a discussion of the pathological case is beyond us here (cf. sec. 20, pp. 108–113, of Boas), we would like to demonstrate the fact that the addition of the requirement that g be bounded on some finite interval is enough to guarantee the uniqueness of the linear solution (cf. Young, 1958).

THEOREM 3.1 Let g be a real function defined on the real line and suppose that g is bounded on some interval, say, $[b,c]$, with $b < c$. Then g satisfies the linear-functional equation

$$g(t_1 + t_2) = g(t_1) + g(t_2) \tag{3.1}$$

for arbitrary real numbers t_1 and t_2, if and only if

$$g(t) = g(1)t \tag{3.2}$$

for all t.

Proof The "if" part is easy, for if g has the form in (3.2), then

$$\begin{aligned} g(t_1 + t_2) &= g(1)(t_1 + t_2) \\ &= g(1)t_1 + g(1)t_2 \\ &= g(t_1) + g(t_2) \end{aligned}$$

and hence g satisfies the linear-functional equation.

The "only if" part takes longer to prove: Suppose that g satisfies the linear-functional equation (3.1). Then we note first that

$$g(0) = 0 \tag{3.3}$$

for it follows directly from (3.1) that if we set $t_2 = 0$, we obtain

$$g(t_1) = g(t_1) + g(0)$$

If then we take $t_1 \in [b,c]$, $g(t_1)$ is bounded by hypothesis and hence may be subtracted from both sides, thus giving (3.3).

Next, we wish to show that if g is bounded on $[b,c]$, then it is also bounded on $[0,\delta]$, where

$$\delta \triangleq c - b \tag{3.4}$$

Suppose that we consider a value of t in $[0,\delta]$. Then $(t + b) \in [b,c]$ and, since g must satisfy (3.1),

$$g(t + b) = g(t) + g(b) \qquad \text{or} \qquad g(t) = g(t + b) - g(b)$$

On taking magnitudes we then get the result

$$\begin{aligned} |g(t)| &= |g(t + b) - g(b)| \\ &\leq |g(t + b)| + |g(b)| \end{aligned}$$

Now g is bounded on $[b,c]$ by hypothesis; that is, there exists a positive number M such that

$$|g(t)| \leq M < +\infty \tag{3.5}$$

for all $t \in [b,c]$. Since for $t \in [0,\delta]$, we have $(t + b) \in [b,c]$ and, in any case, $b \in [b,c]$, it follows from (3.5) that

$$|g(t + b)| \leq M \qquad \text{and} \qquad |g(b)| \leq M$$

and hence that

$$|g(t)| \leq 2M$$

for all $t \in [0,\delta]$.

Suppose now that we consider the function

$$\gamma(t) \triangleq g(t) - mt \tag{3.6}$$

where the constant m is defined by the formula

$$m \triangleq \frac{g(\delta)}{\delta} \tag{3.7}$$

It then follows from (3.6) and from (3.1) that

$$\begin{aligned}\gamma(t_1 + t_2) &= g(t_1 + t_2) - m(t_1 + t_2) \\ &= g(t_1) + g(t_2) - mt_1 - mt_2 \\ &= \gamma(t_1) + \gamma(t_2)\end{aligned} \tag{3.8}$$

Thus we see that γ also satisfies the linear-functional equation (3.1). Further, it follows from the fact that g is bounded on $[0,\delta]$ that γ is also bounded there. For, if $t \in [0,\delta]$,

$$\begin{aligned}|\gamma(t)| = |g(t) - mt| &= \left| g(t) - \frac{g(\delta)}{\delta} t \right| \\ &\leq |g(t)| + |g(\delta)| \left| \frac{t}{\delta} \right| \leq 2M\end{aligned}$$

since $\left| \frac{t}{\delta} \right| \leq 1$ for $t \in [0,\delta]$.

We note, in addition, that γ is a periodic function. For, it follows using (3.7) that

$$\begin{aligned}\gamma(\delta) &= g(\delta) - \frac{g(\delta)}{\delta} \delta \\ &= 0\end{aligned}$$

and hence, using (3.8), that

$$\begin{aligned}\gamma(t + \delta) &= \gamma(t) + \gamma(\delta) \\ &= \gamma(t)\end{aligned}$$

for all t; that is, γ is a periodic function with period δ. Since we have already shown that γ is bounded on $[0,\delta]$, it then follows that γ must be bounded everywhere.

Suppose next that t_0 is some real number such that

$$\gamma(t_0) \neq 0 \tag{3.9}$$

It then follows from (3.8) that

$$\begin{aligned}\gamma(2t_0) &= \gamma(t_0 + t_0) = \gamma(t_0) + \gamma(t_0) \\ &= 2\gamma(t_0)\end{aligned}$$

and hence, by induction, that

$$\gamma(nt_0) = n\gamma(t_0)$$

and therefore that

$$|\gamma(nt_0)| = n|\gamma(t_0)|$$

for $n = 1, 2, 3, \ldots$. If then $\gamma(t_0) \neq 0$ as assumed in (3.9), it follows that we can make $|\gamma(nt_0)|$ as large as we wish simply by choosing n large enough. This, however, contradicts our previous result that γ must be bounded for all t. Thus we see that (3.9) cannot hold true for any t_0, and hence that we must have

$$\gamma(t) = 0$$

for all t. It therefore follows from the definition of γ, that is, from (3.6), that we must also have

$$g(t) = mt$$

for all values of t. On evaluating this result at $t = 1$, we get

$$m = g(1)$$

and hence it finally follows that g must have the form

$$g(t) = g(1)t$$

for all t; thus completing the proof. ∎

4 CHARACTERISTIC FUNCTION

While it is quite sufficient in many applications to know the moment properties of independent increment processes, it is also true that in many other applications we would like to know the probability properties of such processes. In particular, given an independent increment process $\{Y_t, 0 \leq t\}$, we often need to be able to determine the individual and joint-probability distribution functions of the sum random variables Y_t from assumed probability distributions for the increments X_i, where

$$Y_{t_n} = \sum_{i=1}^{n} X_i$$

and hence where

$$X_i = Y_{t_i} - Y_{t_{i-1}}$$

for $t_0 = 0 < t_1 < t_2 < \cdots < t_n$. The basic problem before us then is the determination of the probability distribution functions of sums of independent random variables. This same problem also arises, of course, in many other contexts. For example, consider the sample mean

$$\hat{m}_n = \frac{1}{n} \sum_{i=1}^{n} X_i$$

which we introduced earlier in Sec. 8.2. If the sample random variables

X_i are mutually independent and if we wish to determine the probability distribution function of the sample mean $\hat{m}_n$, then again we are faced with the problem of determining the probability distribution function of a sum of independent random variables.

Now we have, in fact, already studied one method of dealing with this problem. We showed earlier (in Sec. 6.3) that if the random variables X_1 and X_2 are statistically independent and have the probability densities f_{X_1} and f_{X_2}, respectively, then the probability density f_{Y_2} of their sum

$$Y_2 \triangleq X_1 + X_2$$

is given by the convolution integral

$$f_{Y_2}(y) = \int_{-\infty}^{+\infty} f_{X_2}(y - x) f_{X_1}(x)\, dx \tag{4.1}$$

[cf. (6.3.11)]. With this result in mind, suppose that the random variables $X_1, X_2, \ldots, X_n$ are mutually independent random variables, that the random variables X_i has the probability density f_{X_i} (for $i = 1, 2, \ldots, n$), and that we wish to determine the probability density of the sum

$$Y_n \triangleq \sum_{i=1}^{n} X_i \tag{4.2}$$

It follows from this definition that we can write

$$\begin{aligned} Y_n &= X_n + \sum_{i=1}^{n-1} X_i \\ &= X_n + Y_{n-1} \end{aligned}$$

Since the random variables X_n and Y_{n-1} here are independent (cf. Exercise 6.3.2), we can use (4.1) and write

$$f_{Y_n}(y) = \int_{-\infty}^{+\infty} f_{X_n}(y - x) f_{Y_{n-1}}(x)\, dx$$

Successive applications of this recursion formula for values of n going back to $n = 2$ then enable us to obtain the result

$$f_{Y_n}(y) = \int_{-\infty}^{+\infty} \cdots \iint_{-\infty}^{+\infty} f_{X_n}(y - x_{n-1}) \cdots f_{X_2}(x_2 - x_1) f_{X_1}(x_1)\, dx_1\, dx_2 \cdots dx_n \tag{4.3}$$

Thus we see that the probability density of a sum of n mutually independent random variables is given by the $(n - 1)$-fold convolution of their respective probability densities.

While the result expressed by (4.3) does indeed enable us to calculate the probability density of a sum of mutually independent random variables, the evaluation of such a multiple convolution is often exceedingly tedious to carry out, and one is therefore led to ask about the possibility of other methods of attacking this problem. A key to the answer to this question is obtained by recalling that the same sort of problem arose in Chaps. 10 and 11 where we discussed the determination of the output of a cascade of stable, linear, time-invariant systems. We showed there, in fact, that the output of such a cascade could be calculated *either* from a multiple convolution *or* from the inverse Fourier transform of a product of Fourier transforms (cf. Exercises 10.5.6 and 11.3.3). With this fact in mind, let us now apply Fourier transform techniques to the present problem.

In particular, let us consider the Fourier transform ϕ_X of the probability density f_X of the random variable X, where we write

$$\phi_X(v) \triangleq \int_{-\infty}^{+\infty} f_X(x)e^{+ivx}\,dx \tag{4.4}$$

so long as the integral exists, where v is a real parameter and where $i = \sqrt{-1}$. We shall call ϕ_X the *characteristic function* of the random variable X. It then follows from our previous discussion of Fourier transforms that *the characteristic function and the probability density form a Fourier transform pair*, and hence that we may write

$$f_X(x) = \frac{1}{2\pi}\int_{-\infty}^{+\infty} \phi_X(v)e^{-ivx}\,dv \tag{4.5}$$

The factor $1/2\pi$ arises in (4.5) from the fact that we lumped a factor of 2π in with the parameter v in (4.4).

Exercise 4.1 On defining the characteristic functions of the various random variables of concern as in (4.4), show that

$$\phi_{Y_2}(v) = \phi_{X_1}(v)\phi_{X_2}(v)$$

is the characteristic function equivalent of the convolution integral formula in (4.1) □

Exercise 4.2 Again let Y_n be the sum of n mutually independent random variables X_k and let ϕ_{X_k} be the characteristic function of X_k. Show that

$$\phi_{Y_n}(v) = \prod_{k=1}^{n} \phi_{X_k}(v)$$

is the characteristic function equivalent of (4.3).

Thus we see again that the use of Fourier transforms enables us to replace a multiple convolution in one domain by a multiple product in another domain. □

Exercise 4.3 Let X be a real random variable with the probability density f_X and the characteristic function ϕ_X. Suppose that the new random variable Y is defined by

$$Y \triangleq aX + b$$

where a and b are real constants and where a is nonzero. We showed previously (Exercise 6.2.3) that the probability density of Y is given by the formula

$$f_Y(y) = \frac{1}{|a|} f_X\left(\frac{y-b}{a}\right)$$

in this case. Show that it therefore follows that Y has the characteristic function

$$\phi_Y(v) = e^{ivb}\phi_X(av)$$

(We will show a much simpler way of deriving this result in the next section.) □

Exercise 4.4 Let $\hat{m}_n$ be the sample mean:

$$\hat{m}_n \triangleq \frac{1}{n}\sum_{k=1}^{n} X_k$$

and suppose that the random variables X_k are mutually independent and all have the same probability density f_X. Using the result of the preceding exercise, show that

$$\phi_{\hat{m}_n}(v) = \phi_X{}^n\left(\frac{v}{n}\right)$$

where ϕ_X is the characteristic function which corresponds to f_X. □

Exercise 4.5* Let X be a gaussian random variable with a mean $E[X]$ and a variance $\sigma^2(X)$:

$$f_X(x) = \frac{1}{\sqrt{2\pi}\,\sigma(X)} \exp\left[-\frac{(x-E[X])^2}{2\sigma^2(X)}\right]$$

Show that the characteristic function of such a random variable is

$$\phi_X(v) = \exp\left[ivE[X] - \frac{v^2\sigma^2(X)}{2}\right]$$

(*Hint:* Contour integration is required.) □

* Supplementary exercise. We shall, however, often need to use the result of this exercise in the remainder of the text.

Exercise 4.6 Using the result of Exercise 4.5, show that if Y is a sum of n mutually independent gaussian random variables X_k, then the characteristic function of Y has the form

$$\phi_Y(v) = \exp\left[ivm - \frac{v^2\sigma^2}{2}\right]$$

where

$$m \triangleq \sum_{k=1}^{n} m_k \qquad \text{and} \qquad \sigma^2 \triangleq \sum_{k=1}^{n} \sigma_k^2$$

and where m_k and σ_k^2 are the mean and variance, respectively, of the random variable X_k.

Thus we see that a sum of mutually independent gaussian random variables is also a gaussian random variable, that the mean of the sum is the sum of the respective means, and that the variance of the sum is the sum of the respective variances (cf. Exercise 6.3.4). □

Exercise 4.7 Let X be a Cauchy random variable with the probability density

$$f_X(x) = \frac{1}{\pi(1 + x^2)}$$

Show that the characteristic function of X is

$$\phi_X(v) = e^{-|v|}$$

in this case. □

Exercise 4.8 Suppose that the random variables $X_1, X_2, \ldots, X_n$ are mutually independent Cauchy random variables (each with a probability density as given in Exercise 4.7), and suppose $\hat{m}_n$ is the sample mean formed from those n samples.

a. Show that in this case

$$\phi_{\hat{m}_n}(v) = e^{-|v|}$$

that is, the sample mean here has exactly the same characteristic function as does each of its component sample random variables. Thus it follows that the probability density of the sample mean is

$$f_{\hat{m}_n}(z) = \frac{1}{\pi(1 + z^2)}$$

b. How do these results fit into the previous discussion (in Sec. 8.2) of the sample mean made from mutually independent samples? □

A moment's reflection upon the gaussian and Cauchy random variable examples above shows that these are both special cases of the situation in which a random variable (say X) has a characteristic function of the form

$$\phi_X(v) = a^{f(v)} \tag{4.6}$$

where a is some constant and where $f(v)$ is some function of the real variable v. If now we consider a sum Y of n such random variables which are mutually independent and which all have the same probability density, then it follows from the result of Exercise 4.2 that we get

$$\phi_Y(v) = a^{nf(v)} \tag{4.7}$$

as the characteristic function of the sum. That is, the characteristic function of such a sum has, except for a change of scale, the same functional form as does the characteristic function of a component of the sum.

Exercise 4.9 Let X be a random variable with the probability density

$$f_X(x) = \begin{cases} \dfrac{x^{(n-2)/2}e^{-x/2}}{2^{n/2}\Gamma\left(\dfrac{n}{2}\right)} & \text{for } 0 \leq x \\ 0 & \text{for } x < 0 \end{cases}$$

where n is a nonnegative integer and Γ is the gamma function. This probability density is often written as a function of $\chi^2 = x$ and called a *chi-squared* probability density of n *degrees of freedom.*

a. Show that the mean of X is

$$E[X] = n$$

b. Show that the characteristic function of X is

$$\phi_X(v) = (1 - i2v)^{-n/2}$$

in this case. □

Exercise 4.10 Let X_1 and X_2 be statistically independent random variables each with a chi-squared probability density, and let n_1 and n_2 be the numbers of degrees of freedom of X_1 and X_2, respectively. Further, let Y be the sum of X_1 and X_2. Show that the characteristic function of the sum then is

$$\phi_Y(v) = (1 - i2v)^{-n/2}$$

where $n = n_1 + n_2$.

A sum of two independent chi-squared random variables is also a chi-squared random variable, and the number of degrees of freedom of the sum is equal to the sum of the numbers of degrees of freedom of the components. □

Exercise 4.11 Let X be a gaussian random variable with a zero mean and a variance of unity. Suppose that a new random variable is defined as the square of X:

$$Y \triangleq X^2$$

Show that in this case the characteristic function of Y is

$$\phi_Y(v) = (1 - i2v)^{-1/2}$$

It then follows from this result, and from that in Exercise 4.9*b* above, that *the square of a gaussian random variable has a chi-squared probability density with one degree of freedom* (cf. Exercise 6.1.2). □

Exercise 4.12 Let the random variables $X_1, X_2, \ldots, X_n$ be mutually independent gaussian random variables, each with a mean of zero and a variance of unity. Suppose that a new random variable Y is defined as the sum of the squares of the random variables X_k:

$$Y \triangleq \sum_{k=1}^{n} X_k^2$$

Show that the characteristic function of this sum is

$$\phi_Y(v) = (1 - i2v)^{-n/2}$$

Thus we see that *a sum of n mutually independent squared gaussian random variables is a chi-squared random variable with n degrees of freedom.*

□

5 FURTHER PROPERTIES OF THE CHARACTERISTIC FUNCTION

Having defined the characteristic function of a random variable X as the Fourier transform of its probability density f_X, in particular, as

$$\phi_X(v) \triangleq \int_{-\infty}^{+\infty} f_X(x)e^{ivx}\,dx \tag{5.1}$$

and having used it to help determine the probability density of a sum of mutually independent random variables in several specific cases, let us now determine some of its further properties. Much more complete discussions of the characteristic function than we can give here, however, will be found in chap. 6 of Chung and in the monograph by Lukacs.

A moment's reflection upon (5.1) should make one realize that the characteristic function defined there is simply the expectation of e^{ivX}:

$$\phi_X(v) = E[e^{ivX}] \tag{5.2}$$

where, as usual, we assume the parameter v to be real. Now while the Fourier transform definition (5.1) has meaning only in those cases in which the random variable in question has a probability density, the expectation formula (5.2) can obviously be applied in any case. With this fact in mind, *we shall from here on use* (5.2) *as the basic definition of the characteristic function.*

Once we have defined the characteristic function as an expectation, we can not only extend our study to more general cases, but we can, of course, make full use of our previously derived results and, in fact, simplify our derivations in many cases. Consider, for example, the problem of determining the characteristic function of the random variable Y defined in terms of the random variable X by the formula

$$Y \triangleq aX + b$$

where a and b are real constants. Using (5.2) we have

$$\phi_Y(v) = E[e^{ivY}]$$

It then follows from the properties of the expectation that we may write

$$\begin{aligned}\phi_Y(v) &= E[e^{iv(aX+b)}] = e^{ivb}E[e^{i(av)X}] \\ &= e^{ivb}\phi_X(av)\end{aligned}$$

Thus we see that the result previously obtained in Exercise 4.3 for the case in which X has a probability density holds, in fact, in any case. We note, in addition, that our derivation here is much simpler than that required in Exercise 4.3.

One of the apparent advantages in viewing the characteristic function as a Fourier transform is the Fourier transform-pair relation then obtained between the characteristic function and the probability density. The ability to transform back and forth between the probability and characteristic function domains holds in any case, however. Consider, for example, the case in which a random variable X is a discrete random variable with the possible values x_k and the corresponding probabilities $P[X = x_k]$. The characteristic function of X is then given by the formula

$$\phi_X(v) = \sum_k P[X = x_k]e^{ivx_k} \tag{5.3}$$

in this case, where the summation is taken over all values of the index k. The inversion problem here, that is, the problem of going from the characteristic function to the probability distribution, is solved essentially by inspection. If we are given an expression for the characteristic function of a discrete random variable, then the probability distribution of that random variable may be found by expanding the characteristic function expression as a series[1] in e^{iv} as

$$\phi_X(v) = \sum_k a_k e^{ivx_k}$$

and then identifying the coefficients a_k with the probabilities $P[X = x_k]$. This technique is illustrated by the following exercise.

Exercise 5.1 Let X be a Poisson random variable, X having the possible values $k = 0, 1, 2, \ldots$, with the probabilities

$$P[X = k] = \frac{\lambda^k e^{-\lambda}}{k!} \tag{5.4}$$

where λ is a positive constant.

a. Show that in this case

$$\phi_X(v) = \exp[\lambda(e^{iv} - 1)] \tag{5.5}$$

b. Invert (5.5) by expanding $\exp \lambda e^{iv}$ in a power series and thus get (5.4) back again. □

Exercise 5.2 Let Y be a sum of n mutually independent Poisson random variables, each with the same parameter λ. Show that

$$\phi_Y(v) = \exp[n\lambda(e^{iv} - 1)]$$

in the characteristic function of the sum.

Thus we see that *a sum of mutually independent Poisson random variables is also a Poisson random variable.* □

Any time that we define a function in terms of an integral, as in (5.1), or in terms of a series, as in (5.3), the question of existence always arises. With this in mind, suppose that we take the magnitude of both sides of (5.2). We then get

$$\begin{aligned} |\phi_X(v)| &= |E[e^{ivX}]| \\ &\leq E[|e^{ivX}|] \end{aligned}$$

the inequality following from (7.6.4). Since the magnitude of e^{ivX} is unity, and since the expectation of unity is unity, it therefore follows that we obtain the inequality

$$|\phi_X(v)| \leq 1 \tag{5.6}$$

[1] Called a Dirichlet series after the German mathematician Peter Gustav Lejeune Dirichlet (1805–1859).

for all values of v. Thus we see that the characteristic function always exists, no matter whether the random variable X is discrete, has a probability density, or is a mixed random variable. The inequality (5.6) may be rewritten in a more interesting form when we realize that

$$\phi_X(0) = 1 \tag{5.7}$$

We therefore obtain the inequality

$$|\phi_X(v)| \leq \phi_X(0) = 1 \tag{5.8}$$

for all values of v; that is, *the magnitude of the characteristic function $\phi_X(v)$ attains its maximum value of unity at the origin in v.*

Moment-generating property

We shall assume throughout the following discussion that the random variables in question all have probability densities. This is done for convenience only; similar derivations may also be carried out easily for the case of discrete random variables.

If the random variable X has the probability density f_X, the characteristic function of X has the form

$$\phi_X(v) = \int_{-\infty}^{+\infty} e^{ivx} f_X(x)\, dx$$

Suppose now that we may differentiate this integral with respect to the parameter v; we then get

$$\phi_X'(v) \triangleq \frac{d\phi_X(v)}{dv} = i \int_{-\infty}^{+\infty} x e^{ivx} f_X(x)\, dx$$

If we evaluate this derivative at the origin in v, we get

$$\phi_X'(0) = i \int_{-\infty}^{+\infty} x f_X(x)\, dx = iE[X]$$

Thus it follows that we obtain the result

$$E[X] = -i\phi_X'(0) \tag{5.9}$$

We see therefore that *the expectation of a random variable X may be obtained from the derivative of the characteristic function of X whenever that derivative exists for $v = 0$.* This result may easily be extended to higher-order moments.

Exercise 5.3 Let X be a random variable with the characteristic function ϕ_X. Show that

$$E[X^k] = (-i)^k \phi_X{}^{(k)}(0)$$

whenever the kth derivative $\phi_X{}^{(k)}$ exists at the origin in v. □

The result derived in Exercise 5.3 describes the *moment-generating property* of the characteristic function. The calculation of the kth moment of a random variable by taking the kth derivative of its characteristic function is often far simpler to carry out than is the evaluation of the integral which appears in our original definition of that moment.

Exercise 5.4 Let X be a gaussian random variable with a mean of zero and a variance σ^2:

$$f_X(x) = \frac{1}{\sqrt{2\pi}\,\sigma} e^{-x^2/2\sigma^2}$$

Using the result of Exercise 5.3, calculate the first four moments of X (cf. Exercise 7.8.8). □

Suppose finally that the Taylor series expansion of the characteristic function exists throughout some interval in v which contains the origin and that we may write

$$\phi_X(v) = \sum_{k=0}^{\infty} \phi_X{}^{(k)}(0) \frac{v^k}{k!}$$

for all v in such an interval. It then follows from the result of Exercise 5.3 that we may also write

$$\phi_X(v) = \sum_{k=0}^{\infty} E[X^k] \frac{(iv)^k}{k!} \tag{5.10}$$

Thus we see that *if the characteristic function $\phi_X(v)$ of a given random variable X has a Taylor series expansion which is valid in some interval in v which contains the origin, then that characteristic function (and hence the corresponding probability density or probability distribution) is uniquely determined by the moments of the given random variable.* If, on the other hand, only the first n moments of the given random variable exist (if only the first n derivatives of the characteristic function exist at the origin in v), then the characteristic function of that random variable may be expanded in a Taylor polynomial form with a remainder of order n (cf. sec. 6.4 of Chung or sec. 2.3 of Lukacs).

Moment-generating functions

A function which is closely related to the characteristic function, and which often appears in the statistical literature, is the *moment-generating function* M_X defined by

$$M_X(u) \triangleq E[e^{uX}] \tag{5.11}$$

where u is taken to be a real variable. As there is no a priori necessary bound upon the magnitude of the right-hand side of (5.11), a moment-generating function need not exist for every random variable. If, however, all the moments of X do exist, then the moment-generating function also exists.

Exercise 5.5 Let X be a random variable whose moments all exist, and suppose that X has the probability density f_X. Show that

$$E[X^k] = M_X^{(k)}(0)$$

in this case, for all $k = 1, 2, 3, \ldots$. □

6 JOINT–CHARACTERISTIC FUNCTIONS

The results of the preceding two sections concerning the characteristic function of a single random variable may now be extended to the random vector case. Let us consider, for example, an n-dimensional random vector $\mathbf{X}$ with the component random variables $X_1, X_2, \ldots, X_n$. The *joint-characteristic function* of that n-dimensional random vector (and hence of those n random variables) is defined by the formula

$$\phi_{X_1,X_2,\ldots,X_n}(v_1, v_2, \ldots, v_n) \triangleq E\left[\exp\left(i \sum_{k=1}^{n} v_k X_k\right)\right] \tag{6.1}$$

As usual, it is convenient to use matrix notation when dealing with random vectors so as to make our expressions more compact. Thus if we define the column matrices

$$\mathbf{v} \triangleq \begin{bmatrix} v_1 \\ v_2 \\ \cdot \\ \cdot \\ \cdot \\ v_n \end{bmatrix} \qquad \mathbf{X} \triangleq \begin{bmatrix} X_1 \\ X_2 \\ \cdot \\ \cdot \\ \cdot \\ X_n \end{bmatrix}$$

then it follows that we have

$$\mathbf{v}^T\mathbf{X} = \sum_{k=1}^{n} v_k X_k$$

where $\mathbf{v}^T$ is the matrix transpose of $\mathbf{v}$. Thus we may write

$$\phi_{\mathbf{X}}(\mathbf{v}) = E[\exp(i\mathbf{v}^T\mathbf{X})] \tag{6.2}$$

as the matrix equivalent of (6.1).

The properties of the joint-characteristic function generally parallel those of the characteristic function of a single random variable. For example, we note on defining the zero vector $\mathbf{0}$ to be the n-dimensional vector whose components are all zero, that it follows from (6.2) that

$$\phi_{\mathbf{X}}(\mathbf{0}) = 1$$

We also note that, if we take the magnitude of both sides of (6.2), we get

$$|\phi_{\mathbf{X}}(\mathbf{v})| \leq E[|\exp i\mathbf{v}^T\mathbf{X})|] \\ = 1$$

and hence that we may write

$$|\phi_{\mathbf{X}}(\mathbf{v})| \leq \phi_{\mathbf{X}}(\mathbf{0}) = 1 \tag{6.3}$$

paralleling the single variable case. Thus we see that the joint-characteristic function always exists and that it has its greatest magnitude at the origin of the v space.

Suppose next that the n-dimensional random vector $\mathbf{X}$ has the joint-probability density

$$f_{\mathbf{X}} = f_{X_1,X_2,\ldots,X_n}$$

It then follows from the definition of the joint-characteristic function that we may write

$$\phi_{\mathbf{X}}(\mathbf{v}) = \int \exp\,(i\mathbf{v}^T\mathbf{x})f_{\mathbf{X}}(\mathbf{x})\,d\mathbf{x} \tag{6.4}$$

where $\mathbf{x}$ is the n-dimensional vector with components $x_1, x_2, \ldots, x_n$ and where the integration is done over the entire n-dimensional range space of the random vector $\mathbf{X}$. The component version of (6.4) is

$$\phi_{X_1,X_2,\ldots,X_n}(v_1, v_2, \ldots, v_n) = \int_{-\infty}^{+\infty} \cdots \iint_{-\infty}^{+\infty} \exp\left(i \sum_{k=1}^{n} v_k x_k\right) \\ f_{X_1,X_2,\ldots,X_n}(x_1, x_2, \ldots, x_n)\,dx_1\,dx_2 \cdots dx_n \tag{6.5}$$

Thus we see that the joint-characteristic function and the joint-probability density form an n-dimensional Fourier transform pair (cf. art. 10.6 of Cramér). It therefore follows that the joint-probability density is given by the n-dimensional inverse Fourier transform

$$f_{X_1,X_2,\ldots,X_n}(x_1, x_2, \ldots, x_n) = \frac{1}{(2\pi)^n} \int_{-\infty}^{+\infty} \cdots \iint_{-\infty}^{+\infty} \exp\left(-i \sum_{k=1}^{n} v_k x_k\right) \\ \phi_{X_1,X_2,\ldots,X_n}(v_1, v_2, \ldots, v_n)\,dv_1\,dv_2 \cdots dv_n \tag{6.6}$$

which has the matrix version

$$f_{\mathbf{X}}(\mathbf{x}) = \frac{1}{(2\pi)^n} \int \exp\,(-\mathbf{v}^T\mathbf{x})\phi_{\mathbf{X}}(\mathbf{v})\,d\mathbf{v} \tag{6.7}$$

where the integration is performed over the entire n-dimensional $\mathbf{v}$ space.

Exercise 6.1 Let $\mathbf{X}$ be an n-dimensional random vector with the components $X_1, X_2, \ldots, X_n$. Let $\phi_{\mathbf{X}}$ be the joint-characteristic function of $\mathbf{X}$ and suppose that $\mathbf{X}$ has an n-dimensional joint-probability density $f_{\mathbf{X}}$.

Show that the joint-characteristic function satisfies the formula

$$\phi_{X_1,X_2,\ldots,X_n}(v_1, v_2, \ldots, v_n) = \prod_{k=1}^{n} \phi_{X_k}(v_k)$$

if and only if the X_k are mutually independent random variables; thus we obtain another way of characterizing mutually independent random variables. (Note the difference between the above result and that previously obtained in Exercise 4.2.) □

One property of the joint-characteristic function which has no parallel in the single random variable case is the property that the joint-characteristic function of any subset of m of the random variables $X_1, X_2, \ldots, X_n$ $(n > m)$ may easily be obtained from the joint-characteristic function of those n random variables simply by setting equal to zero the particular parameters v_k which correspond to the unwanted random variables. Suppose, for example, that we wish to obtain the joint-characteristic function of the random variables $X_1, X_2, \ldots, X_m$ from that of the random variables $X_1, X_2, \ldots, X_n$, where $m < n$. It now follows from (6.1) that

$$\begin{aligned}\phi_{X_1,X_2,\ldots,X_m,X_{m+1},\ldots,X_n}(v_1, v_2, \ldots, v_m, 0, \ldots, 0) &= E\left[\exp\left(i \sum_{k=1}^{m} v_k X_k\right)\right] \\ &= \phi_{X_1,X_2,\ldots,X_m}(v_1, v_2, \ldots, v_m) \qquad (6.8)\end{aligned}$$

Thus we see that setting $v_k = 0$ in a joint-characteristic function is the Fourier transform equivalent of integrating out the variable x_k from the corresponding joint-probability density—and obviously a lot easier!

Moment-generating properties

Finally, we note that (as in the single random variable case) the joint-characteristic function has moment-generating properties.

Exercise 6.2 Let $\mathbf{X}$ be a two-dimensional random vector with the components X_1 and X_2. Let ϕ_{X_1,X_2} be the joint-characteristic function of X_1 and X_2, and suppose that $\mathbf{X}$ has the joint-probability density f_{X_1,X_2}. Show that

$$E[X_1{}^m X_2{}^k] = (-i)^{m+k} \left. \frac{\partial^{m+k} \phi_{X_1,X_2}(v_1,v_2)}{\partial v_1{}^m \, \partial v_2{}^k} \right|_{v_1=0=v_2}$$

for any nonnegative integers m and k, whenever the indicated derivative exists.

Similar results may be obtained, of course, in the n-dimensional case. □

Taylor series expansions of the joint characteristic function may also be obtained whenever the various joint moments exist. Suppose, for example, that ϕ_{X_1,X_2} is the joint-characteristic function of the pair of random variables X_1 and X_2, and suppose further that the joint moments $E[X_1{}^m X_2{}^k]$ exist for all nonnegative integer values of m and k. Let us suppose also, for convenience only, that the joint-probability density f_{X_1,X_2} exists. We may then write

$$\phi_{X_1,X_2}(v_1,v_2) = \iint_{-\infty}^{+\infty} \exp\,(iv_1x_1 + iv_2x_2) f_{X_1,X_2}(x_1x_2)\,dx_1\,dx_2$$

If now we expand the exponentials in this formula in power-series form, we get

$$\begin{aligned}\phi_{X_1,X_2}(v_1,v_2) &= \iint_{-\infty}^{+\infty} \left[\sum_{m=0}^{\infty} \frac{(iv_1x_1)^m}{m!}\right]\left[\sum_{k=0}^{\infty} \frac{(iv_2x_2)^k}{k!}\right] f_{X_1,X_2}(x_1,x_2)\,dx_1\,dx_2 \\ &= \sum_{m=0}^{\infty}\sum_{k=0}^{\infty} \frac{(iv_1)^m}{m!}\frac{(iv_2)^k}{k!} \iint_{-\infty}^{+\infty} x_1{}^m x_2{}^k f_{X_1,X_2}(x_1,x_2)\,dx_1\,dx_2 \\ &= \sum_{m=0}^{\infty}\sum_{k=0}^{\infty} E[X_1{}^m X_2{}^k] \frac{(iv_1)^m}{m!}\frac{(iv_2)^k}{k!} \end{aligned} \tag{6.9}$$

whenever the various joint moments all exist. This is a two-dimensional Taylor series expansion of ϕ_{X_1,X_2}; similar expansions may also be obtained in the n-dimensional case.

Exercise 6.3 Let **X** be a two-dimensional random vector with the component random variables X_1 and X_2, and suppose that the joint-characteristic function $\phi_{X_1,X_2}(v_1,v_2)$ has a two-dimensional Taylor series expansion which is valid in some nonvanishing region of the v_1v_2 plane which contains the origin. Show that, in this case, X_1 and X_2 are independent random variables if and only if *all* their joint moments factor, that is, if and only if

$$E[X_1{}^m X_2{}^k] = E[X_1{}^m]E[X_2{}^k]$$

for all nonnegative integer values of m and k.

It should be noted when studying this result that saying that X_1 and X_2 are uncorrelated random variables guarantees only that the moment $E[X_1X_2]$ factors (cf. Exercise 7.9.5). □

7 INDEPENDENT INCREMENT PROCESSES (CONTINUED)

Let us now return to our study of independent increment processes. In particular, suppose that the real random process $\{Y_t, 0 \le t\}$ is a random process with stationary and independent increments, that $Y_0 = 0$, that we are given the probability properties of the increments, and that we wish to derive the probability properties of the sum random variable Y_t from those of the increments. Since we are concerned here with sums of mutually independent random variables, it seems natural to make use of characteristic functions.

Suppose that we consider the $n + 1$ time instants

$$t_0 = 0 < t_1 < t_2 < \cdots < t_n$$

The increments

$$X_k \triangleq Y_{t_k} - Y_{t_{k-1}} \tag{7.1}$$

$i = 1, 2, \ldots, n$, are then mutually independent random variables (by hypothesis), and we have

$$Y_{t_n} = \sum_{k=1}^{n} X_k \tag{7.2}$$

as usual (since $Y_0 = 0$). Suppose also that either we are given characteristic functions ϕ_{X_k}, $k = 1, 2, \ldots, n$ or that we are given the probability properties of the increments so that we may calculate those characteristic functions. The characteristic function of the sum random variable Y_{t_n} is, by definition,

$$\phi_{Y_{t_n}}(v) = E[\exp(ivY_{t_n})]$$

It therefore follows from (7.2) that

$$\phi_{Y_{t_n}}(v) = E\left[\exp\left(iv \sum_{k=1}^{n} X_k\right)\right]$$

$$= E\left[\prod_{k=1}^{n} e^{ivX_k}\right]$$

Since the increments X_k are mutually independent here, the expectation of the product factors and thus we finally obtain

$$\phi_{Y_{t_n}}(v) = \prod_{k=1}^{n} \phi_{X_k}(v) \tag{7.3}$$

as a formula for the characteristic function of the sum random variable in terms of the given characteristic functions of the increments.

Suppose next that we wish to determine the joint-characteristic function of the n-dimensional random vector $\mathbf{Y}$ whose components are the sum random variables $Y_{t_1}, Y_{t_2}, \ldots, Y_{t_n}$. Now it follows from (6.1) that

$$\phi_{\mathbf{Y}}(\mathbf{v}) = \phi_{Y_{t_1},Y_{t_2},\ldots,Y_{t_n}}(v_1, v_2, \ldots, v_n)$$
$$= E\left[\exp\left(i \sum_{k=1}^{n} v_k Y_{t_k}\right)\right]$$

If, therefore, we express the sum random variable Y_{t_k} in terms of the increments, we may write

$$\phi_{\mathbf{Y}}(\mathbf{v}) = E\left[\exp\left[i \sum_{k=1}^{n} v_k \left(\sum_{j=1}^{k} X_j\right)\right]\right] \tag{7.4}$$

Let us now expand the summation in the argument of the exponential in (7.4). We get in this way the formula

$$\sum_{k=1}^{n} v_k \left(\sum_{j=1}^{k} X_j\right) = v_1X_1 + v_2(X_1 + X_2) + v_3(X_1 + X_2 + X_3) + \cdots + v_n(X_1 + X_2 + \cdots + X_n)$$

which, on rearranging terms, becomes

$$\sum_{k=1}^{n} v_k \left(\sum_{j=1}^{k} X_j\right) = X_1(v_1 + v_2 + \cdots + v_n) + X_2(v_2 + v_2 + \cdots + v_n) + \cdots + X_nv_n = \sum_{k=1}^{n} X_k \left(\sum_{j=k}^{n} v_j\right)$$

If, therefore, we use this result in (7.4) we get

$$\phi_{\mathbf{Y}}(\mathbf{v}) = E\left[\exp\left[i \sum_{k=1}^{n} X_k \left(\sum_{j=k}^{n} v_j\right)\right]\right]$$
$$= E\left[\prod_{k=1}^{n} \exp\left[i \left(\sum_{j=k}^{n} v_j\right) X_k\right]\right]$$

Again we note that the increments X_k are mutually independent random variables and hence that the expectation of the product factors. Thus we finally obtain the formula

$$\phi_{\mathbf{Y}}(\mathbf{v}) = \prod_{k=1}^{n} E\left[\exp\left[i \left(\sum_{j=k}^{n} v_j\right) X_k\right]\right]$$
$$= \prod_{k=1}^{n} \phi_{X_k}\left(\sum_{j=k}^{n} v_j\right) \tag{7.5}$$

for the joint-characteristic function of the n-dimensional random vector $\mathbf{Y}$ in terms of the given characteristic functions of the increments X_k.

Wiener process

Let us now apply our above general independent increment results to a study of the *Wiener process.* This is the real random process $\{Y_t, 0 \leq t\}$ such that

a. $Y_0 = 0$
b. The process has stationary and independent increments, and
c. For all $t > t' \geq 0$, the increment $Y_t - Y_{t'}$, has a gaussian probability density with a zero mean; that is, the probability density of the increment is[1]

$$f_{Y_t - Y_{t'}}(y) = \frac{1}{\sqrt{2\pi}\,\sigma(Y_t - Y_{t'})} \exp\left(-\frac{y^2}{2\sigma^2(Y_t - Y_{t'})}\right) \tag{7.6}$$

First, we note that since the Wiener process is a random process with stationary and independent increments, it follows from our previous discussion of such processes (Sec. 2) that

$$\text{var}\,(Y_t) = \sigma^2 t \tag{7.7}$$

that for any two time instants t_1 and t_2,

$$\text{var}\,(Y_{t_1} - Y_{t_2}) = \sigma^2 |t_1 - t_2| \tag{7.8}$$

and that

$$\text{cov}\,(Y_{t_1}, Y_{t_2}) = \sigma^2 \min\,(t_1, t_2) \tag{7.9}$$

where

$$\sigma^2 \triangleq \text{var}\,(Y_t)\Big|_{t=1} \tag{7.10}$$

Exercise 7.1 Let $\{Y_t, 0 \leq t\}$ be a Wiener process as defined above. Show that it follows, from the definition of the Wiener process, that Y_t has a gaussian probability density with a zero mean and a variance of $\sigma^2 t$:

$$f_{Y_t}(y) = \frac{1}{\sqrt{2\pi t}\,\sigma} \exp\left(-\frac{y^2}{2\sigma^2 t}\right)$$

where σ is the positive square root of σ^2. □

[1] The validity of this gaussian assumption for the increments of a Wiener process will be discussed later in Sec. 9.

Exercise 7.2 Let $\{Y_t, 0 \leq t\}$ be a Wiener process as defined above, and consider the two time instants t_1 and t_2 such that

$$0 \leq t_1 < t_2$$

Show that the joint-characteristic function of the sum random variable Y_{t_1} and Y_{t_2} is

$$\phi_{Y_{t_1},Y_{t_2}}(v_1,v_2) = \exp\left[-\frac{\sigma^2}{2}(v_1{}^2t_1 + 2v_1v_2t_1 + v_2{}^2t_2)\right]$$

in this case. □

Since we assumed that $t_2 > t_1$ in Exercise 7.2, it follows that

$$\min(t_1,t_2) = t_1$$

there, and hence [from (7.9)] that

$$\operatorname{cov}(Y_{t_1},Y_{t_2}) = \sigma^2 t_1$$

It therefore follows that the joint-characteristic function formula obtained in that exercise may be rewritten in the form

$$\phi_{Y_{t_1},Y_{t_2}}(v_1,v_2) = \exp\left[-\frac{v_1{}^2\operatorname{var}(Y_{t_1}) + 2v_1v_2\operatorname{cov}(Y_{t_1},Y_{t_2}) + v_2{}^2\operatorname{var}(Y_{t_2})}{2}\right]$$

As we shall see in our subsequent discussion of the gaussian random process (Chap. 14), this last result is the joint-characteristic function of two jointly gaussian, zero-mean random variables. An extension of the result in Exercise 7.2 to the case of the n random variables Y_{t_1}, Y_{t_2}, . . . , Y_{t_n} would show that those random variables are also jointly gaussian random variables, and hence (cf. Sec. 9.4) that the Wiener process is a gaussian random process. In particular, *a Wiener process is a nonstationary gaussian random process with a zero mean and a variance which increases linearly with time.*

8 PROBABILITY GENERATING FUNCTIONS

In many problems of interest the random variables of concern can assume only nonnegative integer values, for example, 0 and 1 in the Bernoulli process and 0, 1, 2, . . . in the binomial counting process. In such cases, the characteristic function of a given random variable (say X) has the form

$$\phi_X(v) = \sum_k P[X = k]e^{ivk}$$
$$= \sum_k P[X = k](e^{iv})^k$$

where the summation is over all appropriate values of the integer k. A moment's reflection upon this result shows that the summation on the right-hand side is a power series (or polynomial) in e^{iv}. With this fact in mind, it is convenient to introduce the complex variable

$$z \triangleq e^{iv}$$

and to denote the resultant function of z by $\psi_X(z)$:

$$\psi_X(z) \triangleq \sum_k P[X = k]z^k$$
$$= E[z^X] \tag{8.1}$$

where now, in general, we allow z to assume any complex value. The function ψ_X so defined is the *z transform* of the probability distribution of the integer-valued random variable X and is called the *probability generating function* of that random variable. This latter name arises from the fact that given ψ_X as a function of z, the expansion of that function in a power series (or polynomial) in z gives the probability $P[X = k]$ directly as the coefficient of z^k. It is often analytically more convenient to use the probability generating function than it is to use the corresponding characteristic function.

Exercise 8.1 Suppose that the n random variables $X_1, X_2, \ldots, X_n$ are mutually independent Bernoulli random variables, with

$$P[X_j = 1] = p \qquad \text{and} \qquad P[X_j = 0] = q \triangleq 1 - p$$

for all $j = 1, 2, \ldots, n$.

a. Show that the probability generating function of the random variable X_j is

$$\psi_{X_j}(z) = pz + q$$

for all $j = 1, 2, \ldots, n$.

b. Consider the sum random variable

$$Y_n \triangleq \sum_{j=1}^{n} X_j$$

Show that the probability generating function of the sum random variable is

$$\psi_{Y_n}(z) = (pz + q)^n$$

c. Using the result in (b), show that the sum random variable has the probability distribution

$$P[Y_n = k] = \binom{n}{k} p^k q^{n-k}$$

for $k = 0, 1, 2, \ldots, n$. That is, the sum random variable has a binomial probability distribution (cf. Sec. 9.2). □

Let us now determine a few of the properties of the probability generating function (further developments may be found in chapter 11 of Feller, vol. I, 3d ed.). If we take the absolute value of the probability generating function, we get from (8.1)

$$\begin{aligned} |\psi_X(z)| &= \left| \sum_k P[X = k]z^k \right| \\ &\leq \sum_k P[X = k]|z|^k \end{aligned} \tag{8.2}$$

Now we know that we must have

$$\sum_k P[X = k] = 1$$

in any case, hence it follows that the summation on the right-most side of (8.2) converges (to a value less than or equal to unity) for all values of z such that $|z| \leq 1$. That is, the probability generating function $\psi_X(z)$ of a nonnegative, integer-valued random variable exists at least for those values of z such that $|z| \leq 1$.

Suppose next that we consider differentiation of the probability generating function with respect to the variable z. On taking the first derivative of ψ_X we get, from (8.1)

$$\begin{aligned} \psi'_X(z) &\triangleq \frac{d\psi_X(z)}{dz} \\ &= \sum_k kP[X = k]z^{k-1} \end{aligned}$$

If then we evaluate this result at the point $z = 1$ we obtain

$$\begin{aligned} \psi'_X(1) &= \sum_k kP[X = k] \\ &= E[X] \end{aligned} \tag{8.3}$$

Thus we see that we may calculate the mean value of an integer-valued random variable by taking the derivative of its probability generating function and evaluating that derivative at the point $z = 1$. However:

Exercise 8.2 Let X be a nonnegative, integer-valued random variable with a probability generating function ψ_X. Show that

$$E[X(X-1) \cdots (X-n+1)] = \psi_X^{(n)}(1) = \left.\frac{d^n\psi_X(z)}{dz^n}\right|_{z=1}$$

whenever the indicated derivative exists. It is this property that sometimes gives ψ_X the name *factorial-moment-generating function.* □

Joint-probability generating functions

Now just as we extended the definition of the characteristic function to cover the case of random vectors, so can we similarly extend the definition of the probability generating function. Thus, if $\mathbf{X}$ is an n-dimensional random vector with the components $X_1, X_2, \ldots, X_n$, and if each of the components is a nonnegative, integer-valued random variable, then we define the *joint-probability generating function* of the random vector $\mathbf{X}$ by the formula

$$\begin{aligned}\psi_{\mathbf{X}}(\mathbf{z}) &= \psi_{X_1,X_2,\ldots,X_n}(z_1, z_2, \ldots, z_n) \\ &\triangleq E[z_1^{X_1}, z_2^{X_2}, \ldots, z_n^{X_n}] \\ &= E\left[\prod_{k=1}^{n} z_k^{X_k}\right] \qquad (8.4)\end{aligned}$$

In those cases in which the component random variables X_k are mutually independent, we get

$$\begin{aligned}\psi_{\mathbf{X}}(\mathbf{z}) &= \prod_{k=1}^{n} E[z_k^{X_k}] \\ &= \prod_{k=1}^{n} \psi_{X_k}(z_k) \qquad (8.5)\end{aligned}$$

and, conversely, if (8.5) holds for all values of the z_k, then the X_k must be mutually independent (cf. Exercise 6.1).

Finally, suppose that we set $z_k = 1$ for $k = m+1, m+2, \ldots, n$. We then get from (8.4) the result that

$$\begin{aligned}\psi_{X_1,X_2,\ldots,X_m,X_{m+1},\ldots,X_n}&(z_1, z_2, \ldots, z_m, 1, \ldots, 1) \\ &= E[z_1^{X_1}z_2^{X_2} \cdots z_m^{X_m}1^{X_{m+1}} \cdots 1^{X_n}] \\ &= E[z_1^{X_1}z_2^{X_2} \cdots z_m^{X_m}] \\ &= \psi_{X_1,X_2,\ldots,X_m}(z_1, z_2, \ldots, z_m) \qquad (8.6)\end{aligned}$$

[cf. (6.8) above]. *Thus we see that setting* $z_k = 1$ *in a joint-probability generating function is the equivalent of eliminating the random variable* X_k *in the corresponding joint-probability distribution.*

Exercise 8.3 Suppose that the discrete-parameter random process $\{Y_n, n = 0, 1, 2, \ldots\}$ has independent increments. That is, suppose that given $Y_0 = 0$ we can write for each $n > 0$

$$Y_n = \sum_{j=1}^{n} X_j$$

where the increments X_j, $j = 1, 2, \ldots, n$ are mutually independent, nonnegative, integer-valued random variables. Show that in this case the joint-probability generating function of the random variables $Y_1, Y_2, \ldots, Y_n$ is given by the formula

$$\psi_{Y_1,Y_2,\ldots,Y_n}(z_1, z_2, \ldots, z_n) = \prod_{j=1}^{n} \psi_{X_j}\left(\prod_{k=j}^{n} z_k\right)$$

where ψ_{X_j} is the probability generating function of the increment random variable X_j. □

Exercise 8.4 Suppose that the random process $\{Y_n, n = 0, 1, 2, \ldots\}$ is a binomial counting process. That is, suppose that $Y_0 = 0$, and that for each $n > 0$ we can write

$$Y_n = \sum_{j=1}^{n} X_j$$

where the increments X_j, $j = 1, 2, \ldots, n$, are mutually independent random variables each with the Bernoulli probability distribution

$$P[X_j = 1] = p \qquad \text{and} \qquad P[X_j = 0] = q \triangleq 1 - p$$

for all $j = 1, 2, \ldots, n$.

a. Show that, in this case, the joint-probability generating function of the random variables Y_1 and Y_2 is

$$\psi_{Y_1,Y_2}(z_1,z_2) = p^2 z_1 z_2^2 + pq(z_1 + 1)z_2 + q^2$$

b. Show that it therefore follows that

$$\psi_{Y_1,Y_2}(z_1,z_2) \neq \psi_{Y_1}(z_1)\psi_{Y_2}(z_2)$$

and hence that Y_1 and Y_2 are *not* independent random variables (cf. Exercise 9.2.2). □

9 CENTRAL LIMIT THEOREM

While the use of characteristic functions does indeed provide a means for calculating the probability distribution function of a sum of independent random variables, there are, unfortunately, many situations in which the explicit evaluation of the various transforms involved is not practicable. It can be shown, however, that under a quite reasonable variety of conditions the probability distribution function of such a sum approaches that of a gaussian random variable as the number of independent random variables being summed increases without limit. We shall derive below a particularly simple, but nonetheless instructive, version of this result, called the central limit theorem. Extensive discussions of various forms of the central limit theorem are given in a chap. 7 of Chung, chap. 6 of Loève, and in the monograph by Gnedenko and Kolmogorov.

THEOREM 9.1 CENTRAL LIMIT THEOREM Let $\mathbf{X}$ be a real, n-dimensional random vector with the component random variables $X_1, X_2, \ldots, X_n$ such that: (*a*) the X_k, $k = 1, 2, \ldots, n$, are mutually independent; (*b*) the X_k all have the same probability distribution function, and (*c*) the mean and variance exist for each X_k; in particular, let

$$E[X_k] = m < +\infty \qquad \text{and} \qquad \operatorname{var}(X_k) = \sigma^2 < +\infty \tag{9.1}$$

for all $k = 1, 2, \ldots, n$.

If we define the random variable Y_n to be the sum of the component random variables, that is, if

$$Y_n \triangleq \sum_{k=1}^{n} X_k \tag{9.2}$$

then it follows that the characteristic function of the normalized sum random variable

$$Z_n \triangleq \frac{Y_n - E[Y_n]}{\sigma(Y_n)} \tag{9.3}$$

has the limit

$$\lim_{n\to\infty} \phi_{Z_n}(v) = e^{-v^2/2} \tag{9.4}$$

That is, *the characteristic function of the normalized sum random variable approaches that of a gaussian random variable with zero mean and unit variance as the number of components increases without limit.*

Proof First, we note that it follows from (9.2) and (9.1) that

$$E[Y_n] = nm \qquad \text{and} \qquad \operatorname{var}(Y_n) = n\sigma^2 \tag{9.5}$$

that is, both the mean and the variance of the sum random variable Y_n increase directly with the number of components added and therefore both increase without limit as n approaches infinity. It is for this reason that the normalized sum random variable Z_n is introduced as in (9.3), for

$$E[Z_n] = 0 \qquad \text{and} \qquad \operatorname{var}(Z_n) = 1 \tag{9.6}$$

for all $n = 1, 2, \ldots$.

Let us now consider the characteristic function of the normalized sum random variable Z_n. It follows from (9.3) that

$$\phi_{Z_n}(v) = E[e^{ivZ_n}] = E\left[\exp\left(iv\,\frac{Y_n - E[Y_n]}{\sigma(Y_n)}\right)\right]$$

and hence from (9.2) and (9.5) that

$$\begin{aligned}\phi_{Z_n}(v) &= E\left[\exp\left(iv\,\frac{\sum_{k=1}^{n} X_k - nm}{\sqrt{n}\,\sigma}\right)\right] \\ &= E\left[\prod_{k=1}^{n}\exp\left(i\frac{v}{\sqrt{n}}\,\frac{X_k - m}{\sigma}\right)\right] \\ &= \prod_{k=1}^{n} E\left[\exp\left(i\frac{v}{\sqrt{n}}\,\frac{X_k - m}{\sigma}\right)\right]\end{aligned} \tag{9.7}$$

the last equality following from the hypothesis that the components X_k are mutually independent. Now since (by hypothesis) all the components X_k have the same probability distribution function, it follows that each of the expectations on the right-most side of (9.7) is the same function of v. With this fact in mind, it is convenient to introduce temporarily the generic random variable W which has the same probability distribution function as does each of the normalized components $(X_k - m)/\sigma$ so that we may write

$$E\left[\exp\left(i\frac{v}{\sqrt{n}}\,\frac{X_k - m}{\sigma}\right)\right] = E\left[\exp\left(i\frac{v}{\sqrt{n}}\,W\right)\right]$$

for each $k = 1, 2, \ldots, n$. Having done so, we may then rewrite (9.7) as

$$\phi_{Z_n}(v) = \left\{ E\left[\exp\left(i \frac{v}{\sqrt{n}} W \right) \right] \right\}^n \tag{9.8}$$

We now wish to study the limiting form of ϕ_{Z_n} as $n \to \infty$. To this end, we shall find it convenient first to take the natural logarithm of both sides of (9.8), thus obtaining the formula

$$\log [\phi_{Z_n}(v)] = n \log \left\{ E\left[\exp\left(i \frac{v}{\sqrt{n}} W \right) \right] \right\} \tag{9.9}$$

Next, we shall expand the exponential in this expression, then expand the resulting logarithm, and finally we shall pass to the limit as n approaches infinity.

If we were to use a power-series expansion of the exponential in (9.9), we would get

$$\exp\left(i \frac{v}{\sqrt{n}} W \right) = 1 + \frac{ivW}{\sqrt{n}} - \frac{v^2W^2}{2n} - \frac{iv^3W^3}{3!n^{3/2}} + \frac{v^4W^4}{4!n^2} - \cdots$$

which, on taking expectations, would give

$$E\left[\exp\left(i \frac{v}{\sqrt{n}} W \right) \right] = 1 - \frac{v^2}{2n} - \frac{iv^3E[W^3]}{3!n^{3/2}} + \frac{v^4E[W^4]}{4!n^2} - \cdots \tag{9.10}$$

since W has a zero mean and a unit variance. There is, however, a fundamental objection to this procedure; examination of (9.10) shows that it is implied there that *all* moments of the normalized component random variable W exist. This, of course, violates the fact that we have only assumed the existence of the first and second moments of the components X_k. We can, however, avoid this difficulty by expanding the exponential in a Taylor polynomial with a second-order remainder term (instead of in an infinite series.) Thus it can be shown that we may write

$$E\left[\exp\left(i \frac{v}{\sqrt{n}} W \right) \right] = 1 - \frac{v^2}{2n} - \frac{R_n}{n} \tag{9.11}$$

where the remainder term R_n is such that

$$\lim_{n \to \infty} R_n = 0 \tag{9.12}$$

and, further, where it is not necessary to assume the existence of any moments of an X_k other than its mean and variance.[1]

On using (9.11) back in (9.9) we get

$$\log [\phi z_n(v)] = n \log \left[1 - \frac{1}{n}\left(\frac{v^2}{2} + R_n\right)\right]$$

The logarithm on the right-hand side of this equation is of the form $\log (1 - z)$, where $|z| \to 0$ as $n \to \infty$. We may therefore use the power-series expansion

$$\log (1 - z) = -\left(z + \frac{z^2}{2} + \frac{z^3}{3} + \cdots\right)$$

which is valid for $|z| < 1$. Thus we get

$$\log [\phi z_n(v)] = -n \left[\frac{\frac{v^2}{2} + R_n}{n} + \frac{\left(\frac{v^2}{2} + R_n\right)^2}{2n^2} + \frac{\left(\frac{v^2}{2} + R_n\right)^2}{3n^3} + \cdots \right]$$

$$= -\frac{v^2}{2} - R_n - \frac{\left(\frac{v^2}{2} + R_n\right)^2}{2n} - \frac{\left(\frac{v^2}{2} + R_n\right)^3}{3n^2} - \cdots$$

It therefore follows, on using (9.12), that we get the result

$$\lim_{n\to\infty} \log [\phi z_n(v)] = -\frac{v^2}{2}$$

and hence (since the logarithm here is a continuous function of v) that we finally get

$$\lim_{n\to\infty} \phi z_n(v) = e^{-v^2/2} \tag{9.13}$$

as was to be proved. ∎

Thus we have proved that the characteristic function of the normalized sum tends to that of a gaussian random variable (with zero mean and unit variance) as the number of mutually independent components increases without limit. It therefore also follows (via a nontrivial theorem[2]) that

$$\lim_{n\to\infty} F z_n(z) = \frac{1}{\sqrt{2\pi}} \int_{-\infty}^{z} e^{-u^2/2}\, du \tag{9.14}$$

[1] Compare with secs. 3 and 4 of chap. 10 of Parzen, or pp. 291ff of Uspensky.
[2] See, for example, art. 10.4 of Cramér, or sec. 3.6 of Lukacs.

for all z. That is, *the probability distribution function of the normalized sum tends to that of a gaussian random variable with zero mean and unit variance when the individual components of the sum are mutually independent random variables and all have the same probability distribution function with a finite mean and variance.* This result is known as the *equal components case of the central limit theorem.* It now follows from this theorem, for example, that the probability distribution function of the binomial counting random variable Y_n introduced in Sec. 9.2 tends to a gaussian probability distribution function as the counting index n increases without limit (cf. Sec. 8.4).

It should be noted that (9.14) states that the *probability distribution function* of the normalized sum converges to that of a gaussian random variable; it does not say that the corresponding *probability densities* so converge. Consider, for example, the case in which the individual components are discrete random variables and, hence, so is their sum. Here the probability density of an individual component exists only in the impulse function sense and the same is true for that of the sum. As the number of components increases without limit, the number of impulses in the probability density of the sum also increases without limit. Such a sequence of impulsive densities cannot be said to converge in any ordinary sense, however, to the smooth probability density of a gaussian random variable. The sequence converges, in fact, only in the sense that its integral, that is, the probability distribution function sequence, converges.

As a final point, it should be recalled that the central limit theorem gives only the *limiting* form of the probability distribution function of the normalized sum. It should be realized however, that while that limiting form was found to be gaussian, it will sometimes be found that the gaussian limit gives a relatively poor approximation for the tails of the actual probability distribution function of the sum when n is finite—even though the particular value of n under consideration might seem to be quite large. This point is discussed in some detail in chap. 7 of vol. I of Feller, and in arts. 82 and 89 of Fry.

Wiener process

We can now use the central limit theorem to justify the assumption of gaussian increments in the Wiener process model of Brownian motion (cf. Sec. 7 above): Returning to our introduction to the Brownian motion process $\{Y_t, 0 \leq t\}$ given in Sec. 1, suppose that we observe that process at the $n + 1$ time instants

$$0 \leq t_1 < t_2 < \cdots < t_n$$

If now the time increments $(t_k - t_{k-1})$ are quite long compared to the average time between molecular impacts, then a pollen particle experiences a total displacement

$$x_k = y(t_k) - y(t_{k-1})$$

which is caused by a very large number of molecular impacts. Assuming, as usual, that those collisions are mutually independent, the displacement increment random variable X_k is then the resultant of a large number of independent molecular-impact increments and hence should have a probability distribution function which tends toward the gaussian by the central limit theorem.

10 SUMMARY OF DEFINITIONS AND FORMULAS

Independent-increment process

The real random process $\{Y_t, 0 \leq t\}$ is said to be an independent-increment process if for every set of time instants

$$0 < t_1 < t_2 < \cdots < t_n$$

the increments

$$(Y_{t_1} - Y_0), (Y_{t_2} - Y_{t_1}), \ldots, (Y_{t_n} - Y_{t_{n-1}})$$

are mutually independent random variables. If, in addition,

$$Y_0 \triangleq 0$$

then we may write

$$Y_{t_n} = \sum_{i=1}^{n} X_i \tag{2.2}$$

where

$$X_i \triangleq Y_{t_i} - Y_{t_{i-1}} \qquad i = 1, 2, \ldots, n \tag{2.3}$$

Mean and variances of an independent-increment process

If $\{Y_t, 0 \leq t\}$ is an independent-increment process, then

$$E[Y_t] = mt \tag{2.8}$$

and

$$\text{var}\,(Y_t) = \sigma^2 t \tag{2.10}$$

where

$$m \triangleq E[Y_t]\Big|_{t=1} \tag{2.9}$$

and

$$\sigma^2 \triangleq \text{var}\,(Y_t)\Big|_{t=1} \tag{2.11}$$

Further

$$\text{var}\,(Y_t - Y_{t'}) = \sigma^2|t - t'|$$

and (Exercise 2.2)

$$\text{cov}\,(Y_t, Y_{t'}) = \sigma^2 \min\,(t,t')$$

Characteristic function

Suppose that the random variable X has the probability density f_X. The characteristic function ϕ_X of the random variable X then is defined as the Fourier transform of f_X:

$$\phi_X(v) \triangleq \int_{-\infty}^{+\infty} f_X(x)e^{ivx}\,dx \tag{4.4}$$

so long as that integral exists, where v is a real parameter and where $i = \sqrt{-1}$. It then follows that

$$f_X(x) = \frac{1}{2\pi}\int_{-\infty}^{+\infty} \phi_X(v)e^{-ivx}\,dv \tag{4.5}$$

In general, however, we define the characteristic function of a random variable as the expectation of e^{ivx}:

$$\phi_X(v) \triangleq E[e^{ivx}] \tag{5.2}$$

where again v is a real parameter. If X has a probability density, then (5.2) reduces to (4.4). If X is a discrete random variable with the possible values x_k, then (5.2) becomes

$$\phi_X(v) = \sum_k P[X = x_k]e^{ivx_k} \tag{5.3}$$

where the summation is taken over all values of the index k.

We have the inequality

$$|\phi_X(v)| \leq \phi_X(0) = 1 \tag{5.8}$$

for any characteristic function.

Sums of independent random variables

Suppose that the random variables $X_1, X_2, \ldots, X_n$ are mutually independent and that X_i has the probability density f_{X_i}, $i = 1, 2, \ldots, n$. The probability density of the sum

$$Y_n \triangleq \sum_{i=1}^{n} X_i$$

is then given by the multiple convolution

$$f_{Y_n}(y) = \int_{-\infty}^{+\infty} \cdots \iint_{-\infty}^{+\infty} f_{X_n}(y - x_{n-1}) \cdots f_{X_2}(x_2 - x_1) f_{X_1}(x_1)\, dx_1\, dx_2 \cdots dx_n \quad (4.3)$$

If ϕ_{X_i} is the characteristic function of X_i, then the characteristic function of the sum is given by the formula

$$\phi_{Y_n}(v) = \prod_{i=1}^{n} \phi_{X_i}(v) \qquad \text{(Exercise 4.2)}$$

Linear functions

If the random variable Y is defined in terms of the random variable X by the formula

$$Y \triangleq aX + b$$

where a and b are real constants, and if ϕ_X is the characteristic function of X, then the characteristic function of Y is

$$\phi_Y(v) = e^{ivb}\phi_X(av) \qquad \text{(Exercise 4.3)}$$

Gaussian random variables

Let X be a gaussian random variable. The characteristic function of X is

$$\phi_X(v) = \exp\left(ivE[X] - \frac{v^2\sigma^2(X)}{2}\right) \qquad \text{(Exercise 4.5)}$$

Let Y be a sum of n mutually independent gaussian random variables X_k, then the characteristic function of Y is

$$\phi_Y(v) = \exp\left(ivm - \frac{v^2\sigma^2}{2}\right) \qquad \text{(Exercise 4.6)}$$

where

$$m \triangleq \sum_{k=1}^{n} m_k \qquad \text{and} \qquad \sigma^2 \triangleq \sum_{k=1}^{n} \sigma_k^2$$

and where m_k and σ_k^2 are the mean and variance, respectively, of X_k.

Cauchy random variables

Let the random variable X have the Cauchy probability density

$$f_X(x) = \frac{1}{\pi(1 + x^2)}$$

The characteristic function of X then is

$$\phi_X(v) = e^{-|v|} \qquad \text{(Exercise 4.7)}$$

Chi-squared random variables

Let the random variable X have the chi-squared probability density

$$f_X(x) = \begin{cases} \dfrac{x^{(n-2)/2}e^{-x/2}}{2^{n/2}\Gamma\left(\dfrac{n}{2}\right)} & \text{for } 0 \le x \\ 0 & \text{for } x < 0 \end{cases}$$

where n is a nonnegative integer and Γ is the gamma function. The characteristic function of X then is

$$\phi_X(v) = (1 - i2v)^{-n/2} \qquad \text{(Exercise 4.9)}$$

Poisson random variables

Let X be a Poisson random variable:

$$P[X = k] = \frac{\lambda^k e^{-\lambda}}{k!}$$

for $k = 0, 1, 2, \ldots$, where λ is a positive constant. Then

$$\phi_X(v) = \exp\left[\lambda(e^{iv} - 1)\right] \qquad (5.5)$$

Moment-generating property

Let X be a random variable with the characteristic function ϕ_X. Then

$$E[X^k] = (-i)^k \phi_X^{(k)}(0) \qquad \text{(Exercise 5.3)}$$

whenever the kth derivative $\phi_X^{(k)}$ exists at the origin in v. Thus it follows that we may write

$$\phi_X(v) = \sum_{k=0}^{\infty} E[X^k] \frac{(iv)^k}{k!} \qquad (5.10)$$

if all of the moments of X exist.

Joint-characteristic functions

Let $\mathbf{X}$ be an n-dimensional random vector with the component random variables $X_1, X_2, \ldots, X_n$. The joint-characteristic function of $\mathbf{X}$ is

$$\phi_{X_1,X_2,\ldots,X_n}(x_1, x_2, \ldots, x_n) = E\left[\exp\left(i \sum_{k=1}^{n} v_k X_k\right)\right] \qquad (6.1)$$

If then we define the column matrices

$$\mathbf{v} \triangleq \begin{bmatrix} v_1 \\ v_2 \\ \cdot \\ \cdot \\ \cdot \\ v_n \end{bmatrix} \qquad \mathbf{X} \triangleq \begin{bmatrix} X_1 \\ X_2 \\ \cdot \\ \cdot \\ \cdot \\ X_n \end{bmatrix}$$

we may write

$$\phi_{\mathbf{X}}(\mathbf{v}) = E[\exp i\mathbf{v}^T\mathbf{X}] \tag{6.2}$$

as the matrix equivalent of (6.1), where $\mathbf{v}^T$ is the matrix transpose of $\mathbf{v}$.

Paralleling the single variable case we have

$$|\phi_{\mathbf{X}}(\mathbf{v})| \leq \phi_{\mathbf{X}}(\mathbf{0}) = 1 \tag{6.3}$$

If X has the n-dimensional probability density

$$f_{\mathbf{X}} = f_{X_1,X_2,\ldots,X_n}$$

then we have

$$\phi_{\mathbf{X}}(\mathbf{v}) = \int \exp (i\mathbf{v}^T\mathbf{x}) f_{\mathbf{X}}(\mathbf{x})\, d\mathbf{x} \tag{6.4}$$

where $\mathbf{x}$ is the n-dimensional vector with components $x_1, x_2, \ldots, x_n$, and where the integration is over the entire range space of $\mathbf{X}$. It follows from (6.4) that

$$f_{\mathbf{X}}(\mathbf{x}) = \frac{1}{(2\pi)^n} \int \exp (-i\mathbf{v}^T\mathbf{x}) \phi_{\mathbf{X}}(\mathbf{v})\, d\mathbf{v} \tag{6.7}$$

where the integration here is over the entire n-dimensional $\mathbf{v}$ space. The component versions of these last two formulas are

$$\phi_{X_1,X_2,\ldots,X_n}(v_1, v_2, \ldots, v_n) = \int_{-\infty}^{+\infty} \cdots \iint_{-\infty}^{+\infty} \exp \left(i \sum_{k=1}^{n} v_k x_k\right) f_{X_1,X_2,\ldots,X_n}(x_1, x_2, \ldots, x_n)\, dx_1\, dx_2 \cdots dx_n \tag{6.5}$$

and

$$f_{X_1,X_2,\ldots,X_n}(x_1, x_2, \ldots, x_n) = \frac{1}{(2\pi)^n} \int_{-\infty}^{+\infty} \cdots \iint_{-\infty}^{+\infty} \exp \left(-i \sum_{k=1}^{n} v_k x_k\right) \phi_{X_1,X_2,\ldots,X_n}(v_1, v_2, \ldots, v_n)\, dv_1\, dv_2 \cdots dv_n \tag{6.6}$$

Independent random variables

Let $\mathbf{X}$ be an n-dimensional random variable with the components $X_1, X_2, \ldots, X_n$. We have

$$\phi_{X_1,X_2,\ldots,X_n}(v_1, v_2, \ldots, v_n) = \prod_{k=1}^{n} \phi_{X_k}(v_k) \qquad \text{(Exercise 6.1)}$$

if and only if the X_k are mutually independent random variables.

Moment-generating properties

Let $\mathbf{X}$ be a two-dimensional random variable with the components X_1 and X_2. Then

$$E[X_1{}^m X_2{}^k] = (-1)^{m+k} \frac{\partial^{m+k} \phi_{X_1,X_2}(v_1,v_2)}{\partial v_1{}^m \, \partial v_2{}^k} \bigg|_{v_1=0=v_2} \qquad \text{(Exercise 6.2)}$$

for any nonnegative integers m and k, whenever the indicated derivatives exist. Thus it follows that we may write

$$\phi_{X_1,X_2}(v_1,v_2) = \sum_{m=0}^{\infty} \sum_{k=0}^{\infty} E[X_1{}^m X_2{}^k] \frac{(iv_1)^m}{m!} \frac{(iv_2)^k}{k!} \qquad (6.9)$$

whenever all of the joint moments exist.

Independent-increment processes

Let $\{Y_t, 0 \leq t\}$ be a real random process with stationary and independent increments and let $Y_0 = 0$. If, given the time instants

$$t_0 = 0 < t_1 < t_2 < \cdots < t_n$$

we define the increments

$$X_k \triangleq Y_{t_k} - Y_{t_{k-1}}$$

for $k = 1, 2, \ldots, n$, then the characteristic function of

$$Y_{t_n} = \sum_{k=1}^{n} X_k$$

is given by the formula

$$\phi_{Y_{t_n}}(v) = \prod_{k=1}^{n} \phi_{X_k}(v) \qquad (7.3)$$

while the joint-characteristic function of the random variables $Y_{t_1}, Y_{t_2}, \ldots, Y_{t_n}$ is given by the formula

$$\phi_{Y_{t_1},Y_{t_2},\ldots,Y_{t_n}}(v_1, v_2, \ldots, v_n) = \prod_{k=1}^{n} \phi_{X_k}\Big(\sum_{j=k}^{n} v_j\Big) \qquad (7.5)$$

Probability generating function

Let X be a discrete random variable with nonnegative integer possible values. The probability generating function of X is

$$\begin{aligned} \psi_X(z) &\triangleq E[z^X] \\ &= \sum_k P[X = k]z^k \end{aligned} \tag{8.1}$$

It then follows that

$$E[X] = \psi'_X(1) \tag{8.3}$$

and that

$$E[X(X-1) \cdots (X-n+1)] = \psi_X{}^{(n)}(1) \qquad \text{(Exercise 8.2)}$$

whenever the indicated derivatives exist.

Joint-probability generating functions

Let **X** be an n-dimensional random vector with the component random variables $X_1, X_2, \ldots, X_n$, and suppose that each of the X_k is a nonnegative, integer-valued random variable. The joint-probability generating function of **X** is

$$\begin{aligned} \psi_{\mathbf{X}}(\mathbf{z}) &\triangleq E[z_1{}^{X_1}z_2{}^{X_2} \cdots z_n{}^{X_n}] \\ &= E\left[\prod_{k=1}^{n} z_k{}^{X_k}\right] \end{aligned} \tag{8.4}$$

We then get the formula

$$\psi_{\mathbf{X}}(\mathbf{z}) = \prod_{k=1}^{n} \psi_{X_k}(z_k) \tag{8.5}$$

if and only if the X_k are mutually independent random variables.

REFERENCES

Boas, Ralph P., Jr.: "A Primer of Real Functions," Carus Mathematical Monograph No. 13, John Wiley & Sons, Inc., New York, 1960.

Chung, Kai Lai: "A Course in Probability Theory," Harcourt, Brace & World, Inc., New York, 1968.

Cramér, Harald: "Mathematical Methods of Statistics," Princeton University Press, Princeton, N.J., 1946.

Einstein, Albert: "Investigations on the Theory of Brownian Movement," Dover Publications, Inc., New York, 1956 (reprint of 1926 Methuen edition).

Feller, William: "An Introduction to Probability Theory and Its Applications," 3d ed., vol. I, John Wiley & Sons, Inc., New York, 1968.

Fry, Thornton C.: "Probability and Its Engineering Uses," 2d ed., D. Van Nostrand, Inc., Princeton, N.J., 1965.

Gnedenko, Boris V., and Andrei N. Kolmogorov: "Limit Distributions for Sums of Independent Random Variables," Addison-Wesley Publishing Company, Inc., Reading, Mass., 1964.

Loève, Michel: "Probability Theory," 3d ed., D. Van Nostrand, Inc., Princeton, N.J., 1963.

Lukacs, Eugene: "Characteristic Functions," No. 5 of Griffin's Statistical Monographs, Hafner Publishing Company, Inc., New York, 1960.

MacDonald, David K. C.: "Noise and Fluctuations," John Wiley & Sons, Inc., New York, 1962.

Parzen, Emanuel: "Modern Probability Theory and Its Applications," John Wiley & Sons, Inc., New York, 1960.

Young, Gail S.: The Linear Functional Equation, *Amer. Math. Monthly*, vol. 65, pp. 37–38, 1958.

Uspensky, James Victor: "Introduction to Mathematical Probability," McGraw-Hill Book Company, New York, 1937.

13
The Poisson Process

1 INTRODUCTION

Many physical phenomena are probabilistically the same as the following waiting line, or queuing, situation:[1] A stream of customers arrives at a service point (say, for example, an airline ticket counter or a supermarket checkout station). These customers are found to arrive at the service point independently of one another, that is, independent in the sense that the number of customers which arrive in any given time interval is independent of the number arriving in any other nonoverlapping time interval. Further, the average rate of arrival at the service point is constant (and equal to, say, λ customers per unit time), and the probability that a single customer arrives in any very short time interval of duration Δt is essentially equal to $\lambda\,\Delta t$, while the probability that more than a single customer arrives in that short time interval is essentially zero. Under these conditions, one would like to answer such questions as:

[1] Excellent introductions to the analysis of queues may be found in the monograph by Cox and Smith and in Lee's book.

What is the probability distribution of the number of customers to arrive in an interval of length t?

What is the probability density of the arrival time of the kth customer to arrive?

What is the probability density of the time interval between the arrivals of successive customers?

A different physical phenomenon which has essentially the same probabilistic description as that above is the noise current (called *shot noise*) generated by the randomness of the flow of electrons from the cathode to the anode of a temperature-limited diode.[1] In this case, the cathode-to-anode potential difference is so great that the electrons emitted by the cathode are all pulled to the anode with such high velocities that space-charge effects are negligible. The arrival of an electron at the anode under these circumstances is unaffected by the preceding electron arrivals. Assuming that there are λ electrons emitted per second (on the average), that the probability that a single electron is emitted in any very short time interval of duration Δt is essentially equal to $\lambda\,\Delta t$, and that the probability that more than one electron is emitted in such a short time interval is essentially zero then gives rise to a probabilistic model which has the same properties as does the above queue-arrival system.

We shall show in the following section that physical phenomena of the above type may be modeled by a Poisson random process, [named after the French physicist and mathematician Siméon Denis Poisson (1781–1840)], and we shall then study in some detail various of the statistical properties of the Poisson process. Fuller discussions of the Poisson and closely related processes may be found in chaps. 7 and 9 of Karlin's book and in chaps. 4 and 5 of Parzen's book on stochastic processes.

2 POISSON COUNTING PROCESS

The arrival of customers in a queue (or the emission of electrons from a heated cathode) may be described by a counting function $N(t)$ which changes value by unity every time that a customer arrives (or an electron is emitted), that is, by a function which keeps a running count of the events which occur. We shall assume throughout our following discussion that we measure time from the start of an observation interval and that we only count events which occur after the start of that observation interval. By definition then,

$$N(0) = 0 \tag{2.1}$$

A typical counting function might be that shown in Fig. 2.1. The collection of all possible such counting sample functions may then be repre-

[1] See, for example, secs. 7.1 and 7.2 of Davenport and Root.

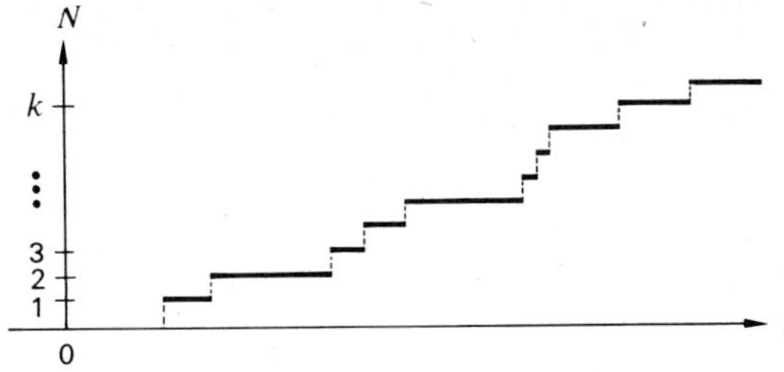

Figure 2.1 A typical counting function.

sented by a nonnegative, integer-valued, continuous-parameter random process.

In order to model phenomena such as the queue arrivals and electron emissions discussed in the preceding section, let us now introduce the *counting random process* $\{N_t, 0 \leq t < +\infty\}$ which we shall assume has the following basic properties: First, it follows from the definition of a counting function that

a. The counting random variable N_t assumes only nonnegative integer values, and

$$N_0 \triangleq 0 \tag{2.2}$$

[Compare with Eq. (2.1).] Next, the assumption of independent customer arrivals (or electron emissions) is modeled by the following hypothesis:

b. The counting random process $\{N_t, 0 \leq t < +\infty\}$ has stationary and independent increments.

Finally, the assumptions that a single event occurs in a very short interval with a probability which is essentially proportional to the length of the interval and that no more than a single event is likely to occur in such an interval are modeled by the hypotheses

c.

$$P[N_{t+\Delta t} - N_t = 1] \triangleq \lambda\, \Delta t + 0(\Delta t) \tag{2.3}$$

and

d.

$$P[N_{t+\Delta t} - N_t > 1] \triangleq 0(\Delta t) \tag{2.4}$$

where λ is a positive constant and where $0(\Delta t)$ is a function of Δt which goes to zero faster than does Δt; that is, where $0(\Delta t)$ is a function such that[1]

$$\lim_{\Delta t \to 0} \frac{0(\Delta t)}{\Delta t} = 0 \tag{2.5}$$

A random process which satisfies hypotheses (*a*) through (*d*) is called a *Poisson counting process* for, as we shall show below, the number of events occurring in any given interval has a Poisson probability distribution.

It should be noted that it follows immediately from hypotheses (*c*) and (*d*) that

$$P[N_{t+\Delta t} - N_t = 0] = 1 - \lambda\,\Delta t + 0(\Delta t) \tag{2.6}$$

that is, the probability that no event occurs in any short interval approaches unity as the duration of that interval approaches zero. In order to derive this result, we note first that since N_t can assume only nonnegative integer values, then the same is true for the counting increment $N_{t+\Delta t} - N_t$. It therefore follows, on summing over all possible values of the increment, that

$$\begin{aligned} 1 &= \sum_{k=0}^{\infty} P[N_{t+\Delta t} - N_t = k] \\ &= P[N_{t+\Delta t} - N_t = 0] + P[N_{t+\Delta t} - N_t = 1] \\ &\qquad + P[N_{t+\Delta t} - N_t > 1] \end{aligned} \tag{2.7}$$

If now we use (2.3) and (2.4) here, we get (2.6). The fact that the increment probabilities given in (2.3), (2.4), and (2.6) are functions of Δt but not of t, is, of course, consistent with the hypothesis (*b*) that the given counting process has stationary increments.

Now let us proceed to determine the probability distribution of the counting random variable N_t. We note, first, that the basic probabilities governing the given process are stated in terms of events which occur in short-duration time intervals and, second, that we are dealing here with an independent-increment process. With these facts in mind, let us consider the time interval $(0, t + \Delta t]$ and split it up into two intervals: one of length t, followed by a second of length Δt, as shown in Fig. 2.2.

[1] We shall use the notation $0(\Delta t)$ throughout to denote *any* function which has the property given in (2.5). Since a constant times such a function and a finite sum of such functions both have this same property, we shall feel free to use the single symbol $0(\Delta t)$ to denote either $a0(\Delta t)$ or any finite sum

$$\Sigma a_i 0(\Delta t)$$

where the coefficients a and a_i are arbitrary finite constants.

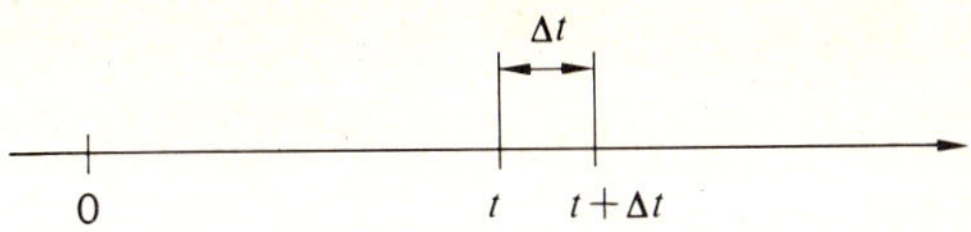

Figure 2.2 Decomposition of the interval $(0,\ t + \Delta t]$.

It is notationally convenient at this time to introduce the symbols

$$p_k(t) \triangleq P[N_t = k]$$

and

$$p_{j,k}(t, t + \Delta t) \triangleq P[N_{t+\Delta t} = k | N_t = j]$$

that is, $p_k(t)$ is the probability that the count is k at time t and $p_{j,k}(t, t + \Delta t)$ is the probability that the count goes from j at time t to k at time $t + \Delta t$. Next, we note that the conditional event $[N_{t+\Delta t} = k | N_t = j]$ is equivalent to the event $[N_{t+\Delta t} - N_t = k - j]$ and hence that their probabilities are equal, that is,

$$P[N_{t+\Delta t} = k | N_t = j] = P[N_{t+\Delta t} - N_t = k - j]$$

Since the given counting process has stationary increments by hypothesis, it follows that

$$\begin{aligned} P[N_{t+\Delta t} - N_t = k - j] &= P[N_{\Delta t} - N_0 = k - j] \\ &= P[N_{\Delta t} = k - j] \end{aligned}$$

the last equality following from the fact that $N_0 = 0$. The transition probabilities $p_{j,k}(t, t + \Delta t)$ thus depend here only upon the time difference Δt, and we shall therefore write

$$\begin{aligned} p_{j,k}(\Delta t) &\triangleq p_{j,k}(t, t + \Delta t) \\ &= P[N_{\Delta t} = k - j] \end{aligned} \tag{2.8}$$

in this case.

Suppose now that we consider the probability that *no* events occur in the overall interval $(0, t + \Delta t]$. This situation can arise in only one way: no events occur in the interval $(0, t]$ and none occur in the interval $(t,\ t + \Delta t]$. Since these intervals are nonoverlapping, the counting increments are independent random variables, and hence we may write

$$p_0(t + \Delta t) = p_0(t)p_{0,0}(\Delta t) \tag{2.9}$$

Now it follows from (2.8) and (2.6) that

$$p_{0,0}(\Delta t) = 1 - \lambda\,\Delta t + 0(\Delta t)$$

On using this result in (2.9), subtracting $p_0(t)$ from both sides, and dividing by Δt, we get

$$\frac{p_0(t + \Delta t) - p_0(t)}{\Delta t} = -\lambda p_0(t) + p_0(t)\,\frac{0(\Delta t)}{\Delta t}$$

If next we pass to the limit as $\Delta t \to 0$, the last term on the right vanishes by virtue of (2.5), and the left-hand side becomes the derivative of $p_0(t)$ with respect to t. Thus we obtain

$$\frac{dp_0(t)}{dt} = -\lambda p_0(t) \tag{2.10}$$

as a differential equation for the probability that no events occur in an interval of duration t. The solution of (2.10) is

$$p_0(t) = ce^{-\lambda t}$$

where c is an integration constant. In order to evaluate c, we note that

$$p_0(0) = c$$

Since zero is the only possible value of N_0, by hypothesis, $p_0(0)$ must be unity and so therefore must c. Thus it follows that

$$P[N_t = 0] = p_0(t) = e^{-\lambda t} \tag{2.11}$$

We note that the probability that no events occur in a given interval becomes vanishingly small as the duration of that interval increases without limit. We note further that *the problem formulation here enabled us to determine the desired probability as a solution of a differential equation.*

Having obtained $p_0(t)$, let us next determine $p_k(t)$ for $k \geq 1$. Again we shall focus upon the intervals $(0,t]$ and $(t, t + \Delta t]$ and make use of the independent increment hypothesis. Starting from $N_0 = 0$, $N_{t+\Delta t}$ can become equal to an integer k in a number of different ways: there may be no events in $(0,t]$ and k events in $(t, t + \Delta t]$; there may be one event in $(0,t]$ and $k - 1$ in $(t, t + \Delta t]$; etc. Thus we may write

$$p_k(t + \Delta t) = \sum_{j=0}^{k} p_j(t)p_{j,k}(\Delta t) \tag{2.12}$$

as these are mutually exclusive events.

Exercise 2.1

a. Show that it follows from (2.4) that

$$p_{j,k}(\Delta t) = 0(\Delta t)$$

when $0 \leq j \leq k - 2$.

b. Show, using (*a*) and the basic hypotheses, that (2.12) leads (on passing to the limit as $\Delta t \to 0$) to

$$\frac{dp_k(t)}{dt} + \lambda p_k(t) = \lambda p_{k-1}(t) \tag{2.13}$$

as a differential recursion equation relating $p_k(t)$ and $p_{k-1}(t)$ for $k \geq 1$. □

There are a number of ways of solving (2.13) for an explicit solution for $p_k(t)$. For example, one could solve the system of differential equations (2.13) recursively by starting with $k = 1$, using the value of $p_0(t)$ obtained previously, and solving (2.13) for $p_1(t)$. That result may then be used back in (2.13) with $k = 2$ to obtain $p_2(t)$, etc. Alternatively, one may derive an integral recursion formula relating $p_{k-1}(t)$ and $p_k(t)$.

Exercise 2.2 Show that the integral recursion formula

$$p_k(t) = \lambda e^{-\lambda t} \int_0^t e^{\lambda \tau} p_{k-1}(\tau)\, d\tau \tag{2.14}$$

is equivalent to the differential recursion equation obtained in Exercise 2.1. [*Hint:* Solve (2.13) by the method of variation of parameters.] □

Exercise 2.3

a. Using either the differential recursion equation derived in Exercise 2.1 or the integral recursion formula derived in Exercise 2.2, show that

$$P_1(t) = \lambda t e^{-\lambda t}$$

and that

$$P_2(t) = \frac{(\lambda t)^2}{2} e^{-\lambda t}$$

b. Show further that

$$P[N_t = k] = P_k(t) = \frac{(\lambda t)^k}{k!} e^{-\lambda t} \tag{2.15}$$

satisfies both of the recursion formulas. We see, therefore, *that the counting random variable N_t has here a Poisson probability distribution.* □

Exercise 2.4

a. Using the Poisson probability distribution in (2.15), obtain an analytic expression for the correction term $0(\Delta t)$ in the equation

$$P[N_{\Delta t} = 1] = \lambda\, \Delta t + 0(\Delta t)$$

[cf. Equation (2.3)].

b. Show that this correction term does in fact have the property

$$\lim_{\Delta t \to 0} \frac{0(\Delta t)}{\Delta t} = 0$$

[cf. Equation (2.5)]. [*Hint:* Use a power-series expansion of the exponential in (2.15).] □

Having shown that the counting random variable N_t has a Poisson probability distribution, it follows from our earlier study of such random variables that the expected value of the counting random variable is

$$E[N_t] = \lambda t \tag{2.16}$$

and hence that

$$\lambda = \frac{E[N_t]}{t} \tag{2.17}$$

[cf. Exercise 7.8.7]. That is, λ *is the average number of events occurring per unit time;* λ, accordingly, is often called the *rate* of the Poisson counting process. It also follows from our earlier Poisson results that

$$\text{var}\,(N_t) = \lambda t \tag{2.18}$$

that is, the variance of the Poisson counting process increases linearly with time. Results of the form given by (2.16) and (2.18) should be expected, of course, as the Poisson counting process is (by definition) a random process with stationary and independent increments (cf. Sec. 12.2).

Exercise 2.5 Let $\{X_t, 0 \leq t < +\infty\}$ and $\{Y_t, 0 \leq t < +\infty\}$ be independent Poisson counting processes with rates λ_x and λ_y, respectively. Let the random process $\{Z_t, 0 \leq t < +\infty\}$ be such that

$$Z_t \triangleq X_t - Y_t$$

for all $t \geq 0$.

Calculate the probability $P[Z_t - Z_{t'} = k]$ for $0 \leq t' < t$ and for $k = 0, \pm 1, \pm 2, \ldots$. Note that the result so obtained shows that *while the sum of two independent Poisson counting processes is a Poisson counting process, the difference is not* (cf. Exercise 12.5.2). □

Exercise 2.6 Let $\{X_t, 0 \leq t < +\infty\}$ and $\{Y_t, 0 \leq t < +\infty\}$ be independent Poisson counting processes with rates λ_x and λ_y respectively. Show that the probability P that exactly k events of the Y_t process occur during the time interval between any two successive events of the X_t

process is given by the formula

$$P = \frac{\lambda_x}{\lambda_x + \lambda_y}\left(\frac{\lambda_y}{\lambda_x + \lambda_y}\right)^k$$

for $k = 0, 1, 2, \ldots$. □

Exercise 2.7 The technique of determining the probabilities which govern a random process as solutions of differential equations may be used in many situations other than just the Poisson. Consider, for example, the following *linear-birth* (or *Yule-Furry*) process (named after G. Udny Yule who introduced the process in 1924 as a model for evolution, and Wendell H. Furry who reintroduced the process in 1937 in a cosmic ray study).

A collection of individuals reproduces according to the following rules:

a. Any individual present at a given time t either remains unchanged in the interval $(t, t + \Delta t]$ for small Δt, or it splits into two individuals during that interval with a probability of splitting given by

$$\lambda\,\Delta t + 0(\Delta t)$$

where the constant λ has the same value for each individual present at time t, and where

$$\lim_{\Delta t \to 0} \frac{0(\Delta t)}{t} = 0$$

as usual.

b. Events (splitting or not) which refer to different individuals are statistically independent.

Let $\{N_t, 0 \leq t < +\infty\}$ be the random process which counts the number of individuals existing at time t, and assume that the process starts with one individual at time zero; that is, $N_0 = 1$ here. Show that

$$P[N_{t+\Delta t} = k | N_t = k] = 1 - k\lambda\,\Delta t + 0(\Delta t)$$

that

$$P[N_{t+\Delta t} = k + 1 | N_t = k] = k\lambda\,\Delta t + 0(\Delta t)$$

and that

$$P[N_{t+\Delta t} > k + 1 | N_t = k] = 0(\Delta t)$$

for $k = 1, 2, 3, \ldots$. □

Exercise 2.8 For the linear-birth random process described in Exercise 2.7:

a. Show that the probability

$$p_1(t) \triangleq P[N_t = 1]$$

satisfies the differential equation

$$\frac{dp_1(t)}{dt} + \lambda p_1(t) = 0$$

and hence that

$$p_1(t) = \begin{cases} e^{-\lambda t} & \text{for } t \geq 0 \\ 0 & \text{for } t < 0 \end{cases}$$

b. Show that the probability

$$p_k(t) \triangleq P[N_t = k]$$

satisfies the differential recursion equation

$$\frac{dp_k(t)}{dt} + k\lambda p_k(t) = (k - 1)\lambda p_{k-1}(t)$$

for $k = 2, 3, 4, \ldots,$ and for $t \geq 0$. □

Alternative derivation of $p_k(t)$

While the above technique of obtaining the Poisson probability

$$P[N_t = k] = p_k(t) = \frac{(\lambda t)^k}{k!} e^{-\lambda t}$$

as a solution of a differential equation is indeed a powerful and widely applicable technique, it is also worthwhile noting that the same result may easily be obtained by using transform techniques:

Since the counting random variable N_t is a nonnegative integer-valued random variable, it has a probability generating function

$$\psi_{N_t}(z) = \sum_{k=0}^{\infty} p_k(t) z^k$$

The partial derivative with respect to t of this generating function is

$$\frac{\partial \psi_{N_t}(z)}{\partial t} = \sum_{k=0}^{\infty} \frac{dp_k(t)}{dt} z^k$$

On using (2.10) and (2.13), then it follows that we get

$$\begin{aligned}\frac{\partial \psi_{N_t}(z)}{\partial t} &= -\lambda p_0(t) - \lambda \sum_{k=1}^{\infty} p_k(t) z^k + \lambda \sum_{k=1}^{\infty} p_{k-1}(t) z^k \\ &= -\lambda \sum_{k=0}^{\infty} p_k(t) z^k + \lambda z \sum_{k=1}^{\infty} p_{k-1}(t) z^{k-1} \\ &= \lambda(z-1) \sum_{k=0}^{\infty} p_k(t) z^k\end{aligned}$$

the last equality following on changing indices in the second sum in the middle line from k to $j = k - 1$. The final summation here is simply the probability generating function of N_t. Thus we obtain

$$\frac{\partial \psi_{N_t}(z)}{\partial t} = \lambda(z-1)\psi_{N_t}(z) \tag{2.19}$$

as a differential equation for the probability generating function of the counting random variable N_t. It should be noted that this result could also have been obtained by directly considering ψ_{N_t} from the start and thus bypassing the derivation of the differential equations for the probabilities $p_k(t)$.

Exercise 2.9 Let $\{N_t,\ 0 \le t < +\infty\}$ be a counting random process which satisfies hypotheses (a) through (d), and let ψ_{N_t} be the probability generating function of the counting random variable N_t:

a. Show that

$$\psi_{N_{t+\Delta t}}(z) = \psi_{N_t}(z)\psi_{N_{\Delta t}}(z)$$

b. Using (a), show that

$$\frac{\partial \psi_{N_t}(z)}{\partial t} = \psi_{N_t}(z) \lim_{\Delta t \to 0} \frac{\psi_{N_{\Delta t}}(z) - 1}{\Delta t}$$

c. Using the series formula for the probability generating function, and (2.2) through (2.6), show for $|z| < 1$ that

$$\lim_{\Delta t \to 0} \frac{\psi_{N_{\Delta t}}(z) - 1}{\Delta t} = \lambda(z-1)$$

and hence, on using this result in (b), that

$$\frac{\partial \psi_{N_t}(z)}{\partial t} = \lambda(z-1)\psi_{N_t}(z)$$

as in (2.19). □

We note that by introducing the probability generating function we have changed our problem from that of solving a system of ordinary differential equations to that of solving a single partial differential equation.

Now for fixed z, Equation (2.19) is an ordinary linear differential equation in the variable t. Its solution is, therefore,

$$\psi_{N_t}(z) = A(z) \exp [\lambda(z - 1)t]$$

where the function $A(z)$ is an, as yet, undetermined function of z. Since we have $N_0 = 0$ by hypothesis, it follows that

$$\psi_{N_0}(z) = E[z^0] = 1$$

With this fact in mind, we also note that it follows from the above formula for $\psi_{N_t}(z)$ that

$$\psi_{N_0}(z) = A(z)$$

Thus we see that $A(z)$ must equal unity and hence that we finally obtain

$$\psi_{N_t}(z) = \exp [\lambda(z - 1)t] \tag{2.20}$$

as an expression for the probability generating function of the counting random variable N_t. The corresponding probability distribution for N_t may now be obtained by expanding the exponential on the right-hand side of (2.20) in a power series in z to give

$$\psi_{N_t}(z) = \sum_{k=0}^{\infty} \frac{e^{-\lambda t}(\lambda t)^k}{k!} z^k$$

It therefore follows that we get, as before,

$$P[N_t = k] = p_k(t) = \frac{e^{-\lambda t}(\lambda t)^k}{k!} \tag{2.21}$$

for $k = 0, 1, 2, \ldots$.

Nonhomogeneous Poisson process

Our previous methods and results can now be easily extended to the case of the nonhomogeneous Poisson process, that is, to the case in which the rate parameter λ is a function of time.

In the present case we change assumption (b) of this section to

b'. *The counting random process* $\{N_t, 0 \leq t < +\infty\}$ *has independent (but nonstationary) increments* and we replace assumption (c), (2.3), by

c'.

$$P[N_{t+\Delta t} - N_t = 1] = \lambda(t)\,\Delta t + 0(\Delta t) \tag{2.22}$$

the assumptions (a) and (d) remaining unchanged. It then follows immediately from our former assumptions (a) and (d) and our new assumptions (b') and (c') that we get the relations

$$P[N_{t+\Delta t} - N_t = 0] = 1 - \lambda(t)\,\Delta t + 0(\Delta t) \tag{2.23}$$

Now as in the constant λ case, we may tackle the problem by splitting up the interval $(0, t+\Delta t]$ into the two nonoverlapping subintervals $(0,t]$ and $(t, t+\Delta t]$ and by again making use of the fact that the counting increments are independent random variables for such intervals. With this in mind we can write

$$P[N_{t+\Delta t} = 0] = P[N_{t+\Delta t} - N_t = 0]P[N_t = 0]$$

which, on using (2.23), leads to the differential equation

$$\frac{dP[N_t = 0]}{dt} = -\lambda(t)P[N_t = 0] \tag{2.24}$$

The solution of this differential equation then is

$$P[N_t = 0] = \exp\left(-\int_0^t \lambda(\tau)\,d\tau\right) \tag{2.25}$$

since $P[N_0 = 0] = 1$. This time-varying rate result reduces to our previously derived result (2.11) when $\lambda(t)$ is taken to be a constant function of time.

We may now similarly obtain a differential equation involving $P[N_t = k]$ by considering all of the different ways that $P[N_{t+\Delta t} = k]$ can be achieved. Thus we get the result

$$P[N_{t+\Delta t} = k] = \sum_{j=0}^{k} P[N_{t+\Delta t} - N_t = k - j]P[N_t = j] \tag{2.26}$$

[cf. (2.12)].

Exercise 2.10

a. Show that it follows from (2.4) that

$$P[N_{t+\Delta t} - N_t = k - j] = 0(\Delta t)$$

for $0 \leq j < k - 2$ (cf. Exercise 2.1).

b. Show, using (*a*) and our present basic assumptions, that (2.26) leads (on passing to the limit as $\Delta t \to 0$) to

$$\frac{dP[N_t = k]}{dt} + \lambda(t)P[N_t = k] = \lambda(t)P[N_t = k - 1] \tag{2.27}$$

as a recursion differential equation relating $P[N_t = k]$ and

$P[N_t = k - 1] \qquad \text{for } k \geq 1$ □

Exercise 2.11 Show that

$$P[N_t = k] = \frac{1}{k!}\left(\int_0^t \lambda(\tau)\, d\tau\right)^k \exp\left(-\int_0^t \lambda(\tau)\, d\tau\right)$$

satisfies the recursion differential equation derived in (*b*) of Exercise 2.10. Thus we see that *the counting random variable N_t has a Poisson probability distribution in the nonhomogeneous case.* □

It now follows immediately from the result of Exercise 2.11 that the mean and variance of the nonhomogeneous Poisson process are given by the formulas

$$E[N_t] = \int_0^t \lambda(\tau)\, d\tau \tag{2.28}$$

and

$$\text{var}\,(N_t) = \int_0^t \lambda(\tau)\, d\tau \tag{2.29}$$

The technique of obtaining the probability distribution by first deriving a partial differential equation for its generating function may also be used in this nonhomogeneous case.[1]

3 ARRIVAL TIMES

It is often necessary in applications to study the length of time required for a given number (say k) of events to occur, as well as to count the number of events which occur in a given time interval (say, of duration t). Assuming that observations start at time zero and that we count the events which occur after the start of the observation interval, we call the time of occurrence t_k of the kth event to occur in the observation interval the kth *arrival time;* we shall also give the same name to the random variable T_k which represents the distribution of possible values of the empirical kth arrival time t_k. Since we are here concerned with continuous-parameter counting processes, it follows that the arrival times are continuous random variables.

[1] For example, see sec. 4-2 of Parzen, "Stochastic Processes."

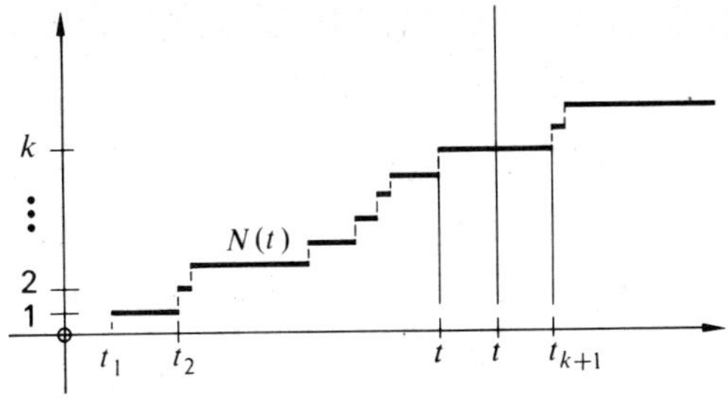

Figure 3.1 A counting function and its arrival times.

Suppose now that we consider a plot of typical counting function $N(t)$, say as shown in Fig. 3.1. A moment's reflection upon this figure shows that there is an explicit relation between the values of a counting function and the corresponding sequence of arrival times. With this observation in mind, let us next determine the relation between the probability distribution function of the kth arrival time T_k and that of the counting random variable N_t.

Now by definition,

$$F_{T_k}(t) = P[T_k \leq t]$$

We note that the events $[T_k \leq t]$ and $[N_t > k - 1]$ are equivalent (and therefore have equal probabilities) since one can occur if and only if the other occurs:

$$[T_k \leq t] = [N_t > k - 1]$$

(cf. Fig. 3.1). Thus it follows that

$$\begin{aligned} P[T_k \leq t] &= P[N_t > k - 1] \\ &= 1 - P[N_t \leq k - 1] \end{aligned}$$

If then we rewrite this result in terms of the corresponding probability distribution functions, we obtain the formula

$$F_{T_k}(t) = 1 - F_{N_t}(k - 1) \tag{3.1}$$

which holds for $k = 1, 2, 4, \ldots$; this is our desired expression. We may, of course, alternatively solve this last result for the probability distribution function of the counting random variable N_t and thus obtain the formula

$$F_{N_t}(j) = 1 - F_{T_{j+1}}(t) \tag{3.2}$$

which holds for $j = 0, 1, 2, \ldots$. Thus we see that given the probability distribution functions of either the counting random variables or the arrival times one can calculate the probability distribution functions of the other. It should be noted that these results hold for *any* counting process $\{N_t, 0 \leq t < +\infty\}$ so long as $N_0 = 0$.

Let us now apply the above general results to the specific case of the Poisson counting process. We showed in the preceding section that

$$P[N_t = j] = \frac{e^{-\lambda t}(\lambda t)^j}{j!}$$

when $\{N_t, 0 \leq t < +\infty\}$ is a Poisson counting process. It therefore follows that, for $k \geq 1$,

$$\begin{aligned} F_{N_t}(k-1) &= \sum_{j=0}^{k-1} P[N_t = j] \\ &= \sum_{j=0}^{k-1} \frac{e^{-\lambda t}(\lambda t)^j}{j!} \end{aligned}$$

On using this result in (3.1), it then follows that we get

$$F_{T_k}(t) = \begin{cases} 1 - e^{-\lambda t} \displaystyle\sum_{j=0}^{k-1} \frac{(\lambda t)^j}{j!} & \text{for } t \geq 0 \\ 0 & \text{for } t < 0 \end{cases} \tag{3.3}$$

as a formula for the probability distribution function of the kth arrival time of a Poisson counting process.

Suppose next that we take the derivative with respect to t of both sides of (3.3) in order to obtain the probability density of the kth arrival time. In this way we get

$$\begin{aligned} f_{T_k}(t) &= -\frac{d}{dt}\left[e^{-\lambda t} \sum_{j=0}^{k-1} \frac{(\lambda t)^j}{j!}\right] \\ &= -\frac{d}{dt}\left\{e^{-\lambda t}\left[1 + \lambda t + \frac{(\lambda t)^2}{2!} + \cdots + \frac{(\lambda t)^{k-1}}{(k-1)!}\right]\right\} \\ &= -\lambda e^{-\lambda t}\left[1 + \lambda t + \cdots + \frac{(\lambda t)^{k-2}}{(k-2)!}\right. \\ &\qquad\qquad \left. - 1 - \lambda t - \cdots - \frac{(\lambda t)^{k-2}}{(k-2)!} - \frac{(\lambda t)^{k-1}}{(k-1)!}\right] \\ &= \lambda e^{-\lambda t} \frac{(\lambda t)^{k-1}}{(k-1)!} \qquad \text{for } t \geq 0 \end{aligned}$$

Thus we see that the probability density of the kth arrival time of a

Poisson counting process is given by the formula

$$f_{T_k}(t) = \begin{cases} \lambda e^{-\lambda t} \dfrac{(\lambda t)^{k-1}}{(k-1)!} & \text{for } t \geq 0 \\ 0 & \text{otherwise} \end{cases} \tag{3.4}$$

This probability density is known both as the *gamma probability law* (with parameters k and λ) and as the *Erlang law* [named after the Danish engineer A. K. Erlang (1878–1929), who derived the formula in a study of congestion in telephone systems.[1]

Exercise 3.1

a. Show, using (3.4), that here

$$E[T_k] = \frac{k}{\lambda}$$

that is, that the expected value of the kth arrival time of a Poisson counting process is just k times the reciprocal of the rate λ of that process.

b. Show also that, in this case,

$$\text{var}\,(T_k) = \frac{k}{\lambda^2}$$

that is, that the variance of the kth arrival time is k times the reciprocal of the square of the rate λ of the Poisson counting process. □

Exercise 3.2 Show that the characteristic function of the kth arrival time of a Poisson counting process is given by the formula

$$\phi_{T_k}(v) = \frac{1}{\left(1 - \dfrac{iv}{\lambda}\right)^k}$$

where $i = \sqrt{-1}$. (*Hint:* Use a table of Laplace transforms if necessary.) □

4 INTERARRIVAL TIMES

Having discussed the arrival times T_k of a counting random process $\{N_t, 0 \leq t < +\infty\}$ in the preceding section, let us next study some of the statistical properties of the intervals between successive arrival times.

[1] See Brockmeyer, Halstrøm, and Jensen.

We shall call the durations Z_k of these intervals the *interarrival times*, where

$$\begin{aligned} Z_1 &\triangleq T_1 \\ Z_k &\triangleq T_k - T_{k-1} \qquad \text{for } k = 2, 3, 4, \ldots \end{aligned} \tag{4.1}$$

A typical sequence of arrival times and the corresponding interarrival times might then be as shown in Fig. 4.1. The sequence of interarrival times thus forms a discrete-parameter, continuous-random-variable random process $\{Z_k, k = 1, 2, 3, \ldots\}$.

Let us now determine the probability distribution function of the kth interarrival time Z_k in terms of the probability distribution of the counting random variable N_t. We note, however, that it is more convenient to first determine the probability

$$\begin{aligned} P[Z_k > z] &= 1 - P[T_k \leq z] \\ &= 1 - F_{Z_k}(z) \end{aligned}$$

It follows from the definition of the interarrival times that the events $[Z_k > z]$ and $[T_k - T_{k-1} > z]$ are equivalent:

$$\begin{aligned} [Z_k > z] &= [T_k - T_{k-1} > z] \\ &= [T_k > T_{k-1} + z] \end{aligned}$$

Suppose now that the observed value of T_{k-1} is t_{k-1}. The event $[T_k > T_{k-1} + z | T_{k-1} = t_{k-1}]$ occurs if and only if the underlying counting process does not change count during the interval $(t_{k-1}, t_{k-1} + z]$:

$$[T_k > T_{k-1} + z | T_{k-1} = t_{k-1}] = [N_{t_{k-1}+z} - N_{t_{k-1}} = 0]$$

Since equivalent events have equal probabilities, we then obtain the result

$$P[Z_k > z | T_{k-1} = t_{k-1}] = P[N_{t_{k-1}+z} - N_{t_{k-1}} = 0]$$

Thus it follows that

$$F_{Z_k}(z | T_{k-1} = t_{k-1}] = 1 - P[N_{t_{k-1}+z} - N_{t_{k-1}} = 0] \tag{4.2}$$

If now the given counting process has stationary increments, then it follows that the probability on the right-hand side of (4.2) is a function only of the difference z; in particular, we have in this case

$$P[N_{t_{k-1}+z} - N_{t_{k-1}} = 0] = P[N_z = 0]$$

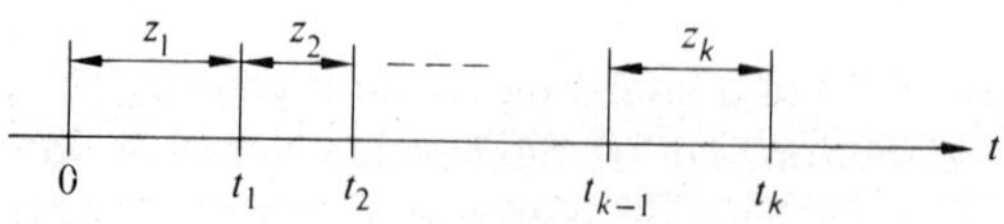

Figure 4.1 Arrival and interarrival times.

since, by hypothesis, $N_0 = 0$ for our basic counting process. The conditional distribution function on the left-hand side of (4.2) is therefore independent of the particular value of k in this case, and hence we finally get the result that

$$F_{Z_k}(z) = 1 - P[N_z = 0] \tag{4.3}$$

for all $k = 1, 2, 3, \ldots$. As this result is independent of the value of the index k, it follows that *if the basic counting process has stationary increments, then the various interarrival times will all have the same probability distribution function.*

Let us now apply the above results to the case of the Poisson counting process.

Exercise 4.1 Let the counting process $\{N_t, 0 \leq t < +\infty\}$ be a Poisson counting process. Show that the interarrival times of such a process have the common probability distribution function

$$F_{Z_k}(z) = \begin{cases} 1 - e^{-\lambda z} & \text{for } z \geq 0 \\ 0 & \text{for } z < 0 \end{cases} \tag{4.4}$$

for all $k = 1, 2, 3, \ldots$, and hence that their corresponding common probability density is

$$f_{Z_k}(z) = \begin{cases} \lambda e^{-\lambda z} & \text{for } z \geq 0 \\ 0 & \text{for } z < 0 \end{cases} \tag{4.5}$$

for all $k = 1, 2, 3, \ldots$.

That is, *the interarrival times of a Poisson counting process are exponentially distributed.* □

Exercise 4.2 Let the counting process $\{N_t, 0 \leq t < +\infty\}$ be a Poisson counting process.

a. Show that for such a process

$$E[Z_k] = \frac{1}{\lambda}$$

for all $k = 1, 2, 3, \ldots$. That is, *the expected value of the interarrival times of a Poisson counting process is equal to the reciprocal of the rate of that process.*

b. Show also that for such a process

$$E[T_k] = kE[Z_k]$$

that is, *the expected value of the kth arrival time of a Poisson counting process is simply k times the common expected value of the interarrival times of that process.* □

Exercise 4.3 Show that the last result,

$$E[T_k] = kE[Z_k]$$

holds true for *any* counting process for which $N_0 \triangleq 0$ and which has stationary increments. □

Exercise 4.4 Let $\{N_t, 0 \leq t < +\infty\}$ be a Poisson counting process and suppose that it is known that an interarrival time, say Z, is greater than some value, say z_1. Show that

$$P[Z > z_1 + z_2 | Z > z_1] = P[Z > z_2]$$

that is, given that we know that the last arrival occurred z_1 seconds previously, the probability that the next arrival will occur at least z_2 seconds in the future is equal to the unconditional probability that the next event occurs at least z_2 seconds in the future. □

Having studied the probability distribution function of the interarrival times of a counting process, it is of interest to know whether different interarrival times are dependent or independent random variables. It can be shown, in fact, that the interarrival times of a Poisson counting process are independent random variables (cf. secs. 4-3 and 4-4 of Parzen, "Stochastic Processes").

5 RENEWAL COUNTING PROCESSES

In our preceding derivation of the properties of the Poisson counting process, we first assumed the probability properties of the counting increments and then we derived the properties of the arrival times and the interarrival times. In many applied problems, however, it seems more appropriate to first assign probability properties to the interarrival times and then to determine the properties of the corresponding counting process than it does to proceed the other way around.

Suppose, for example, that we are concerned with a machine-replacement problem. In particular, let us suppose that having installed some given machine at an initial time (say $t = 0$), we replace that machine when it fails by another machine of the same kind and, further, let us suppose that the lifetime of any one machine is unaffected by the behavior of any other machine. A reasonable model for such a situation might then be a random process in which the interarrival times of the events being studied, for example, the failure of one machine and its replacement by another, are mutually independent random variables, all with the same probability distribution function; we shall call the corresponding counting process a *renewal counting process*. We now wish to determine some of the basic properties of such processes; more complete discussions may be found in Cox's monograph and in chap. 5 of Parzen's book on stochastic processes.

Suppose now that $\{N_t, 0 \leq t < +\infty\}$ is a renewal counting process whose interarrival times Z_k are mutually independent random variables, all with the same probability distribution function F_Z:

$$F_{Z_k}(t) = F_Z(t)$$

for all $k = 1, 2, 3, \ldots$. That the specification of F_Z is enough to enable the determination of the probability distribution function of the renewal-counting random variable may be seen as follows:

First, we recall from the discussion in Sec. 3 that the probability distribution function of the counting random variable N_t may be calculated from that of the arrival times T_j by using the formula

$$F_{N_t}(j) = 1 - F_{T_{j+1}}(t) \tag{5.1}$$

Further, it follows from the definitions of the arrival times T_j and the interarrival times Z_k that

$$T_j = \sum_{k=1}^{j} Z_k$$

and hence (since the interarrival times Z_k are here mutually independent random variables all with the same probability distribution function F_Z), it follows that the characteristic function of the jth time of arrival is given in terms of that of the interarrival times by the formula

$$\phi_{T_j}(v) = \phi_Z{}^j(v) \tag{5.2}$$

If, therefore, we calculate the characteristic function ϕ_Z of an interarrival time using the given probability distribution function F_Z, then we may: (*a*) substitute that result in (5.2), (*b*) calculate the inverse transform f_{T_j} of the resulting ϕ_{T_j}, (*c*) calculate the probability distribution function F_{T_j}, and, finally, (*d*) use that probability distribution function in (5.1) in order to calculate the desired probability distribution function F_{N_t} of the renewal-counting random variable.

Exercise 5.1 Suppose that the interarrival times of a given renewal counting process are exponentially distributed:

$$F_Z(t) = \begin{cases} 1 - e^{-\lambda t} & \text{for } t \geq 0 \\ 0 & \text{for } t < 0 \end{cases}$$

a. Determine the corresponding characteristic function ϕ_Z.

b. Determine the characteristic function ϕ_{T_j} of the jth arrival time of this process.

c. Determine the corresponding probability distribution function of the jth arrival time.

d. Finally, show that the probability distribution function of the renewal-counting random variable N_t of this process is

$$F_{N_t}(k) = \sum_{j=0}^{k} \frac{e^{-\lambda t}(\lambda t)^j}{j!}$$

for $k = 0, 1, 2, \ldots$; that is, show that N_t has a Poisson probability distribution.

It thus follows that we may build up the Poisson counting process either as we did at first by specifying the properties of the counting increments or, as we have done here, by specifying that the interarrival times are mutually independent, exponentially distributed random variables. □

We next wish to derive a much stronger result which shows that it is sufficient to specify the expected value of the counting random variable in order to completely specify the properties of a renewal counting process. We shall call that expected value the *renewal function* and denote it by $m(t)$:

$$m(t) \triangleq E[N_t] \tag{5.3}$$

in addition, we shall call its derivative the *renewal intensity* and denote it by $\lambda(t)$:

$$\lambda(t) \triangleq \frac{dm(t)}{dt} \tag{5.4}$$

With these definitions in mind we note, first, that it follows from the definition of the renewal function that

$$m(t) = \sum_{k=0}^{\infty} kP[N_t = k] \tag{5.5}$$

for all $t \geq 0$, so long as $m(t)$ is finite. Now it also follows from the derivations in Sec. 3 that

$$P[N_t = k] = F_{T_k}(t) - F_{T_{k-1}}(t)$$

for any counting process [cf. (3.1)]. We may therefore rewrite (5.5) as

$$m(t) = \sum_{k=0}^{\infty} k[F_{T_k}(t) - F_{T_{k-1}}(t)]$$

that is

$$m(t) = 0[F_{T_0}(t) - F_{T_1}(t)] + 1[F_{T_1}(t) - F_{T_2}(t)] + 2[F_{T_2}(t) - F_{T_3}(t)] + \cdots$$

Thus we get the result that

$$m(t) = \sum_{k=1}^{\infty} F_{T_k}(t) \tag{5.6}$$

If then we take the derivative with respect to t of both sides of this last result, we get

$$\lambda(t) = \sum_{k=1}^{\infty} f_{T_k}(t) \tag{5.7}$$

for $t \geq 0$, so long as the various derivatives exist.

It is convenient at this time to again introduce the use of Fourier transforms.[1] On denoting the Fourier transform of the renewal intensity $\lambda(t)$ by $\Lambda(v)$

$$\Lambda(v) \triangleq \int_{-\infty}^{+\infty} \lambda(t) e^{ivt}\, dt \tag{5.8}$$

we then get from (5.7)

$$\Lambda(v) = \sum_{k=1}^{\infty} \phi_{T_k}(v)$$

The characteristic function ϕ_{T_k} of the kth arrival time may now be expressed in terms of the characteristic function ϕ_Z of an interarrival time by using (5.2); thus we get

$$\Lambda(v) = \sum_{k=1}^{\infty} \phi_Z{}^k(v)$$

A moment's reflection upon this last result should show that the right-hand side is a geometric series in $\phi_Z(v)$ and hence that we may write

$$\Lambda(v) = \frac{\phi_Z(v)}{1 - \phi_Z(v)} \tag{5.9}$$

or, alternatively,

$$\phi_Z(v) = \frac{\Lambda(v)}{1 + \Lambda(v)} \tag{5.10}$$

Thus we see that the specification of either the renewal intensity $\lambda(t)$ or the probability density $f_Z(t)$ of an interarrival time (or their respective Fourier transforms) enables the determination of the other. It therefore

[1] Depending upon the specific functions involved in any particular problem, ordinary Fourier transforms of the various terms in (5.7) may or may not exist in the usual sense. When they do not, one can often use, however, Fourier transforms of a complex variable or Laplace transforms.

follows further (from our results preceding Exercise 5.1) that the specification of the renewal intensity $\lambda(t)$ is enough to enable the calculation of the probability distribution function of the corresponding renewal counting random variable.

Exercise 5.2 Suppose that the intensity $\lambda(t)$ of a renewal counting process $\{N_t, 0 \leq t < +\infty\}$ is a positive constant; in particular, suppose that

$$\lambda(t) = \begin{cases} \lambda & \text{for } t \geq 0 \\ 0 & \text{for } t < 0 \end{cases}$$

with $\lambda > 0$. Show that the interarrival times of this renewal process then have the probability density

$$f_Z(t) = \begin{cases} \lambda e^{-\lambda t} & \text{for } t \geq 0 \\ 0 & \text{for } t < 0 \end{cases}$$

(*Hint:* Use Laplace transforms rather than Fourier transforms if desired.)

Thus we see that simply specifying the intensity of a renewal counting process to be a constant is enough to determine that the interarrival times of that process are exponentially distributed and hence (using the results of Exercise 5.1) *that that renewal counting process is a Poisson counting process.* □

Now it is often more convenient not to have to deal with the various Fourier transforms discussed above and, instead, to be able to deal directly with the renewal intensity and the interarrival probability density. With this in mind, let us reconsider our results (5.9) and (5.10). It follows from either of these that we may write

$$\phi_Z(v) = \Lambda(v) - \Lambda(v)\phi_Z(v)$$

It therefore also follows, on taking inverse transforms, that

$$f_Z(t) = \lambda(t) - \int_{-\infty}^{+\infty} \lambda(t-u) f_Z(u)\, du \tag{5.11}$$

since the inverse Fourier transform of a product of Fourier transforms is the convolution of their respective inverse transforms. Now since, by definition, a counting process starts at $t = 0$, we have

$$\lambda(t) = 0 \qquad \text{for } t < 0$$

and

$$f_Z(t) = 0 \qquad \text{for } t < 0$$

We may therefore rewrite (5.11) in the more usual form

$$f_Z(t) = \lambda(t) - \int_0^t \lambda(t-u) f_Z(u)\, du \tag{5.12}$$

This result, which gives the probability density of the interarrival times of a renewal process as a solution of an integral equation involving the intensity of that process, is known as a *renewal equation.*

Exercise 5.3 Suppose that the intensity $\lambda(t)$ of a renewal counting process $\{N_t,\ 0 \leq t < +\infty\}$ is a positive constant as in Exercise 5.2. Show, using (5.12), that in this case the probability distribution function F_Z of an interarrival time satisfies the differential equation

$$\frac{dF_Z(t)}{dt} + \lambda F_Z(t) = \lambda \qquad \text{for } t \geq 0$$

which has the solution

$$F_Z(t) = 1 - e^{-\lambda t} \qquad \text{for } t \geq 0$$

that is, that again we find the interarrival time to have an exponential distribution. □

6 UNORDERED ARRIVAL TIMES

Let us now return to the study of the arrival times of a Poisson process which was initiated in Sec. 3 above. Suppose that it is known that exactly k events of a Poisson process occur during an interval of duration t; in other words, suppose that $N_t = k$, where N_t is a Poisson counting random variable. We now wish to study the resultant conditional probability of the times at which these events occur.

To this end, let us first partition the total observation interval $(0,t]$ into M adjacent subintervals by the time instants $t'_0, t'_1, t'_2, \ldots, t'_M$, where we set

$$t'_0 \triangleq 0 \qquad \text{and} \qquad t'_M \triangleq t$$

Then, on defining

$$\tau_m \triangleq t'_m - t'_{m-1} \qquad \text{for } m = 1, 2, \ldots, M$$

we have the partitioning situation pictured in Fig. 6.1. It follows from

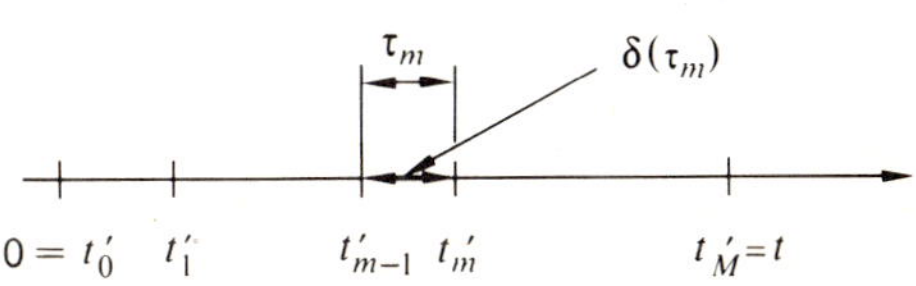

Figure 6.1 The partitioning scheme.

this definition that we may write

$$t = \sum_{m=1}^{M} \tau_m \tag{6.1}$$

It should be emphasized that *there is no a priori relation here between the times at which events occur* (the t_k) *and the partition instants* t'_m.

Next, let us denote the number of events which occur during the subinterval

$$\delta(\tau_m) \triangleq (t'_{m-1},t'_m] \tag{6.2}$$

(where $m = 1, 2, \ldots, M$) by k_m. It then follows that we must have

$$\sum_{m=1}^{M} k_m = k \tag{6.3}$$

since, by hypothesis, exactly k events occurred during the interval $(0,t]$.

The conditional joint probability that k_m events occur during the interval $\delta(\tau_m)$ of length τ_m, $m = 1, 2, \ldots, M$, subject to the hypothesis that k events occur during the entire interval $(0,t]$, is then

$$\begin{aligned} P[N_{\tau_1} = k_1, N_{\tau_2} = k_2, \ldots, N_{\tau_M} = k_M | N_t = k] \\ = \frac{P[N_{\tau_1} = k_1, N_{\tau_2} = k_2, \ldots, N_{\tau_M} = k_M, N_t = k]}{P[N_t = k]} \\ = \frac{P[N_{\tau_1} = k_1, N_{\tau_2} = k_2, \ldots, N_{\tau_M} = k_M]}{P[N_t = k]} \end{aligned} \tag{6.4}$$

The last equality following from the fact that the specification of the M events $[N_{\tau_m} = k_m]$ automatically specifies the event $[N_t = k]$ [cf. (6.1) and (6.3)].

Since the underlying process here is one with stationary and independent increments, it follows that

$$P[N_{\tau_1} = k_1, N_{\tau_2} = k_2, \ldots, N_{\tau_m} = k_M] = \prod_{m=1}^{M} P[N_{\tau_m} = k_m]$$

Since, in addition, the number of events occurring during any interval has a Poisson probability distribution, it further follows that

$$P[N_{\tau_m} = k_m] = \frac{e^{-\lambda\tau_m}(\lambda\tau_m)^{k_m}}{k_m!}$$

and that

$$P[N_t = k] = \frac{e^{-\lambda\tau}(\lambda\tau)^k}{k!}$$

On using these results in (6.4), we then get

$$P[N_{\tau_1} = k_1, N_{\tau_2} = k_2, \ldots, N_{\tau_M} = k_M | N_t = k] = \prod_{m=1}^{M} \frac{\dfrac{e^{-\lambda\tau_m}(\lambda\tau_m)^{k_m}}{k_m!}}{\dfrac{e^{-\lambda t}(\lambda t)^k}{k!}}$$

$$= \frac{e^{-\lambda(\tau_1+\tau_2+\cdots+\tau_M)}}{e^{-\lambda t}} \frac{\lambda^{k_1+k_2+\cdots+k_M}}{\lambda^k} \frac{\prod_{m=1}^{M} \dfrac{(\tau_m)^{k_m}}{k_m!}}{\dfrac{t^k}{k!}}$$

$$= \frac{k!}{t^k} \prod_{m=1}^{M} \frac{(\tau_m)^{k_m}}{k_m!} \tag{6.5}$$

as our desired conditional probability.

Suppose next that we make our partitioning sufficiently fine that no more than one event occurs during any single subinterval. In this case, k of the subintervals will each have *one* event occurring during it; in these k cases we have

$$(\tau_m)^{k_m} = \tau_m \qquad \text{and} \qquad k_m! = 1$$

Each of the remaining $M - k$ subintervals will have *no* events occurring during it; in these $M - k$ cases we have

$$(\tau_m)^{k_m} = \tau_m{}^0 = 1 \qquad \text{and} \qquad k_m! = 0! = 1$$

The product in (6.5) thus consists of a product of k terms (each of the form τ_m) corresponding to the k subintervals during each of which a single event occurs. Let us next (for convenience) re-index those k subintervals so that the subinterval $\delta(\tau_j)$ now contains the jth event to occur. Since, by definition, the time of occurrence of the jth event is the arrival time t_j discussed in Sec. 3, the conditional joint probability in (6.5) becomes now just the joint probability of occurrence of the events $[t_j \in \delta(\tau_j)]$, $j = 1, 2, \ldots, k$, subject to the hypothesis that $N_t = k$. That is, we have thus shown that

$$P[t_1 \in \delta(\tau_1), t_2 \in \delta(\tau_2), \ldots, t_k \in \delta(\tau_k) | N_t = k] = \frac{k!}{t^k} \prod_{j=1}^{k} \tau_j \tag{6.6}$$

where the arrival times t_j are ordered:

$$0 < t_1 < t_2 < \cdots < t_k \leq t$$

by definition.

Having obtained (6.6), we shall next show that, given that k events occur during the interval $(0,t]$, the very same result may be obtained by assuming that the arrival times t_j are the order statistics of *event times* (or *unordered arrival times*) which are mutually independent random variables each of which is uniformly distributed over the interval $(0,t]$.

With this last assertion in mind, consider the k events which occur during the given interval $(0,t]$. Let us now imagine that we could index the objects causing those particular events, for example, index the people which will form a queue or the electrons which will be emitted, by the integers $1, 2, \ldots, k$, and let us then denote by u_i the *event time* that object i causes an event of interest to occur. Note that since there is no a priori guarantee that object number i will give rise to the ith event to occur (in order), there is correspondingly no guarantee that the event time u_i of object number i will be the same as t_i, the arrival time of the ith event to occur. That is, while the arrival times t_j are ordered by definition, the event times u_i are not necessarily so ordered; the situation is essentially that shown in Fig. 6.2. If we denote by U_i the random variable which gives the distribution of values of the empirical event times u_i, we say that the arrival-time random variables T_j are the *order statistics* of the event-time random variables U_i.

Next, let us *assume* that the event times U_i are mutually independent random variables, each of which is uniformly distributed over the interval $(0,t]$; more precisely, let us assume that, given $N_t = k$, the U_i are mutually independent and

$$f_{U_i}(u_i|N_t = k) = \begin{cases} \dfrac{1}{t} & \text{for } 0 < u_i \leq t \\ 0 & \text{otherwise} \end{cases} \tag{6.7}$$

for all $i = 1, 2, \ldots, k$. Since the U_i are independent random variables here by assumption, the conditional joint-probability density of the k

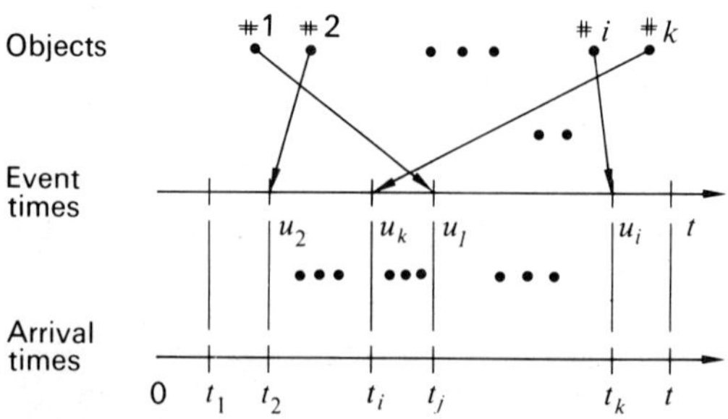

Figure 6.2 The relation between event times and arrival times.

event times U_i is given by

$$f_{U_i,U_2,\ldots,U_k}(u_i, u_2, \ldots, u_k|N_t = k) = \begin{cases} \dfrac{1}{t^k} & \text{for } 0 < u_i \leq t \text{ and } i = 1, 2, \ldots, k \\ 0 & \text{otherwise} \end{cases}$$

Now for a given set of observed arrival times $\{t_j, j = 1, 2, \ldots, k\}$ during $(0,t]$, any one of the k values t_j could be the event time u_1 of object 1; any one of the remaining $k - 1$ values t_j could be the event time u_2 of object 2, etc. Thus we see that there are $k!$ different sets of event times $\{u_i, i = 1, 2, \ldots, k\}$ which could give rise to a particular given set of arrival times. It therefore follows that the conditional joint-probability density of the *ordered* event times (i.e., the arrival times) is given by

$$f_{T_1,T_2,\ldots,T_k}(t_1, t_2, \ldots, t_k|N_t = k) = \begin{cases} \dfrac{k!}{t^k} & \text{for } 0 < t_j \leq t \text{ and } j = 1, 2, \ldots, k \\ 0 & \text{otherwise} \end{cases} \tag{6.8}$$

Now let us return to the k subintervals $\delta(\tau_j)$ referred to in Eq. (6.6). The probability that the arrival time T_1 falls in a subinterval $\delta(\tau_1)$ of duration τ_1, that T_2 falls in a subinterval $\delta(\tau_2)$ of duration τ_2, etc., is then given by the integral of the conditional joint-probability density given in (6.8) over the k given subintervals. That is,

$$\begin{aligned} P[T_1 \in \delta(\tau_1), T_2 \in \delta(\tau_2), \ldots, T_k \in \delta(\tau_k)|N_t = k] \\ = \int_{\delta(\tau_k)} \cdots \int_{\delta(\tau_2)} \int_{\delta(\tau_1)} \frac{k!}{t^k}\, dt_1\, dt_2 \cdots dt_k \\ = \frac{k!}{t^k} \prod_{j=1}^{k} \tau_j \end{aligned}$$

Since this last result is identical to that given by Eq. (6.6), our assertion that, *given that k events occur during an interval* $(0,t]$, *the event times* U_1, U_2, . . . , U_k *are mutually independent random variables* (*each of which is uniformly distributed over the given interval*) is thus proved.

7 FILTERED POISSON PROCESSES

Let us next reconsider the shot-noise process introduced in Sec. 1 of this chapter; this is the random process which characterizes the fluctuating current generated by the flow of electrons from the cathode to the anode of a temperature-limited diode. Now the total current flowing through such a thermionic diode is the resultant of the current pulses produced by the individual electrons which pass through the diode. Since the diode is temperature limited by hypothesis, there is effectively no interaction between the various electrons and, hence, the total diode current is simply the sum of the individual electronic current pulses. In this case, these electronic current pulses all have the same shape and differ only by translations in time caused by the different electron-emission times. In general, the duration of any individual electronic current pulse will be the transit time of the electron passage through the diode; for example, the electronic current pulse $i_e(t)$ caused by the passage of an electron through a parallel-plane, temperature-limited diode has the form shown in Fig. 7.1, where an electron of charge e was emitted at time $t = u$ and where τ_a is the electronic transit time (cf. sec. 7.1 of Davenport and Root).

Suppose now that we start observing the current passing through a particular diode at time $t = 0$ and that we measure the total current i at time t. Assuming that t is greater than the duration of an electronic current pulse (so that the effect of electrons emitted before the start of the observation interval has died out), we may then write for this diode

$$i(t) = \sum_{j=1}^{k} i_e(t - u_j)$$

if exactly k electrons are emitted during the observation interval $(0,t]$. Here u_j is the unordered emission time of the jth electron and $i_e(t - u_j)$ is the electronic current pulse produced by the electron emitted at time u_j. Suppose next that we repeat this observation procedure using a large number of different diodes, all of which have the same physical characteristics. If we did so, we would not be very surprised to find that the number of electrons emitted by each diode in the set varied from diode to diode. With this in mind, we could then describe the ensemble of possible temperature-limited diode-current waveforms by the random process $\{I_t, 0 \leq t < +\infty\}$ in which the total-current random variable I_t is given by the formula

$$I_t = \sum_{j=1}^{N_t} i_e(t - U_j)$$

where N_t is the counting random variable which characterizes the number

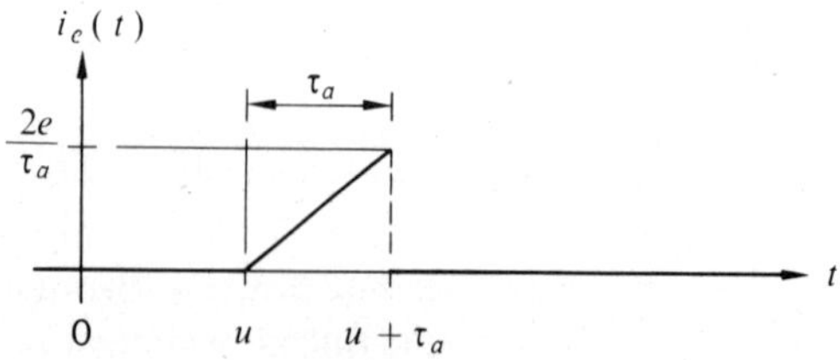

Figure 7.1 A parallel-plane temperature-limited diode electronic current pulse.

of electrons emitted during the interval $(0,t]$, and where the U_j are the unordered emission-time random variables. Assuming that the electron-emission process satisfies the assumptions stated in Sec. 1 above, it then follows that the counting random variable N_t here has a Poisson probability distribution.

Many physical phenomena are probabilistically similar to the above shot-noise process in that the phenomenon of interest may be described as the result of randomly exciting a linear operator with a Poisson process. That is, the random process which describes the phenomenon of interest, say $\{X_t, 0 \leq t < +\infty\}$, is such that we may write

$$X_t \triangleq \sum_{j=1}^{N_t} h(t - U_j) \tag{7.1}$$

where an event which occurs at time u_j generates an outcome $h(t - u_j)$ at time t, where the number of events which occur during the interval $(0,t]$ is described by the Poisson counting random variable N_t, and where the random variables U_j are the unordered arrival times of the events which occur during the interval $(0,t]$. Such a random process is called a *filtered Poisson process* (cf. sec. 4.5 of Parzen's book on stochastic processes). We would like to determine some of the statistical properties of filtered Poisson processes in this section.

First, let us find the expected value of the random variable X_t. We note from (7.1) that X_t is, in fact, a random sum of functions of random variables. With this fact in mind, we shall calculate the expected value of X_t by first assuming a particular value (say k) of the counting random variable N_t, then conditionally averaging the resultant fixed sum with respect to the k unordered emission times, and finally averaging the result over all possible values k of N_t (cf. Exercise 7.11.4). Thus we shall write

$$E[X_t] = \sum_{k=0}^{\infty} P[N_t = k]E[X_t|N_t = k] \tag{7.2}$$

where here N_t has the Poisson probability distribution

$$P[N_t = k] = \frac{e^{-\lambda t}(\lambda t)^k}{k!} \qquad \text{for } k = 0, 1, 2, \ldots \tag{7.3}$$

in which λ is the average number of events occurring per unit time, and where the conditional average $E[X_t|N_t = k]$ is obtained by averaging the sum

$$\sum_{j=1}^{k} h(t - U_j)$$

with respect to the unordered arrival times $U_1, U_2, \ldots, U_k$. Now it follows from the results of the preceding section that (since N_t is Poisson in this case) the U_j are mutually independent random variables, each of which is uniformly distributed over the interval $(0,t]$:

$$f_{U_j}(u|N_t = k) = \begin{cases} \dfrac{1}{t} & \text{for } 0 < u \leq t \\ 0 & \text{otherwise} \end{cases} \tag{7.4}$$

for all $j = 1, 2, \ldots, k$. Thus it follows, since the expectation of a sum is a sum of expectations, that

$$\begin{aligned} E[X_t|N_t = k] &= E\left[\sum_{j=1}^{k} h(t - U_j)\right] \\ &= \sum_{j=1}^{k} E[h(t - U_j)] = \sum_{j=1}^{k} \frac{1}{t}\int_0^t h(t - u_j)\,du_j \\ &= \frac{k}{t}\int_0^t h(u)\,du \end{aligned}$$

where we obtained the last equality by introducing the change of variables $u \triangleq t - u_j$. On substituting this result back into (7.2), we then get

$$\begin{aligned} E[X_t] &= \frac{1}{t}\int_0^t h(u)\,du \sum_{k=0}^{\infty} P[N_t = k]k \\ &= \frac{1}{t}\int_0^t h(u)\,du\; E[N_t] \end{aligned}$$

Since N_t here is Poisson, it then finally follows from (2.17) that we obtain the result

$$E[X_t] = \lambda \int_0^t h(u)\,du \tag{7.5}$$

as a formula for the expected value of a filtered Poisson process.

Suppose, for example, that $\{I_t, 0 \leq t < +\infty\}$ is the shot-noise process discussed above. Since in this case we have (for values of t greater than the duration of an electronic current pulse) the result that

$$\int_0^t i_e(u)\,du = e$$

where e is the charge of an electron, it follows from (7.5) that

$$E[I_t] = e\lambda$$

That is, the average value of the shot-noise current equals the charge of an electron multiplied by the average number of electrons passing through the diode per unit time, a hardly unexpected result!

Exercise 7.1 That the Poisson counting process may be considered to be a special case of the filtered Poisson process can be shown as follows:

Suppose that the response at time t of an event occurring at time u_j is

$$h(t - u_j) = U(t - u_j)$$

where here U is the unit-step function

$$U(t) \triangleq \begin{cases} 1 & \text{for } t \geq 0 \\ 0 & \text{for } t < 0 \end{cases}$$

a. Show, using (7.5), that in this case

$$E[X_t] = \lambda t$$

b. Show that it follows from the definition of h here that

$$X_t = N_t$$

and hence that the result in part (*a*) is to be expected. □

Exercise 7.2 Again let $\{I_t, 0 \leq t < +\infty\}$ be a shot-noise process as described above. Using the averaging technique illustrated by the above calculation of the expected value of a filtered Poisson process, show that the autocorrelation function of the shot-noise process is given by the formula

$$R_{I_t}(\tau) = \lambda \int_{-\infty}^{+\infty} i_e(u) i_e(u + \tau)\, du + (e\lambda)^2$$

and hence that the variance of the shot-noise process is

$$\operatorname{var}(I_t) = \lambda \int_{-\infty}^{+\infty} i_e^2(u)\, du$$

where, in both cases, the assumption is made that t is very much larger than both the time shift τ and the transit time of an electron. (*Hint:* These last assumptions permit making the approximation that $N_{t+\tau} = N_t$.) □

Next, we wish to determine the probability distribution of the filtered Poisson random variable X_t. To this end, we will first determine the characteristic function

$$\begin{aligned} \phi_{X_t}(v) &= E[e^{ivX_t}] \\ &= E\left[\exp\left[iv \sum_{j=1}^{N_t} h(t - U_j)\right]\right] \end{aligned}$$

and then we will invert ϕ_{X_t}. Paralleling our calculation of $E[X_t]$, we have

$$\phi_{X_t}(v) = \sum_{k=0}^{\infty} P[N_t = k]E[e^{ivX_t}|N_t = k] \tag{7.6}$$

where

$$E[e^{ivX_t}|N_t = k] = E\left[\exp\left[iv\sum_{j=1}^{k} h(t - U_j)\right]\right]$$

and where the averaging here is done with respect to the mutually independent, uniformly distributed, unordered arrival times $U_1, U_2, \ldots, U_k$. Since the exponential of a sum is a product of exponentials, we have

$$E[e^{ivX_t}|N_t = k] = E\left[\prod_{j=1}^{k} e^{ivh(t-U_j)}\right]$$

which becomes

$$E[e^{ivX_t}|N_t = k] = \prod_{j=1}^{k} E[e^{ivh(t-U_j)}] \tag{7.7}$$

since the U_j are mutually independent. Since the U_j are also each uniformly distributed over the interval $(0,t]$, we have

$$E[e^{ivh(t-U_j)}] = \frac{1}{t}\int_0^t e^{ivh(t-u_j)}\,du_j = \frac{1}{t}\int_0^t e^{ivh(u)}\,du$$

for all $j = 1, 2, \ldots, k$ (where again we made the substitution $u \triangleq t - u_j$). If now we use this last result back in (7.7), we get

$$E[e^{ivX_t}|N_t = k] = \left[\frac{1}{t}\int_0^t e^{ivh(u)}\,du\right]^k$$

which gives, on substituting back into (7.6), the result that

$$\begin{aligned}\phi_{X_t}(v) &= \sum_{k=0}^{\infty} \frac{e^{-\lambda t}(\lambda t)^k}{k!}\left[\frac{1}{t}\int_0^t e^{ivh(u)}\,du\right]^k \\ &= e^{-\lambda t}\sum_{k=0}^{\infty} \frac{1}{k!}\left[\lambda\int_0^t e^{ivh(u)}\,du\right]^k\end{aligned}$$

since N_t here is a Poisson random variable. The last series may now be summed by remembering the power-series expansion of the exponential function; thus we finally obtain the formula

$$\begin{aligned}\phi_{X_t}(v) &= e^{-\lambda t}\exp\left(\lambda\int_0^t e^{ivh(u)}\,du\right) \\ &= \exp\left[\lambda\int_0^t \left(e^{ivh(u)} - 1\right)du\right]\end{aligned} \tag{7.8}$$

for the characteristic function of a filtered Poisson random process; the probability distribution of X_t may now be obtained by inverting this formula. In some cases the inversion problem is simple to solve (see Exercise 7.3 below), while in others it is quite difficult (cf. Gilbert and Pollak).

Exercise 7.3 Let $\{X_t, 0 \leq t < +\infty\}$ be a filtered Poisson process and suppose that the response function h has the form

$$h(t) = \begin{cases} a & \text{for } 0 \leq t \leq \delta \\ 0 & \text{otherwise} \end{cases}$$

where a is a positive constant and where δ is the duration of the response.

a. Show that, in this case,

$$\phi_{X_t}(v) = \exp\left[\lambda\delta(e^{iva} - 1)\right]$$

b. Show also that it follows from the definition of h, and from the result of part (a), that X_t is a discrete random variable with the possible values $0, a, 2a, 3a, \ldots$, and that X_t has the probability distribution

$$P[X_t = ka] = \frac{e^{-\lambda\delta}(\lambda\delta)^k}{k!}$$

for all $k = 0, 1, 2, \ldots$. Thus we see that here the random variable X_t is itself a Poisson random variable. □

Exercise 7.4 Let $\{X_t, 0 \leq t < +\infty\}$ be a filtered Poisson process. Show, using the moment-generating results of Sec. 12.5, that the formulas

$$E[X_t] = \lambda \int_0^t h(u)\,du \qquad \text{and} \qquad \operatorname{var}(X_t) = \lambda \int_0^t h^2(u)\,du$$

follow directly from the formula for the characteristic function of a filtered Poisson process. □

In many problems of interest, the response function h is nonzero only over a finite interval; say, for example, as in the shot-noise case above or as in Exercise 7.3. In these cases, we have

$$e^{ivh(u)} - 1 = 0$$

for $u < 0$ and $u > \delta$, where δ is the duration of h. With this fact in mind, Eq. (7.8) may be rewritten in the form

$$\phi_{X_t}(v) = \exp\left[\lambda \int_{-\infty}^{+\infty} \left(e^{ivh(u)} - 1\right) du\right] \tag{7.9}$$

for such processes. It should be noted that in this case the characteristic function (and hence the probability distribution) of X_t is *not* a function of t. Since the same result is obtained when the joint-characteristic function of such a process is determined for any n time instants, it follows that such a filtered Poisson process is stationary in the strict sense.

Limiting form as $\lambda \to \infty$

We noted above that while it was not too difficult to obtain a formula for the characteristic function of a filtered Poisson process, it may in fact be quite difficult to invert that formula in order to obtain the corresponding probability distribution. Nonetheless, it is interesting to note also that the characteristic function of the process tends to that of a gaussian random variable as the average number of events occurring per unit time increases without limit. While this result obviously follows in some way from the central limit theorem (since we are dealing here with sums of functions of independent random variables), it is easy enough to derive directly.

Since it follows from the results of Exercise 7.4 that both the mean and the variance of a filtered Poisson process increase linearly with λ, it is convenient to consider the normalized random variable

$$Y_t \triangleq \frac{X_t - E[X_t]}{\sigma(X_t)} \tag{7.10}$$

which has a mean of zero and a variance of unity for all values of λ. If now we define (purely for notational convenience) the parameters α and β by the formulas

$$\alpha \triangleq \int_0^t h(u)\,du \qquad \text{and} \qquad \beta^2 \triangleq \int_0^t h^2(u)\,du$$

then it follows from Exercise 7.4 that we may write

$$E[X_t] = \alpha\lambda \qquad \text{and} \qquad \text{var}\,(X_t) = \beta^2\lambda$$

and hence, using these results in (7.10), we have

$$Y_t = \frac{X_t - \alpha\lambda}{\beta\sqrt{\lambda}} = \frac{X_t}{\beta\sqrt{\lambda}} - \frac{\alpha}{\beta}\sqrt{\lambda}$$

The characteristic function of this normalized random variable may then be expressed in terms of that of X_t as

$$\begin{aligned}\phi_{Y_t}(v) &= E\left[\exp\left[iv\left(\frac{X_t}{\beta\sqrt{\lambda}} - \frac{\alpha}{\beta}\sqrt{\lambda}\right)\right]\right] \\ &= e^{-i(\alpha/\beta)\sqrt{\lambda}v}\phi_{X_t}\left(\frac{v}{\beta\sqrt{\lambda}}\right)\end{aligned}$$

On using (7.8), we then obtain the formula

$$\phi_{Y_t}(v) = \exp\left(-i\frac{\alpha}{\beta}\sqrt{\lambda}\, v + \lambda \int_0^t \left[\exp\left(i\frac{v}{\beta\sqrt{\lambda}}h(u)\right) - 1\right] du\right) \tag{7.11}$$

for the characteristic function of a normalized, filtered Poisson random variable.

Exercise 7.5

a. Using a power-series expansion of the exponential term which is inside the integral with respect to u in (7.11), show that

$$\log \phi_{Y_t}(v) = -\frac{v^2}{2} - \frac{iv^3}{3!\beta^3\lambda^{1/2}}\int_0^t h^3(u)\, du + \frac{v^4}{4!\beta^4\lambda}\int_0^t h^4(u)\, du + \cdots$$

b. Show that it therefore follows from (*a*) that

$$\lim_{\lambda\to\infty} \phi_{Y_t}(v) = e^{-v^2/2}$$

if the integrals of the higher powers of h there all exist. □

Thus we see that *the characteristic function of the normalized, filtered Poisson random variable Y_t tends to that of a gaussian random variable with a zero mean and a unit variance.* Since a similar derivation for the limiting form of the joint-characteristic function of any n random variables $Y_{t_1}, Y_{t_2}, \ldots, Y_{t_n}$ also results in a gaussian characteristic function, it therefore follows that *a filtered Poisson process tends to become a gaussian random process as the average number of events occurring per unit time increases without limit.*

8 RANDOM PARTITIONING

A number of physical phenomena may be modeled by what might be called *random partitioning* of a filtered Poisson process, say, for example, the splitting of a cathode current between the plate and the screen of a pentode (cf. North's paper or sec. 7.8 of Davenport and Root). In such cases, one is given a basic filtered Poisson process, say $\{X_t, 0 \le t < +\infty\}$, and a new process, say $\{Z_t, 0 \le t < +\infty\}$, is then formed by randomly selecting some of the basic events and not others. That is, if we write

$$X_t = \sum_{j=1}^{N_t} h(t - U_j) \tag{8.1}$$

for the basic process as in the preceding section (where N_t is a Poisson counting random variable and where the U_j are the unordered arrival times of the basic events), then we have

$$Z_t \triangleq \sum_{j=1}^{N_t} Y_j h(t - U_j) \tag{8.2}$$

for the new process, where each of the partitioning random variables Y_j has the same Bernoulli probability distribution

$$P[Y_j = 1] = p \qquad \text{and} \qquad P[Y_j = 0] = q \triangleq 1 - p \tag{8.3}$$

where $0 < p < 1$, and where the Y_j are mutually independent random variables which are independent of the event-time random variables U_j. We now wish to show that the partitioned process is a filtered Poisson process with a rate which is p times the rate of the basic process.

The technique which we shall employ here is essentially the same as that used in our study of the filtered Poisson process. Suppose, for example, that we wish to calculate the expected value of the new process. We then write

$$E[Z_t] = \sum_{k=0}^{\infty} P[N_t = k]E[Z_t|N_t = k] \tag{8.4}$$

where here the conditional expectation $E[Z_t|N_t = k]$ must be determined by averaging over *both* the event-time random variables $U_1, U_2, \ldots, U_k$ *and* the partitioning random variables $Y_1, Y_2, \ldots, Y_k$. Now, since the expectation of a sum is a sum of expectations, we have

$$\begin{aligned} E[Z_t|N_t = k] &= E\left[\sum_{j=1}^{k} Y_j h(t - U_j)\right] \\ &= \sum_{j=1}^{k} E[Y_j h(t - U_j)] \end{aligned} \tag{8.5}$$

Then, using the facts that Y_j and U_j are independent random variables and Y_j has the Bernoulli distribution (8.3) while U_j is uniformly distributed over $(0,t]$, we get

$$\begin{aligned} E[Y_j h(t - U_j)] &= E[Y_j]E[h(t - U_j)] \\ &= p\frac{1}{t}\int_0^t h(t - u_j)\,du_j = \frac{p}{t}\int_0^t h(u)\,du \end{aligned}$$

On substituting this result back into (8.5), and thence into (8.4), we obtain

$$E[Z_t] = \sum_{k=0}^{\infty} P[N_t = k] \left[\sum_{j=1}^{k} \frac{p}{t} \int_0^t h(u)\, du \right]$$

$$= \frac{p}{t} \int_0^t h(u)\, du \sum_{j=1}^{\infty} kP[N_t = k]$$

$$= p\lambda \int_0^t h(u)\, du \tag{8.6}$$

the last equality following from (2.17). Since we showed in the preceding section that

$$E[X_t] = \lambda \int_0^t h(u)\, du$$

it therefore follows that

$$E[Z_t] = pE[X_t] \tag{8.7}$$

That is, *the expected value after random partitioning is simply p times the expected value before partitioning.*

Exercise 8.1 Let $\{Z_t, 0 \leq t < +\infty\}$ be a randomly partitioned filtered Poisson random process as defined above. Show that the characteristic function of Z_t is given by the formula

$$\phi_{Z_t}(v) = \exp\left[p\lambda \int_0^t (e^{ivh(u)} - 1)\, du \right]$$

On comparing this result with (7.8) above, we see that the characteristic function of Z_t is that of a filtered Poisson process with a rate equal to $p\lambda$. □

We note that it follows from the results of Exercises 8.1 and 7.4 that the variance of the randomly partitioned process is

$$\text{var}\,(Z_t) = p\lambda \int_0^t h^2(u)\, du$$

and hence that

$$\sigma^2(Z_t) = \text{var}\,(Z_t) = p\,\text{var}\,(X_t) = p\sigma^2(X_t) \tag{8.8}$$

It therefore follows from this result, and from (8.7) above, that

$$\frac{\sigma(Z_t)}{E[Z_t]} = \frac{\sqrt{p}\,\sigma(X_t)}{pE[X_t]}$$

$$= \frac{1}{\sqrt{p}} \frac{\sigma(X_t)}{E[X_t]} \tag{8.9}$$

Since $0 < p < 1$, we see that *the effect of random partitioning of a filtered Poisson process is always to increase the ratio of the standard deviation to the mean, that is, to increase the "noisiness" of the process.*

9 SUMMARY OF DEFINITIONS AND FORMULAS

Poisson process

Let $\{N_t, 0 \leq t < +\infty\}$ be a counting random process such that:

a. N_t assumes only nonnegative integer values and

$$N_0 \triangleq 0 \tag{2.2}$$

b. The process has stationary and independent increments,

c.

$$P[N_{t+\Delta t} - N_t = 1] = \lambda\,\Delta t + 0(\Delta t) \tag{2.3}$$

where λ is a positive constant, and

d.

$$P[N_{t+\Delta t} - N_t > 1] = 0(\Delta t) \tag{2.4}$$

where

$$\lim_{\Delta t \to 0} \frac{0(\Delta t)}{\Delta t} = 0 \tag{2.5}$$

It then follows that

$$P[N_{t+\Delta t} - N_t = 0] = 1 - \lambda\,\Delta t + 0(\Delta t) \tag{2.6}$$

and that

$$P[N_t = k] = \frac{e^{-\lambda\tau}(\lambda t)^k}{k!} \tag{2.10}$$

for $k = 0, 1, 2, \ldots$; that is, N_t is a Poisson random variable. The counting process $\{N_t, 0 \leq t < +\infty\}$ is then called a Poisson counting process.

The expected value of the Poisson process is

$$E[N_t] = \lambda t \tag{2.17}$$

and the variance is

$$\text{var}\,(N_t) = \lambda t \tag{2.19}$$

Arrival times

Let T_k be the random variable which describes the arrival time of the kth event counted by the counting random process $\{N_t, 0 \leq t < +\infty\}$.

Then

$$F_{T_k}(t) = 1 - F_{N_t}(k - 1) \tag{3.1}$$

If $\{N_t, 0 \leq t < +\infty\}$ is a Poisson counting process, then

$$F_{T_k}(t) = \begin{cases} 1 - e^{-\lambda t} \sum_{j=0}^{k-1} \frac{(\lambda t)^j}{j!} & \text{for } t \geq 0 \\ 0 & \text{for } t < 0 \end{cases} \tag{3.3}$$

and hence

$$f_{T_k}(t) = \begin{cases} \lambda e^{-\lambda t} \frac{(\lambda t)^{k-1}}{(k-1)!} & \text{for } t \geq 0 \\ 0 & \text{for } t < 0 \end{cases} \tag{3.4}$$

that is, T_k has an Erlang probability density. In this case:

$$E[T_k] = \frac{k}{\lambda} \qquad \text{(Exercise 3.1)}$$

$$\text{var}\,(T_k) = \frac{k}{\lambda^2} \qquad \text{(Exercise 3.1)}$$

and

$$\phi_{T_k}(v) = \frac{1}{\left(1 - \frac{iv}{\lambda}\right)^k} \qquad \text{(Exercise 3.2)}$$

Interarrival times

Let $\{N_t, 0 \leq t < +\infty\}$ be a counting process with the arrival times $\{T_k, k = 1, 2, 3, \ldots\}$. The durations

$$\begin{aligned} Z_1 &\triangleq T_1 \\ Z_k &\triangleq T_k - T_{k-1} \qquad \text{for } k = 2, 3, 4, \ldots \end{aligned} \tag{4.1}$$

are called the interarrival times of the counting process. We then have

$$F_{Z_k}(\tau) = 1 - P[N_{t_{k-1}+\tau} - N_{t_{k-1}} = 0] \tag{4.2}$$

If the counting process has stationary increments, then, for all k,

$$F_{Z_k}(\tau) = 1 - P[N_\tau = 0] \tag{4.3}$$

Further, if the given counting process is Poisson, then

$$F_{Z_k}(\tau) = \begin{cases} 1 - e^{-\lambda\tau} & \text{for } \tau \geq 0 \\ 0 & \text{for } \tau < 0 \end{cases} \tag{4.4}$$

and hence

$$f_{Z_k}(\tau) = \begin{cases} \lambda e^{-\lambda\tau} & \text{for } \tau \geq 0 \\ 0 & \text{for } \tau < 0 \end{cases} \tag{4.5}$$

In this case

$$E[Z_k] = \frac{1}{\lambda} \qquad \text{(Exercise 4.2)}$$

for all $k = 1, 2, 3, \ldots$. In *any* case

$$E[T_k] = kE[Z_k] \qquad \text{(Exercise 4.3)}$$

Renewal counting processes

Let $\{N_t, 0 \leq t < +\infty\}$ be a counting process. If the interarrival times of this counting process are mutually independent random variables, all with the same probability distribution function, then the given process is called a renewal counting process.

The renewal function $m(t)$ of a renewal counting process is the expected value of that process; that is,

$$m(t) \triangleq E[N_t] \tag{5.3}$$

and its derivative

$$\lambda(t) \triangleq \frac{dm(t)}{dt} \tag{5.4}$$

is called the renewal intensity of the process. It then follows that

$$m(t) = \sum_{k=1}^{\infty} F_{T_k}(t) \tag{5.6}$$

and, if the various derivatives exist,

$$\lambda(t) = \sum_{k=1}^{\infty} f_{T_k}(t) \tag{5.7}$$

On defining $\Lambda(v)$ to be the Fourier transform of the renewal intensity, that is,

$$\Lambda(v) \triangleq \int_{-\infty}^{+\infty} \lambda(t)e^{ivt}\, dt \tag{5.8}$$

it then follows that

$$\Lambda(v) = \frac{\phi_Z(v)}{1 - \phi_Z(v)} \tag{5.9}$$

where ϕ_Z is the common characteristic function of the interarrival times.

Unordered arrival times

Let $\{N_t, 0 \leq t < +\infty\}$ be a Poisson counting process and suppose that $N_t = k$; that is, suppose that k events occur by time t. The unordered arrival times $U_1, U_2, \ldots, U_k$ of those k events are then mutually independent random variables, each of which is uniformly distributed over the interval $(0,t]$:

$$f_{U_i}(u_i|N_t = k) = \begin{cases} \frac{1}{t} & \text{for } 0 < u_i \leq t \\ 0 & \text{otherwise} \end{cases} \tag{6.7}$$

for all $i = 1, 2, \ldots, k$

Filtered Poisson processes

Let $\{N_t, 0 \leq t < +\infty\}$ be a Poisson counting process. The random process $\{X_t, 0 \leq t < +\infty\}$ in which

$$X_t \triangleq \sum_{j=1}^{N_t} h(t - U_j) \tag{7.1}$$

where an event which occurs at time u_j generates an outcome $h(t - u_j)$ at time t and where the random variables U_j are the unordered arrival times of the events which occur during the interval $(0,t]$, is called a filtered Poisson process.

The mean of a filtered Poisson process is

$$E[X_t] = \lambda \int_0^t h(u)\, du \tag{7.5}$$

the variance is

$$\text{var}\,(X_t) = \lambda \int_0^t h^2(u)\, du \tag{7.6}$$

and the characteristic function of the random variable X_t is

$$\phi_{X_t}(v) = \exp\left[\lambda \int_0^t (e^{ivh(u)} - 1)\, du\right] \tag{7.8}$$

Random partitioning

Let $\{N_t, 0 \leq t < +\infty\}$ be a Poisson counting process and let $\{X_t, 0 \leq t < +\infty\}$ be the corresponding filtered Poisson process in which

$$X_t \triangleq \sum_{j=1}^{N_t} h(t - U_j)$$

We say that the random process $\{Z_t, 0 \leq t < +\infty\}$ is a randomly partitioned filtered Poisson random process if

$$Z_t \triangleq \sum_{j=1}^{N_t} Y_j h(t - U_j) \tag{8.2}$$

where the partitioning random variables Y_j are mutually independent random variables which are independent of the unordered arrival times U_j, and where each of the Y_j has the same Bernoulli probability distribution

$$P[Y_j = 1] = p \qquad \text{and} \qquad P[Y_j = 0] = q \triangleq 1 - p$$

where $0 < p < 1$.

In this case,

$$E[Z_t] = p\lambda \int_0^t h(u)\,du \tag{8.5}$$

$$= pE[X_t] \tag{8.6}$$

The characteristic function of the randomly partitioned random variable Z_t is

$$\phi_{Z_t}(v) = \exp\left[p\lambda \int_0^t (e^{ivh(u)} - 1)\,du\right] \qquad \text{(Exercise 8.1)}$$

REFERENCES

Brockmeyer, E., H. L. Halstrøm, and E. Jensen: "The Life and Works of A. K. Erlang," Copenhagen Telephone Company, Copenhagen, 1948.

Cox, David R.: "Renewal Theory," Methuen & Co., Ltd., London (John Wiley & Sons, Inc., New York), 1962.

Cox, David R., and Walter L. Smith: "Queues," Methuen & Co., Ltd., London (John Wiley & Sons, Inc., New York), 1961.

Davenport, Wilbur B., Jr., and William L. Root: "Random Signals and Noise," McGraw-Hill Book Company, New York, 1958.

Gilbert, Edgar N., and Henry O. Pollak: Amplitude Distribution of Shot Noise, *Bell System Tech. J.*, vol. 39, no. 2, pp. 333–350, 1960.

Karlin, Samuel: "A First Course in Stochastic Processes," Academic Press, Inc., New York, 1966.

Lee, Alec M.: "Applied Queuing Theory," Macmillan & Co., Ltd., London (St Martin's Press, New York), 1966.

North, Dwight O.: Fluctuations in Space-Charge Currents, pt. 3, Multi-Collectors, *RCA Rev.*, vol. 5, pp. 244–260, 1940.

Parzen, Emanuel: "Stochastic Processes," Holden-Day, Inc., Publisher, San Francisco, 1962.

14
The Gaussian Process

1 INTRODUCTION

We have seen throughout this text that gaussian random variables, vectors, and processes have come up in our discussions time and time again. The practical reason for this pervasiveness is that the random variables of interest in many physical problems are the results of adding large numbers of independent random variables; the central limit theorem therefore applies in such cases and the resulting sum random variables are gaussian to all practical intents and purposes (as, for example, in the limiting case of the filtered Poisson process discussed in Sec. 13.7). Even though we have already derived many of their most significant properties, we wish to emphasize the great importance of the gaussian random variables and processes and shall do so by repeating here some of our previous results and then deriving some new results in the following sections.

We first defined a *gaussian* (or *normal*) *probability density* to be

$$f(x) = \frac{1}{\sqrt{2\pi}} e^{-x^2/2} \tag{1.1}$$

[cf. (4.5.12)]. We later pointed out that (1.1) was the probability density of a random variable which had a zero mean and a unit variance and that, more generally, a *gaussian random variable* X with mean m and variance σ^2 had the probability density

$$f_X(x) = \frac{1}{\sqrt{2\pi}\,\sigma} \exp\left[-\frac{(x-m)^2}{2\sigma^2}\right] \tag{1.2}$$

[cf. (7.8.15)]. Finally, we said that the random variables X_1, X_2, . . . , X_n were *jointly gaussian random variables* if and only if their joint-probability density had the form

$$f_{X_1,X_2,\ldots,X_n}(x_1, x_2, \ldots, x_n) = \frac{1}{(2\pi)^{n/2}|\mathbf{\Lambda}|^{1/2}} \exp\left[-\frac{1}{2|\mathbf{\Lambda}|}\sum_{j=1}^{n}\sum_{k=1}^{n}|\mathbf{\Lambda}|_{jk}(x_j - m_j)(x_k - m_k)\right] \tag{1.3}$$

where m_j is the mean value of X_j,

$$m_j \triangleq E[X_j] \qquad j = 1, 2, \ldots, n \tag{1.4}$$

and where $|\mathbf{\Lambda}|_{jk}$ is the cofactor of the element λ_{jk} in the determinant $|\mathbf{\Lambda}|$ of the covariance matrix

$$\mathbf{\Lambda} \triangleq \begin{bmatrix} \lambda_{11} & \lambda_{12} & \cdots & \lambda_{1n} \\ \lambda_{21} & \lambda_{22} & \cdots & \lambda_{2n} \\ \cdots & \cdots & \cdots & \cdots \\ \lambda_{n1} & \lambda_{n2} & \cdots & \lambda_{nn} \end{bmatrix} \tag{1.5}$$

in which the elements λ_{jk} are the covariances

$$\lambda_{jk} \triangleq \text{cov}\,(X_j, X_k) \tag{1.6}$$

(cf. Sec. 7.10). While it undoubtedly seems that the form given above for the gaussian joint-probability density was pulled out of thin air, we shall show in the next section that that form is, in fact, a quite logical choice.

In the following sections of this chapter, we shall study the effects of linear transformations on gaussian random vectors; then we shall define the gaussian random process and study a few of its basic properties; finally, we shall introduce the special (but nonetheless practically important) case of the narrowband gaussian process and study some of its properties. More extensive discussions of gaussian random variables and vectors may be found in chaps. 17 and 24 of Cramér's book and in Miller's book, while more complete discussions of gaussian random processes in applied contexts may be found in Rice's papers, in chap. 8

of Davenport and Root, in Blachman's book, and in pt. 2 (chaps. 7 through 11) of Middleton's book.

Exercise 1.1 A probability density closely related to the gaussian or normal is the *log-normal* density which arises in certain wave-propagation problems.

Suppose, for example, that a plane wave propagating in the $+x$ direction passes through a layered medium as indicated in Fig. 1.1. Suppose further that the attenuation constant of the ith layer is α_i and that the thickness of that layer is Δx_i, so that if the amplitude of the wave entering the ith layer is y_{i-1}, then the amplitude of the wave leaving that layer is

$$y_i = y_{i-1}e^{-\alpha_i \Delta x_i}$$

It therefore follows that if the amplitude of the wave initially entering the layered medium is A, then the amplitude of the wave leaving the nth layer is

$$\begin{aligned} y_n &= Ae^{-\alpha_1 \Delta x_1}e^{-\alpha_2 \Delta x_2} \cdots e^{-\alpha_n \Delta x_n} \\ &= Ae^{-z_n} \end{aligned}$$

where we have defined

$$z_n \triangleq \sum_{i=1}^{n} \alpha_i \, \Delta x_i$$

Now in many problems of interest, the attentuation constants and layer thicknesses vary randomly and independently from layer to layer. If the number of layers is very large, then the central limit theorem applies and we may model the situation reasonably accurately by writing

$$Y = e^{-Z}$$

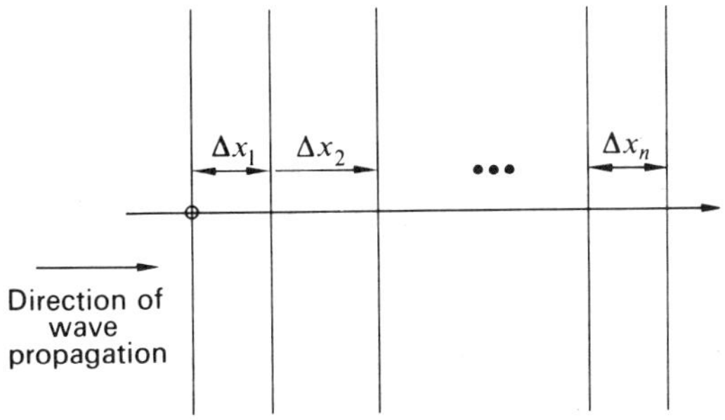

Figure 1.1 Wave propagation through a layered medium.

where Y is the (normalized) random variable which characterizes the final wave amplitude, and where Z has a gaussian probability density, say

$$f_Z(z) = \frac{1}{\sqrt{2\pi}\,\sigma} \exp\left[-\frac{1}{2}\left(\frac{z-m}{\sigma}\right)^2\right]$$

Show that in this case the probability density of the final wave amplitude is given by the formula

$$f_Y(y) = \begin{cases} \dfrac{1}{\sqrt{2\pi}\,y\sigma} \exp\left[-\dfrac{1}{2\sigma^2}(\log y - m)^2\right] & \text{for } y > 0 \\ 0 & \text{for } y \le 0 \end{cases}$$

This density is called the *log-normal* probability density.[1] □

2 GAUSSIAN RANDOM VECTORS*

We now wish to show that the choice of the form of the gaussian joint-probability density given in (1.3) above was not without reason. To this end, let us consider for the moment the n-dimensional random vector $\mathbf{X}$ whose components $X_1, X_2, \ldots, X_n$ are mutually independent gaussian random variables with means $m_1, m_2, \ldots, m_n$ and variances $\sigma_1^2, \sigma_2^2, \ldots, \sigma_n^2$, respectively. It then follows from (1.2) above that the probability density of the kth component is

$$f_{X_k}(x) = \frac{1}{\sqrt{2\pi}\,\sigma_k} \exp\left[-\frac{(x-m_k)^2}{2\sigma_k^2}\right] \tag{2.1}$$

for all $k = 1, 2, \ldots, n$. It further follows, from the result of Exercise 12.4.5, that the characteristic function of the kth component is

$$\phi_{X_k}(v) = \exp\left(ivm_k - \frac{v^2\sigma_k^2}{2}\right) \tag{2.2}$$

Since the components $X_1, X_2, \ldots, X_n$ are assumed to be mutually independent here, their joint characteristic function is the product of their respective individual characteristic functions (cf. Exercise 12.6.1; that is,

$$\begin{aligned} \phi_{\mathbf{X}}(\mathbf{v}) &\triangleq \phi_{X_1,X_2,\ldots,X_n}(v_1, v_2, \ldots, v_n) \\ &= \prod_{k=1}^{n} \phi_{X_k}(v_k) \end{aligned}$$

[1] A comprehensive treatment of the properties of the log-normal density is given by Aitchison and Brown.

* This section will make use of some of the matrix results derived in Sec. 10.3.

Thus it follows from this result, and from (2.2), that

$$\phi_x(\mathbf{v}) = \prod_{k=1}^{n} \exp\left(iv_k m_k - \frac{v_k^2\sigma_k^2}{2}\right)$$
$$= \exp\left(i \sum_{k=1}^{n} v_k m_k - \sum_{k=1}^{n} \frac{v_k^2\sigma_k^2}{2}\right) \tag{2.3}$$

It is convenient at this time to introduce again the use of matrix notation. Suppose, therefore, that we define the matrix

$$\mathbf{v} \triangleq \begin{bmatrix} v_1 \\ v_2 \\ \cdot \\ \cdot \\ \cdot \\ v_n \end{bmatrix}$$

the mean matrix

$$\mathbf{m}_x \triangleq E[\mathbf{X}] = \begin{bmatrix} m_1 \\ m_2 \\ \cdot \\ \cdot \\ \cdot \\ m_n \end{bmatrix}$$

and the covariance matrix

$$\mathbf{\Gamma} \triangleq \begin{bmatrix} \sigma_1^2 & 0 & 0 & \cdots & 0 \\ 0 & \sigma_2^2 & 0 & \cdots & 0 \\ \cdot & \cdot & \cdot & \cdots & \cdot \\ 0 & 0 & 0 & \cdots & \sigma_n^2 \end{bmatrix}$$

It then follows from the rules of matrix multiplication that we may write

$$\sum_{k=1}^{n} v_k m_k = \mathbf{m}_x^T\mathbf{v}$$

(where, as usual, the superscript T denotes matrix transpose), and

$$\sum_{k=1}^{n} v_k^2\sigma_k^2 = \mathbf{v}^T\mathbf{\Gamma}\mathbf{v}$$

If now we use these last two results in (2.3), we obtain

$$\phi_{\mathbf{x}}(\mathbf{v}) = \exp\left(i\mathbf{m}_x^T\mathbf{v} - \tfrac{1}{2}\mathbf{v}^T\mathbf{\Gamma}\mathbf{v}\right) \tag{2.4}$$

as the matrix form of the joint-characteristic function of the n-dimensional random vector **X** whose components are mutually independent gaussian random variables.

Next, let us consider a new m-dimensional random vector **Y** (where $m \leq n$) defined to be a linear transformation of the random vector **X** defined above; that is, suppose that

$$\begin{aligned} Y_1 &= g_{11}X_1 + g_{12}X_2 + \cdots + g_{1n}X_n \\ Y_2 &= g_{21}X_1 + g_{22}X_2 + \cdots + g_{2n}X_n \\ &\cdots\cdots\cdots\cdots\cdots\cdots \\ Y_m &= g_{m1}X_1 + g_{m2}X_2 + \cdots + g_{mn}X_n \end{aligned}$$

where the coefficients g_{jk} are arbitrary real numbers. The matrix form of this transformation is

$$\mathbf{Y} = \mathbf{g}\mathbf{X} \tag{2.5}$$

where **Y** is the column matrix of elements $Y_1, Y_2, \ldots, Y_m$, and where **g** is the transformation matrix

$$\mathbf{g} \triangleq \begin{bmatrix} g_{11} & g_{12} & \cdots & g_{1n} \\ g_{21} & g_{22} & \cdots & g_{2n} \\ \cdots & \cdots & \cdots & \cdots \\ g_{m1} & g_{m2} & \cdots & g_{mn} \end{bmatrix}$$

The joint-characteristic function of this new random vector **Y** is, by definition,

$$\begin{aligned} \phi_{\mathbf{Y}}(\mathbf{u}) &= E\left[\exp\left(i \sum_{j=1}^{m} u_j Y_j\right)\right] \\ &= E[\exp(i\mathbf{u}^T\mathbf{Y})] \end{aligned}$$

[cf. (12.6.2)] where **u** is the column matrix of elements $u_1, u_2, \ldots, u_m$. Then, taking into account the fact that **Y** is defined as a linear transformation of **X**, we may write

$$\begin{aligned} \phi_{\mathbf{Y}}(\mathbf{u}) &= E[\exp(i\mathbf{u}^T\mathbf{g}\mathbf{X})] \\ &= \phi_{\mathbf{X}}(\mathbf{g}^T\mathbf{u}) \end{aligned} \tag{2.6}$$

since

$$(\mathbf{g}^T\mathbf{u})^T = \mathbf{u}^T\mathbf{g}$$

It therefore follows from (2.4) and (2.6) that we obtain

$$\phi_{\mathbf{Y}}(\mathbf{u}) = \exp(i\mathbf{m}_{\mathbf{X}}{}^T\mathbf{g}^T\mathbf{u} - \tfrac{1}{2}\mathbf{u}^T\mathbf{g}\mathbf{\Gamma}\mathbf{g}^T\mathbf{u}) \tag{2.7}$$

as a formula for the joint-characteristic function of the m-dimensional random vector **Y**, when **Y** is defined as a linear transformation of the n-

dimensional random vector $\mathbf{X}$ whose components are mutually independent gaussian random variables.

We now wish to rewrite (2.7) in terms of the mean and covariance matrices of the random vector $\mathbf{Y}$, rather than in terms of those of $\mathbf{X}$. First, we note that it follows from (10.3.2) that

$$\begin{aligned}\mathbf{m}_\mathbf{Y} &\triangleq E[\mathbf{Y}] = \mathbf{g}E[\mathbf{X}] \\ &= \mathbf{g}\mathbf{m}_\mathbf{X}\end{aligned} \tag{2.8}$$

Next, we note that if we denote the covariance matrix of $\mathbf{Y}$ by $\mathbf{\Lambda}$, that is, if we write

$$\mathbf{\Lambda} \triangleq \begin{bmatrix} \lambda_{11} & \lambda_{21} & \cdots & \lambda_{1m} \\ \lambda_{21} & \lambda_{22} & \cdots & \lambda_{2m} \\ \cdots & \cdots & \cdots & \cdots \\ \lambda_{m1} & \lambda_{m2} & \cdots & \lambda_{mm} \end{bmatrix}$$

where the elements λ_{jk} are the covariances

$$\lambda_{jk} \triangleq \operatorname{cov}(Y_j, Y_k)$$

then it follows from (10.3.10) that

$$\mathbf{\Lambda} = \mathbf{g}\mathbf{\Gamma}\mathbf{g}^T \tag{2.9}$$

where $\mathbf{g}$ is the transformation matrix defined above and where $\mathbf{\Gamma}$ is the covariance matrix of the random vector $\mathbf{X}$. On using (2.8) and (2.9) in (2.7), we finally obtain

$$\phi_\mathbf{Y}(\mathbf{u}) = \exp\,(i\mathbf{m}_\mathbf{Y}{}^T\mathbf{u} - \tfrac{1}{2}\mathbf{u}^T\mathbf{\Lambda}\mathbf{u}) \tag{2.10}$$

as the matrix form of the joint-characteristic function of the random vector $\mathbf{Y}$ expressed in terms of its own mean and covariance matrices. Expansion of the matrix products in (2.10) then gives

$$\begin{aligned}\phi_{Y_1,Y_2,\ldots,Y_m}&(u_1, u_2, \ldots, u_m) \\ &= \exp\left\{i\sum_{j=1}^{m} E[Y_j]u_j - \frac{1}{2}\sum_{j=1}^{m}\sum_{k=1}^{m} \operatorname{cov}(Y_j,Y_k)u_ju_k\right\}\end{aligned} \tag{2.11}$$

as the component form of the joint-characteristic function of the m-dimensional random vector $\mathbf{Y}$.

Exercise 2.1 Show that it follows from (2.11) that

$$\phi_{Y_k}(v) = \exp\left\{ivE[Y_k] - \frac{v^2\sigma^2(Y_k)}{2}\right\}$$

for all $k = 1, 2, \ldots, m$.

Thus we see that *if the joint-characteristic function of the random vector* **Y** *has the form given in* (2.10) *or* (2.11), *then the components of* **Y** *are individually gaussian random variables.* □

On comparing the formula in (2.10) with that in (2.4) above, we note that the joint-characteristic function of the random vector **Y** has exactly the same functional form as does the joint-characteristic function of the random vector **X**, when **Y** is defined to be a linear transformation of **X** and when the components of **X** are mutually independent gaussian random variables.

Exercise 2.2 Let **Y** be the m-dimensional random vector defined above and let **Z** be an r-dimensional random vector (where $r \leq m$) defined as a finite linear transformation of **Y**. That is, suppose that

$$\begin{aligned} Z_1 &= h_{11}Y_1 + h_{12}Y_2 + \cdots + h_{1m}Y_m \\ Z_2 &= h_{21}Y_1 + h_{22}Y_2 + \cdots + h_{2m}Y_m \\ &\cdots\cdots\cdots\cdots\cdots\cdots\cdots\cdots \\ Z_r &= h_{r1}Y_1 + h_{r2}Y_2 + \cdots + h_{rm}Y_m \end{aligned}$$

or, in matrix form,

$$\mathbf{Z} \triangleq \mathbf{hY}$$

where **h** is the transformation matrix with elements h_{jk}. Show that the joint-characteristic function of the random vector **Z** has the form

$$\phi_{\mathbf{Z}}(\mathbf{w}) = \exp\,(i\mathbf{m_Z}^T\mathbf{w} - \tfrac{1}{2}\mathbf{w}^T\mathbf{u}\mathbf{w})$$

in this case, where $\mathbf{m_Z}$ is the mean-matrix for **Z** and where **u** is the covariance matrix **Z**.

Thus we see that any further finite linear transformation of the random vector **Y** *still preserves the functional form of the joint-characteristic function.* □

With the results expressed by (2.4), (2.10), and Exercises 2.1 and 2.2 in mind, we define the m-dimensional random vector **Y** to be a *gaussian random vector* (and say that its components $Y_1, Y_2, \ldots, Y_m$ are *jointly gaussian random variables*) if and only if the joint-characteristic function of **Y** has the matrix form

$$\phi_{\mathbf{Y}}(\mathbf{v}) = \exp\,(i\mathbf{m_Y}^T\mathbf{v} - \tfrac{1}{2}\mathbf{v}^T\mathbf{\Lambda}\mathbf{v}) \tag{2.12}$$

where $\mathbf{m_Y}$ is the mean matrix for **Y** and where $\mathbf{\Lambda}$ is the covariance-matrix for **Y**. It should be noted that it follows immediately from this definition of a gaussian random vector, and from the result of Exercise 2.2, that *any finite linear transformation of a gaussian random vector yields a gaussian* *random vector.* This result is a generalization of the special result obtained previously to the effect that a sum of independent gaussian random variables is a gaussian random variable (cf. Exercise 12.4.6).

While its proof is beyond us here, it is worth noting that the converse of our last result is also true. That is, it can be shown[1] that if the random vector **Y** is a gaussian random vector and if **Y** is the result of a linear transformation of the random vector **X**, then **X** must also be a gaussian random vector. Thus it follows that *if the random vector* **Y** *is the result of a linear transformation of the random vector* **X**, *then* **Y** *is a gaussian random vector if and only if* **X** *is a gaussian random vector.* With this last result in mind, it is sometimes found to be convenient to define a random vector to be a gaussian random vector if and only if all linear combinations of its components result in gaussian random variables (see, for example, sec. 2.6 of pt. I of Van Trees).

Exercise 2.3 Let X_1, X_2, X_3, and X_4 be jointly gaussian random variables, each with a zero mean. Show that

$$E[X_1X_2X_3X_4] = E[X_1X_2]E[X_3X_4] + E[X_1X_3]E[X_2X_4] + E[X_1X_4]E[X_2X_3]$$

(*Hint:* Use the moment-generating property of the joint-characteristic function—but be careful of the details!) □

Gaussian joint-probability density

Having defined (2.12) to be the joint-characteristic function of the m-dimensional gaussian random vector Y, we now wish to obtain the corresponding gaussian joint-probability density.

First, we recall that the matrix form (2.12) for the gaussian joint-characteristic function is the same for all gaussian random vectors—whether the individual components are independent random variables or not. Since the joint-probability density here is the m-dimensional Fourier transform of the joint-characteristic function, we might suspect that the matrix form of the gaussian joint-probability density would be the same for all gaussian random vectors, again whether the individual components are independent or not. With this suspicion in mind, let us consider the easy independent-component case first, and then extend our results to the general case.

Suppose that **X** is an m-dimensional gaussian random vector with independent components. The probability density of the ith component then is

$$f_{X_i}(X_i) = \frac{1}{\sqrt{2\pi}\,\sigma_i} \exp\left[-\frac{1}{2}\left(\frac{x_i - m_i}{\sigma_i}\right)^2\right]$$

[1] See, for example, sec. 10.2 of Cramér's tract.

where, as usual,

$$m_i \triangleq E[X_i] \qquad \text{and} \qquad \sigma_i^2 \triangleq \text{var}\,(X_i)$$

Since the components here are independent by hypothesis, the joint-probability density of the random vector $\mathbf{X}$ is the m-fold product of such terms. Thus:

$$\begin{aligned} f_{X_1,X_2,\ldots,X_m}(x_1, x_2, \ldots, x_m) &= \prod_{i=1}^{m} \frac{1}{\sqrt{2\pi}\,\sigma_i} \exp\left[-\frac{1}{2}\left(\frac{x_i - m_i}{\sigma_i}\right)^2\right] \\ &= \frac{1}{(2\pi)^{m/2}\left(\prod_{i=1}^{m} \sigma_i\right)} \exp\left[-\frac{1}{2}\sum_{i=1}^{m}\left(\frac{x_i - m_i}{\sigma_i}\right)^2\right] \end{aligned} \tag{2.13}$$

We would now like to rewrite this result in matrix form. Now the covariance matrix of $\mathbf{X}$ in this case has the form

$$\mathbf{\Gamma} = \begin{bmatrix} \sigma_1^2 & 0 & 0 & \cdots & 0 \\ 0 & \sigma_2^2 & 0 & \cdots & 0 \\ \cdot & \cdot & \cdot & \cdots & \cdot \\ 0 & 0 & 0 & \cdots & \sigma_m^2 \end{bmatrix}$$

It therefore follows that

$$\begin{aligned} \prod_{i=1}^{m} \sigma_i &= \sigma_1\sigma_2 \cdots \sigma_m \\ &= |\mathbf{\Gamma}|^{1/2} \end{aligned} \tag{2.14}$$

Further, since the covariance matrix here is nonsingular, it has an inverse; in fact,

$$\mathbf{\Gamma}^{-1} = \begin{bmatrix} \dfrac{1}{\sigma_1^2} & 0 & 0 & \cdots & 0 \\ 0 & \dfrac{1}{\sigma_2^2} & 0 & \cdots & 0 \\ \cdot & \cdot & \cdot & \cdots & \cdot \\ 0 & 0 & 0 & \cdots & \dfrac{1}{\sigma_m^2} \end{bmatrix}$$

With this result in mind, we can then write

$$\sum_{i=1}^{m}\left(\frac{x_i - m_i}{\sigma_i}\right)^2 = (\mathbf{x}^T - \mathbf{m}_\mathbf{x}^T)\mathbf{\Gamma}^{-1}(\mathbf{x} - \mathbf{m}_\mathbf{x}) \tag{2.15}$$

Thus it follows, from (2.13) through (2.15), that the joint-probability density of the random vector X has the matrix form

$$f_{\mathbf{X}}(\mathbf{x}) = \frac{1}{(2\pi)^{m/2}|\mathbf{\Gamma}|^{1/2}} \exp\left[-\tfrac{1}{2}(\mathbf{x}^T - \mathbf{m}_{\mathbf{X}}{}^T)\mathbf{\Gamma}^{-1}(\mathbf{x} - \mathbf{m}_{\mathbf{X}})\right] \tag{2.16}$$

when $\mathbf{X}$ is a gaussian random vector with independent components. Using a preservation-of-form argument, then, we suggest that the joint-probability density of any m-dimensional gaussian random vector has the form given in (2.16)—so long as the inverse of the covariance matrix exists. That this conclusion is in fact true is shown by the following exercise.

Exercise 2.4 Let $\mathbf{X}$ be an m-dimensional gaussian random vector with independent components so that its joint-probability density is given in matrix form by (2.16). Let the m-dimensional random vector Y be defined as a one-to-one linear transformation of $\mathbf{X}$:

$$\mathbf{Y} = \mathbf{g}\mathbf{X}$$

where g is a one-to-one linear transformation. $\mathbf{Y}$ is therefore a gaussian random vector also, and its joint-characteristic function is given by (2.10).

Show, using the techniques discussed in Sec. 6.4, that the joint-probability density of $\mathbf{Y}$ is given in this case by the formula

$$f_Y(y) = \frac{1}{(2\pi)^{m/2}|\mathbf{\Lambda}|^{1/2}} \exp\left[-\tfrac{1}{2}(\mathbf{y}^T - \mathbf{m}_{\mathbf{Y}}{}^T)\mathbf{\Lambda}^{-1}(\mathbf{y} - \mathbf{m}_{\mathbf{Y}})\right] \tag{2.17}$$

where $\mathbf{\Lambda}$ is the covariance matrix of $\mathbf{Y}$ and $\mathbf{m}_{\mathbf{Y}}$ is the mean matrix of $\mathbf{Y}$. □

Since, by definition, the joint-characteristic function is the Fourier transform of the joint-probability density, it follows that (2.10) and (2.17) form an m-dimensional Fourier transform pair.[1] On comparing (2.16) and (2.17), we conclude that *the joint-probability density of any m-dimensional gaussian random vector is given by* (2.17) *whether the individual components are independent or not.*

The corresponding component form of (2.17) is

$$f_{Y_1,Y_2,\ldots,Y_m}(y_1, y_2, \ldots, y_m) = \frac{1}{(2\pi)^{m/2}|\mathbf{\Lambda}|^{1/2}}$$
$$\exp\left\{-\frac{1}{2|\mathbf{\Lambda}|}\left[\sum_{i=1}^{m}\sum_{j=1}^{m} |\mathbf{\Lambda}|_{ij}(y_i - E[Y_i])(y_j - E[Y_j])\right]\right\} \tag{2.18}$$

where, as usual, $|\mathbf{\Lambda}|$ is the determinant of the covariance matrix $\mathbf{\Lambda}$ of the

[1] Compare with sec. 11.12 of Cramér's book.

random vector $\mathbf{Y}$ and $|\mathbf{\Lambda}|_{ij}$ is the cofactor of the element λ_{ij} of that matrix. Now we know from whence (1.3) came!

3 GAUSSIAN RANDOM PROCESSES

With the preceding gaussian vector results in mind, we say that the real random process $\{Y_t, t \in T\}$ is a *gaussian random process* if for every finite set of time instants $t_j \in T$, the corresponding random variables Y_{t_j} are jointly gaussian random variables (cf. Sec. 9.4). Thus, $\{Y_t, t \in T\}$ is a gaussian random process if, given any positive integer n and any set of n time instants

$$t_j \in T \qquad j = 1, 2, \ldots, n$$

the joint-characteristic function of the random vector

$$\mathbf{Y} \triangleq (Y_{t_1}, Y_{t_2}, \ldots, Y_{t_n})$$

has the (matrix) form

$$\phi_{\mathbf{Y}}(\mathbf{v}) = \exp\left(i\mathbf{m_Y}^T\mathbf{v} - \tfrac{1}{2}\mathbf{v}^T\mathbf{\Lambda}\mathbf{v}\right) \tag{3.1}$$

in which $\mathbf{m_Y}$ is the mean matrix for $\mathbf{Y}$ and $\mathbf{\Lambda}$ is the covariance matrix of $\mathbf{Y}$, that is, if

$$\phi_{Y_{t_1},Y_{t_2},\ldots,Y_{t_n}}(v_1, v_2, \ldots, v_n) = \exp\left[i \sum_{j=1}^{n} E[Y_{t_j}]v_j - \frac{1}{2}\sum_{j=1}^{n}\sum_{k=1}^{n} K_Y(t_j,t_k)v_jv_k\right] \tag{3.2}$$

where K_Y is the covariance function of the given process:

$$K_Y(t_j,t_k) = \operatorname{cov}(Y_{t_j}, Y_{t_k}) \tag{3.3}$$

It is an immediate consequence of this definition, and of the results in the preceding section, that *any finite linear transformation of a gaussian random process results in a gaussian random process.* In addition, it can also be shown that *passing a gaussian random process through a stable linear filter also results in a gaussian random process.*[1] It is for these reasons that the gaussian process is sometimes defined in terms of linear transformations. In particular, it is sometimes convenient to define a random process to be a gaussian process if every linear functional (that is, every linear transformation which results in a single random variable) of that process is a gaussian random variable (see, for example, sec. 3.3.3 of pt. I of Van Trees or sec. 8.1 of Gallager).

[1] See, for example, sec. 8-4 of Davenport and Root.

One further consequence of the form of the joint-characteristic function of a gaussian random process is that *a gaussian process is completely determined* (statistically) *by the specification of the mean* $E[Y_t]$ *and the covariance function* $K_Y(t,t')$ *of the process.* It further follows from this fact that any gaussian process which is stationary in the wide sense is also stationary in the strict sense (cf. Exercise 9.7.6). For, suppose that the random process $\{Y_t, -\infty < t < +\infty\}$ is a wide-sense stationary gaussian random process. In this case we may write

$$m \triangleq E[Y_t]$$

for all t, and

$$K_Y(t,t') = K_Y(t - t')$$

for all t and t'. It then follows that (3.2) becomes

$$\phi_{Y_{t_1},Y_{t_2},\ldots,Y_{t_n}}(v_1, v_2, \ldots, v_n) = \exp\left[i \sum_{j=1}^{n} mv_j - \frac{1}{2} \sum_{j=1}^{n} \sum_{k=1}^{n} K_Y(t_j - t_k)v_jv_k\right] \tag{3.4}$$

when $\{Y_t\}$ is a wide-sense stationary gaussian random process. Suppose now that we translate all of the time instants $t_1, t_2, \ldots, t_n$ by the same amount τ. The joint-characteristic function of the new random variables $Y_{t_j+\tau}$, $j = 1, 2, \ldots, n$, is then

$$\begin{aligned}
\phi_{Y_{t_1+\tau},Y_{t_2+\tau},\ldots,Y_{t_n+\tau}}&(v_1, v_2, \ldots, v_n)\\
&= \exp\left[i \sum_{j=1}^{n} mv_j - \frac{1}{2} \sum_{j=1}^{n} \sum_{k=1}^{n} K_Y(t_j + \tau - t_k - \tau)v_jv_k\right]\\
&= \exp\left[i \sum_{j=1}^{n} mv_j - \frac{1}{2} \sum_{j=1}^{n} \sum_{k=1}^{n} K_Y(t_j - t_k)v_jv_k\right]\\
&= \phi_{Y_{t_1},Y_{t_2},\ldots,Y_{t_n}}(v_1, v_2, \ldots, v_n)
\end{aligned} \tag{3.5}$$

Thus we see that the n-dimensional joint characteristic function (and hence the corresponding n-dimensional joint-probability density) is unchanged by a translation of the time origin. Since this result holds for any positive integer n and any set of n t-values, it follows that we have thus shown that *if a gaussian process is stationary in the wide sense, then it is also stationary in the strict sense;* that is, wide-sense stationarity and strict-sense stationarity are equivalent concepts in the case of gaussian random processes.

Exercise 3.1 Let $\{Y_t, -\infty < t < +\infty\}$ be a zero-mean, stationary, gaussian random process with the covariance function

$$K_Y(\tau) = \begin{cases} 1 - \dfrac{|\tau|}{T} & \text{for } -T \le \tau \le +T \\ 0 & \text{otherwise} \end{cases}$$

A sequence of n samples of the process are taken at the time instants

$$t_j = j\frac{T}{2} \qquad j = 1, 2, \ldots, n$$

a. Determine the mean and the variance of the sample mean

$$\hat{m}_n = \frac{1}{n}\sum_{j=1}^{n} Y_{t_j}$$

b. Determine the probability density of the sample mean. □

Exercise 3.2 Let $\{Y_t, -\infty < t < +\infty\}$ be a zero-mean, stationary, gaussian random process and suppose that this process is the input to a *square-law detector*. Let $\{Z_t, -\infty < t < +\infty\}$ be the detector-output random process, where

$$Z_t \triangleq Y_t^2$$

a. Show that the mean value of the detector output is

$$E[Z_t] = K_y(0) \qquad \text{for all } t$$

and that the covariance function of the detector output is

$$\begin{aligned} K_Z(\tau) &\triangleq K_Z(t + \tau, t) \\ &= 2K_Y^2(\tau) \end{aligned}$$

where, in each case, K_Y is the input covariance function. Note that these results show that *the detector output is stationary in the wide sense.*

b. Let

$$W_t \triangleq Z_t - E[Z_t]$$

be the fluctuation of the detector output about its mean. Show that the spectral density of this output fluctuation is given by the formula

$$S_W(f) = 2\int_{-\infty}^{+\infty} S_Y(f - f')S_Y(f')\,df'$$

where S_Y is the input spectral density. □

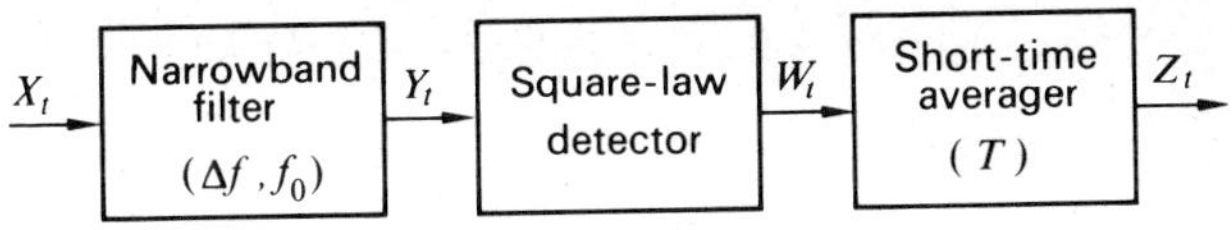

Figure 3.1 The simple radiometer.

Exercise 3.3 Let $\{X_t, -\infty < t < +\infty\}$ be a stationary gaussian random process with a zero mean and a continuous spectral density S_X, and suppose that this process is the input to the *simple radiometer* shown in Fig. 3.1. The radiometer input-filter system function is assumed to have a system function which has a constant magnitude of unity throughout a narrow frequency band of width Δf which is centered on f_0, as shown in Fig. 3.2.

a. Assuming that the radiometer input spectral density is essentially constant over the passband of the input filter, show that the mean value of the radiometer output is

$$E[Z_t] = 2S_X(f_0)\,\Delta f$$

that is, *the mean value of the radiometer output provides a measure of the height of the input spectral density at the center frequency f_0 of the radiometer* [cf. (11.4.7)].

b. Using the results of part (*a*) and Exercise 3.2, determine the spectral density of the fluctuating part of the square-law detector output, and plot your result.

c. Assuming that the integration time T of the short-time averager is very much larger than the reciprocal of the bandwidth of the radiometer input filter, that is,

$$T \gg \frac{1}{\Delta f}$$

show that the variance of the radiometer output is

$$\text{var}\,(Z_t) = \frac{2}{T} S_X^2(f_0)\,\Delta f$$

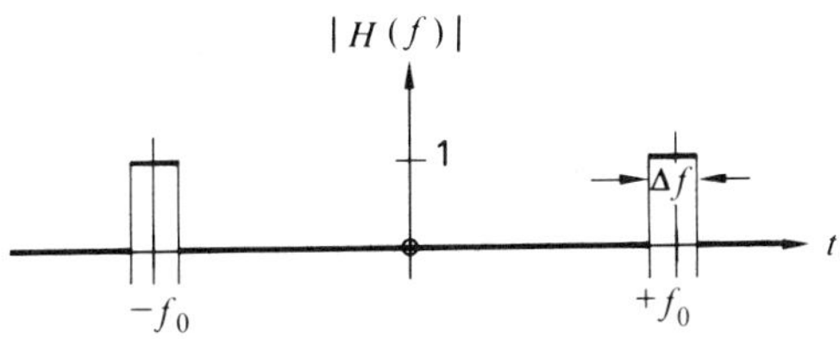

Figure 3.2 The magnitude of the system function of the radiometer-input filter.

Note that it follows from (*a*) and (*b*) here that

$$\frac{[\text{var }(Z_t)]^{1/2}}{E[Z_t]} = \frac{1}{\sqrt{2T\,\Delta f}}$$

that is, *the relative noisiness of the radiometer output decreases as the product of the integration time and the input-filter bandwidth increases.* □

Exercise 3.4 Let Y_1 and Y_2 be jointly gaussian random variables, each with a zero mean and a variance σ^2. Show that the probability that both Y_1 and Y_2 have the same sign is equal to

$$\frac{1}{2} + \frac{1}{\pi}\sin^{-1}\rho$$

where

$$\rho \triangleq \frac{\text{cov }(Y_1, Y_2)}{\sigma^2}$$

(*Hint:* Introduce polar coordinates and integrate first over the radius variable.) □

Exercise 3.5 Let $\{Y_t, -\infty < t < +\infty\}$ be a stationary gaussian random process with a zero mean, a variance σ^2, and a covariance function K_Y. Suppose that this gaussian process is the input to an *ideal limiter* whose output random process $\{Z_t, -\infty < t < +\infty\}$ is such that

$$Z_t \triangleq \begin{cases} +1 & \text{if } Y_t > 0 \\ 0 & \text{if } Y_t = 0 \\ -1 & \text{if } Y_t < 0 \end{cases}$$

Show that the output process has a zero mean and, using the result of Exercise 3.4, that the output autocorrelation function is

$$R_Z(\tau) \triangleq R_Z(t+\tau, t) = \frac{2}{\pi}\sin^{-1}\frac{K_Y(\tau)}{\sigma^2}$$

Note that it therefore follows that the limiter output process is stationary in the wide sense. □

Exercise 3.6* Let $\{Y_t, -\infty < t < +\infty\}$ be a stationary gaussian process with a zero mean, a variance σ^2, and a covariance function K_Y.

* Supplementary exercise (cf. sec. 4.7 of Rice or Sec. 12.4 of Davenport and Root).

Suppose that this gaussian process is the input to a *half-wave linear detector* whose output random process $\{Z_t, -\infty < t < +\infty\}$ is such that

$$Z_t \triangleq \begin{cases} Y_t & \text{when } Y_t \geq 0 \\ 0 & \text{when } Y_t < 0 \end{cases}$$

a. Show that the detector-output mean value is

$$E[Z_t] = \frac{\sigma}{\sqrt{2\pi}} \qquad \text{for all } t$$

b. Show that the detector-output autocorrelation function is

$$R_Z(\tau) = \frac{\sigma^2}{2\pi} \{[1 - \rho^2(\tau)]^{1/2} + \rho(\tau) \cos^{-1} [-\rho(\tau)]\}$$

where

$$\rho(\tau) \triangleq \frac{K_Y(\tau)}{\sigma^2}$$

Hint: It may be shown [cf. eq. (3.5-4) of Rice] that

$$\int_0^{+\infty} du \int_0^{+\infty} dv[uv \exp(-u^2 - v^2 - 2uv \cos \phi)] = \tfrac{1}{4} \csc^2 \phi \, (1 - \phi \cot \phi)$$

c. Using power-series expansions of the square-root and arc-cosine terms in (*b*), show that the detector-output autocorrelation function is given (approximately) by the formula

$$R_Z(\tau) \approx \frac{\sigma^2}{2\pi} + \tfrac{1}{4}K_Y(\tau) + \frac{1}{4\pi\sigma^2} K_Y{}^2(\tau)$$

Note that it follows from the results of parts (*a*) and (*b*) that the detector-output process here is stationary in the wide sense. □

4 NARROWBAND WAVEFORMS

In the remainder of this chapter we wish to study narrowband random processes, and, in particular, narrowband gaussian random processes. First, however, it will be necessary to study narrowband waveforms.

A function of time $x(t)$ is said to be a *narrowband waveform* if the region over which its frequency spectrum

$$X(f) \triangleq \int_{-\infty}^{+\infty} x(t)e^{-i2\pi ft} \, dt \tag{4.1}$$

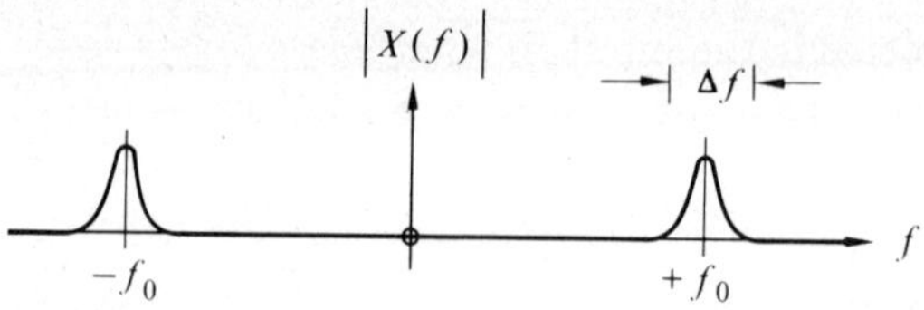

Figure 4.1 The magnitude of the frequency spectrum of a narrowband waveform.

is nonzero is confined to a narrow frequency band of width Δf centered about some frequency f_0, where

$$f_0 \gg \Delta f$$

The magnitude of a typical narrowband frequency spectrum might then have the form shown in Fig. 4.1. If such a narrowband waveform were to be viewed on an oscilloscope screen, it would appear more or less as a sinusoidal wave with a slowly varying envelope function and a slowly varying phase function, for example, as shown in Fig. 4.2. That is, it would appear as though we could write

$$x(t) = v(t) \cos [\omega_0 t + \phi(t)] \tag{4.2}$$

where $v(t)$ is a slowly varying *envelope function* which is always nonnegative, $\phi(t)$ is a slowly varying *phase function*, and

$$f_0 = \frac{\omega_0}{2\pi}$$

is the apparent *carrier* frequency. Here, "slowly varying" means that the respective time variations are slowly varying as compared to those of $\cos \omega_0 t$. The envelope of a narrowband waveform is, of course, given by the output of an envelope detector in response to that waveform, while

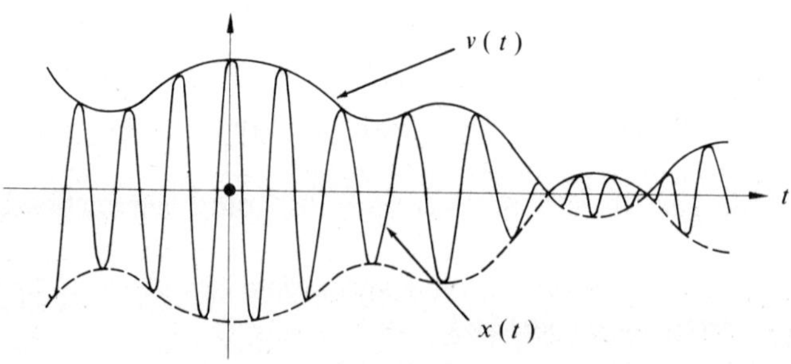

Figure 4.2 A narrowband waveform and its envelope.

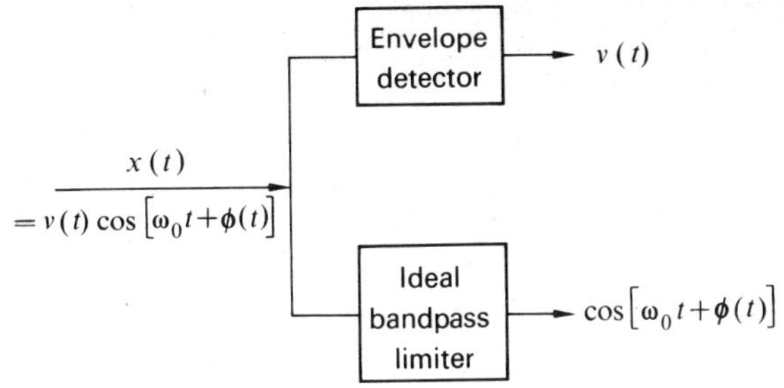

Figure 4.3 Decomposition of a narrowband waveform into envelope and phase-modulated components.

the factor $\cos [\omega_0 t + \phi(t)]$ is given by the output of an ideal bandpass limiter,[1] as indicated in Fig. 4.3.

A useful alternative representation of a narrowband waveform may now be obtained by expanding the cosine in (4.2). Thus we get

$$x(t) = x_c(t) \cos \omega_0 t - x_s(t) \sin \omega_0 t \tag{4.3}$$

where we have defined

$$x_c(t) \triangleq v(t) \cos \phi(t) \qquad \text{and} \qquad x_s(t) \triangleq v(t) \sin \phi(t) \tag{4.4}$$

and where $x_c(t)$ is called the *cosine component* of $x(t)$ and $x_s(t)$ is called the *sine component*. The extension to random processes of the decomposition of a narrowband waveform into its cosine and sine components will be found to be very useful in our subsequent discussions of narrowband random processes when we wish to determine the statistical properties of the envelope and phase of such a process. That we can obtain the envelope-phase form of a narrowband waveform from its cosine-sine component form follows from inverting (4.4) to give

$$v(t) = \sqrt{x_c^2(t) + x_s^2(t)} \qquad \text{and} \qquad \phi(t) = \tan^{-1} \frac{X_s(t)}{X_c(t)} \tag{4.5}$$

A physical interpretation of the cosine and sine components of a narrowband waveform may possibly be gotten from the following discussion: Consider the system shown in Fig. 4.4. It follows from that figure, and from (4.3), that

$$\begin{aligned} w_c(t) &= 2x(t) \cos \omega_0 t \\ &= 2x_c(t) \cos^2 \omega_0 t - 2x_s(t) \sin \omega_0 t \cos \omega_0 t \\ &= x_c(t) + [x_c(t) \cos 2\omega_0 t - x_s(t) \sin 2\omega_0 t] \end{aligned}$$

[1] See sec. 13-2 of Davenport and Root.

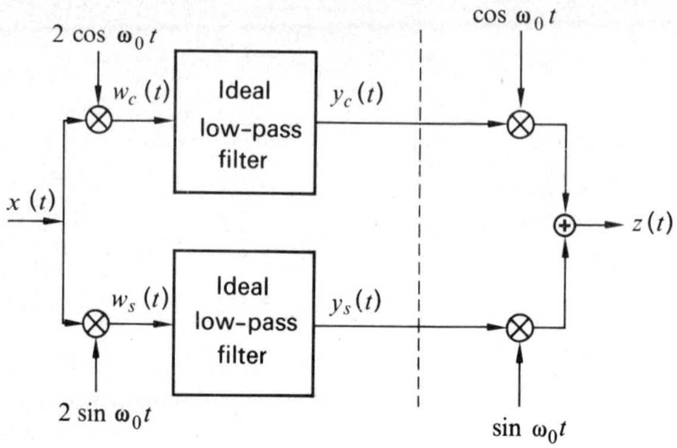

Figure 4.4 Quadrature decomposition of a narrowband waveform.

Since $x_c(t)$ and $x_s(t)$ are both slowly varying functions of time here, we see that the multiplier output $w_c(t)$ has a component centered about zero frequency [that is, $x_c(t)$] and a set of components clustered about twice the carrier frequency. This is the situation pictured, for example, in Fig. 4.5. If now the cutoff frequency of the ideal low-pass filter is chosen so as to permit separation of the two sets of multiplier ouput components, then that filter will pass the terms centered about zero frequency without distortion and will eliminate the terms centered about $2f_0$. We therefore obtain the result that the output of the ideal low-pass filter is

$$y_c(t) = x_c(t)$$

Similarly, it also follows from Fig. 4.4, and from (4.3), that

$$\begin{aligned} w_s(t) &= 2x(t) \sin \omega_0 t \\ &= 2x_c(t) \sin \omega_0 t \cos \omega_0 t - 2x_s(t) \sin^2 \omega_0 t \\ &= -x_s(t) + [x_c(t) \sin 2\omega_0 t + x_s(t) \cos 2\omega_0 t] \end{aligned}$$

and hence that

$$y_s(t) = -x_s(t)$$

Thus we see that beating the original narrowband waveform with in-phase and quadrature phase carriers followed by low-pass filtering gives the

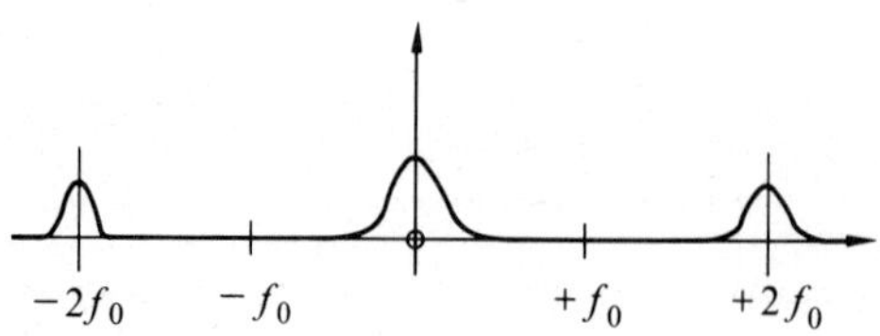

Figure 4.5 Spectral decomposition of the multiplier output.

cosine and sine components of that narrowband waveform. It should be realized that the system shown schematically in Fig. 4.4 is, in fact, widely used in practice for that purpose.

Further examination of Fig. 4.4 now shows that the remainder of the system shown there reconstructs the narrowband input waveform. For, we see from that figure and our previous results that

$$\begin{aligned} z(t) &= y_c(t) \cos \omega_0 t + y_s(t) \sin \omega_0 t \\ &= x_c(t) \cos \omega_0 t - x_s(t) \sin \omega_0 t \\ &= x(t) \end{aligned}$$

that is, the output of the overall system is simply the original narrowband input waveform.

5 NARROWBAND RANDOM PROCESSES

Now let us extend our study of narrowband phenomena to cover random processes. Suppose, therefore, that the random process $\{X_t, -\infty < t < +\infty\}$ is a *narrowband random process;* that is, suppose that $\{X_t\}$ is a zero-mean random process which is stationary in the wide sense with a spectral density S_X which differs from zero only in some narrow frequency band centered about some given frequency, say f_0. Paralleling our preceding discussion of narrowband waveforms, it is sometimes convenient to represent such a narrowband random process in terms of an envelope random process $\{V_t, -\infty < t < +\infty\}$ and a phase random process $\{\Phi_t, -\infty < t < +\infty\}$ by using the relation

$$X_t = V_t \cos (\omega_0 t + \Phi_t) \tag{5.1}$$

where $\omega_0 = 2\pi f_0$. Alternatively, it is also sometimes convenient to represent a given narrowband random process in terms of its cosine and sine component random processes $\{X_{ct}, -\infty < t < +\infty\}$ and $\{X_{st}, -\infty < t < +\infty\}$, respectively, by using the relation

$$X_t = X_{ct} \cos \omega_0 t - X_{st} \sin \omega_0 t \tag{5.2}$$

Again paralleling our development in the preceding section, the relations between the two alternative representations are given either by the formulas

$$X_{ct} = V_t \cos \Phi_t \qquad \text{or} \qquad X_{st} = V_t \sin \Phi_t \tag{5.3}$$

or by the formulas

$$V_t = \sqrt{X_{ct}^2 + X_{st}^2} \qquad \text{and} \qquad \Phi_t = \tan^{-1} \frac{X_{st}}{X_{ct}} \tag{5.4}$$

As in the waveform case, the envelope, phase, cosine component, and sine component are all low-frequency phenomena; that is, their spectral densities are all concentrated about zero frequency and are all essentially zero in value for frequencies greater in magnitude than some small fraction of the "carrier" frequency f_0.

In many practical problems involving narrowband random processes, it is important to know the various correlation functions concerning the cosine- and sine-component random processes.[1] With this fact in mind, we now wish to calculate those correlation functions and, to this end, we shall make use of the system pictured in Fig. 4.4.[2] Consider, in particular, the part of that figure which concerns the cosine component; we have redrawn that part in Fig. 5.1 for convenience. If we denote the weighting function of the ideal low-pass filter by h, it follows from Fig. 5.1 that we may write

$$\begin{aligned} R_c(t+\tau, t) &\triangleq E[X_{c(t+\tau)}X_{ct}] \\ &= E\left[2\int_{-\infty}^{+\infty} h(u)X_{t+\tau-u}\cos\omega_0(t+\tau-u)\,du \right. \\ &\qquad\qquad \left. 2\int_{-\infty}^{+\infty} h(z)X_{t-z}\cos\omega_0(t-z)\,dz\right] \\ &= 4\int_{-\infty}^{+\infty} h(z)\cos\omega_0(t-z)\,dz \\ &\qquad\qquad \int_{-\infty}^{+\infty} h(u)\cos\omega_0(t+\tau-u)\,du\, E[X_{t+\tau-u}X_{t-z}] \\ &= 4\int_{-\infty}^{+\infty} h(z)\cos\omega_0(t-z)\,dz \\ &\qquad\qquad \int_{-\infty}^{+\infty} h(u)\cos\omega_0(t+\tau-u)\,du\, R_X(\tau-u+z) \end{aligned} \tag{5.5}$$

the last equality following from the fact that the given narrowband random process is stationary in the wide sense.

Since the spectral separation property of the ideal low-pass filter is most directly handled in the frequency domain, it is convenient at this time to rewrite (5.5) in that domain. Thus, if we express the autocorrelation function R_X as the Fourier transform of the spectral density S_X,

[1] See, for example, pt. III of Rice or chap. 4 of Blachman.

[2] An alternative derivation which makes use of a Fourier series decomposition of a narrowband random process may be found in sec. 3.7 of Rice and in sec. 8-5 of Davenport and Root.

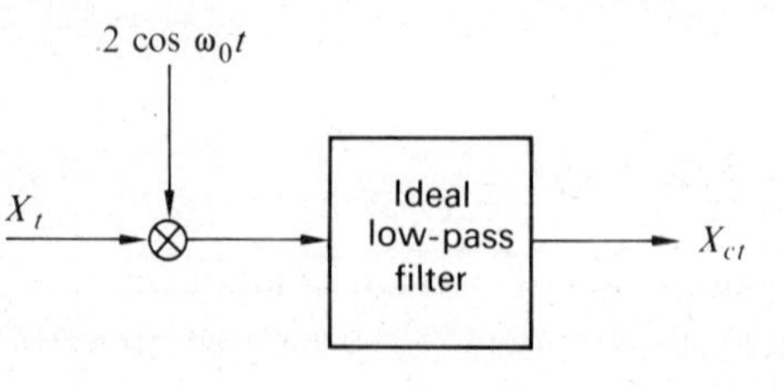

Figure 5.1 Generation of the cosine component of a narrowband random process.

then (5.5) becomes

$$R_c(t+\tau, t) = 4 \int_{-\infty}^{+\infty} h(z) \cos \omega_0(t-z)\, dz$$
$$\int_{-\infty}^{+\infty} h(u) \cos \omega_0(t+\tau-u)\, du \int_{-\infty}^{+\infty} S_X(f) e^{i2\pi f(\tau-u+z)}\, df$$

Suppose now that we expand the cosines in terms of complex exponentials:

$$R_c(t+\tau, t) = \int_{-\infty}^{+\infty} h(z)(e^{i2\pi f_0(t-z)} + e^{-i2\pi f_0(t-z)})\, dz$$
$$\int_{-\infty}^{+\infty} h(u)(e^{i2\pi f_0(t+\tau-u)} + e^{-i2\pi f_0(t+\tau-u)})\, du \int_{-\infty}^{+\infty} S_X(f) e^{i2\pi f(\tau-u+z)}\, df$$
$$= \int_{-\infty}^{+\infty} S_X(f)\, df \left(e^{i2\pi f_0 t} \int_{-\infty}^{+\infty} h(z) e^{+i2\pi(f-f_0)z}\, dz + e^{-i2\pi f_0 t} \int_{-\infty}^{+\infty} h(z) e^{+i2\pi(f+f_0)z}\, dz\right)$$
$$\left(e^{+i2\pi[f_0 t+(f+f_0)\tau]} \int_{-\infty}^{+\infty} h(u) e^{-i2\pi(f+f_0)u}\, du + e^{-i2\pi[f_0 t-(f-f_0)\tau]} \int_{-\infty}^{+\infty} h(u) e^{-i2\pi(f-f_0)u}\, du\right)$$

The integrals of the weighting functions here may now be rewritten in terms of the ideal low-pass system function H, giving

$$R_c(t+\tau, t) = \int_{-\infty}^{+\infty} S_X(f)\, df[e^{i2\pi f_0 t} H^*(f-f_0) + e^{-i2\pi f_0 t} H^*(f+f_0)]$$
$$[e^{i2\pi[f_0 t+(f+f_0)\tau]} H(f+f_0) + e^{-i2\pi[f_0 t-(f-f_0)\tau]} H(f-f_0)] \quad (5.6)$$

Before proceeding further analytically, it is useful to consider the frequency plots of the magnitudes of various terms in (5.6), as shown, for example, in Fig. 5.2. Examination of that figure indicates that when the bracketed terms in (5.6) are multiplied out, the terms containing the factors

$$H^*(f-f_0)H(f+f_0) \qquad \text{and} \qquad H^*(f+f_0)H(f-f_0)$$

vanish because of the nonoverlapping character of $|H(f+f_0)|$ and $|H(f-f_0)|$. Thus it follows that (5.6) becomes

$$R_c(t+\tau, t) = \int_{-\infty}^{+\infty} S_X(f)\, df[e^{i2\pi(f-f_0)\tau}|H(f-f_0)|^2 + e^{i2\pi(f+f_0)\tau}|H(f+f_0)|^2] \quad (5.7)$$

Since this last result is not a function of time, we shall write

$$R_c(\tau) \triangleq R_c(t+\tau, t)$$

from here on. Further examination of Figure 5.2 then shows that we may rewrite (5.7) as

$$R_c(\tau) = \int_0^{+\infty} S_X(f) e^{i2\pi(f-f_0)\tau}\, df + \int_{-\infty}^{0} S_X(f) e^{i2\pi(f+f_0)\tau}\, df$$
$$= \int_0^{+\infty} S_X(f) e^{i2\pi(f-f_0)\tau}\, df + \int_0^{+\infty} S_X(f') e^{-i2\pi(f'-f_0)\tau}\, df'$$

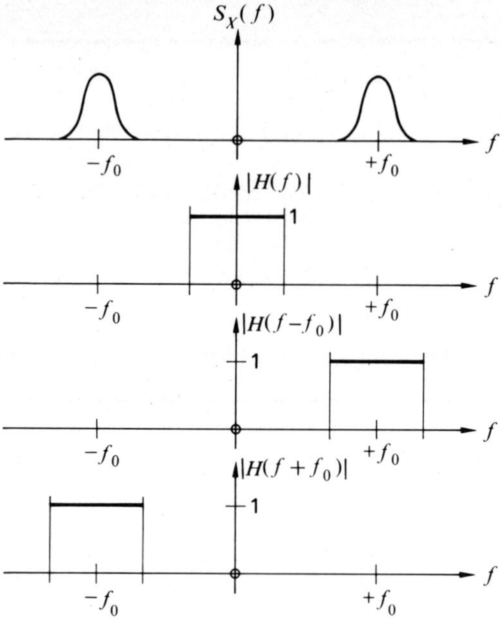

Figure 5.2 Filter passband locations.

the last equality following on using the change of variables $f' = -f$. Thus we finally obtain the formula

$$\begin{aligned} R_c(\tau) &= \int_0^{+\infty} S_X(f)(e^{i2\pi(f-f_0)\tau} + e^{-i2\pi(f-f_0)\tau})\, d\tau \\ &= 2\int_0^{+\infty} S_X(f) \cos\,[2\pi(f-f_0)\tau]\, df \qquad (5.8) \end{aligned}$$

for the autocorrelation function of the cosine component of the given narrowband random process.

Exercise 5.1 By paralleling the derivation leading to (5.8), show that

$$\begin{aligned} R_S(\tau) &\triangleq E[X_{s(t+\tau)}X_{st}] \\ &= 2\int_0^{+\infty} S_X(f) \cos\,[2\pi(f-f_0)\tau]\, df \\ &= R_c(\tau) \qquad (5.9) \end{aligned}$$

where R_S is the autocorrelation function of the sine component of the given narrowband random process. □

Exercise 5.2 By paralleling the derivation leading to (5.8), show that

$$\begin{aligned} R_{cs}(\tau) &\triangleq E[X_{c(t+\tau)}X_{st}] \\ &= 2\int_0^{+\infty} S_X(f) \sin\,[2\pi(f-f_0)\tau]\, df \\ &= -E[X_{s(t+\tau)}X_{ct}] \triangleq -R_{sc}(\tau) \end{aligned}$$

where R_{cs} and R_{sc} are the cross-correlation functions of the cosine and sine components of the given narrowband random process. □

It follows from (5.8), and from the results of Exercises 5.1 and 5.2, that

$$R_c(0) = R_s(0) = 2\int_0^{+\infty} S_X(f)\,df = R_X(0) \tag{5.10}$$

and, further, that

$$R_{cs}(0) = 0 = R_{sc}(0) \tag{5.11}$$

that is, that the cosine and sine component random variables X_{ct} and X_{st}, respectively, are uncorrelated random variables. These various correlation results are conveniently expressed by the covariance matrix $\mathbf{R}(\tau)$ of the random variables X_{ct}, X_{st}, $X_{c(t+\tau)}$, and $X_{s(t+\tau)}$, where

$$\mathbf{R}(\tau) = \begin{bmatrix} R_X(0) & 0 & R_c(\tau) & R_{cs}(\tau) \\ 0 & R_X(0) & -R_{cs}(\tau) & R_c(\tau) \\ R_c(\tau) & -R_{cs}(\tau) & R_X(0) & 0 \\ R_{cs}(\tau) & R_c(\tau) & 0 & R_X(0) \end{bmatrix} \tag{5.12}$$

where R_c is given in terms of the spectral density of the given narrowband process by (5.8) and R_{cs} is similarly given by (5.9).

6 NARROWBAND GAUSSIAN PROCESSES

Having studied narrowband random processes in general, we are finally in a position to study narrowband gaussian processes in particular. Suppose, therefore, that the narrowband random process $\{X_t, -\infty < t < +\infty\}$ is a gaussian random process. Since the corresponding cosine and sine component random processes $\{X_{ct}, -\infty < t < +\infty\}$ and $\{X_{st}, -\infty < t < +\infty\}$, respectively, are obtainable from linear transformations of the given narrowband process (compare with Fig. 5.1), it follows that these component random processes are jointly gaussian random processes.

Suppose now that we consider the component random *variables* X_{ct} and X_{st}. Since these random variables are jointly gaussian, and since it follows from (5.11) that they are uncorrelated, it further follows that they are independent random variables. Then, since both X_{ct} and X_{st} have zero means and each has a variance equal to $R_X(0)$ as shown by (5.10), their joint-probability density is

$$f_{X_{ct},X_{st}}(x,y) = \frac{1}{2\pi R_X(0)} \exp\left[-\frac{x^2+y^2}{2R_X(0)}\right] \tag{6.1}$$

We may now calculate the joint-probability density of the envelope and phase random variables

$$V_t = \sqrt{X_{ct}^2 + X_{st}^2} \qquad \text{and} \qquad \Phi_t = \tan^{-1} \frac{X_{st}}{X_{ct}}$$

respectively, by using the methods of Chap. 6. If we do so, we find that the envelope and phase random variables are statistically independent with the probability densities

$$f_{V_t}(v) = \begin{cases} \dfrac{v}{R_X(0)} \exp\left[-\dfrac{v^2}{2R_X(0)}\right] & \text{for } 0 \leq v \\ 0 & \text{otherwise} \end{cases} \tag{6.2}$$

and

$$f_{\Phi_t}(\phi) = \begin{cases} \dfrac{1}{2\pi} & \text{for } 0 \leq \phi \leq 2\pi \\ 0 & \text{otherwise} \end{cases} \tag{6.3}$$

respectively (compare with Exercises 6.4.4 and 6.4.5). That is, *the envelope of a narrowband gaussian process has a Rayleigh density, and the phase of such a process is uniformly distributed over* $[0,2\pi]$.

Higher-order probability densities of the envelope and phase random processes may also be obtained from the statistical properties of the cosine and sine component processes by using the methods discussed in Chap. 6. Unfortunately, however, the derivations are tedious and the results are complicated.[1] One result which can be so obtained is worth noting here, though, and that is that while the envelope and phase random *variables* are statistically independent, the envelope and phase random *processes* are not statistically independent.[2]

7 SUMMARY OF DEFINITIONS AND FORMULAS

Gaussian random vectors

The m-dimensional random vector $\mathbf{Y}$ is said to be a gaussian random vector (and its components $Y_1, Y_2, \ldots, Y_m$ are said to be jointly gaussian random variables) if and only if the joint-characteristic function of $\mathbf{Y}$ has the matrix form

$$\phi_{\mathbf{Y}}(\mathbf{v}) = \exp\left(i\mathbf{m}_{\mathbf{Y}}^T\mathbf{v} - \tfrac{1}{2}\mathbf{v}^T\mathbf{\Lambda}\mathbf{v}\right) \tag{2.12}$$

where $\mathbf{m}_{\mathbf{Y}}^T$ is the transpose of the mean matrix $\mathbf{m}_{\mathbf{Y}}$ for $\mathbf{Y}$ and where $\mathbf{\Lambda}$ is the covariance matrix for $\mathbf{Y}$. The component form of this joint charac-

[1] Compare with pt. III of Rice or sec. 8.5 of Davenport and Root.
[2] See, for example, prob. 14 in sec. 8.7 of Davenport and Root.

teristic function is

$$\phi_{Y_1,Y_2,\ldots,Y_m}(v_1, v_2, \ldots, v_m) = \exp\left\{i \sum_{j=1}^{m} E[Y_j]v_j - \frac{1}{2}\sum_{j=1}^{m}\sum_{k=1}^{m} \operatorname{cov}(Y_j, Y_k)v_jv_k\right\} \tag{2.11}$$

The corresponding joint-probability density is

$$f_{\mathbf{Y}}(\mathbf{y}) = \frac{\exp\left[-\frac{1}{2}(\mathbf{y}^T - \mathbf{m_Y}^T)\mathbf{\Lambda}(\mathbf{y} - \mathbf{m_Y})\right]}{(2\pi)^{m/2}|\mathbf{\Lambda}|^{1/2}} \tag{2.13}$$

which has the component form

$$f_{Y_1,Y_2,\ldots,Y_m}(y_1, y_2, \ldots, y_m) = \frac{\exp\left\{-\dfrac{\sum_{j=1}^{m}\sum_{k=1}^{m} |\mathbf{\Lambda}|_{jk}(y_j - E[Y_j])(y_k - E[Y_k])}{2|\mathbf{\Lambda}|}\right\}}{(2\pi)^{m/2}|\mathbf{\Lambda}|^{1/2}} \tag{2.14}$$

where $|\mathbf{\Lambda}|$ is the determinant of the covariance matrix $\mathbf{\Lambda}$ and where $|\mathbf{\Lambda}|_{jk}$ is the cofactor of the element λ_{jk} of that matrix.

Gaussian random processes

The real random process $\{Y_t, t \in T\}$ is said to be a gaussian random process if for every finite set of time instants $t_j \in T$, the corresponding random variables Y_{t_j} are jointly gaussian random variables.

Narrowband random processes

The random process $\{X_t, -\infty < t < +\infty\}$ is said to be a narrowband random process if it has a zero mean, is stationary in the wide sense, and if its spectral density S_X differs from zero only in some narrowband of width Δf centered about some frequency f_0 where

$$f_0 \gg \Delta f$$

A narrowband random process may be represented in terms of an envelope random process $\{V_t, -\infty < t < +\infty\}$ and a phase random process $\{\Phi_t, -\infty < t < +\infty\}$ by using the relation

$$X_t = V_t \cos(\omega_0 t + \Phi_t) \tag{5.11}$$

where $\omega_0 = 2\pi f_0$. Alternatively, a narrowband random process may also be represented in terms of cosine and sine component random processes $\{X_{ct}, -\infty < t < +\infty\}$ and $\{X_{st}, -\infty < t < +\infty\}$, respectively, by using the relation

$$X_t = X_{ct} \cos \omega_0 t - X_{st} \sin \omega_0 t \tag{5.2}$$

The relations between these two representations are given by the formulas

$$X_{ct} = V_t \cos \Phi_t \qquad \text{and} \qquad X_{st} = V_t \sin \Phi_t \tag{5.3}$$

which have the inverses

$$V_t = \sqrt{X_{ct}^2 + X_{st}^2} \qquad \text{and} \qquad \Phi_t = \tan^{-1} \frac{X_{st}}{X_{ct}} \tag{5.4}$$

The random variables X_{ct}, X_{st}, $X_{c(t+\tau)}$, and $X_{s(t+\tau)}$ have the covariance matrix

$$\mathbf{R}(\tau) = \begin{bmatrix} R_X(0) & 0 & R_c(\tau) & R_{cs}(\tau) \\ 0 & R_X(0) & -R_{cs}(\tau) & R_c(\tau) \\ R_c(\tau) & -R_{cs}(\tau) & R_X(0) & 0 \\ R_{cs}(\tau) & R_c(\tau) & 0 & R_X(0) \end{bmatrix} \tag{5.12}$$

where

$$R_c(\tau) = 2 \int_0^{+\infty} S_X(f) \cos [2\pi(f - f_0)\tau] \, df \tag{5.8}$$

and

$$R_{cs}(\tau) = 2 \int_0^{+\infty} S_X(f) \sin [2\pi(f - f_0)\tau] \, df \qquad \text{(Exercise 5.2)}$$

Narrowband gaussian processes

The cosine- and sine-component random variables X_{ct} and X_{st} of a gaussian narrowband random process are independent random variables with zero means, each with a variance equal to $R_X(0)$, and a joint-probability density

$$f_{X_{ct}X_{st}}(x,y) = \frac{\exp\left[-\dfrac{x^2 + y^2}{2R_X(0)}\right]}{2\pi R_X(0)} \tag{6.1}$$

The envelope and phase random variables V_t and Φ_t of a gaussian narrowband random process are also independent random variables. The envelope has the Rayleigh probability density

$$f_{V_t}(v) = \begin{cases} \dfrac{v}{R_X(0)} \exp\left[-\dfrac{v^2}{2R_X(0)}\right] & \text{for } 0 \le v \\ 0 & \text{otherwise} \end{cases} \tag{6.2}$$

and the phase is uniformly distributed over $[0,2\pi]$:

$$f_{\Phi_t}(\phi) = \begin{cases} \dfrac{1}{2\pi} & \text{for } 0 \le \phi \le 2\pi \\ 0 & \text{otherwise} \end{cases} \tag{6.3}$$

REFERENCES

Aitchison, John, and James A. C. Brown: "The Lognormal Distribution," Cambridge University Press, London, 1957.

Blachman, Nelson M.: "Noise and Its Effect on Communications," McGraw-Hill Book Company, New York, 1966.

Cramér, Harald: "Random Variables and Probability Distributions," No. 36 of Cambridge Tracts in Mathematics and Mathematical Physics, Cambridge University Press, London, 1937.

———: "Mathematical Methods of Statistics," Princeton University Press, Princeton, N.J., 1946.

Davenport, Wilbur B., Jr., and William L. Root: "Random Signals and Noise," McGraw-Hill Book Company, New York, 1958.

Gallagher, Robert G.: "Information Theory and Reliable Communications," John Wiley & Sons, Inc., New York, 1968.

Middleton, David: "An Introduction to Statistical Communication Theory," McGraw-Hill Book Company, New York, 1960.

Miller, Kenneth S.: "Multidimensional Gaussian Distributions," John Wiley & Sons, Inc., New York, 1964.

Rice, Stephen O.: Mathematical Analysis of Random Noise, *Bell System Tech. J.*, vol. 23, pp. 282–332, 1944, and vol. 24, pp. 46–156, 1945. (Reprinted in Nelson Wax, "Selected Papers on Noise and Stochastic Processes," Dover Publications, Inc., New York, 1954.)

Van Trees, Harry L., Jr.: "Detection, Estimation, and Modulation Theory," pt. 1, John Wiley & Sons, Inc., New York, 1968.

Bibliography

Aitchison, John, and James A. C. Brown: "The Lognormal Distribution," Cambridge University Press, London, 1957.

Apostol, Tom M.: "Mathematical Analysis," Addison-Wesley Publishing Company, Inc., Reading, Mass., 1957.

———: "Calculus," Blaisdell Publishing Company, New York, vol. I, 1961; vol. II, 1962.

Artin, Emil: "The Gamma Function," Holt, Rinehart and Winston, Inc., New York, 1964.

Bartle, Robert G.: "The Elements of Real Analysis," John Wiley & Sons, Inc., New York, 1964.

———: "The Elements of Integration," John Wiley & Sons, New York, 1966.

Beckenbach, Edwin F., and Richard Bellman: "An Introduction to Inequalities," Random House, Inc., New York, 1961.

——— and ———: "Inequalities," Springer-Verlag OHG, Berlin, 2d rev., 1965.

Bennett, William R.: "Electrical Noise," McGraw-Hill Book Company, New York, 1960.

Blachman, Nelson M.: "Noise and Its Effect on Communications," McGraw-Hill Book Company, New York, 1966.

Blackman, Ralph B., and John W. Tukey: "The Measurement of Power Spectra," Dover Publications, Inc., New York, 1958.

Boas, Ralph P., Jr.: "A Primer of Real Functions," Carus Mathematical Monograph No. 13, John Wiley & Sons, Inc., New York, 1960.

Bochner, Salomon: "Lectures on Fourier Integrals," Ann. Math. Studies, No. 42, Princeton University Press, Princeton, N.J., 1959.

Bracewell, Ron: "The Fourier Transform and Its Applications," McGraw-Hill Book Company, New York, 1965.

Brockmeyer, E., H. L. Halstrøm, and E. Jensen: "The Life and Works of A. K. Erlang," Copenhagen Telephone Company, Copenhagen, 1948.

Chambers, R. P.: Random-number Generation on Digital Computers, *IEEE Spectrum*, pp. 48–56, February, 1967.

Chandrasekhar, Subrahmanyan: Stochastic Problems in Physics and Astronomy, *Rev. Mod. Phys.*, vol. 15, no. 1, pp. 1–89, 1943 (reprinted in pp. 3–91 of Nelson Wax, "Selected Papers on Noise and Stochastic Processes," Dover Publications, Inc., New York, 1954).

Chernoff, Herman, and Lincoln E. Moses: "Elementary Decision Theory," John Wiley & Sons, Inc., New York, 1959.

Chung, Kai Lai: "A Course in Probability Theory," Harcourt, Brace & World, Inc., New York, 1968.

Courant, Richard: "Differential and Integral Calculus," Interscience Publishers, Inc., New York, vol. I, 1937; vol. II, 1936.

Cox, David R.: "Renewal Theory," Methuen & Company, Ltd., London (John Wiley & Sons, Inc., New York), 1962.

——— and Walter L. Smith: "Queues," Methuen & Company, Ltd., London (John Wiley & Sons, Inc., New York), 1961.

Cramér, Harald: "Random Variables and Probability Distributions," No. 36 of Cambridge Tracts in Mathematics and Mathematical Physics, Cambridge University Press, London, 1937.

———: "Mathematical Methods of Statistics," Princeton University Press, Princeton, N.J., 1946.

———: "The Elements of Probability Theory and Some of Its Applications," John Wiley & Sons, Inc., New York, 1955.

——— and M. R. Leadbetter: "Stationary and Related Stochastic Processes," John Wiley & Sons, Inc., New York, 1967.

Davenport, Wilbur B., Jr., Richard A. Johnson, and David Middleton: Statistical Errors in Measurements on Random Time Functions, *J. Appl. Phys.*, vol. 23, no. 4, pp. 377–388, 1952.

——— and William L. Root: "Random Signals and Noise," McGraw-Hill Book Company, New York, 1958.

Deutsch, Ralph: "Estimation Theory," Prentice-Hall, Inc., Englewood Cliffs, N.J., 1965.

Doob, Joseph: "Stochastic Processes," John Wiley & Sons, Inc., New York, 1953.

Dwight, Herbert B.: "Tables of Integrals and Other Mathematical Data," The Macmillan Company, New York, 1947.

Einstein, Albert: "Investigations on the Theory of Brownian Movement," Dover Publications, Inc., New York, 1956 (reprint of 1926 Methuen edition).

Evans, John V., and Tor Hagfors: "Radar Astronomy," McGraw-Hill Book Company, New York, 1968.

Feller, William: "An Introduction to Probability Theory and Its Applications," 3d ed., vol. I, 1968; vol. 2, 1966; John Wiley & Sons, Inc., New York.

Fourier, Jean Baptiste Joseph: "The Analytical Theory of Heat," Dover Publications, Inc., New York, 1955 [reprint of an 1878 English translation (by Alexander Freeman) of the 1822 French original].

Fraenkel, Abraham A.: "Abstract Set Theory," North-Holland Publishing Company, Amsterdam, 1953.

Freeman, Harold: "Introduction to Statistical Inference," Addison-Wesley Publishing Company, Inc., Reading, Mass., 1963.

Fry, Thornton C.: "Probability and Its Engineering Uses," 2d ed., D. Van Nostrand Company, Inc., Princeton, N.J., 1965.

Gallagher, Robert G.: "Information Theory and Reliable Communications," John Wiley & Sons, Inc., New York, 1968.

Gel'fand, Izrail Moiseevich, and Georgiĭ Evgen'evich Shilov: "Generalized Functions," vol. I, Academic Press, Inc., New York, 1964.

Gilbert, Edgar N., and Henry O. Pollak: "Amplitude Distribution of Shot Noise," *Bell System Tech. J.*, vol. 39, no. 2, pp. 333–350, 1960.

Gnedenko, Boris Vladimirovich: "Theory of Probability," Chelsea Publishing Company, New York, 1962.

——— and Andrei N. Kolmogorov: "Limit Distributions for Sums of Independent Random Variables," Addison-Wesley Publishing Company, Inc., Reading, Mass., 1964.

Halmos, Paul R.: "Naive Set Theory," D. Van Nostrand Company, Inc., Princeton, N.J., 1960.

Hildebrand, Francis B.: "Methods of Applied Mathematics," Prentice-Hall, Inc., Englewood Cliffs, N.J., 1952.

Hoel, Paul G.: "Introduction to Mathematical Statistics," 3d ed., John Wiley & Sons, Inc., 1962.

Hoffman, Kenneth, and Ray Kunze: "Linear Algebra," Prentice-Hall, Inc., Englewood Cliffs, N.J., 1961.

James, Hubert M., Nathaniel B. Nichols, and Ralph S. Phillips: "Theory of Servomechanisms," McGraw-Hill Book Company, N.Y., 1947.

Jenkins, Gwilym M., and Donald G. Watts: "Spectral Analysis and Its Applications," Holden-Day, Inc., Publisher, San Francisco, 1968.

Karlin, Samuel: "A First Course in Stochastic Processes," Academic Press, Inc., New York, 1966.

Kemeny, John G., Hazleton Mirkil, J. Laurie Snell, and Gerald L. Thompson: "Finite Mathematical Structures," Prentice-Hall, Inc., Englewood Cliffs, N.J., 1958.

Kershner, Richard B., and Lee Roy Wilcox: "The Anatomy of Mathematics," The Ronald Press Company, New York, 1950.

Kolmogorov, Andrei N.: "Foundations of the Theory of Probability," Chelsea Publishing Company, New York, 1956 (English translation of 1933 German edition).

Kreider, Donald L., Robert G. Kuller, Donald R. Ostberg, and Fred W. Perkins: "An Introduction to Linear Analysis," Addison-Wesley Publishing Company, Inc., Reading, Mass., 1966.

Krickeberg, Klaus: "Probability Theory," Addison-Wesley Publishing Company, Inc., Reading, Mass., 1965.

Lee, Alec M.: "Applied Queuing Theory," Macmillan & Co., Ltd., London (St Martin's Press, Inc., New York), 1966.

Lee, Yuk-Wing: "Statistical Theory of Communication," John Wiley & Sons, Inc., New York, 1960.

Loéve, Michel: "Probability Theory," 3d ed., D. Van Nostrand Company, Inc., Princeton, N.J., 1963.

Lukacs, Eugene: "Characteristic Functions," No. 5 of Griffin's Statistical Monographs, Hafner Publishing Company, Inc., New York, 1960.

MacDonald, David K. C.: "Noise and Fluctuations," John Wiley & Sons, Inc., New York, 1962.

Middleton, David: "An Introduction to Statistical Communication Theory," McGraw-Hill Book Company, 1960.

Miller, Kenneth S.: "Multidimensional Gaussian Distributions," John Wiley & Sons, Inc., New York, 1964.

Mood, Alexander M., and Franklin A. Graybill: "Introduction to the Theory of Statistics," 2d ed., McGraw-Hill Book Company, 1963.

National Bureau of Standards: "Tables of Binomial Probability Distributions," Applied Mathematics Series, AMS-6, U.S. Government Printing Office, Washington, 1950.

———: "Tables of Normal Probability Functions," Applied Mathematics Series, AMS-23, U.S. Government Printing Office, Washington, 1953.

———: "Handbook of Mathematical Functions," Applied Mathematics Series, AMS-55, U.S. Government Printing Office, 1964.

Neveu, Jacques: "Mathematical Foundations of the Calculus of Probability," Holden-Day, Inc., Publishers, San Francisco, 1965.

North, Dwight O.: Fluctuations in Space-Charge Currents, pt. 3, Multi-Collectors, *RCA Rev.*, vol. 5, pp. 244–260, 1940.

Papoulis, Athanasios: "The Fourier Integral and Its Applications," McGraw-Hill Book Company, New York, 1962.

Parzen, Emanuel: "Modern Probability Theory and Its Applications," John Wiley & Sons, Inc., New York, 1960.

———: "Stochastic Processes," Holden-Day, Inc., Publishers, San Francisco, 1962.

Pfeiffer, Paul E.: "Concepts of Probability Theory," McGraw-Hill Book Company, New York, 1965.

Polya, George: "Induction and Analogy in Mathematics," Princeton University Press, Princeton, N.J., 1954.

Rice, Stephen O.: Mathematical Analysis of Random Noise, *Bell System Tech. J.*, vol. 23, pp. 282–332, 1944, and vol. 24, pp. 46–156, 1945 (reprinted in pp. 133–294 of Nelson Wax, "Selected Papers on Noise and Stochastic Processes," Dover Publications, Inc., New York, 1954).

Roydan, Halsey L.: "Real Analysis," The Macmillan Company, New York, 1963.

Seeley, Robert T.: "An Introduction to Fourier Series and Integrals," W. A. Benjamin, Inc., New York, 1966.

Titchmarsh, Edward Charles: "Introduction to the Theory of Fourier Integrals," Clarendon Press, Oxford, 1948.

Uspensky, James Victor: "Introduction to Mathematical Probability," McGraw-Hill Book Company, New York, 1937.

Van Trees, Harry L., Jr.: "Detection, Estimation, and Modulation Theory," pt. I, John Wiley & Sons, Inc., New York, 1968.

Wiener, Norbert: "Extrapolation, Interpolation, and Smoothing of Stationary Time Series," John Wiley & Sons, Inc., New York, 1949.

Wilks, Samuel S.: "Mathematical Statistics," John Wiley & Sons, Inc., New York, 1962.

Yaglom, Akiva Moisseevich: "Stationary Random Functions," Prentice-Hall, Inc., Englewood Cliffs, N.J., 1962.

Young, Gail S.: The Linear Functional Equation, *Am. Math. Monthly,* vol. 65, pp. 37–38, 1958.

Index

Index

$Z = X - Y$ $\quad Z = X + (-Y)$

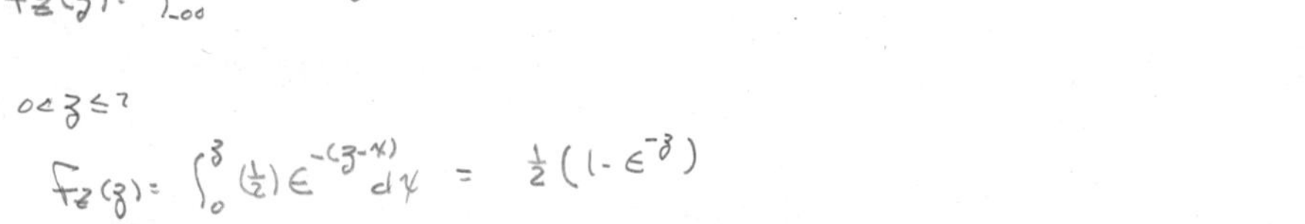

$f_Z(z) = \int_{-\infty}^{\infty} f_X(x) f_Y(z-x)\,dx$ $\quad x<0 \ f_X(x)=0 \quad x>z \ f_Y(z-x)=0$

$0 < z \le 2$

$f_Z(z) = \int_0^z (\tfrac{1}{2}) \epsilon^{-(z-x)}\,dx = \tfrac{1}{2}(1-\epsilon^{-z})$

$2 < z < \infty$

$f_Z(z) = \int_2^z f_Z(2) f_X(x) f_Y(z-x)\,dx$

$= \int_2^z \frac{\frac{1}{2}(1-\epsilon^{-2})}{2} \epsilon^{-(z-x)}\,dx = \frac{1-\epsilon^{-2}}{2} - \epsilon^{-(z-2)}$

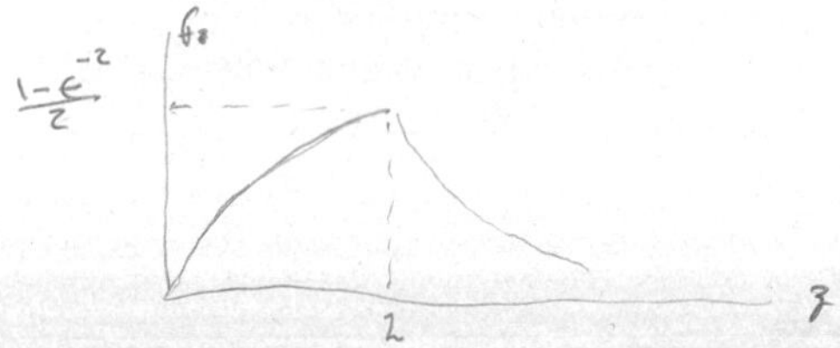

ERROR FUNCTION - GAUSSIAN DISTRIBUTION

$$P(a \le X \le b) = \frac{1}{\sqrt{2\pi}\,\sigma} \int_a^b \epsilon^{-\frac{(x-m)^2}{2\sigma^2}} dx \qquad \text{LET } s = \frac{x-m}{\sigma} \quad \therefore ds = \frac{dx}{\sigma}$$

$$\therefore P(a \le X \le b) = \frac{1}{\sqrt{2\pi}} \int_{\frac{a-m}{\sigma}}^{\frac{b-m}{\sigma}} \epsilon^{-s^2/2} ds = \operatorname{erf}\left(\frac{b-m}{\sigma}\right) - \operatorname{erf}\left(\frac{a-m}{\sigma}\right)$$

$$\operatorname{erf}_1(t) = \frac{1}{\sqrt{2\pi}} \int_{-\infty}^{t} \epsilon^{-s^2/2} ds \qquad -\infty < t < \infty$$

$$\operatorname{erf}_1(x) = \begin{cases} \frac{1}{2} + \operatorname{erf}_2(x) & (x \ge 0) \\ \frac{1}{2} - \operatorname{erf}_2(-x) & (0 \ge x) \end{cases}$$

$$\operatorname{erf}_2(t) = \frac{1}{\sqrt{2\pi}} \int_0^t \epsilon^{-s^2/2} ds \qquad 0 < t < \infty$$

LEIBNIZ'S RULE

$$\frac{d}{dx} \int_a^x f(u)\,du = f(x)$$

$$\frac{d}{dx} \int_{a_1(x)}^{a_2(x)} F(u,x)\,du = \int_{a_1(x)}^{a_2(x)} \frac{\partial F}{\partial x} du + F[a_2(x), x] \frac{da_2}{dx} - F[a_1(x), x] \frac{da_1}{dx}$$

EXAMPLE OF REPEATED TRAILS (LAW OF LARGE NUMBER'S) CHEBYSHEV

CONSIDER TOSSING A COIN. WANT TO FIND n (THE NUMBER OF TOSSES) SUCH THAT THE PROBABILITY THAT THE NUMBER OF HEADS IS BETWEEN $0.45n$ AND $0.55n$ IS GREATER THAN 99.7%

SOLUTION: "HEADS IS THE EVENT AND IS p (SUCCESS)

$$P\left\{\left|\frac{k}{n} - p\right| \le \epsilon\right\} \ge \kappa \qquad \epsilon = 0.55n - 0.45n = 0.05$$

$$1 - \delta = 99.7\%$$

$$\therefore P\left\{\left|\frac{k}{n} - \frac{1}{2}\right| \le 0.05\right\} \ge 0.997 \qquad n > \frac{9pq}{\epsilon^2} = 900$$

$$n > \frac{K^2(\alpha)\,pq}{\epsilon^2} \qquad \operatorname{erf} K(\alpha) = \alpha/2$$

$U = X^2$ TRANSFORMATION PROBLEM

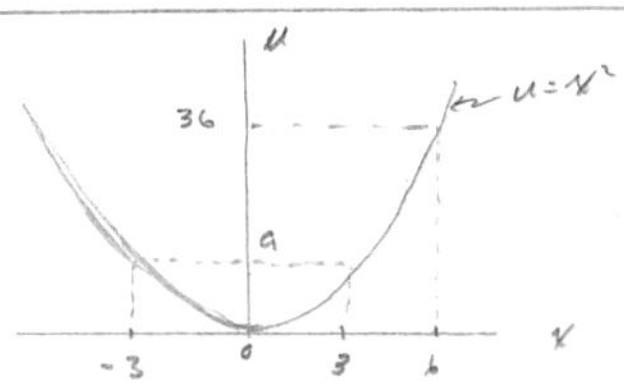

$$f_X(x) = \begin{cases} x^2/81 & -3 < x < 6 \\ 0 \end{cases}$$

$$u = g(x) = x^2 \quad \therefore x = g^{-1}(u) = \pm\sqrt{u}$$

$$F_U(u) = P[U \le u]$$

FOR $0 \le u \le 9$

$$F_U(u) = P[U \le u] = P[X^2 \le u] = P[-\sqrt{u} \le X < \sqrt{u}]$$

$$\therefore F_U(u) = \int_{-\sqrt{u}}^{\sqrt{u}} f_X(x)\,dx$$

FOR $9 < u < 36$

$$F_U(u) = P[U \le u] = P[-3 < X < \sqrt{u}]$$

$$F_U(u) = \int_{-3}^{\sqrt{u}} f_X(x)\,dx$$

$$f_U(u) = \frac{dF_U(u)}{du} \qquad \text{(USE LEIBNIZ RULE)}$$

$$d \int_{-3}^{\sqrt{u}} f_X(x)\,dx = \frac{f_X(\sqrt{u})}{2\sqrt{u}} - f_X(-3)\frac{d(-3)}{du}^{\,0}$$

$$d \int_{-\sqrt{u}}^{\sqrt{u}} f_X(x)\,dx = f_X(\sqrt{u})\,d(\sqrt{u}) - f_X(-\sqrt{u})\,d(-\sqrt{u})$$

$$= \frac{f_X(\sqrt{u})}{2\sqrt{u}} + \frac{f_X(-\sqrt{u})}{2\sqrt{u}}$$

$$\therefore g_U(u) = \begin{cases} \dfrac{f_X(\sqrt{u}) + f_X(-\sqrt{u})}{2\sqrt{u}} & 0 \le u \le 9 \\ \dfrac{f_X(\sqrt{u})}{2\sqrt{u}} & 9 < u < 36 \\ 0 \end{cases} \quad \& \quad g_U(u) = \begin{cases} \sqrt{u}/81 & 0 \le u \le 9 \\ \sqrt{u}/162 & 9 < u < 36 \\ 0 \end{cases}$$

(1.) SHOW THAT FOR THE SPACE $S = \{1,2,3,4,5,6\}$, THE CLASS $\{\phi\}, \{S\}$ $\{1,3,5\}, \{2,4,6\}$ IS A FIELD

CONDITIONS FOR A FIELD ARE:

1) IF $A \in \mathcal{A}$ THEN $\bar{A} \in \mathcal{A}$

2) IF $A \in \mathcal{A}$ & $B \in \mathcal{A}$ THEN $A \cup B \in \mathcal{A}$, $A \cap B \in \mathcal{A}$, $B - A \in \mathcal{A}$ & $A - B \in \mathcal{A}$

3) S & $\phi \in \mathcal{A}$

CHECK FOR $\bar{S}$ & $\bar{\Phi}$

(2) CONDITIONAL PROBABILITY:

GIVEN: A URN: 1 CONTAINING WHITE & RED BALLS (w_1, w_2 w_3 r_1 r_2) REMOVE TWO BALL'S IN SUCCESSION @ random. WHAT IS PROB. THAT FIRST BALL IS WHITE & 2nd red?

$$P[w \cap R] = P[R/w]\,P[w] = (2/4)(3/5)$$

(3) BAYES THEROM:

GIVEN: 2 BOXES: 1 CONTAINS 999 WHITE CARDS & 1 RED CARD
OTHER CONTAINS 999 RED CARDS & 1 WHITE CARD

A BOX IS SELECTED AT RANDOM AND A CARD REMOVED THE CARD IS WHITE. WHAT IS PROB IT COMES FROM 1ST BOX.

$$P[B_1/W] = \frac{\overset{999/1000}{P[W/B_1]}\ \overset{1/2}{P[B_1]}}{\underset{999/1000}{P[W/B_1]}\ \underset{1/2}{P[B_1]} + \underset{1/1000}{P[W/B_2]}\ \underset{1/2}{P[B_2]}}$$

(4) A PAIR OF FAIR DICE IS THROWN. IF THE TWO NUMBERS APPEARING ARE DIFFERENT FIND THE PROBABILITY P THAT a) THE SUM IS SIX, b) AN 1 APPEARS, c) the sum IS 4 OR LESS. $S = \{\quad\}$ 36 combinations less 1,1, 2,2, etc

a) $S = 30$

THE SUM 6 = $\{(1,5)\,(24)\,(42)\,(51)\}$

$\therefore P = 4/30$

b) A 1 APPEARS 10 WAYS $\therefore P = 10/30$

c THE SUM OF 4 or less OCCURS 4 WAYS
(12) (13) (31) (21) $\therefore P = 4/30$.

(5) HOW MANY DICE MUST BE THROWN SO THAT THERE IS A BETTER THAN EVEN CHANCE OF OBTAINING A SIX?

THE PROBABILITY OF not obtaining A SIX = 5/6 on n dice = $(5/6)^n$

$\therefore$ seek smallest n for which $(5/6)^n < \frac{1}{2}$.

EXAMPLE OF STATISTICALLY INDEPENDENT CONDITIONAL DENSITY FUNCTION

$f_X(x) = \epsilon^{-x} \quad x \geq 0, \quad f_{Y/X}(y/x) = x\epsilon^{-xy} \quad y \geq 0,\ x \geq 0$

FIND: $f_Y(y)$

$f_{XY}(x,y) = f_{Y/X}(y/x)\, f_X(x) = x\epsilon^{-xy} \cdot \epsilon^{-x} = x\epsilon^{-(1+y)x} \quad 0 \leq x,\ 0 \leq y$

$f_Y(y) = \int_0^\infty f_{XY}(x,y)\,dx = \int_0^\infty x\epsilon^{-(1+y)x}\,dx = \frac{\epsilon^{-(1+y)x}}{(1+y)^2}\left[-(1+y)x - 1\right]_0^\infty = \frac{1}{(1+y)^2}$

EXAMPLE OF CONDITIONAL DENSITY & DISTRIBUTION FUNCTIONS

LET $f_X(x) = \epsilon^{-x} \quad 0 < x \quad \&\ M = \{1 \leq x \leq 2\}$

$\therefore\ P(M) = P\{1 \leq X \leq 2\} = \int_1^2 f_X(x)\,dx = \epsilon^{-1} - \epsilon^{-2}$

FIND $F_X(x/M)$

$x < 0 \Rightarrow \{X \leq x\} \cap \{1 \leq X \leq 2\} = \phi \Rightarrow P\{x/M\} = 0$

$0 \leq x < 1 \Rightarrow \{X \leq x\} \cap \{1 \leq X \leq 2\} = \phi \Rightarrow P\{x/M\} = 0$

$1 \leq x \leq 2 \Rightarrow \{X \leq x\} \cap \{1 \leq X \leq 2\} = \{1 \leq X \leq x\}$

$P(X \leq x, M) = P\{1 \leq X \leq x\} = \int_1^x \epsilon^{-\alpha}\,d\alpha = \epsilon^{-1} - \epsilon^{-x}$

$2 < x < \infty: \quad \{X \leq x\} \cap \{1 \leq X \leq 2\} = \{1 \leq X \leq 2\}$

$P(X \leq x, M) = P\{1 \leq X \leq 2\} = \epsilon^{-1} - \epsilon^{-2}$

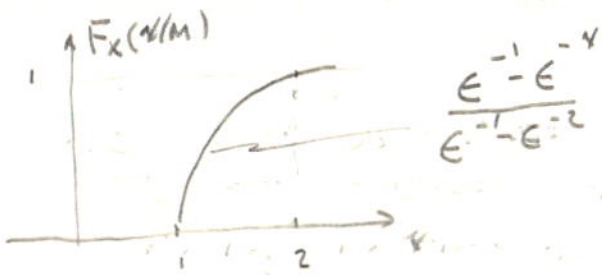

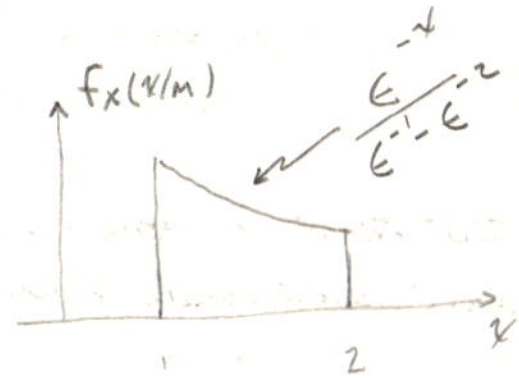

EXAMPLE OF $E[XY]$

A RECTANGLE HAS DIMENSIONS THAT ARE RANDOM VARIABLES. THE BASE X IS A RANDOM VARIABLE UNIFORMLY DISTRIBUTED FROM 0 TO 5. THE HEIGHT Y IS A R.V. UNIFORMLY DISTRIBUTED FROM 0 TO X. FIND E[ ] OF AREA

$f_X(x) = 1/5 \quad 0 < x \leq 5, \quad f_{Y/X}(y/x) = 1/x \quad 0 \leq y \leq x$

$f_{XY}(x,y) = (1/5)(1/x) = 1/5x \quad \therefore\ E[XY] = \int_{-\infty}^\infty \int_{-\infty}^\infty xy\, f_{XY}(x,y)\,dx\,dy$

$= \int_{x=0}^{x=5} \int_{y=0}^{y=x} \frac{1}{5x}\, x y\, dx\,dy = \frac{1}{5}\int_0^5 \frac{x^2}{2}\,dx = \frac{5^3}{30}$

EXAMPLE OF JOINT PROBABILITY

GIVEN: $f_{XY}(x,y) = \begin{cases} K\epsilon^{-(3x+4y)} & 0 < x,\ 0 < y \\ 0 & \text{ELSEWHERE} \end{cases}$

FIND: ① K, ② $F_{XY}(x,y)$ ③ $P\{0 < X \leq 1,\ 0 < y \leq 2\}$ ④ $f_X(x)$

① $\int_{-\infty}^\infty \int_{-\infty}^\infty f_{XY}(x,y)\,dx\,dy = 1 \quad \therefore\ \int_0^\infty \int_0^\infty K\epsilon^{-(3x+4y)}\,dx\,dy = 1$

$K\int_0^\infty \left[-\tfrac{1}{3}\epsilon^{-3x}\Big|_0^\infty\right]\epsilon^{-4y}\,dy = 1$

$K/3 \int_0^\infty \epsilon^{-4y}\,dy = 1$

$K/12 = 1$

$\therefore\ K = 12$

② $F_{XY}(x,y) = \int_{-\infty}^{y}\int_{-\infty}^{x} f_{XY}(\alpha,\beta)\,d\alpha\,d\beta = \int_0^y\int_0^x 12\epsilon^{-(3\alpha+4\beta)}\,d\alpha\,d\beta$

$$= 12\int_0^y\left[-\tfrac{1}{3}\epsilon^{-3\alpha}\Big|_0^x\right]\epsilon^{-4\beta}\,d\beta$$

$$= {}^{12}\!/_{3}\int_0^y\left[\epsilon^{-3x}-\epsilon^{0}\right]\epsilon^{-4\beta}\,d\beta$$

$$= -4(\epsilon^{-3x}-1)\int_0^y\epsilon^{-4\beta}\,d\beta$$

$$= (\epsilon^{-3x}-1)(\epsilon^{-4y}-1)$$

③ $\int_0^2\int_0^1 12\epsilon^{-(3x+4y)}\,dx\,dy = 0.95$

④ $f_X(x) = \int_{-\infty}^{\infty} f_{XY}(x,y)\,dy = 3\epsilon^{-3x}$ $\quad f_Y(y) = 4\epsilon^{-4y}$

EXAMPLE OF POISSON THEOREM

① A SYSTEM CONTAINS 1000 COMPONENTS. EACH COMPONENT FAILS INDEPENDENTLY OF THE OTHERS AND THE PROBABILITY OF ITS FAILURE IN ONE MONTH $= 10^{-3}$

WHAT IS THE PROBABILITY THE SYSTEM WILL FUNCTION (NO COMPONENT FAILURES) AT THE END OF A MONTH?

$p = 10^{-3}$ (PROB. OF A FAILURE = SUCCESS), $k=0$, $n = 10^3$ (NO. OF COMPONENTS)

$np = 10^{-3}\cdot 10^{3} = 1 \qquad \therefore P[k=0] \approx \epsilon^{-np} = \epsilon^{-1}$

---

② RANDOM POINTS IN TIME

A CITY HAD TWO EARTHQUAKES IN A CENTURY. FIND THE PROBABILITY THAT IN THIS CITY THERE WILL BE ONE EARTHQUAKE (a) NEXT YEAR (b) IN THE NEXT 50 YEARS. (c) WHAT IS THE PROBABILITY OF THREE OR MORE INDEPENDENT EARTHQUAKES IN THE SAME MONTH. ASSUME EARTHQUAKES TO BE INDEPENDENT AND EQUIPROBABLE IN ANY MONTH.

(a) $P[\text{1 EARTHQUAKE IN 1 YEAR}] = \dfrac{(\lambda\cdot 1)^{1}}{1!}\epsilon^{-\lambda\cdot 1} = \lambda\epsilon^{-\lambda}$

$k = 1$ EARTHQUAKE $\quad n = 2$

$t_a = 1$ YEAR $\quad \lambda = \frac{n}{T} = \frac{1}{50}$

$T = 100$ YEAR

(b) $P[\text{1 EARTHQUAKE IN 50 YEARS}] = \dfrac{(\lambda\cdot 50)^{1}}{1!}\epsilon^{-\lambda\cdot 50} = \lambda\cdot 50\,\epsilon^{-\lambda 50}$

$t_a = 50$ YRS, $\lambda = 1/50$

(c) $P[\text{3 OR MORE EARTHQUAKES IN 1 MONTH}] = 1 - [P(0)+P(1)+P(2)]$

$P(0) = \dfrac{(\lambda\cdot 1/12)^{0}}{0!}\epsilon^{-\lambda(1/12)} = \epsilon^{-\lambda/12}$ $\qquad P(2) = \dfrac{(\lambda\cdot 1/12)^{2}}{2!}\epsilon^{-\lambda(\frac{1}{12})} = \dfrac{\lambda^2}{288}\epsilon^{-\lambda(1/12)}$

$P(1) = \dfrac{(\lambda\cdot 1/12)^{1}}{1!}\epsilon^{-\lambda(1/12)} = \lambda/12\,\epsilon^{-\lambda(1/12)}$

$t_a = 1/12$

---

③ AN ORDER OF 3000 PARTS IS RECEIVED. THE PROBABILITY THAT A PART IS DEFECTIVE IS 10^{-3}. WHAT IS THE PROBABILITY THAT THERE WILL BE FIVE OR MORE PARTS DEFECTIVE?

$p = 10^{-3}$ (SUCCESS IS A FAILURE) $\qquad np = 10^{-3}\times 3\times 10^{3} = 3 \qquad n = 3\times 10^{3}$

$P\{\text{5 OR MORE PARTS ARE DEFECTIVE}\} = 1 - \{P(0)+P(1)+P(2)+P(3)+P(4)\}$

$P(0) = \epsilon^{-np}\dfrac{(np)^k}{k!} = \epsilon^{-3}\dfrac{(3)^0}{0!} = \epsilon^{-3}$ $\qquad P(1) = \epsilon^{-3}\dfrac{(3)^1}{1!} = 3\cdot\epsilon^{-3}$

$P(2) = \epsilon^{-3}\dfrac{(3)^2}{2!} = \dfrac{9\epsilon^{-3}}{2}$ $\qquad P(3) = \epsilon^{-3}\dfrac{(3)^3}{3!} = \dfrac{27}{6}\epsilon^{-3}$

POISSON THEROM

① A MESSAGE OF 5000 LETTERS HAS BEEN TRANSMITTED THROUGH A CHANNEL WITH AN ERROR RATE OF 10^{-3}. WHAT IS THE PROBABILITY THAT THERE ARE (a) ONE ERROR (b) AT LEAST TWO ERR. (c) TWO OR MORE ERRORS.

$N = 5000, \quad p = 10^{-3} \qquad NP = 5$

$$P_{5000}(K=1) = \frac{(NP)^1}{1!} \epsilon^{-NP}$$

$$P_{5000}(K \geq 1) = 1 - P_{5000}(K=0) = 1 - \epsilon^{-5}$$

$$P_{5000}(K \geq 2) = 1 - P_{5000}(K=0) - P_{5000}(K=1)$$

---

② 26 TUBE'S FAILURE PER YEAR. TUBES REPLACED FROM LOCAL STOCK WHICH IS REPLENISHED EVERY 2 WEEKS. TO WHAT NUMBER MUST THE DEPLETED STOCK ALWAYS BE REPLENISHED TO PREVENT, WITH A PROB. OF 99% or more, THE COMPUTER BEING PUT OUT OF ACTION.

AVG. FREQ $\lambda = \frac{26}{52} = 0.5$ TUBES/WEEK, $t_a = 2$ WEEKS

so $\lambda t_a = 1$, K = Number of failures/week, N = Number of tubes in stock. WHAT $K \leq N$

$$P(K \leq N) = \sum_{K=0}^{N} \frac{(\lambda t_a)^K}{K!} \epsilon^{-\lambda t_a} \geq 0.99$$

$$\text{or} \quad \left(1 + \lambda t_a + \frac{(\lambda t_a)^2}{2!} + \cdots + \frac{(\lambda t_a)^N}{N!}\right) \geq 0.99\, \epsilon^{\lambda t_a} = 2.69$$

ADDING TERMS IN SUM UNTIL IT EXCEEDS 2.69 yields $N \geq 4$

THERE ARE 4 ROADS BETWEEN CITY A & B, EACH CROSSING A RIVER. EACH BRIDGE IS WASHED OUT WITH A PROB. OF 1/3. STAT. INDEPENDENT.

a) NORMAL TRAVEL REQUIRE'S AT LEAST TWO ROADS BE OPEN. WHAT IS PROBABILITY NORMAL TRAVEL CAN OCCUR

$q = 1 - p = 2/3$ (ROAD OPEN), $n = 4 \qquad \sum \binom{n}{k} p^k q^{n-1}$

$P(\geq 2) = 1 - [P(0) + P(1)]$

b) WHAT IS AVERAGE NUMBER OF OPEN ROADS?

$q \times N$